NASA SP-4021

ASTRONAUTICS AND AERONAUTICS, 1976

A Chronology

Eleanor H. Ritchie

The NASA History Series

Scientific and Technical Information Branch   1984
NATIONAL AERONAUTICS AND SPACE ADMINISTRATION
Washington, D.C.

Contents

January	1
February	23
March	43
April	69
May	87
June	111
July	137
August	175
September	203
October	235
November	267
December	285
Appendix A: Satellites, Space Probes, and Manned Space Flights, 1976	307
Appendix B: Major NASA Launches, 1976	337
Appendix C: Manned Space Flights, 1976	341
Appendix D: NASA Sounding Rocket Launches, 1976	343
Appendix E: Abbreviations of References	351
Index and List of Abbreviations and Acronyms	355
Errata in Earlier Volumes	397

Illustrations

Comsats launched by NASA, 1976	2
Concorde arrives at Dulles Airport	103
Third Century America exposition opens at KSC	108
National Air and Space Museum	138
First panoramic view of Mars from *Viking 1*	147
Goldstone antenna used in solar-energy tests	197
Enterprise (Shuttle Orbiter 101) rolled out at Rockwell	223
XV-15 (tilt-rotor research aircraft) rolled out at Bell	258

January 1976

1 January: "An impressive record of cost-cutting that could well set an example to many agencies" had been achieved by NASA since the mid-sixties, said Sen. William A. Proxmire (D–Wis.), one of the agency's severest critics. Chairman of the Senate Appropriations subcommittee that handles NASA's appropriation, Proxmire said that NASA had made its tight budget go farther by acting as its own prime contractor when construction bids were too high; buying electronic parts from a standardized list; providing reimbursable launches to private industry and other outside organizations on an "accelerated" basis, although not—according to Proxmire—charging enough for them; and, in general, making "one dollar do the job that it took two dollars to do in the free and easy money days." (*Today*, 1 Jan 76, 1)

- Failure to reach a satisfactory nuclear arms agreement with the Soviet Union would result in U.S. development of a new land-based intercontinental missile, according to U.S. government sources quoted by Reuters in a *Baltimore Sun* story. The 5-yr defense plan to be submitted to Congress in 3 wk with the new FY 1977 budget would include spending $1.07 billion to start work on a larger and more accurate missile called MX that could be launched from silos, aircraft, or land-mobile platforms. The new missile would be operational by 1985, the report said. (*B Sun*, 1 Jan 76, A6)

- *2 January:* NASA planned 18 major satellite launchings in 1976, 16 from Cape Canaveral and 2 from the Western Test Range at Vandenberg Air Force Base. First of these, on 13 Jan., would be a U.S.–Canadian communications technology satellite; second, on 15 Jan., would be the *Helios* satellite built by West Germany. Other NASA launches during the year would include 2 comsats for the North Atlantic Treaty Organization; 2 maritime comsats (Marisats); *Palapa*, a comsat for the Indonesian government; the second geostationary operational environmental satellite (GOES–B); and *Lageos*, a laser geodynamic satellite to help alleviate earthquake hazards. (*NYT*, 3 Jan 76, 5)

- As he assumed the presidency of the American Association for the Advancement of Science, Dr. William D. McElroy (chancellor of the Univ. of Calif. in San Diego, and former director of the National Science Foundation) said he planned to reorganize the AAAS journal *Science*, involve the association more deeply in government and industry activity, and quadruple the membership. Dr. McElroy, called the "leading world authority on the biochemistry of firefly light," said he hoped to stop the publication of pure research papers in *Science*, as more specialized scientific journals could do this more efficiently, and to promote a free flow of ideas between the universities and the industrial community by

Four of the comsats in NASA's 1976 launch schedule: A, Comsat General's Marisat, launched 19 Feb.; B, NATO III-A, launched 22 Apr.; C, Comsat General's Comstar 1A, launched 13 May; D, Indonesia's Palapa, launched 8 July.

including more articles of interest to users in industry and engineering, and paying more attention to science issues in government policy. He referred to the "unfortunate separation" between industry and university after World War II when science gained new nonindustrial financial support; by increasing the circulation of *Science* magazine, he would expand the AAAS membership, as the magazine is available only to members. (*NYT*, 2 Jan 76, 10)

5 January–4 February: Permission to fly the British- and French-built Concorde supersonic transport into the U.S. was the subject of a hearing called by Transportation Secretary William T. Coleman, Jr., in Washington to help him decide the matter within 30 days. Basic arguments pro and con were not new, observers agreed, as they had all been thrashed out in a bitter 1971 congressional fight that halted U.S. efforts to build an SST after about $1 billion had been spent on the project. The case against the Concorde had been that it was too noisy and that its emissions polluted the atmosphere and depleted the stratospheric ozone shield. Spokesmen for the Concorde cited its benefits in speed of travel and technical achievement. At the hearing, representatives of the Va. state government called for approval of Concorde landings at Dulles, in view of the airport's significance as Virginia's international gateway; these advocates found themselves opposed to the views of the local Va. jurisdictions which did not want the SST landing there. Gov. Hugh Carey of N.Y. opposed the Concorde's coming into JFK Airport because of the noise. On 6 Jan. the Environmental Protection Agency reversed its previous stand, calling for a total ban on commercial operations in the U.S. of the British–French SST. On 13 Jan., the Aerospace Industries Assn.—U.S. manufacturers of aerospace vehicles and components—submitted a letter deploring the cancellation of the U.S. program and supporting Concorde's application to operate. On 29 Jan., the Am. Inst. of Aeronautics and Astronautics, claiming to represent 22 000 engineers and scientists, called for a limited-operations trial period to collect data as the basis for a final decision on giving Concorde access to U.S. airports. The Concorde had begun regular passenger service 21 Jan. between Paris and Rio de Janeiro, and between London and Bahrain, as the climax of 14 yr of technical cooperation between Britain and France and a joint investment of more than $3 billion.

DOT Secy. Coleman announced 4 Feb. that he had granted the Concorde "limited scheduled commercial flights" into the U.S. for a period not to exceed 16 mo under precise limitations. British Airways and Air France could send up to 2 flights per day into JFK and Dulles airports, but the permission could be revoked upon 4 mo notice or immediately in case of emergency. The limitations included prohibitions on landing or takeoff in the U.S. before 7 am or after 10 pm local time, and on flying at supersonic speed over the U.S. or any of its territories. In his 61-page decision with 36 pages of appendix, Coleman said the 16-mo period should be long enough to demonstrate the validity of his judgment. Upon hearing of the secretary's decision, 2 members of the Senate Commerce Committee were narrowly defeated in an attempt to ban the Concorde from landing in the U.S. by attaching such a ban to a bill authorizing funds for airport development. The Concorde's only supersonic competitor was the USSR's Tupolev-144, in service since 26 Dec. 1975 within the Soviet Union as a freight plane. (*NYT*, 5 Jan 76, 1; *Av Wk*, 5 Jan 76, 26; *W Star*, 6 Jan 76, A-3; *W Post*, 6 Jan 76, A-1; AIA Aerospace News Release 76-1; text, AIAA letter to DOT, 29 Jan 76; *NYT*, 21 Jan 76, 16-17; *W Star*, 21 Jan 76, A-3; *B Sun*, 28 Jan 76, A-6; *NYT*, 1 Feb 76, 3-1; DOT Release 09-76, 4 Feb. 76; *NYT*, 5 Feb 76, 1, 16, 32M; *W Post*, 5 Feb 76, A-1, A-18; *B Sun*, 5 Feb 76, A14)

6 January: The Republic of Ireland would become the eleventh member of the European Space Agency, ESA announced today, when parliaments of the existing member states (Belgium, Denmark, France, Germany, Italy, the Netherlands, Spain, Sweden, Switzerland, and the United Kingdom) ratify the agency's convention. Ratification might take 1 to 2 yr, during which time Ireland would retain its observer status. The ambassador of the Republic of Ireland in Paris signed the ESA convention 31 Dec. 1975. (ESA release 5 Jan 76)

7 January: The U.S. Coast Guard began its Project Icewarn field program with the first of a number of daily ice-observation missions by aircraft over the Great Lakes during the winter navigation season. The flights would continue through 2 Feb., when ice conditions were expected to stop all shipping on the Great Lakes, and resume in Mar. with the breakup of the ice and continue through Apr. Objective of the program was to provide Coast Guard personnel with experience in operating a side-looking airborne radar (SLAR) system developed by NASA's Lewis Research Center, which included equipment to convert radar signals to ice-cover images transmitted in real time from the aircraft to the ground through a relay channel on *Sms/Goes 1*. The system would be demonstrated in a simulation during which ice-cover information would be transmitted to captains of lake shipping for use in routing the ships through the ice. (Wkly Briefs for Admr, 12 Jan 76)

- A new computer program developed at Johnson Space Center from Landsat digital data could compile maps on any scale desired showing water surfaces in excess of 0.024 sq km, said specialists in the Earth Observations Division at JSC. Nations needing an inventory of their water resources could obtain maps of their lakes and reservoirs, and state governments could use the service in choosing between recreational and industrial use of available water supplies. The program, called Detection and Mapping (DAM) Package, required only $300 worth of computer time to map more than 33 800 sq km. As the two Landsats cover about 95 percent of the earth's land mass, the system could produce surface-water maps for virtually all populated regions with almost 100% accuracy for water areas as large as 0.04 sq km; position accuracy—the degree to which the maps match the terrain—would be within 90m of dead center. User training would be typically no more than a day, and the system would need no experts to implement it. (NASA Release 76-4; JSC Release 76-01)

- The Peoples Republic of China indicated for the first time that it planned to put a man into space, according to an article in the Peking *Kuang-*

ming Daily entitled "The Launching and the Bringing Back of Artificial Satellites from Earth." As reported by radio from Hong Kong, the article recalled that PRC had put 5 satellites into orbit since 1970, emphasizing that *China 4* (launched 26 Nov. 75) "returned to earth as scheduled after functioning normally." After explaining how satellites were launched and recovered, the article added that recovery was particularly significant in "sending men into space." The radio report said that the article—first of its kind in the official PRC press—was even more surprising because the Chinese supposedly had not heard that man had walked on the moon, as this had never been reported in the Chinese press. (FBIS, Hong Kong AFP in English, by G. Biannic, 7 Jan 76)

8 January: Goes 1, a new geostationary environmental satellite launched in Oct. 1975 for the National Oceanic and Atmospheric Administration (NOAA), became operational, replacing *Sms 1* above the equator over northern Brazil to provide forecasters with visual and infrared pictures of the western hemisphere, to monitor solar-flare activity, and to relay information from data-collection stations in remote areas. *Sms 1* would be moved to standby status over the eastern Pacific south of Mexico; a prototype satellite, some of whose systems have degraded, it could still provide full operational data on an emergency basis. *Sms 2*, launched in Feb. 1975, was moved in Dec. from $115°W$ to $135°W$ to provide improved imagery of the Pacific in the Hawaiian Islands area. (NOAA Release 76-1)

- The Flight Research Center at Edwards, Calif., NASA's prime site for experimental research in aeronautics, was renamed the Hugh L. Dryden Flight Research Center in memory of the pioneer aeronautical researcher who was first Deputy Administrator of NASA, a position he held until his death in 1965. Dr. Dryden's contributions to aeronautical research included investigation of high-speed airfoils, supersonic propeller-tip velocities, boundary layers, and airflow wind turbulence, and development of high-speed wind tunnels. He was director of the former National Advisory Committee for Aeronautics from 1947 until that agency became NASA in 1958. (NASA Release 76-7; FRC Release 1-76; *Goddard News,* Feb 76, 6)

- In observing the results of *Salyut 4's* first year of operation in space, Soviet cosmonaut Konstantin Feoktistov in an article in *Pravda* reviewed the research and experiments conducted on board the space station, calling it "a major achievement for Soviet space navigation." Special mention was given to the solar radiation studies and environmental photography done from *Salyut 4;* the first expedition photographed a million sq km of Soviet territory, and the second crew also took pictures to enable scientists to follow developments in hydrology, vegetation, and climate. Medical studies of the crews centered on the effects of long space flights and the mechanism of the body's adaptation to weightlessness. *Salyut 4*, the article noted, was built with an eye to longer activity for the station and better facilities for prolonging manned missions. (FBIS, Tass in English, 8 Jan 76)

9 January: Geos 3, the geodynamics experimental ocean satellite launched 9 Apr. 1975, completed 9 mo in orbit and the project was judged successful with respect to the prelaunch mission objectives. Problems with orbit definition and revised criteria for data distribution had delayed the release of large quantities of data, said a report by Charles W. Mathews, Associate Administrator for Applications at NASA, but the first batches would be shipped 30 Jan. to principal investigators, and subsequent batches would be released monthly thereafter. (MOR E-855-75-01 [postlaunch], 25 Feb 76)

10 January: An unmanned ground terminal that could operate with several satellites in synchronous orbit above the earth was patented by three engineers for the Communications Satellite Corporation. Previously, a separate earth terminal was needed for each satellite. William K. Sones, Laurence F. Gray, and Louis Pollack, of the ComSat staff in Washington, D.C., invented the new facility, considered a major advance that would add reliability and reduce costs. The structure included a single reflector about 9.7m by 15.4m with enough traveling-wave tubes, transmitters, receivers, and amplifiers to handle two or more satellites, plus monitors and controls; the reliability feature was a system that switched in another tube if one became defective. Possible uses would be on offshore rigs or at oil-pipeline installations that were unmanned but required constant communication by satellite. (*NYT,* 10 Jan 76, 31)

11 January: The French government halted operations of sounding-rocket and scientific satellite launchings from its base at Kourou, French Guiana, after 7 yr of activity that included launches of 275 rockets or balloons, many of them under U.S. government or National Science Foundation programs. The base had become a victim of European economic problems that had led to the abandonment of ELDO—the European Launcher Development Organization formed by several countries in July 1966—when the first launching in 1971 failed. The head of the operations division at Kourou said there was a complete lack of coordination among the countries, although France and its partners had spent about $500 million on the site and support facilities at Kourou. Space research would resume in about 4 yr with the launch of a 3/4-ton telecommunications satellite sponsored by the French government, under a new system in which each project would be under the management of one country. Ten European governments would contribute money, and the funds would be allocated in the form of contracts for each project. (*W Post,* 11 Jan 76, E 7)

12 January: Use of dirigibles instead of drifting stations in the Arctic Ocean for polar exploration and scientific studies was advocated by Soviet polar explorer Nikolay Blinov, a staff member of the Leningrad Arctic and Antarctic Institute, as reported by Tass. The USSR had been using observatories set up on ice floes to forecast weather and ice formations on the northern sea route for about 40 yr; the 23rd North Pole station was established on a large iceberg in the last quarter of 1975. Blinov pointed out that the continuously drifting stations, never safe from breakup and melting away upon encounter with warm water, offered less reliability than dirigibles, which could remain over a location for an indefinite time to register data on the "hydrological regime" of a certain area and would be durable and less expensive to maintain. (FBIS, Tass in English, 12 Jan 76)

13 January: Eighty percent of the 114 federal research laboratories had not adopted cost-cutting activities recommended 6 yr ago by the General Accounting Office. A new GAO survey found that agencies had not made periodic inspections of equipment to find unused items that could be sold; had not established equipment pools to eliminate duplication and underuse of costly scientific tools; and had not started using "time meters" as a check on how often the equipment was used. Agencies following the three recommendations could save millions of dollars, said Rep. Les Aspin (D–Wis.) in releasing the GAO report, citing an atomic research facility that saved $24.2 million by disposing of unneeded equipment. Aspin also referred to purchases of expensive sophisticated equipment for a single experiment that was never used again, when it could have been sold or passed along to another agency. (*W Star,* 13 Jan 76, C8)

- The largest optical telescope in the Southern Hemisphere began operation on a mountaintop about 483 km north of Santiago, Chile, to give astronomers their best look at objects such as the Magellanic Clouds and the brightest globular star clusters visible only from that hemisphere. The 13.7m-long telescope's steerable portion weighed 300 tons, so delicately balanced that one person could move it by hand. Its mirrors were made of Cervit, an optical material insensitive to temperature changes; its 15-ton main mirror was 4 m in diam. and 61 cm thick, and the secondary mirror was 1.3 m in diameter. The new Cerro Tololo Inter-American Observatory would be run by the Association of Universities for Research in Astronomy (AURA), Inc., under contract to the National Science Foundation and in cooperation with the Univ. of Chile at Santiago; Cerro Tololo and its sister institution, Kitt Peak National Observatory near Tucson, Ariz., would be national research centers with 60% of all telescope time available to qualified visiting scientists who would otherwise lack access to instruments capable of frontier research in astronomy. (NSF Release PR 76-4)

14 January: The French government had earmarked a preliminary budget of $3.6 million for development of a spy satellite that would survey the earth's surface with infrared cameras, the *N.Y. Daily News* said. Weighing several hundred kg and orbiting at 500 to 600 km altitude, the satellite would use several lenses in visible and infrared light to detect details at sizes ranging from 10 to 100 m. The government's representative for armaments, J.-L. Delpech, said that France had "no military space policy" because reconnaissance satellites were "not essential" to France's "deterrent strategy." (FBIS, in French, 13 Jan 76; *NY News,* 14 Jan 76, 58)

15 January: Helios 2, third* cooperative project of the U.S. and the Federal Republic of Germany, was launched from Complex 41 of the Eastern Test Range at 12:34 am EST (0534 GMT) on a Titan–Centaur (TC-5) into

---

*First FRG–U.S. venture was the German research satellite GSR-A, launched in Nov. 1969 and titled *Azur* in orbit.

a solar orbit that would put it between 44 million and 144 million km from the sun, 3 million km closer than its twin, *Helios 1*, launched 10 Dec. 1974. Each Helios carried 7 experiments of West German scientists and 3 from the U.S. The experiments would investigate solar processes and events such as coronal, solar-wind, and interplanetary fields and waves; cosmic rays, both solar and nonsolar; dust particles and zodiacal light; and celestial mechanics. *Helios 2* was the first spacecraft carrying a detector for gamma-ray bursts in space, whose cause and source had not been identified since discovery in 1969. Combining Helios data with information from other satellites might pinpoint the direction of the sources and permit their identification with visible celestial objects. The Helios spacecraft weighed 376 kg, with a 1.75 m cylinder carrying conical solar arrays on each end that gave it a spool shape; with deployable booms extended, *Helios 2* would measure 32 m tip to tip. The solar arrays would provide a minimum of 240 w at aphelion—much more when closer to the sun—to power the data handling and transmission. Scientists also hoped for more information on the unexpected concentration of micrometeorites found by *Helios 1*; about 15 times more of these particles were detected within 53 million km of the sun than had been observed near the earth. Cost of the 2 Helios missions was about $260 million, of which the German share was about $180 million for spacecraft units, 7 experiments, and command and data-acquisition expenses; mission control would be at the German Space Operations Center near Munich. The U.S. paid for the 2 launch vehicles and the 3 U.S. experiments, plus support services, a total of about $80 million. A third Helios had been considered for launch in 1980 to measure solar activity at the height of the 11-yr cycle and to study Comet Encke. (NASA Releases 75-317, 76-2; MOR S-823-76-02 [prelaunch] 7 Jan 76, [postlaunch] 23 Jan 76; *NYT*, 16 Jan 76, 27)

- The USSR's supersonic passenger plane TU-144, in service between Moscow and Alma Ata in Kazhakstan, had flown further on the special test bench still being used to operate its engines under varying flight conditions than it had in the skies, according to a story in the newspaper *Pravda*. The bench, designed simultaneously with the aircraft itself, could accommodate the 65-m-long plane and simulate conditions such as outside-temperature changes from −60°C to 150°C; oncoming and vertical air currents with a velocity of 30 to 50 m per sec; and air-current impact on skin and control mechanisms in TU-144 takeoff and landing. The plane's wings and fuselage were "wrapped" in 5000 steel rods with sensors attached to 12 000 points that signaled the slightest change to computers; the bench included 8000 thermometers, as well as 3300 km of wires from the TU-144 to instruments and computers, "equal to the distance between Moscow and Alma Ata," the report said. During the tests, the cabin retained normal atmospheric pressure and temperature even when exterior conditions imitated those at 18 to 20 km in the stratosphere where the plane would be flying. (FBIS, Tass in English, 14 Jan 76)
- Scientists attempting to find a source of unlimited energy like that which powers the sun had taken two potentially significant directions, Walter

Sullivan reported in the *New York Times*. Soviet researchers had shifted from emphasis on laser beams for crushing nuclear fuel to superdensity, to use of electron beams for that purpose. Although 90% of U.S. effort had been on converging pulses of laser light, a new system of using ion beams had shown advantages over the electron-beam method from which it evolved: the apparatus would deliver a vast amount of energy to a fuel pellet, causing the shell to explode both inward and outward. The inward blast, crushing the pellet core to 1000 times its original density, would produce a fusion reaction resulting in helium, and a small amount of mass would be converted into a large amount of energy. The U.S. electron-beam study was based at Sandia Laboratories, in N.M., one of three research centers operated by the Energy Research and Development Administration. Although the U.S. use of ions was the chief novelty in the fusion-energy field, laser fusion was still the front runner, Sullivan noted. (*NYT*, 15 Jan 76, 22)

16 January: An impartial "science court" to weigh controversial national issues such as pesticide use or nuclear-reactor safety was one of several ideas under consideration by two advisory groups appointed by the President last year to give the proposed White House office of science and technology a head start in its task of making major policy decisions. Dr. Simon Ramo, chairman of the advisory group on technological contributions to economic strength, pointed out that no procedure existed for dealing with scientific and technical portions of important issues. Dr. William O. Baker, president of Bell Laboratories, would chair the other advisory group on advances in science and technology. A 2-day meeting of the groups in Washington ended with the plan to test the science-court idea in an experiment, choosing a controversial issue in which a clearcut statement of scientific facts seemed feasible, then arguing the case with advocates and cross examinations for both sides, with an impartial panel of scientific judges to make the decision. Dr. Ramo said the experiment should teach the advisory groups about the issue they chose, help the government agency that must deal with the issue, and demonstrate whether the court idea would work. Major issues on which the advisory groups would attempt to assist the White House included the world problem of food and nutrition, the issue of technological innovation and its effect on productivity, and the effect of government regulation on advancement in science and technology. (*NYT*, 16 Jan 76, 32)

• The board of directors of Communications Satellite Corporation (ComSat) declared a quarterly dividend of 25 cents per share payable 15 Mar. to all shareholders of record at close of business 13 Feb. The group's 22nd consecutive quarterly dividend would be the 7th at the 25-cent rate. (ComSat Release 76-1)

17–29 January: The Communications Technology Satellite designed by U.S. and Canadian technicians—world's most powerful comsat—was launched from Complex 17, Eastern Test Range, at 6:27:54 pm EST (2328 GMT) on a Delta 2914 vehicle, delayed from its scheduled date of 13 Jan. The 346-kg space craft cost $60 million in Canadian funds and $22 million in U.S. funds. The design departure used on *Cts* was the system for providing solar power to the transmitter: a pair of accordion-pleated solar arrays to be extended by thin steel booms to a wingspan of 7.5 m each, that could provide 1 kw of power over *Cts*'s 2-yr lifetime. A 3-axis stabilization system would keep the panels pointed toward the sun for power while the satellite antennas aimed accurately at the center of target transmission areas, about a timezone wide. The large-diameter low-profile spacecraft would be turned over to the Canadian Department of Communications for operation after the spacecraft reached synchronous orbit. Apogee-motor firing at 3:41 pm EST 20 Jan. put *Cts* into a synchronous orbit at 35 888 km altitude, with final station at 116°W over the equator reached 29 Jan.

Cts was the result of a 5-yr international program of cooperation between NASA and CDC to pioneer in new methods of providing communications services by transmitting high-quality color TV and other data to small user-operated ground stations in remote areas. It carried a high-efficiency 200-w traveling-wave tube amplifier developed at LeRC that would operate at power levels 10 to 20 times higher than those of previous satellites, permitting the use of smaller and less expensive ground equipment, and in a new frequency band providing microwave signals in the 12-ghz region. In 1971, the World Administrative Radio Conference had begun assigning frequencies between 11 and 14 ghz and 18 and 30 ghz to signal-relay and broadcast satellites of the future, in hopes of averting a communications jam in orbit in the 1980s. The satellites could provide any country with a means of transmitting its own TV programs and setting up medical and educational consulting services in sparsely settled regions by using small, inexpensive—even portable—ground stations. (NASA Release 76-9, 75-316; MOR E-610-76-01 [prelaunch] 22 Dec 75, [postlaunch] 28 Jan 76, [postlaunch #2] 13 Feb 76; ESA news releases 16 Jan 76, 6 Feb 76; *W Star*, 18 Jan 76, A-2, 21 Jan 76, A-11; *NYT*, 30 Jan 76, C 11)

18 January: Ats 3, launched 5 Nov. 1967 primarily as a comsat, went into semiretirement when the National Oceanic and Atmospheric Administration turned off its multicolor spin-scan cloud camera that had taken the first color photograph of the earth from space. During its 8 yr of orbiting 35 800 km in space synchronously with the earth's rotation, *Ats 3* helped prove the theory that continuous viewing of earth's cloud-cover would provide meaningful weather information. *Ats 3* transmitted thousands of photographs of hurricanes, tornado-bearing thunderstorms, and other life-threatening weather phenomena, including the timely warning of Hurricane Camille before its assault on the Gulf Coast in Aug. 1969. Besides weather pictures, *Ats 3* had relayed live TV coverage of events such as the Apollo missions, Pope Paul's visit to Colombia, and the 1968 Olympic games in Mexico City. Although superseded by more sophisticated geostationary spacecraft such as *Sms 1* and *2* and *Goes 1*, which offered better visual transmission and provided nighttime imagery through infrared sensors, *Ats 3* would continue to relay data as part of NOAA's worldwide weather-facsimile broadcast system, and would be used by NASA in communications demonstrations such as medical and educational experiments. (NOAA Release 76-5)

- Reviewing the Energy Research and Development Administration's first year of operation, ERDA Administrator Robert C. Seamans Jr. said the U.S. had taken major steps to assure adequate energy supplies in the future. Establishment of ERDA on 19 Jan. 1976, bringing together energy research functions formerly located in other agencies, was itself a milestone. Specific steps cited by Seamans included a major solar-heating demonstration and completion of two heating-and-cooling projects; development of a national energy R&D plan emphasizing conservation of energy and development of new sources; demonstration and testing of more efficient recovery and use of coal, oil, and shale; biohazard screening of 150 compounds in fossil fuels to determine which might cause cancer or genetic mutations; and identifying problems in the nuclear-fuel cycle, including reprocessing of fuel and disposal of wastes. Describing fossil and nuclear energy as the near-term and midterm resources of the nation, Seamans said that conservation should enable them to carry the U.S. into the next century, when essentially inexhaustible energy sources should be available: solar electricity, breeder reactors, and fusion power. (ERDA Release 76-5)
- The Apollo lunar scientific experiments package (ALSEP) set up on the moon in Feb. 1971 by astronauts Alan B. Shepard and Edgar D. Mitchell during the third manned exploration of the moon ceased transmission, probably because of failure of electronic components, engineers at Johnson Space Center said. The *Apollo 14* instrument, one of five stations on the moon, had provided scientists with lunar seismic data in combination with stations set up during the *Apollo 12, 15, 16,* and *17* missions. Originally designed with a 1-yr lifetime, the ALSEP 14 was first of the stations to fail completely, although it had lost its ability to receive commands a year previously. The remaining stations continued to provide data on the moon's seismic activity, heat flow, interactions with earth's magnetic field, and the solar-wind and cosmic particles that continuously bombard the lunar surface. Findings based on ALSEP data included measurement of moonquakes, mostly at about 4 on the Richter scale; indications of a lunar core at or near the melting point; a thick lithosphere that probably prevented mountain-building on the moon; and a tenuous atmosphere deriving from solar-wind particles. The seismic instruments on the ALSEPs were the most sensitive in existence, and the remaining stations were estimated to last for another 3 yr. (*JSC Roundup*, 13 Feb 76, 1)
- Food preparation and packaging techniques developed by NASA to feed Apollo and Skylab crews during space flight had been used in a pilot program to provide balanced nutrition to elderly persons living alone, the Johnson Space Center announced. The program, called Meal Systems for the Elderly, was part of JSC's Technology Utilization program to apply space-developed technology to solution of earthbound problems. The goal was to provide a meal system that could be opened, cooked, eaten, and cleaned up by older people living alone; each meal would provide at least a third of the daily diet allowance for an older person, and a field demonstration had been scheduled that would evaluate the meal design and delivery methods by the end of 1976. The JSC team developing the

program would aim at a shelf-stable, multimeal package that could be distributed by several methods—even parcel post—to senior citizens living beyond the range of hot-meal delivery or to those in areas where meals were not provided. (NASA Release 76-6; JSC Release 76-02)

20 January: Dr. Wernher von Braun, who had retired in 1972 as Deputy Associate Administrator for Long-Range Planning at NASA Hq, said in a *Washington Star* interview that he had been brought to Washington in 1970 to "lend clout to NASA's ability to get its appropriations." His assignment had been to present to Congress the program of a manned expedition to Mars; "so many national problems" had arisen that Congress would not commit itself to another multibillion-dollar space program, and the reusable space shuttle was the only element of the concept that survived. Dr. von Braun recalled that NASA's resources were cut to about a third during his last 2 yr at NASA, and the new organization with which he was associated—the National Space Institute—was created to "reach large numbers of people" and broaden public support of basic science, "where it is much more difficult to predict the payoff." "Remember," he added, "nuclear energy, which ultimately was used in the hydrogen bomb, was discovered because some astrophysicists were interested in what keeps the sun hot." (*W Star,* 20 Jan 76, A-1)

21 January: President Ford sent a $394.2-billion FY 1977 budget request to Congress, an increase of $44.8 billion over the FY 1976 request; however, comparisons were distorted by the switch this year to a new federal budget calendar. (FY 1976 would end as usual on 30 June, but FY 1977 would not start until 1 Oct.; for the "transition quarter," July through Sept., the budget would run about $98 billion, revenues at $81.9 billion, and the deficit at $16.1 billion. The change appeared in the Budget Reform Act of 1974, to give the Congress more time after it convened in Jan. each year to adopt a budget for the coming fiscal year.)

The budget attempted to explain "what went wrong" when an estimated national deficit of $9.4 billion turned out as an actual deficit of $43.6 billion; in FY 1975, the last completed fiscal year, spending totaled $324.6 billion ($20 billion more than anticipated), whereas receipts were only $281 billion, compared with an original estimate of $295 billion, the difference partly owing to an antirecession tax rebate enacted toward the end of the fiscal year. The biggest single factor—$6.5 billion spent in unemployment compensation—resulted from the recession unforeseen early in 1974; apart from higher unemployment payments under the existing program, Congress had extended benefits and coverage, accounting for $1.5 billion of the increase. The budget said that most of the discrepancy "can be explained by differences between actual and assumed economic conditions and the effects of new legislation." The *NYT* commented that "total figures for outlays, receipts, and the deficit in the [budget] document submitted at the beginning of each year for the fiscal year to follow have become increasingly meaningless."

In a press briefing on his budget, President Ford stated his intent to "restrain the growth of federal spending and restore the vitality of the private economy." This was expected to be the keystone of his campaign for reelection. Total federal outlays under the proposed budget would increase 5.5% over the current fiscal year, less than half the average annual growth in federal spending over the past 10 yr. Largest percentage increase in the budget again would be in the field of energy research; aside from defense, increased spending in areas such as mass transit and water pollution would represent work on programs already begun rather than new starts.

Highlights of the budget were:
- Increased defense spending of about $10 billion, exceeding $100 billion for the first time.
- Increased Social Security payments by employers and employees, to close the gap between income and outgo.
- A proposal for consolidating about 60 current federal programs—in areas such as health, education, child nutrition, and social services—into "block grants" to the states, with few strings attached.

Proposed spending for defense represented a turnaround of sorts: defense spending had fallen for a while after Vietnam, then rose again in FY 1975 with the end of the draft and conversion to higher priced all-volunteer armed forces with an upward pressure on wages. The outlay in FY 1977 would be for weapons; more than half the recommended increase would go for procurement, research, and development. Defense at $101.1 billion would constitute a fourth of the total budget, as in recent years, and Social Security about another fifth. Interest on the public debt would add a tenth; these three items would account for more than half. Five other basic budget items—Medicare-Medicaid, unemployment compensation, civil service retirement, and veterans benefits—would make up three fourths of the total, leaving only a fourth of the total budget for all remaining government spending.

Science, represented by $23.5 billion in federal R&D spending proposed for FY 1977, would use more than half that amount for military projects; the space program would get $3.6 billion, only 4% more than the current year and much less than price escalation. Highest priorities for increases in civilian spending on science would be colleges and universities, which would get $2.6 billion in research funds, and the National Science Foundation, which would get $812 million. NASA, under the proposed budget, would postpone purchase of one of the Space Shuttles for a year, and would drop its plan for a Mars–Jupiter–Saturn explorer. At the height of spending for the Apollo program 10 yr ago, civilian and nonspace science got 20% of R&D funding; in FY 1977, it would get 37%. (*W Star*, 21 Jan 76, A-1, A-10, A-11; *W Post*, 22 Jan 76, A-1, A-14, A-15, A-16; *NYT*, 22 Jan 76, 1, 24, 25, 26)

- At a hearing on solar power from satellites conducted by the subcommittee on aerospace technology and national needs of the Senate Committee on Aeronautical and Space Sciences, NASA Deputy Administrator George M. Low and Dr. John M. Teem, ERDA's Assistant Administrator for Solar, Geothermal, and Advanced Energy Systems, were questioned closely by Sen. Wendell H. Ford (D–Ky), subcommittee chairman, on the ERDA and NASA responsibilities for the program. Testimony revealed that ERDA had prime responsibility for solar, as well as other, power-generating concepts, and that NASA had not requested funds for developing an independent program but was working on solar-power ideas to use in satellite missions. Primary goal of ERDA's program, Dr. Teem testified, was development and demonstration of terrestrial solar-energy applications that were commercially attractive and environmentally acceptable: solar thermal electric, photovoltaic, wind, and ocean-energy conversion. Dr. Teem pointed out that the FY 1977 budget included no direct funding for energy R&D for NASA, as the basic responsibility rested with ERDA. Dr. Low added that, when ERDA decided that research should proceed, NASA would do the work with ERDA funding. (Text, pp. 192-227)

- Flight tests by the Air Force had confirmed the concept called LATAR (laser-augmented target acquisition and recognition) that would permit a pilot in a high-performance single-seat fighter to fly the aircraft and operate its weaponry at the same time. The LATAR pod, mounted under the fuselage of an F-5E directly beneath the cockpit, contained a laser-target designator, a spot tracker, and an electro-optical sensor. The spot tracker was a device that finds a laser "spot" from another target designator operated from another aircraft or on the ground; initial target identification could be either visual, by radar, or by a helmet-mounted sight system that integrated sight control with movement of the pilot's head. The flight tests were conducted at the Air Force Flight Test Center, Edwards AFB, Calif. (OIP Release 308.75)

22 January: The Aerosat Space Segment Board, representing the European Space Agency, Comsat General Corp., and the government of Canada, decided to issue requests for proposals for a supply of Aerosat spacecraft beginning 1 Mar., with target date for award of contract 15 Nov. The Aerosat program, to be carried out by ESA, the U.S. Federal Aviation Administration, and the government of Canada, would set up an experimental satellite-communications system between transoceanic aircraft and the ground to test functions and timing of an operational system for the International Civil Aviation Organization. Two satellites designed for 7-yr lifetime would be launched into geostationary orbit over the Atlantic Ocean, probably in 1979-80. (ESA release 23 Jan 76; Comsat General Release 76-103)

- Commenting on the first 3 mo of operation of the USSR Venus orbiters, Tass reported that optical instruments had yielded new data on the planet's cloud layer, much more transparent than earth clouds, and on luminescence of the dark side of Venus, formed in a relatively narrow layer at high altitudes. The orbiters were measuring magnetic fields connected with the solar wind in the vicinity of Venus, but had not registered a magnetic field of the planet itself, the report said. (FBIS, Tass in English, 22 Jan 76)

23 January: Three space scientists at Johnson Space Center would begin a week-long round-the-clock test of experiments and procedures in a mockup of the Spacelab designed for the Space Shuttle. The team, headed by astronaut-physician Dr. Story Musgrave, would eat, sleep, and perform space-related duties in a mockup of the Shuttle orbiter crew compartment, and would carry out more than 20 experiments in space medicine and one in space physics inside the 6.8-m-by-4.06-m Spacelab mockup. The team inside would be in constant communication with flight operations engineers and scientists at control points outside the mockup, which had been equipped with both instruments and experiments just as the Spacelab would be. The team would demonstrate 14 operational tests whose results would serve in planning inflight crew activities, procedures, and scheduling, and in studying items such as personal hygiene, general housekeeping and special-purpose cleaning and maintenance, and functional utility of the orbiter aft deck from which many of the Spacelab experiments would be monitored. The simulation was also designed to evaluate ground-support procedures and data-handling techniques. (JSC Release 76-04)

- INTELSAT—the International Telecommunications Satellite Organization—awarded a 9-mo $75 000 contract to Hughes Aircraft Co. for development of improved vibration-test techniques for testing future Intelsat spacecraft, leading to reduced spacecraft weight and a lessening of spacecraft fatigue. (INTELSAT Release 76-1-M)
- The Soviet Union would use artificial satellites and manned orbital stations to investigate dynamic processes in the ocean during the new 5-yr plan period, Tass reported. Soviet oceanologist Leonid Brekhovskikh in an interview forecast a growing number of experiments conducted jointly with other countries, among them a "polymode" experiment with the U.S. to study the nature of huge vertical formations in the ocean discovered by Soviet scientists in 1970. The USSR also planned a global satellite experiment called "pigap" for concurrent study of dynamic processes in the atmosphere and the ocean. (FBIS, Moscow Tass, U 1, 29 Jan 76)

24 January: The National Transportation Safety Board reported that U.S. commercial airlines had the best safety record in 1975 since 1957, with only 42 accidents and a total of 124 deaths, 112 of those in the major air disaster of the year: the crash of an Eastern Airlines flight short of the runway at Kennedy International Airport in New York City on 24 June 1975. The fatality rate—statistically 0.001 per million air miles flown—was especially encouraging as it came after 467 persons had been killed a year earlier in commercial aviation accidents. John H. Reed, acting chairman of the safety board, commented that it was "as difficult to explain a good year as it is a bad one . . . we'd like to feel that all the effort toward safety is a contributing factor." (*W Post*, 25 Jan 76, A3)

25 January: The Great Galactic Ghoul struck again, reported *Washington Post* staff writer Thomas O'Toole, when one of the unmanned Viking spacecraft (*Viking 2*) approached the orbit of Mars and lost one of its three ovens designed to heat samples of the Martian surface and look in the gases for signs of Martian life. Space scientists found that almost all the accidents with Mars-bound spacecraft—three failures and four near-failures—had occurred in the same region, which had no unusual features such as extra cosmic dust or increased solar wind, magnetic field, or background radiation. First "victims of the Ghoul" were Soviet spacecraft—*Zond 2* in 1964, and *Mars 1* a year later—whose batteries died when they crossed the area and never came on again. One Mariner lost its radio in the Ghoul's orbit but came back on when it reached Mars; another lost one or two instruments to cosmic dust; two others lost their guidance stars when they reached the Ghoul's orbit, but locked on them again when they left it. Most costly casualty of the Ghoul was *Mariner 7*, whose battery exploded the day in 1969 when it crossed the Ghoul's orbit, damaging the rest of the spacecraft so that it was useless by the time it flew past Mars. The area was about 56 million km from earth and 209 million km from the sun, and *Viking 2* would not leave it until some time in February; scientists at the Jet Propulsion Laboratory were hoping for no further damage to it or to the other Viking, which was coming along a month behind the first. (*W Post*, 25 Jan 76, A-3)

- Establishment of an Office of Science, Engineering, and Technology Policy in the White House, a goal of U.S. scientists since 1973, would become effective within 6 wk or so, the *New York Times* reported. Since President Nixon had abolished a similar office, the scientific community had been concerned over the lack of provision at the top level of government for sophisticated advice on science and technology affecting U.S. policies on world food supplies, environmental pollution, energy, transportation, disarmament, and similar problems. The Ford Administration had encouraged Congress to establish such an office by law rather than set one up by executive order, and the House of Representatives had passed an Administration-backed bill in Nov. 1975. A Senate bill, "roughly comparable" according to the *NYT*, had been agreed upon by several committees and was on its way to passage; this measure would establish a White House office with a director and as many as 4 associates, who would work closely with the Office of Management and Budget on science and technology budgets and would prepare an annual report for the President to send to Congress on science, engineering, and technology, with options on federal investment and priorities in the three related fields. The Senate bill would also call for authorization of funds to establish science and technology offices in each state; the House bill did not have similar provisions. (*NYT*, 25 Jan 76, 26)

- The Federal Aviation Administration's flight service station at Tucumcari, N.M., was receiving reports of cigar-shaped flying objects with pulsating colored lights dominating the night skies over Clovis, N.M. The objects were reported by policemen and residents for 4 nights, and a reporter photographed one of the objects showing a curved strip of light against a black background. FAA radar had shown nothing unusual, a spokesman said. (W *Star*, 25 Jan 76, A-7)

26 January: Prompted by the crash of a TWA flight near Dulles Airport that took 92 lives in Dec. 1974, the Federal Aviation Administration hired 6 veteran airline pilots to analyze air traffic-control procedures, but the findings were so scathing that the report had not been released, said an editorial in the *Washington Star*. The National Transportation Safety Board this week issued its findings in the case with a "rare dissent" in assigning blame for the crash. The *Star* pointed out that, although the Air Line Pilots Assn. disagreed with the majority finding of pilot error and the Professional Air Traffic Controllers disagreed with the minority finding of ground-control error, leaders of both groups—and members of the board—all questioned FAA handling of its responsibilities. The *Star* called on new FAA administrator John McLucas, sworn in in Nov., to make air safety his highest priority. (*W Star*, 26 Jan 76, A-14)

- The National Oceanic and Atmospheric Administration (NOAA) had joined in research in Antarctica on stratospheric pollution with a team from the Univ. of Wyo. supported by the National Science Foundation. NOAA's Aeronomy Laboratory would launch 2 balloons carrying probes with containers that would open at different altitudes to collect a vertical profile of fluorocarbons and nitrogen oxides in the relatively clean stratosphere over Antarctica. The measurements would be used in testing two theories of ozone destruction in the stratosphere, the first being the production of destructive chlorine through breakdown of manmade fluorocarbon aerosols migrating into the upper atmosphere. The other theory posed a natural ozone-limiting cycle, involving dissociation in the stratosphere of bacterially produced nitrous oxide that became nitric oxide harmful to the ozone layer. The Antarctic launches would help to explain variations in nitrous oxide measurements as a function of latitude, season, or some other cause. (NOAA Release 76-17)

27 January: Laser beams had become the latest major tool in manufacturing aircraft, the Aerospace Industries Association announced, reporting on a year-long series of tests of the numerically controlled laser cutting-arm technique. The test objective was to define effects of a laser on physical characteristics of the parts produced, such as corrosion and fatigue. The cutting arm was coupled with a computer data bank that chose the best method of positioning parts on flat thin aluminum sheets, saving both time and money and reducing aluminum scrap by 30 to 50%. Usual methods such as blanking, routing, and sawing had been used since World War II. The Air Force was funding a program to use thicker sheets of metal and higher power lasers in an application of new technology to aircraft manufacture. (AIA Release 76-4)

28 January: The *Viking 1* on its way toward Mars might have lost the use of one of its ovens just as *Viking 2* was reported last week to have done, project officials said. Each of the Mars landers carried three ovens to heat surface samples for analysis by a gas-chromatograph mass spectrometer that would determine atmosphere elements and search for organic material that would indicate biological or nonbiological activity. Loss of one oven on each craft would not affect the operation of the instruments, but would mean that only two instead of three soil analyses would be done. A separate biology instrument on each lander would search directly for life forms in soil samples. A monitoring device on the ovens was suspected as the cause of the test-data anomaly. (NASA Releases 76-14, 76-15)

- The 12th annual meeting of the American Institute of Aeronautics and Astronautics opened in Washington with the theme "Aerospace and Public Policy," offering 9 panel discussions beginning with one on the Space Shuttle during which John F. Yardley, NASA Deputy Administrator for Space Flight, would give the keynote address on shuttle program status and flight operations plans. Other panels would discuss domestic direct-broadcast satellites; inflation, capital formation, and the aerospace industry; national transportation policy; military aircraft selection; exporting aerospace technology; aerospace and energy; transition to the future; and the environmental impact of aerospace operations. The AIAA convention was preceded by a 2-day aerospace sciences meeting during which about 200 papers were presented at 75 technical sessions. The AIAA's von Kármán lecture would be presented by I.E. Garrick, distinguished research associate at NASA's Langley Research Center, speaking on aeroelasticity.

At the honors night banquet 29 Jan., 14 major AIAA awards would be presented; fellows and honorary fellows of 1975 would be honored; and Edgar M. Cortright—former director of LaRC and previously Chief of Advanced Technology, Assistant Director of Lunar and Planetary Programs, and Deputy Associate Administrator for Space Science and Applications and later for Manned Space Flight at NASA Hq—would be installed as AIAA president. Dr. Cortright had left NASA in Sept. 1975 after 17 yr to become vice president and technical director of Owens-Illinois Inc. Speakers at the honors banquet would be Clarence "Kelly" Johnson, chief of Johnson Space Center's spacecraft design division, who would receive the spacecraft design award of the year, and astronaut Neil A. Armstrong, who commanded *Apollo 11* and took the first walk on the moon. (AIAA releases 6 to 16 Jan 76)

- A communications satellite system to serve more than 200 U.S. public television stations within the next 2 yr was proposed by the Public Broadcasting Service and the Corporation for Public Broadcasting in a report prepared for managers of PBS member stations before the annual meeting in Los Angeles next month. If the member stations approved the proposal and the Federal Communications Commission followed with its approval promptly, the system could be in operation within 21 mo, putting PB ahead of commercial networks by renting circuits on satellites operated by Western Union Telegraph Co. instead of leasing land lines from American Telephone & Telegraph at $12 million a yr. The $38-million satellite system would offer 3 or 4 channels instead of the single land line; expense of additional land lines had precluded additional transmissions to special audiences, such as Spanish-speaking populations in major cities and the U.S. Southwest; the cost of the land lines would increase yearly; and the cost of adding other stations to the PB satellite system would be limited to the cost of a simple ground receiver. (*W Star*, 28 Jan 76, A-1; *W Post*, 29 Jan 76, A-7)

29 January: Intelsat IV-A F-2, second in a series of improved commercial comsats, was launched from Cape Kennedy on an Atlas-Centaur at 6:56 pm EST (2356 GMT) into a transfer orbit at about 31 000-km apogee, aimed at a station over the Atlantic Ocean at 29.5°W which it should reach by April. Working as a backup to *Intelsat IV-A F-1*, the newly launched spacecraft would provide almost two thirds more communications capability than previous Intelsats, by increasing the number of transponders from 12 to 20 and by using an improved antenna system that permitted frequency reuse by beam separation. The east-pointed am used for transatlantic service would illuminate Europe and Africa;

the same 320-mhz bandwidth would be reused in the west beam aimed at North and South America, doubling the usability of the frequency and the satellite's capacity. The cylindrical spacecraft was about 7 m high and 2.38 m in diameter, with a solar panel about 2.8 m long; liftoff weight was 1515 kg, and in-orbit weight 825.5 kg. (NASA Release 76-8; MOR E-491-633-76-02, 12 Feb 76 [prelaunch], 25 Feb 76 [postlaunch])

- Soon after Rockwell International Corp. was named prime contractor on a $5-billion Space-Shuttle program, Rep. Olin E. Teague (D-Tex.), chairman of the House Science and Technology Committee, had accepted an invitation for a trip to Rockwell's resort in the Bahamas. Teague's name was added to a list of senators and representatives who had accepted invitations from Rockwell, a prime government contractor; most went to hunt at the company's lodge in Maryland. Three Senate staff members had accepted invitations to hunt at the Maryland lodge of another contractor, Northrop Corp. During the previous week, the Pentagon had rebuked 38 civilian and military officials, including admirals and generals, for accepting similar invitations from Northrop. House rules prevented members, officers, or employees from accepting gifts "of substantial value, directly or indirectly, from any persons, organization, or corporation having a direct interest in legislation before the Congress;" the Senate had no similar written regulation. (*W Post*, 29 Jan 76, F-5)

- A vacuum test facility built near Philadelphia in 1962 by General Electric to test satellites would be used to freeze-dry most of New York City's air-pollution control records, damaged in a 16th-floor records office on Wall Street by flooding from a ruptured check valve in the 17th-floor airconditioning system, the *New York Times* reported. Hundreds of thousands of cards and pages had been shipped in a refrigerator truck to Pa. to be dried before mold could form and consume the cellulose fibers in the paper. After placement in the chamber, the records would undergo a vacuum so that the water would turn into vapor, move to a condenser, and turn to ice; after being frozen and dehydrated, the papers would be rewarmed. Use of the facility would save "hundreds of thousands of dollars" and untold time in attempting to duplicate the records, the Commissioner of Air Resources said. (*NYT*, 29 Jan 76, 35)

30 January: Col. Stuart A. Roosa, 42, an astronaut since 1966, announced in Houston that he would retire from the Air Force and from NASA on 1 Feb., one day after the 5th anniversary of the launch of his only space flight, *Apollo 14*. Roosa had piloted the command module while Capt. Alan B. Shepard and Cdr. Edgar D. Mitchell had made the third moon landing, and had previously been on the *Apollo 9* backup crew. Roosa told associates he was exploring "companies involved in new procedures." (*NYT*, 30 Jan 76, 34; *W Star*, 30 Jan 76, A-2)

- *San Marco C-2*, launched from the San Marco platform 18 Feb. 1974 into an elliptical orbit to record day-by-day variations in the equatorial atmosphere's density, composition, and temperature, was declared a success after 23 mo in orbit during which it obtained more than 17 diurnal cycles of data. The data had been correlated with data from *Explorer 51* (AE-C) in studying the physics and dynamics of the thermosphere, and these studies would continue as more data from the two satellites were processed and analyzed. Two U.S. scientific instruments were still operating normally, but the Italian air-drag balance instrument (which malfunctioned shortly after launch) had not performed properly. (MOR S-894-74-04, 30 Jan. 76)

- INTELSAT—the International Telecommunications Satellite Organization—awarded an 18-mo $275 000 fixed-price contract to TRW Systems, Inc., for a skewed reaction-wheel thruster attitude-control and stabilization system. A system based on the reaction-wheel principle would offer a combination of light weight, high reliability, and accurate performance for future comsats. External disturbances of satellite stability would be canceled by spinning up the reaction wheels to create offsetting torques, then stopping the spin by use of thrusters. (INTELSAT Release 76-4-M)
- INTELSAT—the International Telecommunication Satellite Organization—awarded a 12-mo $22 000 fixed-price contract to the Republic of China's Government Radio Administration for collection and study of ionospheric scintillation data at 4 ghz, to be used as a data base for studying satellite-signal fluctuations caused by ionospheric effects. (INTELSAT Release 76-5-M)

During January: Using Landsat imagery of an 85 000-sq-km region from the western edge of the Nile to the border of Libya along the Mediterranean coast, the Egyptian Academy of Scientific Research and Technology at the request of President Anwar Sadat made a study of the region's geology, drainage, and ground water. The study resulted from problems of a rising ground-water table and salinity on newly reclaimed areas west of the Nile Delta. A copy of the report, sent to NASA by the Egyptian academy's remote-sensing research project, said in part, "The new maps prepared from Landsat satellite images are more detailed and much more elaborate as compared to previous maps prepared by the traditional methods . . . It has been possible to construct on the Landsat images 14 geological and environmental units in the investigated area . . . This work helps to decipher the geological history of northern Egypt, the Mediterranean Sea and the Nile Delta . . . It is emphasized that additional irrigation waters should not be introduced in the area of investigation until detailed geological and hydrogeological studies have been carried out." (Wkly Briefs for Administrator, 12 Jan 76)

- Two scientific experiments carried on last July's Apollo-Soyuz (ASTP) mission had produced spinoffs that would benefit sufferers from phlebitis and leukemia. Each of the experiments had used electrophoresis—passing an electric current through a solution to separate differing types of cellular matter—under the weightless conditions of space flight to produce results unobtainable in earth's gravity. One experiment used human kidney cells in isolating pure samples of the 5% of cells that manufactured an enzyme (urokinase) effective in removing blood clots from veins and arteries, with major potential for treating persons with conditions such as phlebitis. The separated cells, frozen and returned for use as starters in a culturing process, provided 6 to 7 times more urokinase than the original sample. The other spinoff was the successful testing of a preservative medium developed for use on the ASTP mission that would make it possible to perform transfusions of a certain type of white blood cells in the treatment of leukemia; the preservative, used to freeze granulocytes on the mission, actually improved the survivability of the cells so that a stock of the material could be kept frozen for use upon demand, instead of searching for donors with resultant delay in emergencies. (NASA Releases 76-3, 76-5)

- Communications via satellite was a science-fiction writer's dream 30 yr ago, but within 2 decades after Arthur C. Clarke predicted it, the first comsat—*Early Bird*, launched 6 Apr. 1965 into synchronous orbit—began operations 28 June 1965 and opened a new chapter in communications history. In an article in the *Telecommunications Journal*, Santiago Astrain, secretary general of the International Telecommunications Satellite Organization (INTELSAT), viewed the past 10 yr, up to the new Intelsat IV-A, largest operational comsat to date.

  The brief history of INTELSAT summarized the organization's development from 1960, when President Eisenhower called for the creation of a global communications system; after the United Nations passed a resolution supporting the concept, the U.S. Congress passed the Communications Satellite Act of 1962 that brought the Communications Satellite Corporation (ComSat) into being. On 20 Aug. 1964, 11 countries (representing 85% of the world's telecommunications traffic) created an international consortium to be managed by ComSat on an interim basis. On 12 Feb. 73, the Definitive Agreements were signed; by the time they took effect, the number of member nations had increased to 80. Membership was open to all countries that were members of the International Telecommunications Union (ITU); any country, whether or not a member of INTELSAT, could have access to the system on a nondiscriminatory basis.

  Current planning was geared to meet a traffic demand for capacity to handle about 70 000 telephone channels by 1984; this would represent a 6-fold increase within 10 yr. Intelsat V satellites due to enter service in 1979 would provide more than 23 000 channels. As INTELSAT had the object of providing high-quality communications service, its policy had been to maintain a "spare" comsat in each ocean to make additional capacity always available; this had enabled the organization to provide domestic services at a reasonable cost. Since operations began in 1965, the number of countries actually participating in operations had increased from 5 to 64; the number of earth stations, from 5 to 90; and the number of antennas, from 5 to 115. (*Telecommunications Jl*, vol 42, 672-677)

- NASA issued 2 major reports—SP-386, "Outlook for Space," and SP-387, "A Forecast of Space Technology 1980-2000"—prepared by a task group initiated by Administrator James C. Fletcher in June 1974 to examine the civilian role of the U.S. space program during the coming 25 yr. The group included 20 representatives of NASA and 1 from the USAF. "Outlook for Space," according to the foreword by Fletcher, was "an impressive analysis of the services that space systems and technology might provide the world of today and tomorrow," and related this analysis directly to national needs and human purposes.

"A Forecast of Space Technology," according to a preface by study director Donald P. Hearth, "was an important element of the study and provided key inputs to the study and its conclusions." The technology forecast was conducted by a team from the Jet Propulsion Laboratory, supported by representatives of other NASA centers; Jack N. James and Rob Roy McDonald of JPL led the technology team. This report forecast developments in technology of acquiring, processing, transferring, and storing information, energy, and matter.

"Outlook," on the other hand, identified future objectives of space programs and prepared background information that could serve as a basis for development of program plans. Objectives were established as earth-oriented or extraterrestrial; categories of need were either physical (provision of food, shelter, health, security, education, good environment, and the work necessary to obtain these) or needs of mind and spirit (the quest for knowledge, the need to explore the unknown, the sense of accomplishment in the face of challenge). The 3 conclusions of the report were that a space program could help fill the need to improve food production and distribution, to develop new energy sources, to meet new challenges to the environment, and to predict and deal with natural and manmade disasters; that the space program could answer the need for intellectual challenge, exploration, and knowledge that would help humanity understand its relation to the universe; and that those in charge of the space program must make it recognized as meeting these public needs. Efforts recommended by the study group would be in data management, predictive modeling of future missions, advanced communications, space processing, and a permanent space station. (Text)

February 1976

1 February: Scientists at Mass. Institute of Technology, using NASA's *Sas 3* x-ray astronomy satellite, found a stellar object unlike any other and were unable to say what it was. The discovery, located by Dutch astronomers in the center of the galaxy in the constellation Sagittarius, gave off x-ray bursts that were unevenly spaced, although pulse after pulse proved to be identical. The object might be a single new source or a combination of old ones, but its distance was millions of light years and the source might be nonexistent now, the investigators said. (*W Star*, 1 Feb 76, A-6)

2 February: A "so far secret NASA material" in outfits worn by 3 U.S. skaters was credited by an Olympic games official with the clean sweep of the international women's speed-skating championships at Davos, Switzerland. Sheila Young of Detroit, Mich., set a world record of 40.91 sec for the 500-m race; Leah Poulos of Northbrook, Ill., set a U.S. national record of 2:13.98 for the 1500-m event, followed by Young and Nancy Swider of Park Ridge, Ill. The press chief of the Innsbruck organizing committee said a flood of records could be expected because skating garments "contain an aerodynamic property which is better than any used before," and attributed the U.S. victories to a "frog suit" made of the new material. U.S. Olympic Committee officials said they had no knowledge of such suits, and team members were not available for comment. (*W Star*, 2 Feb 76, D-1)

3 February: Tests on the Lageos—a geophysical research satellite expected to remain in orbit several million years—concluded at Goddard Space Flight Center at the end of Jan., NASA announced, and the satellite would be shipped to the Western Test Range in California for launch in late April. Called the "cosmic golfball" because of its 426 special reflectors designed to return laser pulses to their exact point of origin on the earth, Lageos would provide a stable point in the sky to permit measurement of the relative locations of participating ground stations within a few centimeters; these measurements would provide scientists with models of earth's crustal motion, useful in predicting earthquakes. Manager of the Lageos project would be the Marshall Space Flight Center at Huntsville, Ala. (MSFC Release 76-17)

- A team of engineers and scientists at NASA's Jet Propulsion Laboratory had devised a way to transfer blood without contamination and lengthen its safe storage time, NASA announced. Using a concept developed in spacecraft sterilization, the team—under contract to the National Heart and Lung Institute of the National Institutes of Health—worked out an aseptic fluid-transfer system on which patents had been applied for by the Calif. Institute of Technology, which operates JPL for NASA. Two dual-walled tubes, fused and penetrated by heat, would permit clean transfer of blood from one container or bag to another; the unique connector would be manufactured as part of the container. Outer portions of the tubing would be polyvinyl chloride, and the inner parts of heat-resistant kapton; application of heat at 200°C through a metal clamp to the flat end links of each tube would effectively fuse the tube and sterilize the adjoining areas. Heat could be applied by a portable device no larger than an ordinary hair dryer. The method was reported to kill 99.999% of all bacteria and spores, even when surfaces had been purposely contaminated. (NASA Release 76-20)

- *Pioneer 10*—on its way out of the solar system—would cross the orbit of Saturn nearly a billion miles from earth on 10 Feb., NASA announced. As it crossed Saturn's orbit, the Pioneer would be 1 384 600 000 km from the sun and 1 435 807 000 km from earth; however, the big-dish antennas of NASA's Deep Space Network should be able to hear the spacecraft as far out as the orbit of Uranus—3.2 billion km from earth—which it should reach in 1979. Systems aboard the 256-kg spacecraft were still operating, and the Pioneer was returning valuable information on the character of the interplanetary medium in the unexplored space beyond the orbit of Jupiter, which it flew past in Dec. 1973. Project officials said that communication might be possible well beyond Uranus. *Pioneer 10* carried a message for any intelligent beings who might retrieve it on its wanderings through the galaxy; scientists calculated that it might encounter a star system once every million years, and should remain in good condition even though its nuclear power source would die in a few decades. (NASA Release 76-21; ARC Release 76-04)

- An environmental satellite operated by the National Oceanic and Atmospheric Administration (NOAA) had used its capability of measuring ocean temperatures to help fishermen on the Calif. coast find tuna and salmon. A pilot project using satellite imagery to locate areas of "upwelling" off the coast—areas where cold nutrient-rich water would rise from the bottom to the surface—had shown that these areas were favored by the fish, and the fishermen could save time and fuel by using the data to locate them. NOAA's polar-orbiting satellite passing over the coast twice daily would use visual and infrared sensors to relay data including sea-surface temperatures to earth, where images of thermal fronts indicated upwelling areas. The fronts transferred to navigation charts would be available to fishermen at northern Calif. ports. NOAA's National Environmental Satellite Service conducted the research through a NOAA-supported sea grant program at Humboldt State Univ.; the technique would be applicable to fishing industry in many parts of the world. (NOAA Release 76-10)

- A new manmade fiber stronger than nylon for possible use in drogue parachutes had been tested in the 5-m transonic tunnel at the Arnold Engineering Development Center in Ohio, the Air Force announced. The material, called Kevlar, had been developed by DuPont for use in tires, and could be woven into parachute materials twice as strong as nylon with only half the weight and volume of nylon counterparts. The AEDC tests used 4 ribbon-type parachutes 2 m in diameter, tested to destruction; one was 100% nylon, another 100% Kevlar, and the remaining two were composed of different blends of the two substances. The tests measured dynamic loads as the chutes were deployed at about 966 km per hr at simulated altitudes from 15 to 115 km; steady-state loads were also measured at those altitudes for speeds of from 640 to 1440 km per hr. The material-evaluation tests were a novelty for the tunnel, normally used for aerodynamic studies of large-scale models or engine compatibility with flight hardware. (AF OIP 003.76)

4 February: First mainstage test of the Space Shuttle main engine occurred at the National Space Technology Laboratories in Hancock County, Miss., Marshall Space Flight Center announced. The engine—called the Integrated Subsystem Test Bed—was fired for 3.38 sec, reaching and stabilizing at its minimum power level: 50% of its rated power level of 1 668 075 newtons at sea level, or 2 090 654 newtons at altitude. Tests were conducted by the prime contractor for the engine, Rocketdyne Div. of Rockwell International Corp., under MSFC direction. (MSFC Release 76-29; Rockwell Release RD-2)

- An asteroid more than 3 km across was photographed near earth 27 Jan., about 7.5 million km distant, by the Mt. Palomar Observatory, the Calif. Institute of Technology reported. The asteroid, of the type that made giant craters on planets of the solar system, was closer to earth than any known celestial body other than the moon. The institute said there was no chance that the asteroid—called 1976 AA—would collide with the earth on its present path, as the orbits of the earth and the asteroid did not touch. (*NYT,* 4 Feb 76, 21)

5 February: The "Historic Redstone Test Site" at MSFC, neglected since the final Redstone test firing in Oct. 1961, would be restored to its original appearance as an exhibit for visitors during the Bicentennial celebration, MSFC announced. The interim test stand, as it was called, was built by the U.S. Army in 1953 and used to test moderate-range Redstone rockets. On 31 Jan. 1958, a modified Redstone renamed Jupiter-C launched into orbit the first U.S. satellite, *Explorer 1.* Another modification of the Redstone tested on the stand was the Mercury-Redstone, one of which powered the first U.S. manned space flight—that of Alan Shepard in May 1961. The stand was the site of 364 test firings between 1953 and 1961, after which it was retired from a space program that had become much more sophisticated. Painted in its original colors and designs, and with a Redstone borrowed from the Army Missile Command installed on it, the site would appear much as it did 15 yr ago in its new role as an historical exhibit. (MSFC Release 76-32)

- Physicists at the Fermi National Accelerator Laboratory in Batavia, Ill., discovered a new elementary particle, heaviest ever observed, the annual meeting of the American Physical Society was told. The investigators from Columbia Univ., the State Univ. of N.Y. at Stony Brook, and the Fermi lab said the "upsilon" particle raised the possibility that physics might never be able to identify any ultimate or basic building blocks of matter. The upsilon was discovered in debris of particles created by acceleration of a proton beam aimed at a beryllium target; the resulting collision produced a cluster of electron-positron pairs with a mass of 6 billion electron volts, more than 6 times that of the proton and 1.5 times that of any other particle. The report on the upsilon particle was a last-minute addition to the annual meeting's agenda because the data were available only in the last few wk; first observed about 5 mo ago, the upsilon had been seen only 12 times so far, but was expected to require new thinking in theoretical physics. (*B Sun*, 6 Feb. 76, A–3; *W Post*, 9 Feb 76, A–3)

• The accepted explanation of solar energy—fusion of hydrogen atoms into helium at the sun's core—had been challenged by more than one recent discovery, the *Christian Science Monitor* reported. Measurements of natural vibrations in the sun, published in the journal *Nature* by a Soviet and British group, had revealed that the sun pulsed steadily every 2 hr 40 min. Such regularity had not been predicted by the fusion theory. Another group, reporting in the journal *Science* on a 3-yr experiment directed by the Brookhaven National Laboratory, found fewer neutrinos created by thermonuclear reactions at the sun's center than was "consistent with standard ideas of stellar evolution." Suggested explanations were a tiny black hole at the sun's center, which could radiate energy outward without creating neutrinos, or a shutoff of the sun's thermonuclear furnace, which would lower the number of neutrinos long before any decrease in heat or light would become apparent. (*CSM*, 5 Feb 76, 16; *B Sun*, 23 Jan 76, A–3; *NYT*, 23 Jan 76, 1; 18 Feb 76, 32)

• Thermal protection fabrics developed for use in the Apollo and Skylab programs would be the basis for research into improved clothing and equipment for firefighters, under a contract signed between NASA and the Department of Commerce's National Fire Prevention and Control Administration. The agreement assigned management responsibility to Marshall Space Flight Center for a 3-yr program funded jointly by NASA and NFPCA at a cost of $300 000 in the first year. The program would emphasize weight reduction, performance, and cost of equipment to improve the chances of a firefighter's surviving any fire uninjured, in a job described as the most hazardous in the U.S. (MSFC Release 76–31)

6 February: U.S. airlines suffered a near-record deficit in 1975, according to year-end statements recently released that showed declines for nearly all the 11 trunk carriers and a total deficit of $87 million, second only to 1970's loss of $100.8 million. In 1974, the group had earned $250.8 million. Reasons given included rising fuel prices, costly strikes, and proliferating fare discounts. The recession in the first part of 1975 cut passenger traffic for all lines, and fuel costs continued to soar another 30% in 1975 after doubling the previous year. Increasing numbers of discount fares resulted in an average of 7.04 cents of revenue per domestic passenger mile for TWA, down from 7.51 cents for the same month a year earlier. Although January 1976 traffic had increased about 11%, and predicted profits from an increase of 2% in unit revenues might produce as much as $140 million for the 11 major carriers, this would represent only a 6% return on equity compared to the 12% considered fair by the Civil Aeronautics Board. (*NYT*, 6 Feb 76, 39)

- A small-scale model of the Space Shuttle and its launch pad would be used in a test program begun at MSFC to reduce noise during a Shuttle launch. Sound energy could damage sensitive instruments or other payloads carried in the Shuttle orbiter, and reducing launch noise by modifying the launch facility would be more economical than redesigning the launch vehicle itself or its associated payloads to tolerate anticipated levels of noise. The model would test various designs that sprayed water into, above, and below the rocket exhaust; firing the engines into a water spray would convert acoustic (sound) waves into thermal energy, which would dissipate as steam. Use of the water technique had significantly reduced—but not eliminated—the damaging noise. (MSFC Release 76-33)

9 February: An ad hoc committee of 9 prominent astronomers, assisted by a group of experts from NASA, ERDA, and the Air Force and the atmospheric and astronomical science communities, reviewed the program of solar research being conducted by the U.S. Air Force's Sacramento Peak Observatory in N.M., and recommended that the National Science Foundation assume responsibility for its operation after 30 June 1976. The Air Force had established the observatory near Alamogordo in 1952 to research methods of predicting solar geophysical disturbances that might affect military responsibilities of the AF; reduced manpower had forced the AF to earmark the observatory for phaseout. Dr. H. Guyford Stever, director of NSF, announced that NSF would take over the SPO operation and continue its role in solar research at a productive level. Constructed at a cost of about $8 million, SPO had an estimated replacement value of about $20 million. (NSF Release PR76-15)

10 February: Atmosphere Explorer D, second in a series of 3 maneuverable unmanned spacecraft designed to explore a specific area of earth's outer atmosphere, had ceased functioning, spacecraft controllers at GSFC said. *AE–D* had stopped working 29 Jan. when a power-supply electronics malfunction disabled its solar array. Launched 6 Oct. 1975, *AE–D* had transmitted important data on energy transfer in the upper atmosphere and on processes critical to the atmospheric heat balance; although it did not last its design lifetime of 1 yr, it did transmit a full set of lower thermosphere data on the daylight side from North to South Pole at constant times during its 4–mo operation. *AE-C*, launched in 1973, was still functioning normally, returning upper-atmosphere data in the region above 128 km; *AE-E*, launched in Nov. 1975, had a payload similar to that of the others but also carried equipment to measure earth's ozone layer between 20°N and S. (NASA Release 76-24)

12 February: The three U.S. astronauts who participated in last summer's Apollo-Soyuz project—Thomas P. Stafford, Vance Brand, and Donald K. Slayton—left Saudi Arabia after a 2-day visit in Riyadh and headed for Abu Dhabi, capital of the United Arab Emirates, on a goodwill tour of the Middle East. (*NY News,* 12 Feb 76, 96)

- NASA announced award of an $8.8-million extension for 1 yr of a contract with Northrop Services, Inc., of Houston for operation and maintenance of lab and test facilities at Johnson Space Center. Northrop had provided these services to JSC for the past 3 yr; the extension would bring the value of the contract to $37 416 742. Northrop had used 425 employes to maintain and operate the life sciences and engineering labs and the lunar curatorial lab.

NASA also announced award of a $6.8-million extension for 2 yr of a contract with the Charles Stark Draper Labs., Cambridge, Mass., for technical support of Space Shuttle orbiter avionics software development. Draper had provided this support to JSC since 1974. The effort, employing 55 people, would include software design, verification, simulation, requirements formulation, and analysis as required for programming the guidance, navigation, and control computer for the Shuttle orbiter. (JSC Releases 76-12, 76-14)

- Louis Morton, one of the foremost U.S. military historians and chairman of NASA'S historical advisory committee from 1970 to 1973, died at the age of 63 in Burlington, Vt., after surgery. A member of the National Archives advisory council since 1968, he had worked strongly for separation of the National Archives and Records Service from the General Services Administration, and had advocated a coordinating office for the federal government's entire historical program with a chief government historian. Based at Dartmouth College, where since 1960 he had been teacher, writer, and administrator, Morton was active in national organizations related to history and had served on history and biography juries for the Pulitzer prize. He had been with the Army's office of military history from 1946 to 1959, serving as historian, then as deputy chief and chief of the Pacific branch; he was editor of an 11-vol. history of the U.S. Army in the Pacific during World War II, and general editor of the 17-vol. "Wars and Military Institutions of the U.S." (*W Post*, 15 Feb 76, B 12)

13 February: Ceremonies at GSFC honored 45 members of the team responsible for NASA'S Delta rocket, the "workhorse" of the agency's most dependable launching systems. NASA Administrator James C. Fletcher and Deputy Administrator George M. Low participated in the event. Delta placed NASA'S first communications satellite (*Echo 1*) into orbit 12 Aug. 1960; through January 1976, 119 Deltas had been launched. The Delta's reliability and economy had made it the primary vehicle for commercial spacecraft, international users, and operational meteorological systems. The number of reimbursable launches for non-NASA users had increased over the years; of the 11 Delta missions scheduled for 1976, all but 2 would be launched for foreign or commercial users on a reimbursable basis, better than 80% of the total. Delta's payload capability had risen from 68 kg to 910 kg. As Dr. Fletcher remarked, "Delta is becoming a standard against which to measure the conduct of NASA'S business in the future." (NASA Release 76-27)

- Dr. Alexander M. Lippisch, designer of the first operational rocket-powered fighter aircraft, died in Cedar Rapids, Ia., at the age of 81. As chief of design after 1939 at Germany's Messerschmitt Company, his speed research led to development of the ME 163B, world's fastest airplane at that time, which had a speed of nearly 1015 km per hr and could climb to 9 km in 3 min. As director of research at the Airplane Research

Institute in Vienna after 1943, he was rounded up in Operation Paperclip by the Allies, who put him to work for the U.S. Air Force technical intelligence section and sent him to the U.S. in 1946 to work at Wright-Patterson Field in Ohio. He had left government service in 1957 and gone to work for Collins Radio, where he headed the company's aeronautical laboratory. In 1965 he left Collins to form his own group, the Lippisch Research Corp., and at his death was developing an airfoil boat and an aerodyne (wingless aircraft) for the West German government. (*NYT*, 13 Feb 76, 34)

15 February: The long-secret site of USSR space launches—previously identified by U.S space experts as Tyuratam, for the name of the local railway station—was identified in the newspaper *Kazakhstanskaya Pravda* as "Leninsk," in a dispatch on construction of electric power-transmission lines. The dispatch, stating that the Leninsk area had been connected to the Central Asian power grid, was believed to be the first in which the name appeared in a Soviet publication. U.S. astronauts who visited the space complex last year during Apollo-Soyuz mission preparations had reported that a city of about 50 000 population was associated with the launch complex, and was known as Leninsk. Publication of the name—which had not appeared on Soviet maps or reference books—was considered either an oversight by the censor or a decision to make the name public. (*NYT*, 15 Feb 76, 12)

- An around-the-world auto race planned for mid-1976 as part of the U.S. Bicentennial celebration would feature a 1914 Model T Ford, tracked by a NASA satellite from the starting point in Paris to its termination in New York City, NASA announced. The car, driven by Goddard Space Flight Center employee Robert H. Pickard, would carry a 13.6-kg electronics package that would transmit the car's ground speed to the random-access measuring system (RAMS) carried on NASA'S *Nimbus 6* meteorological satellite that covers the entire globe once every 12 hr. *Nimbus 6* would relay information to GSFC through a ground station in Alaska. The auto race would demonstrate use of the RAMS system for ground tracking applications where transmissions could encounter various kinds of interference; normally, RAMS would collect data from instruments on moving platforms such as balloons or buoys to allow calculation of wind or sea-surface movements. Pickard, a car enthusiast, had worked in the space program since the Vanguard days and was presently project manager for the comsat-metesat system called GOES (geostationary operational environmental satellite). (NASA Release 76-23)

17 February: Under an agreement between NASA and ERDA—the Energy Research and Development Administration—in the initial phase of ERDA'S 5-yr program for solar heating and cooling, Marshall Space Flight Center would assist in managing the commercial demonstration projects portion that would show the nation's businesses and industries that solar heating, cooling, and hot-water systems could be used economically to relieve the growing demand for fossil fuels. MSFC would evaluate responses to Program Opportunity Notices (PONs)—ERDA's equivalent of NASA's Requests for Proposals—and would administer the demonstration projects resulting from contracts awarded by ERDA. The PONs solicited a variety of commercial projects that would use solar-based systems in commercial buildings; of the 308 proposals received from industry, ERDA expected to award from 8 to 20 contracts during 1976. (MSFC Release 76-42)

18 February: A "house of the future" incorporating technology from aerospace research would be open to the public at Langley Research Center after July 1, NASA announced. The Tech House, a project of NASA's technology utilization program, would demonstrate how an average family could cut fuel consumption by two thirds and water consumption by one half, using innovative energy and water management systems integrated with building designs and materials. Building started in late Jan.; all equipment and features in the house would be available to the public within 5 yr, or are available now. A family selected by NASA would live in the house for at least a year, beginning early in 1977; a systems engineer would monitor all use of the systems in the normal life of the family and record day-to-day savings. The contemporary one-story house would have an enclosed living space of about 500 sq m; its major feature combined solar collectors with night radiators and heat pump for one of the most cost-effective heating and cooling systems now available. Contributing to the NASA project were the National Association of Home Builders, the National Bureau of Standards, and the Department of Housing and Urban Development. (NASA Release 76-26)

- The Communications Satellite Corporation (ComSat) reported 1975 net income of slightly more than $46 million, compared to nearly 145 million for 1974, attributing the increase to additional leasing of circuits to the corporation's customers—3833 at the end of 1975, compared to 3510 in 1974. The report noted the FCC decision of 4 Dec. 1975 ordering lower rates for services provided through ComSat's global system, and warned that the decision would have "a substantial adverse impact" on future earnings unless modified by judicial review. ComSat had appealed the ruling and delayed filing a schedule of lower rates. (ComSat Release 76-3)

- More than 25% of all scientists and engineers in the U.S. and the USSR were engaged in weapons work of some kind, whereas fewer than one hundredth of 1% were engaged in arms control or disarmament, said William Epstein, former director of the U.N. Disarmament Division, in a letter to the *New York Times*. Warning of the "terrible doomsday weapons that scientists may yet develop" in the spiraling arms race, Epstein called on scientists and engineers to establish nationally and internationally a code of conduct that would include educational work on the perils of the arms race and political efforts to achieve arms control and disarmament. "Science may be neutral and amoral, but scientists are not . . . They have a moral duty to use their capabilities for the benefit of humanity and not for its destruction." Uniting their efforts behind a "Hippocratic oath" not to engage in means of mass murder, associations of scientists and other professional bodies could provide moral support and tangible assistance to those "even in dictatorial countries" where imprisonment or harassment would be the result of such action. (*NYT*, 18 Feb 76, 33)

19 February: Marisat 1—first satellite of a privately owned $100-million system to provide rapid high-quality communications between ships at sea and shore stations—was launched on a Delta for Comsat General Corp. from Cape Canaveral at 5:32 pm local time (1032 GMT). The 655-kg craft was headed for a stationary orbit 35 788 km above the Atlantic Ocean at 15°W, about 547 km southwest of the coast of Liberia. A second satellite, Marisat B, would be stationed over the Pacific later this year. *Marisat 1* carried 2 channels each about 4 mhz wide, operating in the L and C bands; one would translate shore-to-ship signals from 6 to 1.5 ghz, the other would translate ship-to-shore signals from 1.6 to 4 ghz. Ground stations at Southbury, Conn., and Santa Paula, Calif., would provide earth-satellite communications links, using 12.8-m-dia. antennas to relay tracking, telemetry, and command information between the satellites and the Comsat General control center in Washington, D.C. The U.S. government would use three UHF channels on the satellite, completely separate from the L- and C-band channels, through its own terminal facilities. The satellite system, owned and operated by a consortium headed by the Comsat General Corp., would be used by commercial shipping lines, as well as by the U.S. Navy pending completion of its own Fleet Satellite Communications System. By early April, commercial telephone, telex, and data communications would be available to link ships and offshore facilities with shore stations connected into domestic and international communications networks worldwide. Marisat mobile terminals had been purchased or leased from ComSat and installed on ships of a number of nations. (NASA Release 76-22; MOR M-492-205-76-01 [prelaunch] 26 Jan 76, [postlaunch] 13 Apr 76; ComSat Release CG 76-107; *W Star*, 20 Feb 76, A-6)

- A new comet now approaching the sun—named Comet West, after Richard M. West of the European Southern Observatory at Geneva, Switzerland, who discovered it in Nov. 1975—would be the object of extensive space and ground-based study to identify and measure its constituents, NASA announced. Comet West would come closest to the sun on 25 Feb., but would not be readily visible until about 2 March because of solar brightness; it would provide the first opportunity for extensive comet study since the appearance in 1973 of Comet Kouhoutek (a "visual disappointment" from which more had been learned about comets than in all the time that had gone before). NASA would participate in the study along with the Naval Research Laboratory, and with the Univ. of Colo. at Boulder and the Johns Hopkins Univ. (NASA Release 76-31)

- Work began on the world's largest radiotelescope—the "Very Large Array," consisting of 27 dish antennas each 25 m wide and weighing 160 tons, plus a Y-shaped layout of rail tracks and underground tubing along three 21-km legs—and was scheduled for completion in 1981 at a cost of $76 million. The plains of St. Augustine, N.M., isolated and desolate, had been chosen because a ring of mountains would prevent radio interference. The VLA would probe objects such as quasars that appeared to speed away at more than half the speed of light, as well as other puzzles such as black holes, star formation, galactic structures, and interstellar molecules. Its design would improve on three weaknesses of current telescopes: poor resolution, poor sensitivity, and inability to make images or maps quickly. Antenna signals routed through underground waveguides to a control center would be amplified and fed into a computer system that could turn out high-resolution images of the radio source in 8 to 12 hr. The array would be supervised by officials of the National Science Foundation, Associated Universities, Inc., and the National Radio Astronomy Observatory. (CSM, 19 Feb 76, 16)

- The universe had been analyzed and found to be open, a group of astronomers told the annual meeting of the American Association for the Advancement of Science in Boston. The director of Cornell Univ.'s national astronomy and ionosphere center, Dr. Frank Drake, said that the most recent astronomical data indicated that in about 30 billion years the universe would have changed, and no stars would appear in a night sky, having gathered into "Island Universes"—clusters of galaxies widely separated from each other. Computer analysis of recent observations indicated that separation of galaxies into islands had not yet begun, which meant that the universe might be comparatively young. The conclusion that the universe was open had been based on factors such as expansion rate, deceleration, and density of the universe. Evidence today would support the "big bang" theory which stated that the universe had exploded in the beginning from a single solid mass, and would go on expanding. (*B Sun*, 19 Feb 76, A-3)

- U.S. observers were concerned about spiraling strategic arms programs in the USSR, the *Christian Science Monitor* reported. Heavy payload in the Soviet nuclear system and an expanded program of deploying multiple warheads had aroused fears of a possible "first strike" by the USSR against U.S. intercontinental ballistic missile (ICBM) systems. Intelligence experts estimated the USSR would have 1500 land-based missiles by summer, compared with 1054 for the U.S.; an article in *Foreign Affairs* magazine by Paul Nitze—Secretary of the Navy in the Johnson administration—had stated the Russians believed they could "win" a nuclear war, whereas popular feeling in the U.S. was that all participants in a nuclear exchange would be destroyed. A Russian "cold launch" system for missile launching that did not damage the silo was contrasted to the U.S. "hot launch" system that damaged the silo during launch. The Pentagon had sought $78 million for development of a new ICBM called "MX," and a FY 1977 budget of $9.4 billion for strategic forces. (*CSM*, 19 Feb 76, 6)

- Extinction of several species of life forms resulted from a weakening of the earth's ozone shield nearly a million years ago, according to a study of recorded atmospheric occurrences that had not previously been correlated. In the Jan. issue of the British scientific journal *Nature*, Drs. George C. Reid and I.S.A. Isaksen of NOAA and Thomas E. Holzer and Paul J. Cruzen of the National Center for Atmospheric Research, Boulder, Colo., stated that fears about man-caused destruction of stratospheric ozone might be well founded. Examination of earth-core samples showed extinction of species of microscopic sea animals during a reversal in polarity in earth's magnetic field that weakened magnetic forces shielding the atmosphere from solar and cosmic radiation. The resultant heavy bombardment of radiation formed nitric oxide in the high atmosphere, in turn destroying part of the ozone shield. Similar increases in radiation, recently recorded, would inevitably affect present organisms, the report said; aside from direct effects on higher forms of life, indirect damage would come from the consequent upset of the ecological life cycle. (*NYT*, 19 Feb 76, 21)

20 February: Astronomical observations and laboratory experiments spanning 16 yr had failed to detect any signals of life in outer space, the 142nd annual meeting of the American Association for the Advancement of Science was told. The session in Boston heard reports on a 4-galaxy scan using the largest radiotelescope in the world—the 600-m-dia. antenna at Arecibo, P.R.—from Dr. Carl Sagan of Cornell Univ. and his Cornell colleague, Dr. Frank Drake, director of the National Astronomy and Ionosphere Center at Arecibo. "Of the trillion stars we looked at in four galaxies," Dr. Sagan said, "there is not a single one that at the time of our observation was devoting a major effort to communicate with us." Another effort—a 4-yr scan by the National Radio Astronomy Observatory at Green Bank, W.Va., of 659 stars in the region of the Milky Way—had yielded negative results after analysis of 90% of the observations. However, fewer than a millionth of the stars in the Milky Way had been scanned, for very brief periods, at only a few wavelengths. Negative results did not rule out the possibility of receiving the looked-for signals some day, the meeting was told. (*NYT*, 21 Feb 76, 40; *B Sun*, 21 Feb 76, A-3; *W Star*, 20 Feb 76, A-3)

- NASA announced selection of the Boeing Commercial Airplane Co. of Seattle for award of a $20-million contract to build an experimental flight research aircraft for use in noise-reduction technology. Boeing would modify a government-furnished C-8 Buffalo transport aircraft into a quiet short-haul research airplane (QSRA) for the Ames Research Center's noise-reduction research program, aimed at reducing aircraft-noise impact on U.S. communities and congestion at major airports. The redesign to meet QSRA concepts would enhance aircraft performance and control at low speeds, and would demonstrate propulsive-lift technology that would permit civil transport aircraft to achieve a short takeoff and landing (STOL) capability. STOL aircraft could use runways at existing smaller airports to relieve traffic at large metropolitan airports, as well as to reduce noise impact. (NASA Release 76-33)

- A simple, inexpensive, and effective method of suppressing noise in wind-tunnel testing of space vehicles would be patented by two MSFC engineers, MSFC announced. While developing an acoustic environment for reentry of the Space Shuttle solid rocket booster, Paul W. Howard and Luke A. Schutzenhofer of MSFC's Systems Dynamics Lab. found their acoustic data distorted by background noise reflected at transonic speeds from the walls of the wind tunnel, causing shock waves. The primary aerodynamic cause of the background noise was found to be the edgetone effect of high-velocity windflow over holes in the perforated walls of the tunnel;

covering the perforated walls with wire screening minimized direct contact between the airstream and the edges of the holes, and the noise was greatly reduced. The temporary fix proved so successful that the screens were permanently installed, and the patent should be issued to the public by early summer. (MSFC Release 76-44)

21 February: A simple device to reorient an orbiting satellite toward the earth stations with which it would communicate had been used successfully on *Satcom 1*, RCA-owned comsat launched by NASA in December 1975. Patented by two engineers at the RCA Electronics Div. in N.J., the new equipment took advantage of the gyroscopic effect of a wheel normally rotated to stabilize a satellite in synchronous orbit; only a simple start signal from the earth could activate it. Previous systems needed precision controls involving delicate sensors and thrust motors manipulated from the ground. *Satcom 1*, in synchronous orbit at about 35 000-km altitude, had been rotated $90°$ by the new system to point its instrumentation at the earth; it would begin regular transmission in March. (*NYT*, 21 Feb 76, 31)

- A Russian system for intercepting and destroying satellites apparently was unsuccessful this week, U.S. intelligence sources said. The test involved Russian satellites only, and posed no threat to any U.S. satellites in space. A target spacecraft (*Cosmos 803*) launched from Tyuratam 12 Feb. was the object of the test, carried out by an interceptor spacecraft (*Cosmos 804*) launched 4 days later. Observers differed on whether the interceptor had been maneuvered close enough to destroy the target, although a routine announcement by the USSR said the second launch had successfully completed its mission. In 5 tests between 1968 and 1971, USSR interceptor satellites had blown up on radio signal after approaching the targets in orbit, and the explosions had destroyed both interceptors and targets; *Cosmos 804* had not blown up. U.S. Secretary of State Henry Kissinger had said that the 1972 nuclear arms limitation agreement with Russia forbade the 2 countries from interfering with each other's satellites, but did not prohibit tests of an antisatellite system. (*B Sun*, 21 Feb 76, A-1; *W Star*, 21 Feb 76, A-3)

22 February: Study of earth's climate to permit predictions of global climatic changes by the end of the century should be the top space priority, said a study group formed by NASA last year to assess U.S. progress in space and where it should go next. The 3-vol report of the study group did not mention manned exploration of Mars, which had been the prime recommendation of a forerunner study 7 yr ago. Citing the dependence of the world's population on a complex system producing food and fiber where climate is favorable, the report said the entire system was predicated on a constant climate. Understanding of climatic processes would help solve earth's most pressing problems: food shortages, and worldwide weather changes caused by pollution. The report estimated that a 1-degree cooling of earth's annual temperature would mean loss of $1 billion in grain production, $2.2 billion in timber and fiber output, and $1.4 billion in fish catches; the drop would also increase the demand for electrical energy by $700 million, with a $2.4 billion rise in health costs other than treating skin cancers caused by alterations in earth's atmosphere. The group recommended development of 6 satellites to identify and assess worldwide crop conditions; other satellites to study chemical changes in the atmosphere and to watch for solar changes affecting earth's weather; and a network of 4 to 6 large geostationary satellites to observe ice and ocean conditions and keep tabs on the earth's radiation balance. Manned flight should develop along the lines of Space Shuttle and space station—to be used partially for climate study—and instrumented craft should be sent to explore every solar-system planet except Pluto, the report said. Cost of the weather-satellite program was set at $1.7 billion, with operating costs over 20 yr estimated at $4.6 billion; the planetary exploration program would cost a total of $5.1 billion to the year 2000. (*W Post*, 22 Feb 76, A-3)

- NASA announced plans for a seventh Lunar Science Conference to be held 15 to 19 March at JSC, bringing together scientists in geology, chemistry, physics, astronomy, engineering, and biology under joint sponsorship of JSC and the Lunar Science Institute of Houston. Discoveries about the moon would be applied to problems of the origin and early history of the solar system, for instance using lunar data to interpret craters and volcanoes photographed on the surfaces of Mercury and Mars. Other papers would discuss meteorites, satellites of Jupiter (about the size of earth's moon), and use of earth-based telescopes to measure chemical composition of asteroids. Lunar science conferences had been held yearly since 1970, when the first such assembly heard scientific results from the *Apollo 11* moon-landing mission. More than 700 scientists from as far away as Australia attended the 1975 conference. (NASA Release 76-25)

- Steam might be the power behind rockets of tomorrow, the annual meeting of the American Association for the Advancement of Science was told by Dr. Freeman Dyson of the Princeton Institute for Advanced Study. Steam engines of the future would be powered not by coal but by laser beams so powerful that they could instantly turn water into superheated high-velocity steam that could carry a 1-ton spacecraft into earth orbit. Basic research on laser propulsion had already been done by Dr. Arthur Kantrowitz of the Avco Everett Research Lab. in Cambridge, Mass. A laser beam with energy of 1 million kw would be needed to carry a rocket and its spacecraft away from earth; this laser would be 10 times more powerful than any developed so far. Ideally, the laser and rocket would be located on a mountain at 3-km altitude, where air would be free of water vapor that would reduce laser efficiency. Development of laser propulsion would take money, Kantrowitz said, but not as much as other rocket engines have cost. (*W Post*, 23 Feb 76, A-2)

23 February: Western Union filed with the Federal Communications Commission a tariff designating 11 more major metropolitan areas as "Satellite Access Cities." WU had already placed on line to Westar, the first U.S. domestic comsat system, 9 major metropolitan areas—New York, Washington, Atlanta, Chicago, Dallas, Los Angeles, San Francisco, Houston, and Pittsburgh—served by 5 ground stations, located near

New York, Chicago, Atlanta, Dallas, and Los Angeles. The Glenwood, N.J., station (near N.Y. City) would serve Boston, Buffalo, Philadelphia, Baltimore, and Wilmington under the new access. The station at Lake Geneva, Wis. (near Chicago), would serve Detroit, Milwaukee, Cleveland, Cincinnati, and St. Louis; the station at Cedar Hill, Tex., would serve Kansas City. In operation for more than a year, the 2-comsat Westar system had been used by several hundred corporations for both voice and data transmission; Westar also included TV centers in New York, Los Angeles, Chicago, and Dallas that had been transmitting increasing amounts of U.S. video news and sports coverage since July 1975. (WU news release 24 Feb 76)

- In a close race to tame the hydrogen-bomb reaction process—fusion—to make electricity, the U.S. and USSR were both ready for larger engineering units to demonstrate the practical production of electricity by the mid-1980s, reporter Robert Toth said in the Los Angeles *Times*. The Kurchatov Institute in Moscow had announced earlier in Feb. a "breakthrough" in the first successful test conducted in its Tokamak-10, the last purely experimental machine intended to prove feasibility of the approach. Last November, U.S. had announced a comparable achievement with the Alcator Tokamak at the Mass. Institute of Technology, described as a major development in fusion research. Although U.S. scientists claimed a fusion process had been first tested at Los Alamos 20 yr ago, it was not seriously pursued, whereas the late Dr. Lev Artsimovich of the USSR had worked along the same line independently and given the machine its name—a Russian contraction for current-machine chamber—upon achieving the first success.

Advantages of the fusion process were that it was the most efficient reaction known, producing 180 times more energy than consumed; fuel, the heavy forms of hydrogen, would be as inexhaustible as the seas; and the process left no radioactive waste like that from atomic-power plants based on the fission reaction. Fusion would require containment of the reaction in a magnetic field, which had been accomplished, and continuous production of energy would require temperatures of a million degrees Centigrade and a density of 100 trillion ions for a second—conditions beyond the capability of present machines. Both the U.S. and USSR results had been 5 to 10 times less than needed, although they were 5 to 10 times greater than previously achieved. (*W Post*, 23 Feb 76, A-1)

24 February: The Apollo lunar scientific experiment package (ALSEP) left on the moon's surface during *Apollo 14*, reported dead [see 18 Jan.] when transmission ceased last month from unknown causes, came to life just as unexpectedly 19 Feb. with transmitter, receiver, and experiments functioning very well. One experiment—the charged-particle lunar environment experiment—had previously been unable to perform during lunar daylight because temperature variations had degraded its power supply; since the revival, the experiment had been sending good data during lunar daylight. JSC scientists and Bendix engineers who designed the ALSEPs had no idea what happened to the *Apollo 14* station, one of

5 transmitting data from the moon; one possibility cited was a relay in the power system that had stuck and then became unstuck. Estimated remaining life of the *Apollo 14* station was once again 2 to 3 yr; all the stations had performed well, with the oldest—from *Apollo 12*—nearing its 7th year. ALSEP experiments were still providing data on conditions inside the moon and on moonquakes; scientists were still hoping for seismic signals from a large meteor impact that would identify the moon's core. (NASA Release 76-34; JSC Release 76-17)

- Pure carbon materials originally developed as ablative heatshields on the Apollo spacecraft were being used at medical centers across the U.S. in research on artificial limbs and implants in human tissue. The inert tendencies and high degree of purity exhibited by the carbon materials made them suitable for implantation without causing infection or rejection. Northwestern Univ. in Ill. and the Rancho Los Amigos hospital center run by the Univ. of Southern Calif. at Los Angeles had studied the materials in skeletal fixation devices that used carbon-button implants to connect the devices to nerves. Other projects included carbon implants to control pain (especially in the lower back) and neuroelectric stimulators to relax contracted muscles in paralyzed patients. Much of the technology used in purifying the carbon materials originated at NASA's Marshall Space Flight Center. (MSFC Release 76-45)

25 February: Fred W. Haise, Jr., astronaut on the *Apollo 13* lunar mission of April 1970 that nearly ended in disaster, would command the first free-flight test of the Space Shuttle, Johnson Space Center officials announced. A research pilot for NASA since 1959, Haise worked at Lewis Research Center and the Flight Research Center before his selection as a civilian astronaut in 1966. He was technical assistant to the manager of the Shuttle orbiter project from April 1973 to January 1976. With Haise on the test Shuttle flight would be Air Force Lt. Col. Charles G. Fullerton, appointed astronaut in 1969, who had never flown in space. The Shuttle was scheduled to fly piggyback in 1977 on a modified Boeing 747 jet transport, then in free-flight testing and landing after release at about 8-km altitude. Haise and Col. Fullerton would guide the Shuttle to an unpowered landing at the Dryden Flight Research Center in Calif. to demonstrate handling and reusability. (*B Sun*, 25 Feb 76, A-7; *W Post*, 25 Feb 76, B 5; NASA Release 76-35; JSC Release 76-17)

- Comsat General Corp. announced an agreement with ARCO—the Atlantic Richfield Co.—to install a Marisat terminal on the S/S *Arco Prudhoe Bay*, a tanker commissioned in Baltimore in 1971, to evaluate satellite communications with an eye to equipping other Arco ships if the new medium should prove reliable. The tanker would transport crude oil from Alaska to U.S. west-coast ports, and communications via satellite would be valuable during periods when the aurora borealis (northern lights) made high-frequency communication difficult. Comsat facilities had been installed previously on 15 commercial vessels; the Marisat system would be the 16th, and would begin operations in commercial service about 1 April. (Comsat Release 76-105)

26 February: A radio sensing system deployed by the federal government in remote areas of the U.S. since the spring of 1975 to warn of impending natural disasters had already passed one test, reported David S. Johnson, director of the National Environmental Satellite Service. An impending flood upstream from the town of Deming, Wash., had been detected by the sensors, and the river had crested an inch above its flood stage as predicted. The radio sensors, attached to transmitters, collected and relayed data from NOAA geostationary satellites (*Goes 1*, watching the East Coast, and *Sms 2* watching the Pacific Coast); the satellites relayed the information to NOAA computers in the World Weather Building at Suitland, Md. Initial work on the system began in the 1960s, and it was already operational. By summer of 1976, 150 to 200 sensor stations would be in place; a total of about 10 000 would eventually be situated to warn of floods, earthquakes, forest fires, or tidal waves that began in unfrequented areas. The system could also be used for keeping track of ships at sea or of trucks on the highway. (*B Sun*, 26 Feb 76, A-1; NOAA Release 76-32)

- Recovery of the unmanned *Soyuz 20* spacecraft that landed in Kazakhstan 16 Feb. carrying a "comprehensive biology payload"—including both plant and animal life—concluded what the *Christian Science Monitor* described as an impressive 3-mo biology experiment. Results from *Soyuz 20* would be compared with those of similar experiments—including some from the U.S.—carried on the Vostok-type biosat *Cosmos 782* which stayed 19.4 days in space last year. *Cosmos 782* had carried a small centrifuge to test effects of artificial gravity on flora and fauna; comparison of the previous tests with those from *Soyuz 20* would help decide whether—and how much—artificial gravity would be needed in future space stations. *Soyuz 20* had docked with the *Salyut 4* space station 19 Nov. 1975, an example of a self-docking biolab that demonstrated the USSR plan for assembling complex space objects from separately launched modules. (*CSM*, 26 Feb 76, 6)

27 February: Beginning with the 1976 forest-fire season, a new satellite-linked monitoring system of 23 NASA-designed ground stations would monitor forest conditions throughout thousands of sq km in Calif.'s Region One redwood area, providing data every 3 hr to foresters in Sacramento by means of a geostationary weather satellite. The 90-kg self-powered stations developed by NASA's Ames Research Center in cooperation with the state Division of Forestry would transmit continuous reports on wind speed and direction, air temperature, solar radiation, relative humidity, and the moisture content of forest litter such as pine needles and grass. Sensors to be added would measure rainfall and air pollution, including particulate matter and ozone concentration. The Synchronous Meteorological Satellite 2 (*Sms 2*) operated by NOAA from its vantage point over the equator would receive and relay the data. (*JSC Roundup*, 27 Feb 76, 3; NASA Release 76-28; ARC Release 76-07)

- Charles W. Mathews, NASA Associate Administrator of Applications, retired after 33 yr of government service with NASA and its parent organization, NACA. As a member of the science staff at the Langley Research Center since 1943, he was chairman of the group that developed specifications for the Mercury spacecraft before NASA was established. When Mercury became an official program in 1958, Mathews transferred to Houston (now the Johnson Space Center), serving as manager of the Gemini program. Upon completion of Gemini, Mathews went to NASA Hq as director of the Skylab program (then known as Apollo Applications Program); in 1968, he became Deputy Associate Administrator for Manned Space Flight, and remained in that position until 1971. Awarded the NASA Distinguished Service Medal by President Johnson in 1966 for contributions to the manned space program, Mathews received the award a second time in 1969 for contributions to the first manned lunar landing. At his retirement, he stated that he would "continue to devote effort to the further realization of the tremendous potential of space . . ." (NASA Release 76-38)

29 February: The concept of a "science court"—an impartial quasi-judicial body of scientists to conduct inquiries into conflicting scientific claims—was advanced by Dr. Arthur Kantrowitz, director of Avco-Everett Laboratories, a high-technology industrial firm in Mass. The proposal would be a means of assuring a rational and orderly assessment of scientific facts underlying highly complex issues of public concern. Although hearings had been held by Congress and federal agencies on issues such as radiation, automotive safety, air pollution, drugs, and pesticides, these had frequently been superficial and led to no resolution of facts. A recent Harvard University book on the subject said in part: "The principal shortcoming of such hearings can be revealed simply by asking how 'adversary' these adversary hearings are . . . direct battle over facts, if it occurs at all, takes place as a free-for-all in the mass media." Conflicting statements by experts had added to confusion and paralyzed decision-making processes. A test of the "science court" concept had been recommended as part of the new White House office of science and technology policy. (*NYT*, 29 Feb 76, 8)

- Information of value to the U.S. Pioneer mission to Venus scheduled for 1978 was published by the USSR in *Pravda* after 4 mo of analysis of the 2 Soviet spacecraft that reached Venus in Oct. 1975. Each of the Russian craft, weighing about 5000 kg, separated into an orbiting section and a landing capsule; the two landers transmitted the first panoramic photographs of the Venus surface to earth. U.S. scientists had debated the angle at which landers should enter the Venus atmosphere; the Russian report stated that *Venera 9* entered at 20.5° and *Venera 10* at 22.5° to the horizon. Besides the entry angle, the report included details of descent, amount of sunlight on the surface, wind speeds, rock chemistry, and character of surface erosion. Although the 2 landers failed under surface temperatures of 482°C and pressure 90 times that of earth—*Venera 9* lander operated 53 min., *Venera 10* 65 min—the orbiters continued to transmit data on cloud cover and upper atmosphere. The U.S. planned to launch 2 spacecraft, an orbiter and a lander, toward Venus in 1978, the lander separating into 4 or 5 probes to relay data via the orbiter to the earth. (*NYT*, 29 Feb 76, 34)

- Grover Loening, pioneer aeronautical inventor and first person to receive an M.A. degree in aeronautics from an American university, died at the age of 87 at his home in Fla. after a long illness. In 1913, after working for an aeroplane builder in N.Y., Loening became assistant to Orville Wright and manager of the Wright brothers' factory in Dayton, O. In 1914 he was appointed chief aeronautical engineer of the Army Signal Corps aviation section. As vice president of Sturtevant Aeroplane Co., he pioneered the first U.S. steel-frame airplane in 1915; later he formed his own company and developed the M8 2-seat Pursuit monoplane using rigid strut bracing, which he patented. After World War I, Loening produced the Flying Yacht which established world records and won him the Collier Trophy in 1921; his next success was the Loening Amphibian, with the first practical retractable carriage. A member of the Aviation Hall of Fame, Loening was one of the first directors of Pan American Airways and a member of the Smithsonian Institution board for about 20 yr. Awarded the Presidential Medal of Merit and nearly every U.S. aviation medal, he had been selected 3 wk ago to receive the Smithsonian's Langley gold medal for aerodromics. (*W Star*, 3 Mar 76, B-5)

During February: Water hyacinths thought to be environmentally affected by the location of NASA's National Space Technology Laboratories in Miss. proved to have unexpected benefits for space-age processes and byproducts, when NASA scientists monitoring the effects of wastes discharged into nearby streams found that the water plants had a peculiar ability to absorb and concentrate toxic metals, and to metabolize various other chemical pollutants, while they continued to produce massive amounts of plant material. For a year, biochemists at NSTL had worked with the plants as a filtration system for purifying polluted waters, as a source of biogas for fuel, as a protein and mineral additive to cattle feed, and as a soil fertilizer and conditioner. The vascular plants could absorb and metabolize large quantities of nutrients and pollutants from domestic sewage waste waters. An installation of the size needed to meet pollution standards at the nearby community of Orange Grove would have cost about $500 000; stocking the sewage lagoon with enough water hyacinths to purify up to a half million gallons of sewage outflow daily cost only a few thousand dollars.

Plants taken from sewage lagoons had been dried and ground into feeding rations for beef cattle, producing a meal rich in minerals and protein, and added to corn silage fed to a herd of steers at the nearby agricultural experiment farm operated by Miss. State Univ. A 4-mo study resulted in weight gain comparable to that on a diet fortified with cottonseed or soybean meal. The high cost of fuel for drying the plants after harvesting led to construction of a prototype solar dryer at NSTL that was designed to dry 18 tons of wet material each 36 hr; the dryer might prove a solution for agricultural problems other than drying of grains and forage.

Another use of the plants was as a source of methane gas and fertilizer, using an anaerobic fermentation process with a yield of nearly 57 000 cu.m. (2 million cu. ft.) of gas from an acre of water hyacinths. A state like La., with more than a million acres of unwanted water hyacinths, might produce more than enough methane gas to fuel

2 million homes in the New Orleans region annually. Another study showed a remarkable ability of the plants to filter out "heavy" metals (cadmium, mercury, nickel, lead, silver) and other toxic organic substances common in industrial waste waters. Although plants harvested from these sites could not be used for animal feed or fertilizer, they could be used to produce methane gas; researchers found that biogas generated from plants containing trace metals yielded a higher percentage of methane than plants free from metals. Spinoff projects resulting from worldwide interest in the NASA findings included a survey by NASA investigators of water hyacinths blocking up the White Nile, requested by the Sudanese government; a plan to use hyacinth plantings as a detection system for assessing heavy-metal pollution in waters near thermonuclear installations; and use of the plants in a system to recover millions of tons of gold "tailings" left over from mining and present in streams near disused mines in the western U.S. (NASA Release 76-36)

- A 7.6-m radio-controlled model of the dirigible *Hindenburg*, star of a recent motion picture on the last days of the world's largest airship, was given to the National Air and Space Museum for display along with a full-size control car used in interior and exterior filming of the picture. Other full-size sets had been constructed to represent the passenger compartments, internal engineering and cargo areas, and a complete bow section. The radio-controlled model could perform 28 separate functions, from dumping water ballast to operating a complex system of interior lighting. (NAA newsletter Feb 76)

- The USSR had made good use of observations from its manned space stations, Soviet Academician Leonid Sedov said, citing photographs taken by *Salyut 3* showing 67 locations where oil and natural gas could be found in the Caspian Sea region and 84 in Uzbekistan. The satellite did in 3 mo a job that had taken 60 yr of ground prospecting to reveal 102 deposits, he said. Published in the Feb. issue of the British magazine *Spaceflight*, the report by Sedov mentioned contributions to agriculture from a space survey that revealed millions of gallons of water "just below the surface" of arid lands near the Caspian Sea, showing areas that might be suitable for "oasis farming" or grazing livestock. Discussing changes in the environment made possible by space technology, Sedov said such influences should "bear a global character," to avoid a repetition of the "notorious experiment" (the so-called West Ford) in which the U.S. released a cloud of copper fibers in orbit in 1963 for communications experiments; West Ford caused worldwide complaints by radio astronomers. (*Spaceflight*, Feb 76, 54)

March 1976

1 March: NASA's cancellation of a mission to Uranus by withholding funds in a severely restricted budget for FY 1977 had produced a "Uranus option" plan, sponsored by engineers at the Jet Propulsion Laboratory, that would take a spacecraft destined for Saturn and give it the ability to go the extra distance to Uranus as well. In an interview with the *New York Times*, John R. Casani—deputy project manager of the Mariner project at JPL—said the necessary trajectories had been plotted, and modification of the spacecraft remote-sensing instruments to use at Uranus was already under way.

Of the two Mariner missions scheduled for launch in Aug. and Sept. 1977, to fly by Jupiter in 1979 and Saturn in Nov. 1980 and Jan. 1981, the first mission would remain unchanged. The second Mariner, trailing the first by 9 mo, would fly a lower speed trajectory to allow a bigger slingshot effect from Saturn's gravity for propulsion toward Uranus; the slower speed would also allow flight controllers to decide on the basis of the first Saturn flyby whether to use the second Mariner for a closer look at Saturn or to go for Uranus. The second choice would mean that, by 1985, U.S. spacecraft would have looked closely at all the planets of the solar system except Neptune and Pluto; a Mariner flyby of Uranus between Nov. 1985 and Jan. 1986 would take the spacecraft further out into the solar system, toward Neptune. Obtaining data on Neptune would take "a miracle," Casani said, although the JPL engineers would not rule out the possibility. The infrared spectrometer on the second Mariner would be improved in sensitivity to detect the temperature of Uranus, twice as far from the sun as Saturn is, and 20 times as far as earth; also, the spacecraft would be able to photograph Uranus, 4 times the diameter of earth, and gather data on its magnetic field, atmosphere, and other physical properties. The Mariner missions to Jupiter and Saturn would cost about $305 million; addition of the trip to Uranus would add about $21 million to the cost of the mission. (*NYT*, 1 Mar 76, 11)

- An all-systems simulation of landing *Vikings 1* and *2* on the surface of Mars, to exercise the entire 750-person Viking flight team so that everyone knew exactly what to do when the time came later this year, began 20 Feb. and built up to a "Mars landing" 22 Feb., concluding today as a "fantastic success" according to James S. Martin, Jr., project manager at Viking Control in the Jet Propulsion Laboratory installation in the San Gabriel Mts. near Pasadena. The computer program had to be redesigned to facilitate uplink command to the spacecraft cameras, and more time had to be allowed for certain ground-control and data-processing procedures, but no major problems developed. In the room where a full-scale model of the Viking lander complete with operational cameras stood on a simulated Martian surface, prankish team members had put three tiny marine fossils from earth's Paleozoic period on the sand within the camera field-of-view. When the camera came on to transmit after the simulated landing, the first picture was startling, but the fossils were immediately recognized, and the joke demonstrated that even tiny objects near the landing site would appear in the pictures transmitted from Mars. (*NYT*, 4 Mar 76, 32)

- Lt. Col. Michael A. Love, test pilot of the successful NASA lifting-body program, was killed in the crash of an F–4C fighter on a dry lake bed at Edwards AFB, Calif., shortly after takeoff in a proficiency flight. Love's navigator, Maj. E.B. Underwood, Jr., ejected from the plane before it crashed and was hospitalized in stable condition. Love, 37, was chief AF pilot assigned to the X–24B program that tested a wingless aircraft to develop a space vehicle that could be flown to earth and landed like a plane. After serving in the lifting-body program as a chase pilot for various M–2 and X–24A flights, Love made his first X–24B flight on 4 Oct. 1973 and had piloted the plane to its fastest speed—better than 1860 kph—before terminating the program 20 Aug. 1975 with a hard-surface landing of the X–24B on the runway at Edwards. (*W Star*, 2 Mar 76, A–5; *W Post*, 3 Mar 76, A–20; *NASA X-Press*, 12 Mar 76, 2)

- At 0300 GMT (12:30 pm local time) the Japanese ionosphere sounding satellite *Ume* was launched from the Tanegashima site on the 3-stage liquid-fuel N rocket to an orbital altitude of about 1000 km. The cylindrical satellite weighed 139 kg, had a diameter of 94 cm, and was 82 cm high. Its estimated lifetime was 1.5 yr. This was the second successful launch with the new N rocket. (FBIS, Kyodo 29 Feb 76)

- A technical management team called SPICE (for Spacelab payload integration and coordination in Europe) would be established by the European Space Agency as the result of a planning meeting in Paris, ESA announced. The team would work at the Federal Republic of Germany's technical center at Porz-Wahn, with half of its 20 members from the center's staff; the ESA announcement called the center "the European organization that has devoted the most effort to forecasting, studying and preparing the utilisation of Spacelab." Max Hauzeur, manager of the SPICE team, would be responsible for a number of items for the first Spacelab payload: coordination of instrument and experiment development; data management; design and interface specifications; schedule monitoring; acceptance of experiments, and approval of integration tests; and coordination of training programs. (ESA release, 1 Mar 76)

- The European Space Agency (ESA) announced an agreement on loan of a radiometer for 2 yr to the Iranian College of Science and Technology at Tehran for use in propagation experiments in the 12-ghz band, to be used by European relay satellites for long-distance public telecommunications. The experiments would be carried out in "particularly interesting conditions," ESA said, because of Iran's geographical and climatic characteristics. (ESA release, 1 Mar 76)

2 March: Discovery of a new kind of photosynthesis—a bacterial system using purple instead of green pigment to convert sunlight into chemical energy and food—was announced by a team of scientists from NASA's

Ames Research Center and the Univ. of Calif. medical center at San Francisco. Contained in cell membranes of *Halobacterium halobium*, found in salty waters, the pigment when energized by sunlight could transfer protons across the membrane. It appeared to increase the evaporation rate of salt, with possible applications in desalination of sea water; it also seemed to resemble rhodopsin, the little-understood visual pigment of the eye, and might aid in understanding the evolution and operation of vision. The purple pigment, which acted as a proton pump, might also explain the key process of ion transport in all cells. Deposited on a film and exposed to sunlight, it had already been tried out in crude solar cells. Scientists were excited by the possibility of practical uses of the purple pigment because—although less efficient than chlorophyll—it could be purified and was stable over a broad range of temperatures and activities. The newly discovered process was the first instance of a system other than the one based on green chlorophyll in which cells could use light energy to survive, said Dr. Walter Stoeckenius of the University of Calif., chief researcher. Bacteria used in the research came from salt flats near the Mediterranean, where they had been known for a century to cause pinkeye in salted fish and red herrings to become red. The study in which the discovery was made was part of research into earth organisms living in extreme environments like those expected to exist on other planets. (NASA Release 76-30; ARC Release 76-12; *W Post*, 3 Mar 76, A-1; *B Sun*, 3 Mar 76, A-3)

- A cooperative use of spaceflight techniques by U.S. and USSR scientists might verify the existence of "gravitational waves" that could be used to probe quasars and other explosive cosmic events, said an article in the *Astrophysical Journal*. The new technique—proposed by Dr. Kip S. Thorne of Calif. Institute of Technology and Prof. Vladimir B. Braginsky of Moscow State University—would use a net of radio signals between earth and interplanetary spacecraft to monitor sudden unexplained fluctuations in the returned radio frequencies. Analysis of changes in the length and shape of the waves would reveal what had happened to the matter that generated the wave, according to the theory. The idea of gravitational waves had grown by analogy with the three types of manifestations of electrical force: the static electric field, static magnetism, and radiated waves that took form as visible light, radio waves, x-rays, etc. As the extremely precise clocks needed for the experiments had only recently become available, the search for gravity waves might not be successful for another decade, the scientists said. (*NYT*, 2 Mar 76, 17)

2-9 March: Comet West, one of the few such objects visible to the unaided eye in daylight, reached maximum visibility in the eastern sky before sunrise this week. Having skirted the sun a week ago, the comet had become less brilliant as its distance from the sun increased, but was more easily seen as it moved away from the sun. The comet was said to be the brightest such object since Comet Bennett's appearance in 1970. The comet had appeared in photographs taken in August 1975 at an observatory in the Chilean Andes, but was not identified until Richard West, a Dane working in Geneva, studied the photographs. Comets are of special interest because they are thought to be composed of materials from the outer fringes of the solar system. Reports from Italy, Switzerland, and elsewhere described the comet's tail as 2 to 4 times the apparent diameter of the moon, unusually short for so bright a comet. NASA reported extensive observations of the comet using rockets, high-flying aircraft, and ground-based instruments; smog or clouds might make ground observation difficult. (*NYT*, 2 Mar 76, 17; 9 Mar 76, 43; *B Sun*, 3 Mar 76, A-3)

3 March: Women would definitely be admitted to the U.S. space program, the *Chicago Tribune* reported, and NASA would make a formal announcement in July. Only one woman—Russia's Valentina Tereshkova, who orbited for 3 days on *Vostok 6* in June 1963—had been in space so far. The U.S. space women would fly on the Space Shuttle scheduled to be operational by 1980; the exact number had not been determined, but they would form part of a group of 15 mission specialists who would be chosen within the coming year. Screening, evaluation, and physical examination of applicants would be completed by Dec. 1977, and selectees would join NASA as candidates for space flight by July 1978. (*B Sun*, 3 Mar 76, A-3)

5 March: First microwave observations of carbon monoxide in the upper atmospheres of earth, Venus, and Mars were reported by a team of scientists at NASA's Jet Propulsion Laboratory at Pasadena. Establishment of carbon monoxide as a common component of the upper atmospheres of the planets should lead to a better understanding of the delicate balances in earth's atmosphere, said Dr. Joe W. Waters, who headed the team. Carbon monoxide at lower altitudes on Venus and Mars, as well as on earth, had previously been measured by infrared techniques that had also detected CO on Jupiter last year, according to JPL's Dr. Reinhold Beer. JPL microwave observations used a radiotelescope at Aerospace Corp.'s electronic research laboratory to detect the carbon-monoxide absorption frequency at high altitudes, well above the smog layer over Los Angeles; calculations indicated that the same technique could measure CO in the upper atmospheres of the other planets. The Mars experiment used the larger radiotelescope at Kitt Peak, Ariz., to record the identifying signal. CO occurred in 20 parts per million at 80 km altitude on earth, and about 1000 parts per million at 100 km above Venus; the Mars readings were still being analyzed. (NASA Release 76-40)

- Simultaneous measurement of ocean-surface conditions from an aircraft and from an altimeter on NASA's *Geos 3* satellite 840 km above the surface was the mission of a NASA-Navy team just back from a month's expedition to Newfoundland. Correlation of aircraft data on actual surface conditions with remotely sensed data from *Geos 3*, launched 9 April 1975, would demonstrate satellite capability of making accurate and rapid sea-state measurements. Newfoundland was chosen for the experiments because it exhibited a wide range of climatic and oceanographic conditions within a relatively short time period. Measurements taken were sea ice, sea state, surface-wind field, and determination of atmospheric refractivity. Each pass of the satellite—8 to 10 times a day—would produce 10 to 15 min of data relayed to Goddard Space Flight Center via tracking stations at Bermuda, Madrid, or Winkfield, England.

Satellite observations plotted for the North Atlantic would be compared with aircraft data and with daily Navy–NOAA forecast maps. The team consisted of 8 persons from the Naval Research Laboratory and 11 representatives of Wallops Flight Center and Lockheed contractors. (NASA Release 76–39; WFC Release 76–2)

- NASA selected Grumman Aerospace Corp. and McDonnell Douglas Astronautics Corp. to negotiate fixed-price contracts for parallel systems analyses of concepts for space stations in low- and synchronous-altitude earth orbits, including orbit-to-orbit transportation. The 2 studies, each to last 18 mo and cost about $700 000, would be managed by MSFC for Grumman and by JSC for McDonnell Douglas; the work would begin 1 Apr. The studies would define an operational base in space to serve as a space laboratory, of modular construction with potential for growth over the years. Proposals were also received from Rockwell International Space Div. and Boeing Aerospace Co. (NASA Release 76–41; JSC Release 76–19; MSFC Release 76–50)

- The Dept. of Defense announced its first successful powered flight test of a bomber-launched cruise missile, prototype of a weapon the USSR had sought to curb in strategic arms talks. The Boeing-made missile, basically a pilotless jet drone resembling a small plane with airbreathing engines, could carry nuclear warheads at low altitudes to evade detection by enemy radar. The successful test, in which the robot-like craft was launched from a B–52 at an altitude of about 3 km and flew for 11 min. over the White Sands range, was viewed as a step in final development of a highly strategic weapon for the U.S. Control of cruise missiles—both air- and submarine-launched types—had been a major issue in negotiations with the USSR. The Air Force missile just tested would be an aid to bomber penetration of enemy defenses, whereas the Navy's cruise missile would be a tactical type for use against enemy ships rather than a strategic type aimed at land targets. The first test scheduled for 26 Feb. had been canceled because of problems with the B–52's airconditioning unit. Later tests would try out the programmed guidance system. (*W Post*, 6 Mar 76, A–12; *W Star*, 6 Mar 76, A–2; *B Sun*, 6 Mar 76, A–1; *NYT*, 6 Mar 76, 18)

- A new device called an ultrasonic bolt-stress monitor—using ultrasonics to measure stress on a bolt—had been developed by LaRC physicist Joseph S. Heyman, the *Langley Researcher* announced, and would be submitted to *Industrial Research* magazine as a candidate for selection among the 100 most significant new products developed during the year. The bolt-stress monitor would replace equipment now in use to monitor a multitude of bolts in a facility or a piece of equipment, as it would detect a 1-psi change in 50 000 psi in measuring stress. Heyman had invented a continuous-wave ultrasonic microemboli monitor, used to observe blood circulating outside the body during surgery and detect larger than normal particles, that was selected by the magazine as an IR 100 winner in 1974. (*Langley Researcher*, 5 Mar 76, 3)

6 March: U.S. failure to renew the Indian government's access to the ATS satellite relaying birth-control and agricultural information to 2400 villages might intensify tensions between Washington and New Delhi, the

Baltimore Sun said in an editorial. An alliance of U.S. scientists and environmentalists had already protested the sale to India of uranium for a power station near Bombay; the group had asked the Nuclear Regulatory Commission to require India to demonstrate that the substance would not be used to make bombs, as India had done in 1974 with plutonium derived from Canadian material supplied for a joint atomic-energy project. Combined with Ford Administration efforts to use development aid to force India into a more friendly posture, the *Sun* said that withholding advanced technology used by India for humanitarian purposes would create "an unnecessary blur" between humanitarian and development aid and would exacerbate rather than relax existing tensions. (*B Sun*, 6 Mar 76, A-12)

7 March: Dr. James B. Pollack, research scientist at Ames Research Center, had been named to receive the H. Julian Allen award for 1975, for major findings about the atmosphere of Venus. Dr. Pollack and coworkers used airborne infrared telescope observations in 1974 to show that the top 20 km of Venus's heavy cloud cover consisted of a water solution of concentrated sulfuric acid. The heat absorption peculiar to sulfuric acid might account for conditions of extreme heat near the surface of Venus. Dr. Pollack would share honors with 8 other Ames employees; the award, given for outstanding papers written by members of the Ames staff, would be accompanied by an honorarium of $1000. (ARC Release 76-11)

8 March: Jan A. Bijvoet (pronounced byfoot) of the Netherlands arrived at Marshall Space Flight Center where he would on 1 April become the new representative of the European Space Agency (ESA) in the Spacelab program office. Most of his career had been with transatlantic cooperative programs; from 1961 to 1970, he had worked for NATO as a senior scientist in the evaluation methods section of SHAPE's technical center. Before coming to MSFC he had spent 6 yr in ESA's aeronautical satellite program, a joint activity of ESA and the Federal Aviation Administration. At MSFC Bijvoet would be responsible for technical and managerial liaison between NASA and ESA on the Spacelab program. He succeeded Robert Mory, who had been at MSFC since the beginning of the Spacelab program in 1973 and had been reassigned to the Paris office of ESA. (MSFC Release 76-70)

9 March: Former astronaut William A. Anders, serving as chairman of the Nuclear Regulatory Commission, was appointed U.S. ambassador to Norway by President Ford. Anders, 42, had served at NRC since Dec. 1974 and had told the President he wished to remain in the post for only a yr. Selected as an astronaut by NASA in 1963, Anders was one of the 3 crewmen of *Apollo 8* to orbit the moon at Christmastime 1968. Anders left the corps of astronauts in 1969 to become executive secretary of the National Aeronautics and Space Council, a White House post abolished by President Nixon 4 yr later; he then was appointed head of the Atomic Energy Commission (whose functions were reorganized in 1974 into the Energy Research and Development Administration and NRC). He had turned down an offer of reappointment at NRC for a 5-yr term. (*WSJ*, 9 Mar 76, 3; *W Post*, 8 Mar 76, A-20)

12 March: Tests completed in flight on the two Mars-bound Viking spacecraft showed that each had 2 ovens of the organic chemistry experiment in good working order. JPL spokesmen said they were confident the instruments would carry out the investigations planned on the Mars surface later in the year. Earlier data indicated that one oven on each spacecraft might have failed, but investigators would not be certain until the Mars landings, scheduled for early July and Sept. The ovens—3 on each Viking lander—were designed to heat surface samples to 500°C to release organic matter in the soil for analysis by the gas chromatograph mass spectrometer on each lander. (NASA Release 76-43)

- Dr. Bruce T. Lundin, director of Lewis Research Center, received the 1976 Astronautics Engineer Award from the National Space Club at its 19th annual Dr. Robert H. Goddard memorial dinner in Washington. The award, made by a group of judges including former NASA scientist Dr. Wernher von Braun, was for outstanding leadership in development and operation of the Centaur high-energy rocket stage, and the Atlas-Centaur and Titan-Centaur launch vehicles. The Atlas-Centaur launched all NASA's Surveyor spacecraft in the 1960s and many others such as OAOs, the recent Intelsats, Mariners, and Pioneers. The Titan-Centaur, largest U.S. rocket, launched West Germany's Helios spacecraft toward the sun and the Viking spacecraft to Mars. A NASA employee since 1943, Dr. Lundin became director of LeRC in 1969 after a year and a half at NASA Hq as Deputy Associate Administrator and Acting Associate Administrator for Advanced Research and Technology. As leader in the development and operation of the Centaur vehicle, he built the Centaur staff at Lewis and made the decisions that set the character and style of the whole project. (*Lewis News*, 19 Mar 76, 1; LeRC Release 76-11)

- First of 4 U.S.-West German rocket launches to investigate the source of the aurora borealis, in a study called Project Porcupine, would be fired to an altitude of 500 km from northern Sweden, NASA announced. Two identical 250-kg payloads would be flown at twilight between 17 Mar. and 4 Apr. when moonlight interference would be at a minimum; a second pair of rockets would be launched in 1977. The instrument packages would have 12 quill-like booms sticking out to gather data for aurora-probe experiments beyond the orbits of the U.S. Skylab and USSR Salyut space stations. At about 450 km altitude, canisters of barium would be exploded to identify magnetic-field lines and plasma drifts; a NASA jet from Ames Research Center would be in flight over Greece to photograph the clouds released from the rockets. Data from Project Porcupine would be compared with energetic-particle and magnetic-field data gathered by NASA's *Ats 6*. The payloads, largest and most complex to be fired on sounding rockets, would be the first flown on the new Aries rocket, modified from Minuteman intercontinental ballistic missile parts provided by the U.S. Air Force. Project Porcupine would be part of the 1976-78 International Magnetospheric Study, a 40-nation effort to explore the upper atmosphere. (NASA Release 76-44; *NYT*, 28 Mar 76, 21)

14 March: U.S. worldwide predominance in science and technology had eroded over the past 15 yr, said a National Science Foundation report transmitted to Congress by President Ford. Based on a review of 492 major technological innovations, the report said the U.S. share of the total sank from 75% in 1953-55 to 58% in 1971-73. U.S. spending on research and development declined from a peak of 3% of gross national product in 1963 to 2.3% in 1974. Since 1960, U.S. receipts for use of American inventions abroad had tripled, but payments the other way had increased 4.5 times. Foreign inventors receiving U.S. patents now accounted for more than 30% of all those issued by the Patent Office. The study was the most specific compilation to date on changes in relative support for science in the U.S. and other nations; the USSR, West Germany, France, and Japan had improved their inventiveness and worker productivity faster than the U.S. The message to Congress did not mention the international comparisons, but said inflation and recession had adversely affected science and technology in the U.S. (*NYT*, 14 Mar 76, 1)

15 March: A Titan III-C rocket launched from Kennedy Space Center at 2:55 am local time (0755 GMT) carried into stationary orbit at about 36 000 km 2 experimental nuclear-powered comsats for the U.S. Air Force; 23 min later, the Navy's *Solrad 11-A* and *11-B* riding piggyback on the same rocket were sent toward an orbit more than 120 000 km up and at opposite sides of the earth, where they would monitor solar flares that disrupt communications and navigation systems on earth. The Air Force's *Les-8* and *Les-9* were designed to guard against space jamming of U.S. military communications; spokesmen said the satellites were "right on the money" but would undergo various tests before they began their experimental message traffic. (*NYT*, 17 Mar 76, 39)

• Insulation systems developed under contract at MSFC for the liquid-hydrogen tanks of NASA's Saturn V launch vehicle had found new uses in the shipping industry, MSFC announced. Two firms—McDonnell Douglas and Rockwell International Corp.—had developed the polyurethane-foam systems for efficient cryogenic (low-temperature) insulation. The McDonnell techniques had been used to insulate storage tanks on liquefied natural-gas carriers in conjunction with a French naval engineering firm; the system, to be marketed jointly by the 2 firms, had the advantages of quick and easy installation with high efficiency that would reduce boil-off (loss through evaporation) for a given thickness of insulation. The Rockwell system was adapted for tuna-boat insulation; more than 40 boats had been equipped with the new insulation before being sold at prices ranging from $1 million to $4 million. (*Marshall Star*, 31 Mar 76, 4; MSFC Release 76-51)

• Astronaut Eugene A. Cernan, last man to leave footprints on the moon's surface when he commanded *Apollo 17* in 1972, would leave government service 1 July. Cernan, a Navy captain, would complete 20 yr in the Navy in June and would retire from both the Navy and the astronaut corps. Selected as a NASA astronaut in 1963, Cernan had had 3 space missions (including 2 flights to the moon) and was the second American to walk in space, when he was pilot of *Gemini 9* in June 1966. As lunar-module pilot of *Apollo 10* in May 1969, he and Astronaut Thomas P. Stafford had flown the module within about 15 km of the moon's surface in a full-scale rehearsal of all but the final moments of the first lunar-landing mission 2 mo later. In Dec. 1972, Cernan and Astronaut Harrison H. Schmitt landed at Taurus–Littrow during the last scheduled manned lunar mission. In 1973, Cernan became special assistant to the Apollo program manager, working on the joint U.S.–USSR Apollo–Soyuz Test Project; in Sept. 1975, he became chief of training operations in the astronaut office. (NASA Release 76–47; JSC Release 76–20; *W Star*, 16 Mar 76, A–2)

16 March: The 50th anniversary of Dr. Robert Goddard's successful launch of the world's first liquid-fueled rocket at his aunt's farm near Auburn, Mass., was observed in a re-enactment at the launching site—now marked by a granite monument—and in 15th anniversary ceremonies at the Goddard Space Flight Center, which also launched a commemorative rocket. Dr. Goddard had written on space navigation while still in high school; his early theoretical writings prompted the *New York Times* to print an article on its editorial page in 1920 saying that Dr. Goddard was ignorant of elementary physics if he thought a rocket would work in the vacuum of space. The *Times* ran a correction on 17 July 1969, the morning after the Apollo astronauts were launched toward the first lunar landing.

The original rocket launched by Dr. Goddard in 1926 was about 3 m high, traveled a little more than 12 m upward, and attained a speed of 56 m in 2.5 sec. powered by a combination of gasoline and liquid oxygen. By contrast, the Saturn V rocket with its Apollo payload stood over 110 m tall and could cover 25 km in 2.5 sec powered by a mixture of kerosene and liquid oxygen, a fuel remarkably similar to that used by Dr. Goddard. Although, as GSFC Director Dr. John F. Clark pointed out at the Center re-enactment, "Not a line appeared in a single paper about the Auburn flight . . . In fact, the attention he did get in the early 1920s was adverse . . .," Dr. Goddard continued his work in New Mexico in the 1930s and flew his rockets to heights over 3 km and at speeds up to supersonic.

Of Dr. Goddard's concepts, Wernher von Braun said: "Goddard did most of the basic research and development that made possible rockets such as the Saturn V." Dr. Goddard had died in 1945, but his widow had attended the dedication of the Center when it opened in 1961; she had planned to attend the re-enactment in Auburn, but a heavy snowstorm swept New England 16 Mar. and seriously curtailed attendance at the event. Astronaut Eugene Cernan, principal speaker at the Mass. ceremonies, had reached Boston's Logan Airport only an hour before it was closed down by the storm, and was taken by car to the scene as the scheduled helicopter could not fly. (*NYT*, 16 Mar 76, 16; *W Post*, 17 Mar 76, C–2; GSFC *News*, Mar 76, 1; Worcester *Telegram*, 17 Mar 76, 1; report of commemorative committee, Auburn Rotary Club, 19 Mar 76)

- Timing the rebound of laser pulses aimed at reflectors left on the moon by Apollo astronauts had served to reinforce Einstein's theory of relativity, according to a study in *Physical Review Letters* reported by the *New York Times*. The equivalence principle—that objects of different weight fell at essentially the same speed—was tested in the 16th century by Galileo, who showed that all bodies responded similarly to gravity regardless of size or composition. Einstein's theory was based on a related assumption, that the mass responsible for an object's inertia was equivalent to the mass responsible for the gravity that it generated. However, if the gravitational pull of the earth were influenced by the sun's gravity, the relationship between a body's inertia and its gravity would not always be uniform, and the report described experiments to test a theoretical departure from equivalence. An eclipse of the sun in 1919 had been used to test Einstein's theory, and showed that light from distant stars was bent—as expected—as it passed through powerful gravity near the sun. Other experiments—bouncing radar beams off Venus and Mercury to see if the beams were bent or lengthened by the sun's gravity, or monitoring round-trip signals to *Mariner 9* as it sailed behind the sun or radio emissions from quasars in that direction—had narrowed any deviations from Einstein's formulation to a small margin.

    The experiments reported recently were independent analyses of 1389 measurements of round-trip travel times of laser pulses aimed at moon reflectors, to detect changes in earth-moon distance to within 127 mm. If earth gravitational energy were influenced by the sun's gravity, the effect would alter the earth's gravity more than that of the moon, and the moon's motion about the earth would deviate by as much as a meter from conventional predictions. The experiments showed no such effect. (*NYT*, 16 Mar 76, 17)

- The Energy Research and Development Administration requested proposals for what the *Wall Street Journal* called a "scaled-down" Institute for Solar Research, a modest facility with a budget of $4 to $6 million in its first year. The National Academy of Sciences months ago recommended an institute with a wide research role, a staff of about 1500, and an annual budget of $50 million; unsolicited proposals had been arriving from nearly every state, with many politicians interested in having the institute locate in their constituency. ERDA would operate the facility through contracts with any organization, including nonprofit groups, corporations, universities, or state or local governments, and operations should begin by early 1980, according to Robert Hirsch, an acting assistant administrator of ERDA, at the press conference. The institute was a requirement of a 1974 law calling for various solar-energy demonstrations to be financed by the federal government; ERDA wanted initial proposals for a 5-yr operation, with future functioning dependent on performance and availability of funds. As ERDA currently contracted out most of its solar research to private facilities, a reporter asked why its work called for a special institute instead of expanding the current research program; Hirsch responded, "because the law requires it." (*WSJ*, 16 Mar 76, 3)

17 March: NASA asked scientists to submit proposals for scientific experiments to be carried on the first Spacelab mission in late 1980, according to Dr. Noel W. Hinners, Associate Administrator for Space Science. Primary objective of Spacelab 1 would be to verify performance of systems and subsystems and to measure the environment surrounding NASA's Space Shuttle; secondary objectives would be to obtain scientific, applications, and technology data and to demonstrate Spacelab's ability to perform space research. The first flight would emphasize stratospheric and upper atmosphere research, but proposals from other scientific disciplines were also solicited.

Objectives of the Spacelab 1 mission had been planned jointly by NASA and the European Space Agency (ESA); the NASA announcement of opportunity would be sent to U.S. scientists and those in all other nations not members of ESA, and ESA would make a separate solicitation to its 11 member states. Responses would be coordinated between the two agencies. (NASA Release 76-49)

- The U.S. Navy Dept. selected General Dynamics Corp. to develop its cruise missile, under a $34.8-million contract that would cover costs of integrating systems into the prototype. GD won the award after 2 yr of competition with Vought Corp., a unit of LTV; the Navy recently finished a series of tests comparing prototypes made by the 2 companies. Last week the Navy had ordered Vought to stop work on its version, after the company had reported cost overruns and the missile performed badly under testing. The cruise missile, a jet-powered weapon with an advanced guidance system, would fly at low altitudes to make detection difficult. The sea-launched version could be put into submarine torpedo tubes and fired from under water as well as from ships. In both the Navy version and the Boeing missile being tested by the Air Force [see 5 Mar.], McDonnell Douglas Corp. was maker of the guidance system and Williams Research Corp. the engine manufacturer. (*WSJ*, 18 Mar 76, 2)

18 March: Donald K. (Deke) Slayton, pilot of the docking module on the Apollo–Soyuz Test Project, had been appointed to a newly created position as Deputy Director of Flight Operations for approach and landing tests of the Space Shuttle at Johnson Space Center, the *Spaceport News* of KSC reported. Slayton would be responsible for planning and implementing the shuttle approach and landing test project, under the Director of Flight Operations for the Space Shuttle project. (*Spaceport News*, 18 Mar 76, 5)

- NASA announced it had completed negotiations with Owens–Illinois, Inc., for replacement of a mirror blank for the Infrared Telescope Facility being built at Mauna Kea, Hawaii. The 126-cm-dia. replacement would be ready for processing at the Kitt Peak National Observatory that would turn it into a telescope mirror. Opticians at Kitt Peak had cut a center hole in the original disc preparatory to grinding it when a fracture was detected last Sept.; the crack, extending radially more than 86 cm from the center hole, rendered the mirror blank useless. The original and replacement were cast from Cervit, a transparent ceramic material obtained by NASA for testing as part of the Space Telescope. When a different material was decided on for the Space Telescope, the Cervit blank was transferred to the infrared telescope project. Cervit had practically zero thermal expansion, making it ideal for telescope materials, and had been used in 2 of the world's 5 largest telescopes. Under the new agreement, Owens-Illinois would provide a replacement lens for $200,000 and return of ownership of the damaged blank to the Ohio firm. (NASA Release 76-42)

19 March: Dr. George M. Low, Deputy Administrator of NASA, announced that he would leave government service at the end of June to become president of Rensselaer Polytechnic Institute, where he received his bachelor's degree in aeronautical engineering in 1948 and master's in 1950. He had been a trustee of the institute since 1971. In 1949 Low became an aeronautical research scientist at NACA's Lewis Flight Propulsion Laboratory in Cleveland. When NASA was formed 9 yr later, he came to Hq. as Chief of Manned Space Flight and was chairman of the committee that planned the Apollo lunar landing program. He became Deputy Associate Administrator for Manned Space Flight before transferring to the NASA Manned Spacecraft Center at Houston in Feb. 1964 as Deputy Director. In April 1967 after the Apollo 204 fire, Low became manager of the Apollo spacecraft program; under his leadership, the spacecraft was redesigned, manned space missions resumed 23 mo later, and *Apollo 11* made the first manned lunar landing in July 1969. Appointed Deputy Administrator of NASA by the President in Dec. 1969, Low negotiated the agreements with the USSR that led to the Apollo-Soyuz flight and other cooperative space projects. (NASA Release 76-54; Admr's letter, 19 Mar 76; *NYT,* 20 Mar 76, 17; *Marshall Star,* 24 Mar 76, 1)

- A press briefing following the 7th lunar science conference at Johnson Space Center [see 22 Feb.] featured the cochairmen, Dr. Larry Haskins, chief of planetary and earth sciences at JSC, and Dr. Robert Pepin, director of the Lunar Science Institute, and Dr. Deter Hayman of Rice Univ. Dr. Haskins said the conference aimed at broadening the planetary view of people working with lunar samples and data; a substantial interest was developing in the lunar polar-orbiter mission among people more interested than before in planetology, and aware that orbiting and remote-sensing missions would be the principal source for studying geoscientific evolution of these bodies. As an example, long-range operation of the lunar-surface packages had produced information that resulted in a lowered value for lunar heatflow; this in turn "constrained" the composition of the moon and its evolutionary history—"Whatever history it has, the amount of heat in the interior has to match the present heat loss from the body."

    Also, said Dr. Haskins, lunar data were being used widely by scientists not previously concerned with lunar research: for example, archived data would be used to tell whether the moon was differentiated only in its outer extent, or more than 50% of it had undergone chemical separation or even developed a core. Another example would be the matching of seismic with magnetic data to get a better overall picture of lunar evolution.

Dr. Pepin noted that 740 people had attended the conference, many more than the previous year; the program committee had purposely included discussions of other planetary and meteorite investigations that revealed much about the first 500 million yr of solar-system history, whose clues still exist in the moon (although hard to interpret) but no longer exist on earth. Experience with other planets had been insufficient to say if they had preserved early solar-system records; "best chances are not," he added. Early solar-system history, especially of the inner planets, was characterized by enormous bombardment of all solid surfaces by objects now referred to as meteorites, but on a scale in those days "that almost suggests we should call them something else." These huge objects clearly left their mark, with evidence that the bombardment took place more than 4 billion yr ago, then rapidly tapered off. Data from Mercury, Mars, and Venus would be expected to show that the historical record had been erased by time, constant impact, and geological activity in which the surface was pounded, dispersed, or remelted.

The intense bombardment had wiped out earth's early geological record too; terrestrial geologists had tried to put the age of the oldest earth rocks farther and farther back, getting in gradual stages as far back as 3.7 billion yr in a large anorthosite deposit in Greenland. This boundary condition for earth geology was one reason for the growing interest in lunar and general planetary studies: the flow of information had been not only from planetary to terrestrial scientists, but also in the other direction. Lunar scientists had begun to examine terrestrial craters such as the 60-km impact phenomenon called Manicouagan: what would have been referred to 5 yr ago as remnants of a Manicouagan volcano had now been established as an impact, mainly because the ease of sampling (greater there than on the moon) had permitted study of energy partition and formation of rock, glass, and melt material in a large-scale impact. Comparative planetology means applying results of studying one planetary surface to explain phenomena on other planetary surfaces. Earth, unlike other planets we know about, continued as a thermally active planet with volcanism, mountain upheaval and wearing down, and had obliterated much of its earlier record. The early stages of the lunar and planetary program focused attention outward; now these studies had provided knowledge applicable to earth, and the generation and development of lunar-surface morphology as a result of external bombardment was only one example of this application.

Over the next 2 to 3 yr, a group of 50 to 100 scientists would be studying in detail the phenomenon of volcanism on all the terrestrial planets as a fundamental stage of planetary development, in a pilot project to define comparative planetology and relate it to a specific aspect. Mercury apparently had basalt flows; *Venera* pictures suggest recent activity on the surface of Venus that produced what seem to be basalts. Mars clearly had basalts: it had the largest volcano in the solar system, "an enormous structure that would . . . cover the state of Kansas if . . . plunked down there." In 2 or 3 yr, Dr. Pepin said, we should be able to give an integrated picture of the issuing of basalts from the interior onto the surface, an extremely important process in the development and evolution of a planet.

Life in the solar system was an exciting subject now, said Dr. Hayman, and discoveries were coming in at a pace that would make today's theories outmoded by the time they got into print 6 mo from now. The task of scientists, he said, was to include the general public in the excitement, and to make it filter through to the colleges and universities, to high schools, and even to grade-school levels.

One thing that had changed in the past 20 yr since the days of Harold Urey and Hans Seuss was the concept of the solar nebula and the formation of planets and satellites by accretion. The problem was the noncondensable material in the nebula, and where it went: the solar physicists explained its absence by the theory of a solar upheaval that cleaned out the inner parts of the solar system by a superstorm of ions. Applying thermodynamics to this theory made it possible to calculate which compounds would condense out at what temperature. The Allende meteorite that fell in Mexico in 1969 contained compounds similar to what had been predicted as a result of condensation of the solar nebula at high temperatures. Then, 2 or 3 yr ago the idea arose that the sun and planets had formed not only from gases but also from so-called presolar grains, which had not been specifically identified; their identity on the earth and moon would have been lost over time because of geological processes. So the search shifted to meteorites.

Although none of the grains had been definitely identified, the conference had heard 2 reports that might be interpreted as finding presolar grains. Metallic nuggets in the Allende meteorite were found to consist of platinum-group metals in extraordinarily small fragments, a millionth of a meter, that did not vaporize or melt readily and might be presolar grains (although it would take a lot of work to settle the question, Dr. Hayman said). Also, a group from Berkeley had dissolved carbonaceous meteorites in hydrofluoric acid to get rid of the silicates; the carbon residue was found to be full of gas called "planetary" because its composition differed from that of the sun and was much more like that of earth. The question was how carbon material could have been of planetary origin under the solar-nebula theory: speculation was that the carbonaceous material might predate the formation of the sun and have originated elsewhere. It could not have undergone great heat without being destroyed, so the objects containing the material must have been created farther out, not in the inner region of the sun and terrestrial planets but in the region of the giant planets.

If the heavier elements were synthesized in the stars by nuclear processes, especially in explosions of novas and supernovas that produced all sorts of elements as they blew away their outer envelopes, all the elements are probably represented in this expanding envelope. With rapidly dropping temperatures, the condensation of presolar grains would occur within 10 yr. It would be possible to look at the product of a single event to extrapolate the total of many such events of which all the elements around us would be the product.

Dr. Haskins said in summary that developments in methodology and the availability of more precise measurements would make it possible to supplant earlier measurements. Geophysicists were learning to extract information from bits of material instead of bucketsful or large chunks as they had done in the past. The Apollo program had been responsible for a steady progress in the field of measurements: "there simply was not the funding and interest correlated together to do it prior to that time."

Asked what the scientific community had learned about the moon in the 3 yr 3 mo since the last moon mission, Haskins said they were beginning to understand the major stages in the evolution of every planetary body composed of rocky material; the moon had passed through stages that the earth had not reached, and had preserved a record of what had happened—for example, a preview of the kind of rock that would be formed later in earth history. The findings provided a broad framework in which more and different questions could be asked, not just investigation of the more obvious characteristics of the many materials brought back. Queried on the lunar-sample curatorial facility, Dr. Haskins said that moon rocks from the Apollo expeditions were housed in a leaky, flimsy facility that would not protect them for use of future generations of scientists. A new facility was desperately needed, because the present building was not fireproof and the roof leaked; within the building, the rocks had been kept in nitrogen-filled cabinets and handled only in clean-room conditions, to prevent contamination. However, moisture in the surroundings might have led to erroneous deductions regarding the moon environment when scientists analyzed the lunar samples. (Transcript, 7th lunar science conf. press briefing, 19 Mar 76; NASA Release 76-25; JSC Release 76-15; *Science*, vol. 185, 346; *W Post*, 21 Mar 76, A-3)

- Congress's Joint Committee on Defense Production, inquiring into government–contractor activities involving hospitality or gratuities toward federal employees, revealed that Rockwell International Corp. had entertained Eugene A. Cernan and Ronald E. Evans, NASA astronauts, at a facility on Bimini in the Bahamas, in addition to 11 other NASA employees who had enjoyed hospitality at a hunting lodge in Md. Both Rockwell and Northrop Corp., another NASA contractor, ran hunting lodges on the eastern shore of Md. where 13 NASA employees and one former employee had acknowledged acceptance of entertainment. The DOD had announced it would reprimand its chief of research, Dr. Malcolm R. Currie, and dock his pay, for accepting a similar weekend from Rockwell. (*W Star*, 19 Mar 76, A-1; *WSJ*, 17 Mar 76, 17)

22 March: A General Accounting Office review of the Seasat oceanographic spacecraft program revealed a lack of formal agreements between NASA and potential Seasat users on application of resources, said an article in *Aviation Wk and Space Technology*. GAO said there "should have been top-level agreements among NASA, the National Oceanographic and Atmospheric Administration, and the Defense Dept. concerning Seasat participation prior to project approval within NASA." NASA responded that it was "of the opinion that there are clear understandings . . . regarding roles and responsibilities in the Seasat-A project and that formalized agreements are being prepared." GAO also reported increased costs during Seasat spacecraft definition; objections from the user community to changes in instrumentation had prompted NASA to raise the cost ceiling on the spacecraft rather than fly hardware that would not meet user requirements. GAO asked Congress to require that NASA provide specific data on how Seasat would produce $350 million yearly in benefits; why NASA, rather than users, should pay for Seasat improvements to benefit the user community; and what agency responsibilities would be in future Seasat operations. (*Av Wk & Sp Tech,* 22 Mar 76, 44)

- In a special message to Congress, President Ford requested an increase in federal spending to $24.7 billion for scientific research and development in the energy, defense, and space programs. Ford said his request for the 1977 fiscal year, an 11% increase over 1976 estimates, included $2.5 billion for basic research. The budget asked $2.6 billion for energy R&D, a 35% increase over 1976, including funds for nuclear power, development of solar and geothermal energy, and fusion power; major increases were for energy conservation and for research on fossil fuels to improve direct combustion of coal for production of oil and gas. The President also asked congressional conferees to act quickly on bills to establish the Office of Science and Technology. (*W Star,* 22 Mar 76, 7)

- *Pioneer 10,* on its way out of the solar system, spent 24 hr in an enormous magnetic "tail" area of the planet Jupiter, the Ames Research Center reported, registering zero on its solar-wind detector during that time because the magnetic envelope completely shut out the solar wind. Dr. John H. Wolfe, project scientist, said it was conceivable that the solar wind "could have died completely for a whole day without our being in the tail . . . But we believe we've found that Jupiter has a very stretched-out magnetic envelope . . ." Exact shape and size were not known, but were estimated as conical and better than 800 billion km long, spanning the distance between the orbits of Jupiter and Saturn. Earth's magnetic tail had been measured at more than 6 billion km by previous spacecraft; the magnetic tails of the 2 planets had been shown to extend to great distances by the force of the solar wind, a supersonic charged-particle stream constantly emitted from the sun. (*NYT,* 26 Mar 76, 19; *W Post,* 27 Mar 76, A-5; NASA Release 76-55; ARC Release 76-22)

- Cost of the Air Force B-1 bomber might run more than $11 billion over original estimates because of inflation and higher prices, said Rep. Les Aspin (D-Wis.), adding that total cost increases on the B-1 would be around $1 billion while inflation would be responsible for another $10 billion increase. Three research contracts on the B-1 had already run $129 million higher than expected, he said, and those represented only money spent for models, engines, and electronics. By the end of 1976, Congress would have to decide whether to go ahead with the bomber, intended as a replacement for the aging B-52. (*LA Times,* 22 Mar 76, 11C)

23 March: The House passed, 330 to 35, and sent to the Senate an authorization bill of $3.7 billion for the nation's space programs for FY 1977, slightly less than the administration request, but $133 million more than

for FY 1976. Research and development authorizations totaled $2.77 billion, nearly $10 million more than requested. The Space Shuttle would use $1.29 billion in R&D money, plus $40 million for 8 construction projects. (*W Post*, 23 March 76, 6)
- A new network of space satellites would replace the existing hotline between Washington and Moscow later in 1976, said Willis K. Naeher, Deputy Assistant Secretary of State for Communications, on a weeklong working visit of U.S. communications specialists to Russia with Soviet counterparts to check Soviet facilities and transmit test messages over both the Molniya and the Intelsat networks. The new system was planned to provide more reliable emergency communications between the two capitals than had been available for a dozen years using a transatlantic cable and radio system. The overland line had been disrupted when parts of the cable had been damaged by fire, stolen, and once plowed up by a Finnish farmer. Coordination was necessary because the U.S. and Russian systems functioned differently: whereas Intelsat used a "necklace" of satellites around the equator maintaining a fixed attitude toward earth, the Molniya traveled in an elliptical north-south orbit and would be visible from both countries for periods of 8 hr at most, so that 4 satellites would be needed instead of the 3 originally planned. Failure of a Molniya III launch caused the delay from 1974, original starting date. Final test of the Intelsat portion would begin in Apr. (*NYT*, 23 Mar 76, 14)
- The 2 USSR spacecraft orbiting Venus had completed their program, the Tass news agency announced, stating that further scientific experiments would be made "under an additional program." Tass did not elaborate on the further experiments, nor did it indicate how long the two craft would remain in operation. *Venera 9* had completed 75 orbits since its arrival at Venus on 22 Oct. 75, and *Venera 10* had completed 71 since its arrival 3 days later, Tass said. (*NYT*, 23 Mar 76, 3)

24 March: Travelers aboard NASA's Space Shuttle orbiter in the 1980s would use a unique space suit and rescue system, developed at Johnson Space Center. The Shuttle suit, a departure from the customized astronaut suits, would consist of a two-piece upper and lower torso cover in small, medium, and large sizes to accommodate all astronaut candidates or crews, including females. Only the pilot and mission specialist would be outfitted with the space suit; the commander and payload specialists would be provided with a personal rescue system nicknamed the "cosmic soccerball," consisting of a container nearly a meter in diameter constructed of three layers—urethane, Kevlar, and an outside thermal protective layer, with a small viewing post of tough Lexan—containing its own simplified life-support and communications systems. During a rescue operation in space, a space-suited astronaut could transfer the rescue balls in one of three ways: using the handle to carry the enclosure like a suitcase from one vehicle to another; hooking a device like a clothesline between two spaceships and passing the rescue ball with its passenger from the disabled ship to the other; or using the remote-manipulator arm in the orbiter's cargo bay to pluck the rescue ball with its passenger from the disabled ship and put it aboard the rescue ship.

Materials to be used in the rescue ball and the Shuttle suit would provide longer shelf life, according to technicians who ran pressure and abrasion tests on them; use of the fabric to make joints in the space suits (instead of the Apollo and Skylab suit use of neoprene rubber molded into convolutes containing cables) would provide better mobility and reduce cost and weight of each suit. The new suit also featured a module construction that closed with a body seal at the waist, considered more reliable than the pressure-seal zippers used in the previous suits. The new suits also were designed with an integral portable life-support system as part of the rigid upper torso, instead of the bulky 34-kg package that previously had to be unpacked and connected to the suit. (NASA Release 76-56; *JSC Roundup*, 26 Mar 76, 1)

25 March: The first satellite in the 2-ocean Marisat system inaugurated communications services to U.S. Navy ships. However, the scheduled 1 Apr. start of service to commercial shipping was canceled because of malfunctions in C- and L-band transmissions, involving random variations in signal strength. The Navy operations used UHF frequencies. *Marisat 1* was launched 19 Feb. into synchronous orbit over the Atlantic; Comsat General Corp.—86% owner and manager of the Marisat system—said tests indicated that the problems could be solved "with respect to the [second] spacecraft" awaiting launch at ETR, in time to permit its scheduled launch 27 May for service in the Pacific Ocean. The difficulty appeared to be associated with a despun C-band triplexer, arising from loose metal particles. (*Av Wk & Space Tech*, 29 Mar 76, 14; Comsat 1975 rept. to stockholders, 9)

• Space-related technology would serve to monitor municipal water pollution and dispose of solid wastes, NASA announced, describing systems for electronic monitoring of water quality in cooperation with the Gulf Coast Waste Disposal Authority and for sewage treatment at a plant midway between Los Angeles and San Diego. Johnson Space Center at Houston developed under contract with Boeing a trailer-mounted automated system to process data from up to 40 different water sensors for rapid indications of pollution, temperature, turbidity, and similar items. The current procedure was to sample city water at regular intervals and send the samples to a laboratory for analysis, with results available days later. JSC and the Goddard Space Flight Center developed a biosensor to give total bacteria count directly, by adding chemicals that caused bacteria to radiate light; this sensor could both detect and quantify living or dead bacteria in a continuous-flow water sample.

At Langley Research Center, scientists working on Skylab environmental control had developed an electronic sensor that could detect human or nonhuman fecal coliform bacteria in a few hours, rather than a few days; the device would permit public health authorities to act quickly if large quantities of disease-producing bacteria entered a water supply. The automated monitoring system would eventually include a device to detect known carcinogenic chemicals—chloroform and carbon tetrachloride, for instance—already found in the drinking water of cities surveyed for the project. Some of the pollutants were thought to be byproducts of chlorine added to city water to guard against waterborne diseases. A gas-chromatograph technique developed at Ames Research Center to extract minute samples of organic materials from the atmospheres of other planets would be valuable to concentrate the harmful chemicals detected in city water for rapid onsite analysis. Another Ames system, attaching fluorescent dyes to bacteria so that their presence could be recorded by electronic sensors, could be applied to detection of waterborne viruses.

As for sewage disposal, the Jet Propulsion Laboratory used for the new million-gallon-per-day treatment plant at Huntington Beach, Cal., a process invented by Marshall Humphrey, a JPL engineer working on lightweight materials for insulating rocket motors. Using a pyrolytic converter to produce activated carbon by chemical changes induced by heat, he found that sewage solids were excellent raw material for production of carbon. The new process would convert solid sewage to activated carbon, which could then be used to remove impurities from incoming waste water and recycled and reactivated with new incoming sewage. Gases generated from sewage solids would serve as fuel for the converter. Dry black odorless powder of carbon and ash would be the only residue from the process. Even the billion-dollar secondary sewage-treatment plants in compliance with EPA standards had not solved the problem of solid-waste disposal; primary treatment of sewage had left about 40% solid waste in water leaving the plants and discharged into rivers and offshore waters. The JPL process would exceed EPA standards for ocean discharge and reduce capital costs of processing systems as much as 25%. (NASA Release 76-57; *KSC News*, 4 Mar 76, 2)

- NASA's reusable Space Shuttle program would produce significant benefits for civil aviation, the agency reported in announcing the first operational tests of an advanced flight-control system for civil aircraft to be conducted by the Shuttle, together with advances in structural materials technology for aerospace manufacturers.

    Standard heavyweight mechanical backup flight-control systems had been required when reliability of electronic control systems was in question; however, aerodynamic requirements for the use of mechanical systems had hampered development of flight-efficient aircraft designs. Use of the advanced electronic controls would reduce weight, resulting in less fuel consumption, and a computer tie-in with ground navigation and mission controls would provide constant reliable communication and reduce traffic delays. Use of composites—strong lightweight combinations of metals and plastics—would reduce structural weight by more than 30% compared to aluminum. Developed by NASA and the Air Force in collaboration with industry, the composites would demonstrate in Shuttle testing the weight reduction and cost effectiveness needed for future designs. (NASA Release 76-58)

26 March: RCA-B, called *Satcom 2* in orbit, was launched from the Eastern Test Range at 5:42 pm EST (2242 GMT) on a Delta 3914 vehicle into a synchronous transfer orbit at 35 753-km apogee and 185-km perigee, and inclination of 27.25°. Second in a series of 3 large 24-transponder comsats to be launched by NASA on a fully reimbursable basis for the Radio Corporation of America, *Satcom 2* would transmit voice, data, telex, and facsimile messages to and from the continental U.S., Alaska, and Hawaii from its geosynchronous orbit at about 35 800-km altitude above the equator at 120°W. The spacecraft was box-shaped, $1.2 \times 1.2 \times 1.6$ m, weighing at launch about 990 kg; it was 3-axis-stabilized, carrying 2 rectangular solar panels about $7m^2$ continuously oriented to the sun to provide electric power. After dropping the apogee motor, *Satcom 2* would weigh about 463 kg in orbit. The communications system included a fixed 4-reflector antenna assembly and a lightweight transponder of traveling-wave-tube amplifiers; the 24 input and output filters were of a graphite-epoxy composite. All 24 channels were simultaneously operable throughout a minimum 8-yr lifetime, each having a 36-mhz bandwidth within the 500-mhz allocation to RCA's Globcom Inc. (NASA Release 76-37; MOR M-492-206-76-02 [prelaunch] 18 Mar 76, [postlaunch] 13 Apr 76)

- The Hugh L. Dryden Flight Research Center, formerly the Flight Research Center at Edwards, Calif., was formally dedicated in a ceremony by NASA Administrator James C. Fletcher; Dr. T. Keith Glennan, first administrator of NASA from 1958-1961; Sen. Frank E. Moss (D-Utah), chairman of the Senate Committee on Aeronautical and Space Sciences; and Mrs. Dryden, guest of honor, who unveiled a bust of her late husband that would be placed in the Center's lobby. After the ceremony, visitors were invited to a display of aircraft in the main hangar, including the 747 aircraft that would be used for the Space Shuttle orbiter approach and landing tests next year. Center efforts to transfer space technology to industry were represented by a solar-energy display and low-drag truck exhibits. Dr. Dryden in 1947 had been named to the newly created post of director of the National Advisory Committee for Aeronautics; when Dr. Glennan, then head of the Case Institute of Technology, was selected as NASA's first administrator, he insisted on Dr. Dryden as his deputy. When James E. Webb was chosen NASA's second administrator, he also made the condition that Dr. Dryden remain as his deputy; he had served in that position until his death in 1965. (*NASA X-Press*, 26 Mar 76; program)

- Marshall Space Flight Center forecast completion by 1 April of a unique facility—one of the largest and most significant construction jobs undertaken in the past 10 yr—at the Center, an x-ray test facility costing about $4 million. The new facility included a 305-m-long stainless steel x-ray path guide tube nearly a meter in diameter, connecting the x-ray source with a 6-m-diameter instrument chamber housing the telescopes and other instruments to be tested. Work began in January 1975, and would be completed on schedule. The facility, the only one of its size and type, would be used first to test instruments for NASA's High Energy Astronomy Observatory (HEAO) program; later it would be used in x-ray verification testing and calibration of mirrors, telescope systems, and instruments, for rocket payloads in x-ray stellar studies and similar projects. After checkout, the facility would be turned over to MSFC's

Science and Engineering Directorate. (*Marshall Star*, 31 Mar 76, 4; MSFC Release 76-53)

- NASA's Technology Utilization Office, publisher since 1963 of more than 6000 single-page Tech Briefs on new techniques and innovations resulting from advanced research and technology programs in NASA, announced it would begin publication of *NASA Tech Briefs* in April 1976. The new document would be a two-color journal designed for non-aerospace users of NASA technology, containing information on as many as 600 items annually that previously appeared in the single-page briefs as well as new data that previously appeared in Technology Compilation booklets. The document would sort the information into 9 technical categories to make it accessible as a current-awareness medium, and a cumulative index would appear once a year. The journal would be free to U.S. citizens. (NASA Release 76-60)

27-28 March: A series of rocket experiments funded by the National Science Foundation and NASA was launched from Poker Flat Range, Alaska, to measure magnitude and direction of electric fields and neutral winds in the auroral atmosphere. Spectrographic observations of the releases of barium vapor on 2 successive days indicated that a pulsating aurora was induced by vapor releases near 250 km altitude, but only when the explosions occurred in the path of precipitating electrons associated with the visible aurora. Previous experiments had produced no definite evidence of pulsations, and the experiments would be repeated to clarify the results. (*Nature*, 12 May 77, 135)

28 March: A reorganization of the NASA office (N₄PO) located at the contractor-operated Jet Propulsion Laboratory at Pasadena became effective. N₄PO would no longer report to the Associate Administrator for Center Operations at NASA Hq in Washington, D.C.; a resident legal office would report to the NASA General Counsel, and a resident procurement office would report to the NASA Assistant Administrator for Procurement at Hq. E.S. Groo, the Associate Administrator for Center Operations, said the change was made to relate the N₄PO functions more directly to the functions performed by their Hq counterparts, or by certain NASA field centers. N₄PO's cryogenics procurement activities and personnel would be relocated to the MSFC resident office at Canoga Park, Calif., and would be functionally responsible to MSFC; the Delta procurement activities and personnel would remain at JPL but would be functionally transferred to GSFC at Greenbelt, Md. (NASA Release 76-50; Groo anno. 15 Mar 76)

- A component of the backpack worn by Neil A. Armstrong on the moon's surface had been used by the former astronaut—now professor of aerospace engineering at the Univ. of Cincinnati—in bioengineering experiments testing better blood pumps for heart-lung machines and artificial hearts or kidneys. The Apollo double-diaphragm pump that circulated cooling water through plastic tubes in the astronaut's space suit proved in the UC tests to be 10 or more times less damaging to red blood cells than pumps previously used. Blood, a delicate substance easily damaged by prolonged mechanical pumping, had lost critical amounts of hemoglobin in procedures such as open-heart surgery when pumps had taken over the heart's function for several hours. Armstrong, working on a team at the university's Institute of Engineering and Medicine, recalled the Apollo pump's efficiency from a power and weight standpoint and borrowed one from NASA for experiments with animals. The pump, designed to run on minimum power, created little turbulence in the flow of fluid and had no moving parts to damage the cells. It would be further modified for medical use. (*NYT*, 28 Mar 76, 64)

30 March: No immediate serious problem appeared to exist in modification of the stratosphere, according to a report released by an interagency task force on inadvertent modification of the stratosphere (IMOS) prepared by the Interdepartment Committee for Atmospheric Sciences (ICAS). Assessing presently postulated manmade modifiers of the stratosphere other than fluorocarbons, the report considered 6 kinds of possible hazards: nitrogen fertilizers, brominated compounds, other chlorinated compounds, particles, the Space Shuttle, and carbon monoxide. The report concluded that concern over modification of the stratosphere was so far speculative, and that it was based on compounds not yet released to the upper atmosphere in quantities believed sufficient to produce a hazardous effect. Although a large number of substances had been considered, only those appearing most potentially harmful were singled out. Also, cumulative effects from several substances might become significant later, the report said, even if the effect of any single substance were relatively slight. The report urged increased research into stratosphere modifiers and prevention of inadvertent modification now and in the future. (IMOS report, 30 Mar 76)

31 March: First test firing of the second Space Shuttle main engine—Engine 0002, first to have a flight-type engine-mounted controller—achieved the programmed 1.5-sec duration, firing through a diffuser used for altitude simulation when production engines were throttled in testing. Engine 0002, a flight configuration that would not be flown, was a developmental engine instrumented for test purposes. It was fired on the National Space Technology Lab's Stand A-2, used in Saturn V tests during the 1960s and modified and reactivated for the Space Shuttle program. NASA planned to fire each engine on the A-2 throttling test position before using it on a Shuttle flight. Stand A-1, the sea-level test position, had been used for about a year to test the first SSME, a test version known as the integrated subsystem test bed; A-1 was also developed for the Saturn program and modified for Shuttle tests. (MSFC Release 76-57; Rockwell Release RD-7)

- Dr. William H. Pickering, director of the Jet Propulsion Laboratory at the California Institute of Technology since 1954, retired, announcing that he would return to Caltech as a professor of engineering. He had earned all his degrees at that institution and had joined the faculty in 1936, becoming a member of the JPL staff in 1944. The staff at that time numbered a few hundred; when he became director, the total had reached 1000, and numbered about 4000 when he announced his retirement. Born in New Zealand in 1910, Dr. Pickering had become one of the world's foremost experts in space technology. Under his leadership,

JPL had designed *Explorer 1*, first successful U.S. satellite, and went on to develop and operate the Ranger, Surveyor, and Mariner space probes. Dr. Pickering was first president of the American Institute of Aeronautics and Astronautics in 1963, and president of the International Astronautical Federation in 1965–66. NASA's announcement of his retirement said he would work on gaining worldwide scientific support for the proposed International Solar System Decade of research and space exploration of all aspects of the solar system. (NASA anno. 23 June 75; *NYT*, 21 June 75, 41; *LA Times*, 31 Mar 76, 1; *Pasadena Star-News*, 1 Apr 76, 1)

- Gerald A. Mossinghoff, NASA's Assistant General Counsel for general law since Jan. 1974, was appointed Deputy General Counsel effective this date. He joined NASA in 1963 as a patent attorney in the Office of General Counsel; in 1966 he was named director of the Office of Legislative Planning at the U.S. Patent Office, and returned to NASA in Dec. 1967 as Director of Legislative Liaison. He received NASA's Exceptional Service Medal in 1971. A registered patent attorney, Mossinghoff had received a B.S. in electrical engineering from St. Louis University and a J.D. with honors from George Washington University. (Admr.'s announcement, 31 Mar 76)

During March: Launch of RCA's *Satcom 2* this month was the second launch of a Delta vehicle using 9 large Castor IV solid rocket boosters instead of the standard 9 Castor IIs. (First to use the larger rockets was the Delta that launched *Satcom 1* on 12 Dec. 75.) RCA had underwritten development expenses of adapting the NASA "workhorse" Delta to use the larger rockets, in a unique NASA–customer–launch vehicle contractor arrangement that had enabled RCA to design its domestic comsat to be the heaviest to date without having to use a larger and more expensive launch vehicle. The Castor IV boosters would permit a Delta to put into synchronous orbit a payload weighing about 900 kg, compared to about 700 kg for a vehicle equipped with the Castor IIs. NASA was thinking of the more powerful Delta as a standard launch vehicle for other customers who could use the increase in permissible weight of about 29% in many areas of spacecraft design: weather forecasting, scientific exploration, and communications. (*Spaceport News*, 18 Mar 76, 1)

- More new discoveries would be made in space than on the ground during the first 10 to 15 yr of Shuttle operations, a colloquium on bioprocessing in space was told by James H. Bredt, NASA program manager for space processing. The meeting of more than 200 industrial and academic bioscience researchers at Johnson Space Center heard reports on Space Shuttle and Spacelab capabilities and status. Noting that Spacelab would be able to carry 300 to 400 experiments, more on each flight than all other previous manned space missions combined, Bredt urged the audience to submit experiment programs that would make use of the new national resource. Participants recommended formation of a small group of bioscientists to advise NASA in discussions of space processing suggestions, and scheduling of followup meetings to define areas of interest so that programs could be developed before instrumentation development began. NASA spokesmen said more than 1400 experiment proposals for Shuttle flights were on hand, and Shuttle flight opportunities covering the first 2 Spacelab missions would be announced later this year. (*Av Wk & Sp Tech*, 29 Mar 76, 53)

- Endorsed by the state of Colo. as a bicentennial project, a mobile space museum prepared by the High Flight Foundation—a nonprofit religious organization founded in 1972 by *Apollo 15* astronaut Col. James B. Irwin—would tour the U.S. to give residents of 30 cities a chance to see a large exhibit on national accomplishments in space. The exhibit would include tools and photographs, miscellaneous objects and films. During the tour, various state governors participating in bicentennial celebrations would receive from the astronauts state flags that had been flown to the moon. With Col. Irwin on the board of the High Flight Foundation were astronauts Col. William R. Pogue *(Skylab 3)* and Col. Alfred M. Worden *(Apollo 15)*. (*Spaceport News*, 18 Mar 76, 8)
- NASA announced selection of the initial crew of the Boeing 747 shuttle-carrier aircraft that would carry and launch the Space Shuttle orbiter in approach and landing tests. Pilots would be Fitzhugh L. Fulton, Jr., and Thomas C. McMurtry, both of Dryden Flight Research Center; flight-test engineers would be Victor W. Horton, also from Dryden, and Louis E. Guidry, Jr., of Johnson Space Center. All the men chosen were civilians. The ALT flights, scheduled for early 1977, would carry the orbiter to an altitude of 7.5 km, where it would separate from the 747 and the orbiter crew would pilot it to a glide landing. Unmanned and manned captive flights would precede initial free flights. (DFRC Release 3-76)
- A device used in training Apollo astronauts was adapted to assist persons incapable of supporting their entire weight with their legs. Developed at Langley Research Center, the flying lunar-excursion experimental platform would exert a constant weight-relieving force during any vertical movement by the patient; the supporting force would be selected for a particular patient, and could be changed as the patient's condition improved. Besides its application as a device for exercise during rehabilitation, the apparatus showed potential for use in warehouses and storage facilities where heavy equipment was handled. The project was directed by LaRC's Technology Utilization Office. (*Langley Researcher*, 5 Mar 76, 7)
- The Federal Aviation Administration would use 15 new twin-jet Sabreliners to do a nationwide airways navigation-system checking job that formerly took about 50 older aircraft to do, according to a report by Rockwell International Corp., makers of the Sabreliner. The old Civil Aeronautics Administration (predecessor to FAA) in 1932 had hired 2 pilots in single-engine planes to inspect beacon markers—then the only night-navigation aids for air travel—as well as to check safety at emergency landing fields. The report reviewed the advent of radio-range transmitters, omnidirectional signals, and instrument-landing systems, for which the FAA operated a Flight Inspection National Field Office based in Oklahoma City with field offices in the continental U.S., the North Atlantic, and the Caribbean. Until recently, safety checks had been the job of a fleet of military-surplus C-47s (commercially called DC-3s)

known in World War II as Gooney Birds. These became unsatisfactory with the advent of the jet age, and Sabreliners were added to the fleet in 1968. FAA officials noted that the new twin-jet could cruise at 435 knots, compared to 150 knots for the DC-3s scheduled for retirement in June 1976. (*Skyline*, Spring 76, 34)

- Observance of 1976 as the 50th anniversary of U.S. commercial aviation would be established by a resolution introduced in the Senate by Sen. Howard Cannon (D-Nev.). The Air Mail Act of 1925—known as the Kelly Bill—gave the Postal Dept. authority to contract airmail carriage to private carriers, but the first contract airmail flight did not take place until 15 Feb 1926 in Detroit, Cleveland, and Chicago. The Postal Dept. had laid out a transcontinental route, and the contracted routes acted as feeder lines that greatly expanded airmail service around the country. Passenger airlines existed, but passengers were few, so that carrying the mail could mean the difference between success and failure for the airlines of the period. Each of the 46 cities that inaugurated commercial airmail service 50 years ago would cancel its special commemorative cover on the date of the first airmail flight to that city. Covers were issued by the Aviation Historical Foundation in cooperation with the U.S. Postal Service. (*NAA News*, Mar 76, 2)

- On a visit to the ERDA Model Zero 100-kw wind generator installed at NASA's Plum Brook Station near Sandusky, O., a reporter from *Popular Science* magazine climbed the steel tower to get a close look at the monster propeller measuring more than 398 m and rotating at maximum speed of 40 rpm. The flexible aluminum propeller blades operated downwind of the 30 m steel tower similar to utility rigs, set on a raised concrete foundation designed to withstand high wind and rotor-thrust loads and to be accessible for maintenance. The NASA approach to generating power had been to use large generators rather than groups of small ones; the power from wind was said to increase with the square of the rotor diameter, so that doubling the rotor diameter would produce 4 times the power, besides keeping costs down. Design improvements for future wind turbines would include use of composite materials to reduce cost and weight and improve reliability, and development of a different hub to reduce bending moments of the rotor-blade roots and make them less likely to break. (*Popular Science*, Mar 76, 73)

- A new technique of using aircraft to map surface temperatures of bodies of water with an accuracy close to 1°C was announced by Calspan Corp. The company said the method could be used to monitor the discharges of large volumes of heated water into rivers and lakes by power plants, steel mills, and other industries. Ecologists had sought regular monitoring of such flows, said to be potentially dangerous to fish and other marine life. Like the conventional method of taking water temperatures from aircraft, the Calspan method used an infrared thermal mapper to record infrared radiation from the surface, but did not require "ground truth" readings taken simultaneously at the sites being scanned, using boats, equipment, ground personnel, and correlations that were both time-consuming and expensive. The new system used a radiometer to track along the image produced by the thermal mapper, extrapolating true ground temperatures from a series of passes over sites with identical temperature and noting changes in infrared intensity. Maps using this technique had been within 1 °C of agreement with temperatures checked at the water's surface. (*Calspan News*, Second Quarter 76, 3)

- An FAA test of a new method of conserving energy, to be used when weather or other factors caused delays in aircraft landings, saved nearly 2.5 million liters of fuel in 1 day. The test consisted of holding Chicago-bound flights on the ground at 150 airports until they could be accepted at Chicago's O'Hare International Airport with a minimum of airborne delay. (*Av Wk & Space Tech.*, 22 Mar 76, 37)

- A 3-yr International Magnetospheric Study that began 1 Jan., sponsored by the International Council of Scientific Unions, would be coordinated by a steering committee set up by the Special Committee on Solar-Terrestrial Physics, chaired by Juan G. Roederer, professor and senior research physicist at the University of Denver. The 20 or more countries participating in the study had already submitted more than 1000 individual research programs, including ground-based, balloon, rocket, aircraft, and spacecraft experiments as well as networks of magnetic, auroral, and other geophysical observatories. Among major spacecraft projects planned by the European Space Agency, Japan, the Soviet Union, and the U.S. would be use of ESA's Geos spacecraft in geostationary orbit. ESA and NASA would join in an International Sun-Earth Explorer (ISEE) program to measure properties of different parts of the magnetosphere simultaneously. About 40 other spacecraft launched before and during the study period were expected to contribute sun-earth physics data to the study. Roederer told the National Academy of Sciences that the international coordination differed from that used in the International Geophysical Year, relying on interlocking spacecraft missions and quick and active exchange of information and scheduling. (*NAS News Report*, Mar 76, 1)

April 1976

1 April: Dr. Bruce C. Murray, professor of planetary science at California Institute of Technology and geologist by training, succeeded Dr. William H. Pickering as director of Caltech's Jet Propulsion Laboratory. After earning bachelor's, master's, and doctor's degrees at Massachusetts Institute of Technology, Dr. Murray became a geologist and later served in the Air Force. At the USAF Cambridge Research Lab., trying to expand knowledge of the earth's shape and gravity field by the use of earth satellites, he became interested in space technology. First faculty appointee at Caltech in the field of planetary astronomy in 1960, Dr. Murray and Prof. James Westphal were in 1962 the first observers of infrared emission from an object outside the solar system (a star) and did infrared mapping of the moon, Venus, and Jupiter. Dr. Murray was the author or coauthor of 4 books and more than 60 scientific papers, including the recent book "Navigating the Future." (*Pasadena Star-News*, 1 Apr 76, 1; Glendale *News-Press*, 1 Apr 76, 1; Montrose *Ledger*, 1 Apr 76, 1)

- Images from NASA's *Landsat 1* earth-resources survey satellite helped Alaskan Indians to choose the best areas of timberland and mineral deposits in compliance with the Alaska Native Claims Settlement Act of 1971, NASA announced. The act set aside 99 million acres from federal public lands so that more than 200 native village corporations and 12 native regional corporations—representing about 100,000 Indians, Eskimos, and Aleuts—could settle claims going back to the U.S. purchase of Alaska from Russia in 1867. A regional corporation—Doyon, Inc.—asked the Univ. of Alaska to recommend the best land out of inaccessible, irregularly shaped tracts scattered over the huge state. Though known to be rich in minerals and commercially profitable forests, the area had no detailed maps and few settlements, roads, or airfields. Scientists at the university's Geophysical Inst. used Landsat images, combined with limited ground and aerial data, to map 7 million acres of which Doyon chose 2 million. The Doyon selections were based heavily on the Landsat maps; the university reported that application of Landsat data "at least doubled the value of the land selected . . ." (NASA Release 76-63)

- Egypt and the Federal Republic of Germany had agreed that Germany would supply Egypt with a ground station for receiving transmissions from the *Symphonie 1* and *Symphonie 2* comsats jointly operated by FRG and France, the MENA news service from Cairo announced. The stations would use 200 lines initially for telephone contacts between Egypt and "each of Germany and France," as well as between Egypt and any other Arab state in which a similar station would be established. The stations would also be used for television transmissions between France or Germany and Egypt. Egyptians would be sent to FRG for training in operating the proposed station, and steps toward establishing the stations would begin in 1976. (FBIS, MENA in English, 1 Apr 76)

2 April: Gerald D. Griffin, Deputy Associate Administrator for Operations in NASA Hq's Office of Space Flight, had been named Deputy Director of the Dryden Flight Research Center effective 1 May, NASA announced. Griffin joined NASA in 1964 and served as a flight controller at Johnson Space Center during the Gemini program before being named an Apollo flight director in 1968; he served in that position on all 11 Apollo missions, and was lead flight director on Apollo missions 12, 15, and 17. He went to NASA Hq in 1973 and had previously been Assistant Administrator for Legislative Affairs. He had served in the Air Force and was an aerospace engineer with Lockheed and with General Dynamics before coming to NASA. (NASA Release 76-66; DFRC Release 5-76)

• Two engineers at Marshall Space Flight Center had developed a solar-energy collector using air rather than water for heat transfer; this might be the key to low-cost solar-energy systems. The collector consisted of 3 parts: a rigid foam structure, a metal collector plate, and a transparent cover. The device needed no tools or fasteners, but could be fitted together, and required no insulation; it measured 1 by 2 m and weighed about 17 kg. The inventors said their collector had shown itself in tests to be as efficient as other more costly and complicated collectors, and the chief patent counsel at MSFC was considering it for patenting. An average house of about 140 m^2 would require about 67 m^2 of collector area, or a 40-collector system, to provide sufficient heating. A solar-energy system could save a household about 75% of its normal winter heating expense, and could be modified to heat water year round. (NASA Release 76-56)

• The largest American flag ever painted would be exhibited on an outside wall of the Vehicle Assembly Building at Kennedy Space Center as part of KSC's preparation for the Bicentennial exposition Third Century America, scheduled from 20 May through Labor Day 1976. The flag, measuring 64 m by 33.5 m, would be painted on the VAB—world's second largest building—along with a huge Bicentennial symbol 33.5 m in diameter. More than 1900 liters of specially formulated paint had been donated by the Montgomery Ward paint laboratories, and the Bicentennial administration would pay for the painting. The 160-m-high south wall of the VAB would be washed down, rinsed, and primed with white before the finishing coat of white could be added and the design overlaid in red and blue. KSC had been chosen by President Ford to house the Third Century America exposition, featuring exhibits by 16 federal agencies and many U.S. industries, the only such exposition to be sponsored by the government during the Bicentennial year. (KSC Release 112-76)

• Pluto—outermost of the 9 planets circling the sun—might have a surface of frozen methane, putting its temperature in its near-vacuum environment at close to absolute zero (total absence of heat) and indicating that never since the ice formed had the sun's heat been able to boil off the covering, according to a team at the Kitt Peak National Observatory in Ariz. Dr. David Morrison of the University of Hawaii said that Pluto might look and behave just the way it did when it was formed during the creation of the solar system; nowhere but on Pluto had methane been found as a solid, and nowhere else was it present in such abundance. The frozen surface also meant that astronomers might have been fooled since Pluto's discovery in 1930 into thinking it larger than it really was; instead of being the size of the earth, Dr. Morrison said, "it might be only half our size or even smaller." Pluto, 4.8 billion km distant, was the last planet to have scientists define its surface composition. (*NYT*, 2 Apr 76, 37; *W Post*, 2 Apr 76, A-1; *Science*, 23 Apr 76, 362)

4 April: Engineers at the Jet Propulsion Lab. at Pasadena had used equipment identical to that used by NASA's Deep Space Network to maintain comfort conditions in a 6-story building at JPL with almost no operation of the building's conventional heating and cooling equipment, the *L.A. Times* reported. Adaptation of a simple 8-station remote-control panel allowed remote control of building ventilation, heating, and cooling; adaptation of a highly sensitive resistance device to measure return air on one of 4 air handlers permitted constant monitoring of the building's average comfort level. Operators demonstrated maintenance of the comfort level by using the thermal storage of the building structure, outside weather, heat from lights, and heat generated by occupants. Over a month, the building had used less than half the energy required in pre-energy crisis days. The program was expected to save more than 10 percent of JPL's annual energy costs. (*L.A. Times*, 4 Apr 76, 1)

5 April: Harris M. (Bud) Schurmeier, manager since 1972 of the Mariner Jupiter/Saturn project at Jet Propulsion Laboratory, was named Assistant Laboratory Director for Civil Systems, JPL Director Bruce C. Murray announced. The new post would direct Civil Systems activities at JPL, begun in the 1960s to apply specific JPL capabilities in technical and management areas to critical problems of society, including technological transfer to energy, environment, biomedical, and transportation systems. Replacing Schurmeier as manager of the Jupiter/Saturn project that would send 2 instrumented Mariner spacecraft to the giant outer planets next year would be John R. Casani, former manager of JPL's Guidance and Control Division. While serving as Mariner Jupiter/Saturn project manager, Schurmeier had also acted as Deputy Assistant Laboratory Director for Flight Projects for the past 6 yr; the latter position would not be filled immediately. (JPL release 5 Apr 76)

• Howard R. Hughes, 70, aviation pioneer and founder of a financial empire that included aerospace and aviation holdings, died en route to Houston. He was president of Hughes Aircraft and sole trustee of its owner, the Howard Hughes Medical Institute. His estate was estimated to be worth between $1.5 and 2 billion. Newspaper accounts of his wealth and years of withdrawal from public contact told of his early achievements in flying, sparked by his learning to fly during his production in 1930 of "Hell's Angels," then the most expensive motion picture ever made. On 13 Sept. 1935 he set a world land-plane speed record, flying the H-1 (his own design) at 566 kph; in the same plane, he established a transcontinental speed record 19 Jan. 1937 of 7 hr 28 min. On 10 July 1938, he took off in another Hughes plane from N.Y. and flew around the world with 4 associates in 91 hr. Although he had no degree, he studied at Caltech and Rice Institute and had helped to design several aircraft, including the Lockheed P-38 fighter and the same company's triple-tailfin Constellation transport. (*NYT*, 6 Apr 76, 1; *W Post*, 6 Apr 76, A-1; *Av Wk*, 12 Apr 76, 23)

6 April: Marshall Space Flight Center announced award of a contract to Ivey's Plumbing and Electrical Co. of Kosciusko, Miss., for construction of a test facility at MSFC for the solar heating and cooling development program MSFC is directing for the Energy Research and Development Administration. The $647,243 contract was awarded by the U.S. Army Corps of Engineers, which would monitor construction activities; completion date set by the contract was 1 Nov. 1976. The facility would be located in an area formerly used as a swing-arm test facility for Saturn; data-acquisition equipment—instrument terminals and cabling from pad to blockhouse—used for Saturn would be available for the heating and cooling tests. The facility would consist of 4 test areas, 2 designed for complete system or subsystem tests; a third would be used to test liquid energy storage subsystems, and the fourth for passively testing solar collectors. A solar simulator constructed in a nearby building would be linked to the test facility for data collection.

Subsystems and components to be tested would include solar collectors, thermal-energy storage equipment, and solar heating and cooling devices. Testing the hardware against predetermined performance standards would produce data for input to analytical programs; the analyses in turn would be used to evaluate total systems performance under environmental conditions other than those simulated during tests. Components of the facility—heat exchangers, cooling towers, chillers, fans, pumps, and control valves—would be arranged to permit maximum flexibility for future modifications and use of other test positions or procedures. (MSFC Release 76-59)

- New Mexico State University and NASA would sponsor a 3-day symposium aimed at motivating Hispanic and Native American college and high-school students to follow science and engineering as career fields, the Johnson Space Center announced. The symposium sent invitations to faculties of 14 colleges and universities with high numbers of Hispanic and Native American students, as well as local area high schools, to hear speakers and view exhibits from NASA field centers and from the aerospace industry. The session from 21 to 23 April would highlight career opportunities in the aerospace field, including a workshop session on the need for additional aerospace courses in the school curriculum and a job fair providing information on placement. (JSC Release 76-23)
- The Michoud Assembly Facility operated for NASA by MSFC at New Orleans was nearing production capability for the Space Shuttle external tank, MSFC announced. Delivery of all required items to Michoud would require an estimated 225 trucks, about three fourths of them carrying loads classified as oversize. By the time all 225 trucks arrived, total "train" length would have reached more than 3 km. Getting the equipment from

the suppliers located at faraway points such as Dallas, San Diego, Baltimore, and Nashville had presented problems because of the size of the loads; special escorts were required for oversized loads, and travel time at New Orleans for oversize loads was restricted to the period between 10 pm and 4 am. Tooling and fixtures being installed at Michoud would give Martin Marietta, prime external tank contractor, the capability needed to meet flight schedules; each Shuttle flight would require a new external tank, and NASA had planned more than 400 flights between 1979 and 1989. Launches should number 60 per year by 1984. (MSFC Release 76-60)

- Every ruble spent on space research had been returned many times to the Soviet national economy, said Cosmonaut Aleksey Leonov in a Tass interview during observations of the 15th anniversary of Yuri Gagarin's space flight. Leonov was one of the two USSR participants in the Apollo-Soyuz Test Project, and had been the first man to walk in space during a *Voskhod 2* flight 18 Mar. 1965. He stated that 496 spacecraft had been launched from Soviet cosmodromes in the last 5 yrs, 35 of them communications satellites, 13 Meteor weather satellites, and several Prognoz science satellites, as well as about 400 in the Cosmos series. The long-functioning Salyut stations were the main line of Soviet space activity, Leonov said, carrying out research in cartography, geology, agriculture, forestry, and hydrology as only a few of the many fields of knowledge expanded by work on orbital stations. Leonov also stressed the importance of the Soviet-American Apollo-Soyuz test project. (FBIS, Tass in English, 6 Apr 76)

6-9 April: Signatories of the International Telecommunications Satellite Organization (INTELSAT) held their 4th annual meeting in Singapore, with 68 of the 93 signatories attending. The meeting adopted the board of governors' recommendation to increase the capital ceiling for the global satellite system from $500 million U.S. to $900 million U.S., to allow financial flexibility for the Intelsat V spacecraft program. (The capital ceiling was defined as the net capital contributions of signatories and all outstanding contractual capital commitments.) The meeting also approved requests from Nigeria and Zaire for use of the Intelsat space segment for domestic public telecommunications on the same basis as for international services. Intelsat satellites were providing fulltime services through 137 antennas at 109 earth stations in 73 countries. (INTELSAT Release 76-12-I)

7 April: The Rocketdyne Division of Rockwell International Corporation was awarded a $6.9 million contract by Goddard Space Flight Center to produce RS-27 engine systems for the Delta launch vehicle. The contract called for delivery of 10 engines between March and Sept. 1977. NASA had returned 12 Rocketdyne-built H-1 engines for use in producing the new RS-27 systems. (Rockwell Release RD-6)

- Tass announced that the USSR expedition Polar Experiment North-76 had begun work in the high latitudes of the Arctic. Using 10 scientific ships, as well as planes, radar, and installations for rocket probing of upper atmosphere layers, a team of scientists would observe ice floes in the polar basin, including the geographic North Pole. The expedition, said to be the largest in the history of Arctic exploration, would simultaneously determine the heat balance deep in the ocean, in the atmosphere, and in floating ice in the North Pole area. Ice floes in the polar basin had been found to forecast climate changes in the densely populated and economically developed areas of the northern hemisphere. From the beginning of the century to the 1940s, the floes had receded from continental coastlines, indicating a relative rise in global temperature; they had begun to appear further south again, indicating a cold phase, and the appearance had been correlated with difficulties in northern navigation and fisheries and unusually snowy winters in western Europe, droughts in Asia, and unexpected cyclones and hurricanes. (FBIS, Tass in English, 7 Apr 76)

9 April: The Johnson Space Center announced award of a 1-yr extension of the Pan American World Airways contract for plant maintenance and operations services at the Center. Pan American had been selected 13 Feb. 1974 to receive a cost-plus-award-fee contract for operation of all utility systems and maintenance of utilities, buildings, roads, ditches, and special equipment at JSC; the current award represented the second of 2 planned extensions and would extend the effective period through 12 Feb. 1977. Annual estimated amount of the contract was $7.4 million. (JSC Release 76-25)

- A "mysterious force" on the moon had made one of the remote-controlled stations set up by Apollo astronauts behave peculiarly, Boyce Rensberger reported in the *N.Y. Times*. After operating without interruption since its placement on the lunar surface in Feb. 1971, the station went dead, returned to life spontaneously a few weeks later, then went dead again. Engineers at JSC were waiting to see if the unknown influence would switch the station on again. Called ALSEPs, for Apollo lunar scientific experiment packages, the five stations deployed on the moon at intervals since Nov. 1969 had been sending a steady stream of scientific data on moonquakes, heat flowing from the moon's interior, and the nature of particles in the solar wind; the atomic power supplies, designed for a 1-yr lifetime, had proved more durable. The faulty station, left by the *Apollo 14* mission, stopped receiving ground commands in March 1975 when its receiver failed; on 18 Jan. 1976, its transmitter stopped; on Feb. 19, the entire station came back to life—both receiver and transmitter working—and one experiment that had never operated during lunar daytime began working flawlessly night and day. Then, exactly a month later, the entire station shut down again. One theory was that the starts and stops might relate to extreme temperature changes on the moon, varying from 121°C at lunar noonday to −184°C at lunar midnight. However, as day and night had not originally affected the station's operation, scientists could not explain why temperature change did not produce failure from the start. (*NYT*, 9 Apr 76, 4)
- Astronomers at the University of Massachusetts had clocked pulsars— superdense remnants of collapsed stars—at speeds over 2 million kmh, making them the fastest movers in earth's galaxy, the National Science Foundation reported. Some of the objects exceeded galactic escape

velocity, meaning they could pull free from the Milky Way and spin off into intergalactic space. First discovered in 1968, pulsars were defined as dense bodies in which a sun-size mass had been compressed into a sphere 16 km in diameter. The material was said to be so dense that a teaspoonful would weigh a billion tons. With its collapse, the body would spin increasingly faster and its magnetic field would increase in intensity; the newly formed pulsar would send out electromagnetic signals that pulsed like a radio beacon with each turn (hence the name pulsar). The high velocities were thought to be caused by the intense magnetic fields, up to a trillion times that of earth's magnetic field: a slight displacement of the pulsar's magnetic field could release large amounts of electromagnetic energy in one direction, upon which the pulsar would move in the opposite direction as would a rocket. (*Pasadena Star-News*, 9 Apr 76, 5)

12 April: Louis Mogavero was appointed Director of NASA's Technology Utilization Office in the Hq Office of Industry Affairs and Technology Utilization, the agency announced. Mogavero, who joined NASA in 1966, had held management and planning positions in the former Office of Manned Space Flight, and in 1972 had worked with a White House study group evaluating new opportunities in technology. Before joining NASA he was operations manager for new product development with the Boeing Co., and received a master's degree in engineering administration in 1970 from George Washington University. (NASA Hq announcement, 12 Apr 76)

- Electricity from outer space to power homes and industry would be not only possible, but entirely probable, soon after the turn of the century, said Ralph I. LaRock of NASA Hq at a recent meeting held at Marshall Space Flight Center. LaRock, director of the Energy Technology Applications Division in NASA's Office of Energy Programs, described a proposed satellite power system under study by NASA, industry, the Energy Research and Development Administration, and other groups. MSFC's share of the study was aimed at defining overall systems requirements (including satellite structure, economical size and materials for the solar array, new technology needed in satellite and support systems, and transporting, erecting, and maintaining the system in space). LaRock said predictions of technology advancement during the 1980–1990 time period indicated a good chance of "our having a few solar-power satellites in operation by the turn of the century." (MSFC Release 76-75)

13 April: The House of Representatives passed H.R. 13172, authorizing a supplemental $16 800 000 for NASA for the period 1 July 1976 through 30 Sept. 1976 (the so-called "transition period" after which the new government fiscal year would begin on 1 Oct.). The authorization, supplementing NASA's research and program management request, was $3 186 000 less than the original request of $19 986 000. (NASA Ofc. of Budget Operations, Chron. Hist. FY 76, 16 June 76)

- Calling the Space Shuttle "clearly . . . the next logical step" in tapping the scientific and commercial possibilities of space, the *Baltimore Sun* said in an editorial that criticisms of the Shuttle program from an economic point of view should be balanced with the rewards that would come only from experience. The first Shuttle was never intended to be self-supporting, the editorial said, and the rewards would include increased knowledge to serve as the basis for more sophisticated programs and the increased likelihood of constructing a permanently staffed permanently orbiting space station. Complaints from prospective users—ComSat and DOD—that Shuttle fees would be too high "before NASA even has announced these fees" were an obvious ploy to keep the fees as low as possible, the editorial said. (*B Sun*, 13 April 76)

- A 2-yr comparative study of data on dust storms on Mars and earth revealed that Martian dust storms were much like severe ones on earth, "only more so," the Jet Propulsion Lab. announced. JPL investigator Peter M. Woiceshyn used Lowell Observatory data to conclude that a wall of dust more than 50 km high swept down the slopes of the Hellas area on Mars in July 1971 at speeds greater than 480 km per hr. *Mariner 9* occultation data verified that winds of extreme velocity would be required to raise surface dust in the low atmospheric pressure of Mars, where the air density was only 1/100th that on earth. When *Mariner 9* arrived at Mars in Nov. 1971, another dust storm had been raging for several weeks, and dust-cloud tops were estimated to be 50 to 70 km above the surface. The JPL report said the two 1971 storms and another detected by astronomers in 1956 began in the same location on the slopes of Hellas, triggered by a cold jet stream from the Martian north pole funneling down a long valley across the planet's equator. Similar storms on earth in Russia, in Iran, and on the U.S. plains east of the Rockies caused great damage from soil erosion, similar to that on Mars revealed by *Mariner 9*. More storm data from Mars would be sought by the Viking spacecraft scheduled to land there in July and Sept. 76. (JPL release 13 Apr 76)

- A celebration in Moscow of the 15th anniversary of Yuri Gagarin's flight into space heard a speech by A.P. Aleksandrov, president of the Academy of Sciences, reviewing Soviet firsts in space: launch of the world's first artificial satellite, first manned flights and walk in open space, first flight to the moon and delivery to earth of lunar soil samples by automatic devices, creation of orbital stations, and unique experiments in the study of solar-system planets. Aleksandrov recalled that the first manned space flight occurred only 3.5 yr after launch of the first artificial satellite, which he said made the Russian word "sputnik" common to all peoples in all languages of the world. He reviewed Soviet progress in space research and applications, especially in communications, mentioning that the number of Orbita network ground stations to reach remote parts of the Soviet Union had now climbed to 68. He referred to the success of the Apollo–Soyuz joint experimental flight as "a significant contribution to the progress of world cosmonautics." (FBIS, Tass report 13 Apr 76)

14 April: A NASA earth-survey aircraft—one of two U-2 high-altitude research planes flown by the agency—had been studying the stratosphere over Central America, South America, Canada, the Caribbean, and the Pacific Ocean to measure ozone, nitric oxide, and manmade pollutants as part of a long-term global effort to understand the effects of pollutants on global climate over a long period of time. The data flights used 4 scientific instruments: a stratospheric air sampler, to measure ozone and nitrogen ozide; an aerosol particle sampler, to measure minute aerosols; a stratospheric cryogenic sampler to measure halocarbons (freons) and methane levels; and a foil sampler, to measure aerosols and trace gases for the National Center for Atmospheric Research. The other 3 studies were under the direction of NASA's Ames Research Center. Data from the flights would be available to the other governments involved in the studies. (NASA Release 76-60)

15 April: Alternative "super safe" landing sites for the second Viking spacecraft scheduled to arrive on Mars were being investigated by NASA, Dr. Harold Masursky told the annual meeting of the American Geophysical Union in Washington. Masursky, an astrogeologist with the U.S. Geological Survey, said the prime landing sites had been carefully selected and should present no problem. However, in case of trouble with the first lander, scheduled to touch down early in July, the second lander now scheduled to descend in a less well known area would proceed to a less challenging site to ensure salvaging the mission. The prime site, in the Chryse region at the mouth of the largest channel system identified on Mars, was of special interest because scientists thought areas with highest probabilities of water would be most likely to harbor life as known by man. The scheduled second landing area, called Cydonia, had the highest recorded atmospheric water content during the season when the Viking would land, but was in an area not accessible by radar from earth and was therefore not as well known. Giant radar telescopes in the U.S. and at Arecibo, P.R., were scanning areas near the Mars equator that would be safer than the prime sites, even if not as scientifically productive. (*NYT*, 18 Apr 76, 25)

16 April: Dr. John F. Clark would retire 1 July after 10 yr as director of the Goddard Space Flight Center, NASA announced. His successor would be Dr. Robert S. Cooper, deputy director of GSFC. Dr. Clark had come to NASA in 1958, serving from 1963 to 1965 as Deputy Associate Administrator for Space Sciences and Applications, after which he was appointed to his post at GSFC. An authority in atmospheric and space sciences, Dr. Clark while director at GSFC had been responsible for major advances in communications, weather and climate, earth resources, space physics, and space astronomy. (NASA Release 76-71)

• Marshall Space Flight Center announced award of 2 contracts totaling $486 388 for fabrication and installation of 2 test towers to be used in structural load tests of the Space Shuttle external tank. One contract, awarded to the Lucey Boiler Co. of Chattanooga for $344 660, was for pickup from MSFC of government-furnished construction materials and fabrication and delivery by July 1976 of all sections and components of the towers. The other contract, awarded to Teledyne Brown Engineering of Huntsville for $141 728, was for installation of the towers by 30 Sept. (MSFC Release 76-69)

• The Communications Satellite Corp. reported net income of $11 041 000 for the first quarter of 1976, compared with $12 692 000 for the same period of 1975. Although total income of $37 276 000 for the first quarter was 2% higher than that for the same period in 1975, higher depreciation charges associated with the launch of additional satellites—*Intelsat IV-A F-1* and *F-2*, and the first Marisat—and higher operating costs, together with expenses of operating Satellite Business Systems—a partnership formed by subsidiaries of ComSat, IBM, and Aetna— accounted for the reduction in income. ComSat had appealed the FCC decision of 4 Dec. 75 reducing its rates, and FCC had delayed the effective date of the lower rates until the court had ruled on the appeal. (ComSat Release 76-7)

17 April: The $6.2-billion Space Shuttle, "America's only remaining space spectacular," might come apart at the seams because of unsafe nuts and screws, Jack Anderson and Les Whitten reported in the *Washington Post.* Discussion at NASA as early as Apr. 1973 had centered on the menace of substandard screws; in July 1973, Johnson Gage—a Bloomfield, Conn. firm—had sent NASA a warning based on tests of individual screw and nut threads stating that the standards "provide for a loophole that allows [NASA to] accept outright junk." Computer tests had revealed that millions worth of faulty threads reached Rockwell International Corp., principal Shuttle contractor, because of low standards and it was feared the inferior screws had gotten into Shuttle equipment. Although company engineers were worried about the standards and the resultant products, the company bought fasteners from outside suppliers and could not control quality from 6000 vendors in 47 states, company minutes said. Industry had fought to keep the low standards rather than pay the estimated $120 million cost of retooling; the American National Standards Institute, led by industry, had blocked every move for tighter standards, Anderson and Whitten charged. (*W Post*, 17 Apr 76, B-11)

20 April: India's second artificial earth satellite would be launched in 1978 from a USSR cosmodrome, Tass announced. The agreement had been made at a Moscow meeting of a working group of Soviet and Indian specialists. (FBIS, Tass in English, 20 Apr 76)

21 April: Kennedy Space Center announced award to Pan American Technical Services, Inc., Cocoa Beach, Fla., of the first contract to modify KSC facilities for Spacelab processing as part of the Space Shuttle program. The $129 627 contract included architect and engineer services to design the modifications of the Operations and Checkout Bldg., including changes in utilities (gaseous nitrogen, helium, high-pressure air, water and airconditioning) and adaptation of Apollo equipment acceptance and checkout rooms to make them compatible with ground-support equipment provided by the European Space Agency for automated Spacelab testing. Spacelab—to be carried into space by the Space Shuttle orbiter—was being designed, developed, and built by 9 member nations of ESA at a cost of $300 to $400 million. At the end of each mission, the Spacelab would be removed from the landed orbiter and prepared for the next mission. The engineering model to be used for facility and orbiter checkout was scheduled to arrive at KSC in July 1978; the first flight model was due in June 1979. (KSC Release 128-76)

- William A. Anders, former *Apollo 8* astronaut who became the first chairman of the Nuclear Regulatory Commission in 1974, was sworn in as U.S. ambassador to Norway [see 9 Mar.]. (NRC Release 76-93)
- Joseph A. Zinno made aviation history "of a sort," the National Aeronautic Association reported, when his man-powered 68-kg airplane swooped off the ground and flew more than 24 m for 5 sec at Quonset Point, R.I. In 1959 a British industrialist had offered a prize of $24,000 (now grown to $92,500) for flight in a machine powered solely by the crew that could fly a figure-eight course around 2 pylons set 0.8 km apart and cross the starting and finish lines at least 3 m above the ground. So far 16 Europeans and Zinno had succeeded in flying man-powered planes. However, aeronautical experts believed the altitude needed to make a figure-eight maneuver (more than 12 m) would undo any aviator striving for the prize. (NAA newsletter, June 76, 3)
- Tass announced the return to Riga of the 20th Soviet Antarctic expedition after completing "its successful 15-month program of research." For the first time under Antarctic conditions, scientists had operated automatized processing of data from rocket soundings of the atmosphere; also for the first time, a shelf glacier had been drilled with 3 holes, one of them 358 m deep (entirely through the glacier). The Soviet scientists reported increasing cooperation with colleagues; the 20th expedition included geologists and medical workers from the German Democratic Republic, and a U.S. weather specialist spent the winter at the Soviet station while a Soviet glaciologist spent the winter at the U.S. McMurdo station. (FBIS, Tass in English, 21 Apr 76)

22 April: Nato IIIA, first in a series of NATO–USAF military comsats, was launched from Eastern Test Range at 3:46 pm EST (1046 GMT) by a Delta vehicle into a synchronous transfer orbit at about 35 000 km altitude before being positioned above the equator off West Africa at about 15°W. The comsat would transmit voice, data, facsimile, and telex messages among military ground stations; a NATO IIIB spacecraft was scheduled for launch later in 1976 into geosynchronous orbit above the Indian Ocean. *Nato IIIA*, weighing 310 kg in orbit and designed for a 7-yr lifetime, was drum-shaped, 2.2 m in diameter and 2.2 m long, and extended about 3 m with antennas out. The eventual 3-satellite system including 16 major ground terminals had been estimated to cost about $340 million. (NASA Release 76-46 and 76-46A; MOR M-492-207-76-01 [prelaunch] 19 Apr 76, [postlaunch] 20 May 76; *W Post*, 23 Apr 76, A-21)

- NASA's Wallops Flight Center received a National Safety Council award for its performance in on-the-job safety. The Council selected Wallops for a third-place award in its annual contest honoring the lowest disabling-injury frequency achieved by NSC member organizations. Wallops qualified for its award with a recorded rate of zero lost-time injuries per million man-hours worked, as compared to a rate of 1.4 injuries for all competing units in the Aerospace Section, Research and Development Division. (WFC Release 76-4)

- The European Space Agency's launch vehicle Ariane during its 5-yr development period would use hydrazine fuel supplied by the USSR, reported the British journal *New Scientist*. The U.S. hydrazine plant near Baltimore had been closed down because of contamination by nitrosamines, hydrazine sources suspected of causing cancer. Hydrazine, highly corrosive and difficult to handle, had not been manufactured in Western Europe, and no one wanted to do so, according to the report. Ariane's discontinued predecessor Europa used a less powerful fuel—liquid oxygen and kerosene—and the USSR had used hydrazine from the plant near Moscow to power the large rockets that launched the Salyut space stations and large interplanetary probes. The report said that the hydrazine deal stemmed from France's longstanding cooperation with the USSR in aerospace technology; if the Soviets should restrict hydrazine supplies, France would build a European plant because its commitments would override environmental objections. (*W Post*, 22 Apr 76, A-52)

23 April: Addressing the Utah Air Force Association's Bicentennial program in Salt Lake City, NASA's Deputy Administrator George M. Low gave a status report on ongoing projects such as Landsat, Viking, and the Space Shuttle, then speculated at length on accomplishments in space that might be celebrated on the occasion of the U.S. Tricentennial. Referring to "the beginning of the industrialization of space," using the first-generation reusable space-transportation system called Shuttle to set up the first factories in space, he said "the real breakthrough" would come with energy collected from the sun and beamed down with microwaves to provide all the electrical needs of a major city. None of this technology, he added, was beyond today's knowledge. He then forecast a cleanup of earth's environment, communications applications of all kinds, and exploration of outer space made possible by discoveries using the Space Telescope, ending with descriptions of space colonies that would "become the real frontier." He concluded by stating that "all of my projections are overly conservative" and could all become reality long before the end of America's third century. (Text)

26 April: "No useful data" appeared in photographs taken during the artificial-eclipse experiment on the Apollo-Soyuz Test Project, reported *Aviation Week and Space Technology* magazine, quoting U.S. scientists who examined the photographs. This view contrasted with earlier Soviet statements. The experiment consisted of Apollo's blocking out the solar disk while pictures of the corona were made from the Soyuz. According to the magazine, many NASA officials and U.S. scientists "believed privately before the flight that the Soviet experiment would produce only marginal results." (*Av Wk*, 26 Apr 76, 11)

- A $4.5-million structural test data-acquisition system to be used for tests on Space Shuttle components was nearing completion of installation and checkout at Marshall Space Flight Center, MFSC announced. Developed by Avco Electronics, the system included a central facility with a computer system, 6 line printers, 5 graphics display units, a printer-plotter, 2 card readers, and 4 video hard-copy units; 4 data-selector units; and 24 static-input units. The central facility, to be located in MSFC's Test Laboratory, was designed for a lifespan of 10 yr and would combine extreme accuracy with high reliability and ruggedness. Testing the external tanks, solid-fuel rocket booster, and other parts of the Shuttle for structural integrity would require monitoring up to 6000 channels of data simultaneously, fed into the computers from test hardware elsewhere at the Center. The Avco contract also provided for training MSFC personnel to operate the system when completed late in 1976. (MSFC Release 76-73)

- Rockwell International awarded subcontracts totaling nearly $2 million for equipment and material for the U.S. Air Force's new B-1 strategic bomber development effort. Rockwell was awarded the B-1 system contract in 1970, and the first prototype flew for the first time 23 Dec. 1974. A second B-1 equipped with full offensive avionics made its first flight 1 April 1976; a third prototype nearing completion was scheduled to begin flight tests this year. Contracts just awarded to 5 Ohio companies were for manufacture of the fourth prototype. The companies were: Cleveland Pneumatic Co., main landing gear, $1,270,000; TRW Inc., fuel pumps, $195,800; Westinghouse Aerospace Electric Div., generator and controls system, $73,000; RMI Co., titanium, $165,000; and Titanium Metals Corp., titanium, $106,000. The titanium would form major B-1 components such as the wing carry-through structure. (Rockwell Intl. Release LA-2)

- NASA was beginning to make headway in efforts to solve basic problems of U.S. aircraft manufacturers caused by a long-term dearth of low-speed aeronautical research, reported *Aviation Week & Space Technology* magazine. Industry engineers reporting on discussions at a recent Wichita meeting of the Society of Automotive Engineers said use of the workshop technique had been effective in uncovering primary problem areas by crosstalk with such specialized segments as aerodynamicists and propulsion engineers. NASA research being further refined in industry product development included general-aviation aircraft airfoils, aircraft drag reduction, propeller efficiency, and reduction of airflow separation on wing trailing edges. Lewis Research Center had begun exploration of alternatives to conventional reciprocating engines, such as diesel and rotary concepts. NASA had also begun studying the application of advanced technology to agricultural aircraft, said to be responsible for increasing farm production by about $10 billion annually, to solve problems such as corrosion of aircraft structures by agricultural chemicals and control of chemical drift. (*Av Wk*, 26 Apr 76, 56)

- Employment in the U.S. aerospace industry continued downward, the annual employment survey by the Aerospace Industries Association said. Employment figures would fall below 900 000 by the end of 1976, compared with 925 000 at the start of the year. Peak employment in the industry was 1 500 000 in 1968. The predicted decline, amounting to 3.5%, would be spread among all sectors of the industry; greatest decline—about 6000—would be in employment on aircraft programs, much of it in plants producing transport aircraft. The continuing erosion of the high-technology manpower base was principally due to slackening demand for commercial jetliners, uncertainties in export markets, the level of government commitment to new or replacement military aircraft, and reduced investment in the space program, the report said. The forecast was based on data provided by 48 major firms representing more than 65% of total aerospace employment. (*Aerospace News*, 26 Apr 76)

- Comsat General Corp. signed an agreement with the Water Resources Division, U.S. Department of Interior's Geological Survey, to conduct a 6-mo joint evaluation of the use of satellites to transmit data from remote hydrological sensors to a central facility. Comsat General would provide 11 small data-collection platforms to receive and transmit the data via satellite to a central receive facility. The DCPs would be located near Survey sensors in the continental U.S.: 5 in the Pacific Northwest, 5 in eastern Pa., and one near Survey headquarters at Reston, Va. Central receiving station would be the ComSat station at Southbury, Conn., where the data would be stored on magnetic tape and accessed by the Survey on interconnecting terrestrial lines. The program would provide a chance to evaluate collection of environmental data by satellite under operational conditions; it would begin later in 1976 and conclude within 6 mo, subject to FCC approval. (ComSat Release CG 76-112)

- The vast spaces between galaxies previously thought to be empty contained large clouds of gas, 2 scientists from the University of Arizona said. The National Science Foundation, which funded the research using the 2.3m telescope at Kitt Peak, announced the discovery as the most solid evidence to date for the presence of substantial amounts of matter between the galaxies. The discovery, expected to affect current theories on the evolution of galaxies, might contribute to learning whether the universe would continue to expand indefinitely; more mass in the universe, with resulting greater gravitational forces, would eventually halt and reverse the expansion. Dr. Robert E. Williams and Dr. Ray J. Weymann studied clouds of gas in the vicinity of certain quasars and found that the clouds were associated not directly with the quasars but with the cluster of galaxies that contained the quasar. (NSF Release PR76-35)

27 April: Marshall Space Flight Center announced shipment of the first production segment of a motor case for the Space Shuttle solid-fuel rocket motor from the manufacturer (Ladish Co. of Cudahy, Wis.) to the Cal-Doran Metallurgical Services plant at Los Angeles, where it would undergo heat treatment and cleaning before being forwarded to Rohr Industries at Chula Vista, Calif. Rohr—subcontractor to prime contractor Thiokol Corp.'s Wasatch Div. in Utah—would process the cylindrical segment, with a finished weight of about 5060 kg, for delivery to Thiokol in Sept. 1976. Thiokol would load the solid propellants into 4 motor-case segments over 3 m in dia. and 45.5 m long, each holding about 500 000 kg of propellant. The 4 segments joined would stretch to more than 35 m and constitute about three fourths of the solid-fuel rocket booster. Each launch of the Shuttle would use 2 boosters, burning from ignition on the pad to burnout at an altitude of about 42 km; at that point, the SRBs would be separated and descend by parachute into the ocean for recovery, refueling, and reuse. (MSFC Release 76-76)

- Observations from NASA's airborne Kuiper observatory by a group of Cornell University scientists reported the first occultation, or eclipse, of a star by a planet observed from above the lower layers of earth's atmosphere. The occultation was observed from an altitude of 12.4 km over the Atlantic near Bermuda, using the 1-m telescope to view the passage of Mars between the earth and the star Epsilon in the constellation Gemini. Light from the star passed through the Martian atmosphere before reaching earth; changes in the light after it passed through the planet's atmosphere would give astronomers new information about the density and composition of that atmosphere, thought to consist of argon rather than water and carbon dioxide which had been frozen into the Martian polar caps. The two unmanned Viking spacecraft scheduled for Mars landing later this year would also measure the argon present in the Mars atmosphere. (NSF Release PR76-36)

28 April: A model of a nuclear reactor developed by NASA for space applications, using a gaseous rather than a solid nuclear fuel, had begun tests at the Los Alamos Scientific Laboratory, NASA announced. A reactor using gaseous fuel could operate at temperatures that would melt solid-fuel rods; the higher operating temperature would make the gaseous reactor potentially more efficient than conventional solid-core reactors. Also, the gaseous fuel would permit continuous reprocessing of the fuel, eliminating the need for a separate reprocessing plant required for solid fuel; the predicted efficiency would translate into reduced mass and weight in space applications. Nuclear-energized laser research by NASA indicated that power from a gaseous reactor could be generated as laser beams, offering the prospect of a new space technology by which energy from a nuclear-power station in space might be transmitted over large distances, using laser beams, to users on space platforms, lunar bases, or space ships for propulsion. The tests were designed to use hardware salvaged from an earlier NASA-Atomic Energy Commission nuclear-rocket program. (NASA Release 76-76)

29 April: Rear Admiral Stuart J. Evans had been named NASA's Assistant Administrator for Procurement, effective 1 June, NASA announced. Adm. Evans, the Deputy Chief of Naval Procurement, would assume his new post after retiring from the Navy 31 May 1976. He would succeed Rear Admiral Kenneth L. Woodfin (Ret.), present Assistant Administrator for Procurement, who had announced his plans to leave NASA after a year's service to join a private firm. Adm. Woodfin had been Deputy Chief of Naval Material (Procurement and Production) before coming to NASA. (NASA Releases 76-68, 76-78)

• Tass reported a new development in U.S.-Soviet scientific and technical cooperation in a joint oceanographic experiment called Polymode to study vortex formations in the ocean. Professor Alan Robinson of Harvard University, heading a delegation of U.S. oceanographers to a plenary meeting of the Polymode organizing committee, said the mission would help crack the secrets of the ocean and place its resources at the service of mankind. Scientists had theorized that ocean vortexes, like cyclones in the atmosphere, had substantial influence on weather all over the globe. Work on Polymode would begin in mid-1977, with the northern area of research in the North Atlantic, 1000 km west of the Azores, and the southern area 200 to 300 km north of the Antilles. (FBIS, Tass in English, 29 Apr 76)

30 April: Venus—whose dense atmosphere had prevented observation of the surface, except by radar—might be as tectonically active as Mars, and possibly as active as the earth, researchers at JPL's Goldstone tracking station announced. As reported in *Science* magazine, new high-resolution maps from 2 different radiotelescopes disclosed the planet's surface with enough detail to permit a study of its geology. A series of 8 maps taken by the 64-m steerable dish at Goldstone revealed a 1500-km-long trough near Venus's equator comparable to large earth rift systems such as the East African, giving "strong evidence of extensional tectonic activity" on Venus. Evidence of a fault movement seemed to appear in another Goldstone map showing a long arc-shaped mountain range crossed and bowed by another linear feature. A high-resolution image obtained at the Arecibo observatory in P.R. showed a large part of the Venus northern hemisphere including a basin-like feature about 1000 km across with a bright sharp rim; "the shape is wrong" for an impact basin, according to the Arecibo report. The new maps raised as many questions as they answered, *Science* commented. (*Science*, 30 Apr 76, 454)

- Long-term directions and opportunities in civil aviation—including passenger air travel by rotorcraft from small urban-center airports, or intercontinental air transportation on hypersonic craft rated environmentally acceptable—were projected in a NASA study, "Outlook for Aeronautics," announced by the agency as the result of its study on the role it should play in research and development of aviation and the technical advances that might be needed. Relatively few major new developments could be expected through the early 1980s because of economic setbacks and environmental pressures, the agency said, but, with adequate research and technology investments, new opportunities should arise in the period 1985–2000. Demand for passenger transportation should grow from 250 million annually to about 1 billion annually by the year 2000; by the late 1980s, air traffic should exceed the capacity of the present airport system. Congestion should lead to development of short-haul aircraft using smaller airports; increasing costs would result in greater efficiency and economy with improved safety in subsonic aircraft, the report said. (NASA Release 76-78)

During April: Air Force cold-weather tests of an air-cushion landing system for use on large transport aircraft were completed after 4 wk at a site in Canada. The device, which resembled an upside-down life raft installed under the fuselage, was made of rubber and nylon; it measured about 10 m long and about 4 m wide. The elastic container was filled with air from two engines mounted under the wings; air forced out through more than 6800 holes on the bottom surface created an air bearing between the landing surface and the trunk. The cushion system, successfully demonstrated in the late 1960s on a smaller single-engine aircraft, had the advantage of exerting a very small amount of pressure over the entire landing surface as contrasted to conventional landing systems. Further tests of the system would be conducted on the twin-engine short-takeoff-and-landing (STOL) aircraft. (*AFSC Newsreview*, Apr 76, 15)

- Bicentennial year 1976 also marked the 50th birthday of scheduled-airline service in the U.S., the National Aeronautic Association reported. On 13 April 1925, Henry Ford had started an air-freight service between Detroit and Chicago, first such commercial flights on a regular schedule. Upon passage of the Air Commerce Act of 1926, one of NAA's founding members, William P. MacCracken, Jr., took office as the first U.S. Assistant Secretary of Commerce for Aeronautics. By 1931, airlines were spanning the country; by 1950, U.S. airlines carried 19 million passengers 16 billion km, and in 1975 they carried more than 200 million passengers 263 billion km—amounting to almost 80% of intercity public passenger travel in the U.S. Airlines also accounted for 93% of travel to overseas destinations. In 1976, the U.S. scheduled-airline fleet included 2200 jet aircraft serving communities nationwide with 13 000 daily flights. (NAA *News*, Apr 76, 3)

May 1976

1 May: NASA announced issuance of a patent to Richard T. Whitcomb, an aerospace technologist at Langley Research Center, for an aircraft wing to be used at subsonic speeds. The new wing, called a supercritical or "upside down" wing, was described as increasing speed without need for additional power and as suitable for any aircraft flying near the speed of sound. Whitcomb last year received $25 000 for the invention from NASA's Inventions and Contributions Board. (*NYT*, 1 May 76, 31)

3 May: The European aerospace industry was frustrated in attempting to increase its share of commercial and military aircraft sales, said Robert B. Hotz, editor-in-chief, in an editorial in *Aviation Week* magazine. Although maintenance of a viable European aerospace industry would be in the best interest of the Western world, Hotz said, attempts to organize it on a truly international competitive scale had emphasized inherent weaknesses and failed to achieve fiscal success. Recent developments—nationalization of the British aircraft industry, and French moves toward a transatlantic rather than a cross-channel cooperation—had intensified the problems, characterized by higher wage scales and lower productivity. Suggestions for improvement included organizations based on commercial and technical considerations, rather than on political boundaries and restrictions; marketing goals based on the broadest markets possible, rather than on narrow domestic requirements; and controlling labor costs that had priced European products out of the export market and had held production to a rate below that needed to fill international market requirements. (*Av Wk*, 3 May 76, 5)

4 May: NASA launched *Lageos*—a laser geodynamic satellite—from Western Test Range on a Delta rocket at 3 am CDT (0800 GMT) into a near-circular orbit with 5940-km apogee, 5845-km perigee, inclination of 109.8°, and period of 225.5 min. *Lageos*, resembling a "cosmic golfball" 60 cm in diameter and weighing 411 kg, would act in space as a sophisticated mirror reflecting laser beams directed at it by stations on the ground; scientists timing the round trip of the laser beams would be able to detect movement of the earth's surface as small as 2 cm. The 3-yr phase 1 of the Lageos mission would validate laser-ranging techniques as compared with long-baseline interferometry and lunar laser ranging; phase 2 would consist of application of phase-1 techniques continuously for the usable life of the passive satellite. Although *Lageos* was not expected to fall back to earth for at least 8 million yr, and contained no moving parts or electronics to wear out, the 426 laser reflectors on its surface would be eroded in the space environment probably within 50 yr. (NASA Release 76-67; MSFC Release 76-82; MOR

E-639-76-01 [prelaunch] 28 Apr 76, [postlaunch] 27 May 76; *NYT*, 5 May 76, 30; *W Post*, 5 May 76, A-8)

5 May: Oso 8, an orbiting solar observatory launched 21 June 1975, had performed without failure of any subsystem except a redundant solar sensor, and was judged successful, according to a postlaunch mission operations report. Primary mission objectives were high-resolution spatial and spectral observations of the solar chromosphere and transition regions, obtained by the pointed experiments aboard the observatory; secondary mission objectives were met with "substantial numbers" of observations of solar x-rays, earth airglow, and cosmic x-ray background radiation. At launch, *Oso 8* carried the most comprehensive package of cosmic x-ray experiments ever included in a single payload, including the first satellite instrument containing large-area thin plastic window detectors to measure ultrasoft x-rays; the first high-sensitivity crystal spectrometer flown to measure cosmic x-ray sources with extremely high spectral resolution; and the first satellite experiment flown to measure x-ray polarization of cosmic sources with high sensitivity. (MOR S-821-75-09 [postlaunch], 10 May 76)

- Agreements between the French government and the European Space Agency for use of launch facilities at Kourou, French Guiana, were signed in Paris by Michel d'Ornano, minister for industry and research, and Roy Gibson, director general of ESA. The Kourou facilities include the Guiana Space Centre belonging to France's Centre National d'Études Spatiales (CNES) and the Ariane launching site belonging to ESA that replaced the former equatorial base belonging to the defunct ELDO organization [see 11 Jan.]. The meeting of the European Space Conference in April 1975 had decided that all ESA member states would contribute to the costs of the Guiana space center until the end of 1980. Ariane (the ESA heavy launcher now under development) would provide Europe with a satellite launch capability of its own after 1980. (ESA newsletter, Aug 76, 2)

- The Energy Research and Development Administration announced plans to launch a giant manned and instrumented balloon, *Da Vinci II*, between 24 May and mid-June to determine physical and chemical changes in air pollutants over distances of several hundred km from their source. The helium-filled balloon more than 70 m high—equaling, with its gondola, the height of a 15-story building, and carrying the same crew of 4 that flew on *Da Vinci I* in November 1974—would fly for up to 36 hr at altitudes from 3 to 9 km from St. Louis, Mo., to a point in Ill., Ky., or O. *Da Vinci I* flew from Las Cruces to Wagon Mound, N.M., in 12 hr to prove the feasibility of using a manned instrumented balloon for lower atmosphere research. Crew members were Dr. Rudolf J. Engelmann of NOAA; Otis Imboden, photographer from the National Geographic Society; Jimmie Craig, pilot from the U.S. Naval Weapons Center; and Mrs. Vera Simons, project consultant and experienced balloonist, who originated the idea for the project. The crew would use a variety of instruments to conduct more than 20 experiments to show what happens to a plume of polluted air as it moves across several states. (ERDA Release 76-128)

6 May: NASA announced award of a grant to Rensselaer Polytechnic Institute, Troy, N.Y., to establish a center for the study of new composite materials and their application to the development of more efficient commercial transport aircraft. The grant would continue a relationship that began with establishment of a NASA program in materials engineering at RPI in 1961. The new grant for research into composites—very strong lightweight materials combining various substances, such as fiberglass bonded with epoxy—would emphasize more exotic composites using carbon fibers that exhibited high tensile properties. The potential structure-weight reduction of up to 30% in future aircraft using composites could translate into a 10 to 15% reduction in aircraft fuel consumption. First-year funding for the RPI center would be $300 000. (NASA Release 76-82)

- The FAA announced plans to test a computer-based system for predicting potential hazards from wake vortices along aircraft approach and departure paths. The tests would take place at Chicago's O'Hare International Airport and would run for several mo in an attempt to validate results of thousands of measurements previously taken at Kennedy Airport in New York, London's Heathrow Airport, and Denver's Stapleton Airport, to establish how long vortices linger on runways in relation to wind speed and direction. Additional tests at O'Hare would check out a system for detecting and tracking wind shear, another hazard to landing and departing aircraft. (FAA Release 76-43)

7 May: NASA announced plans to launch a second maritime satellite (Marisat) for the Comsat General Corp. later this month, as part of a system to provide communications to the U.S. Navy, commercial shipping, and offshore industries. The first satellite of the system, *Marisat 1*, was successfully launched 19 Feb. into an orbit over the Atlantic at 15°W longitude, where it had provided UHF service to the Navy since 25 Mar. NASA hoped to inaugurate fulltime commercial voice and data communications, using both Marisats, by 1 July. A third satellite had been constructed as a spare. (NASA Release 76-83)

- A launch-abort system for the Space Shuttle, for use in case of malfunction during the first 2.5 min of flight, was "quietly" removed by NASA 3 yr ago although such a system had been designed into the Shuttle late in 1971, according to the *Washington Post*. (In 2.5 min the Shuttle would reach a 40-km altitude from which it could "fly" to earth.) Staff writer Thomas O'Toole said that in 1973 NASA "reversed itself and dropped the launch-abort system . . . a decision understood to have met with dissent inside the space agency." The abort system designed for the Shuttle consisted of 2 huge solid-fuel rocket motors, one on each side of the Shuttle tail, that could be fired to separate the spacecraft and its occupants from the booster engines and main-engine fuel tanks in case of trouble; however, the abort motors weighed 43,500 kg—half as heavy as the entire 68 000-kg Shuttle carrying an average 18 000-kg payload—and even after they fired, the Shuttle would fall for 2 or 3 sec before being lifted away from the boosters.

Elwood W. Land, director of system operations for the Shuttle program, defended the decision to remove the abort system, saying that it was not needed because of redundancy built into the spacecraft and its engine. Land noted that, in 58 manned space flights, neither the U.S. nor the USSR had resorted to a launch-abort system to rescue spacecraft crews, notwithstanding 2 close calls: *Gemini 6* astronauts had almost fired their ejection seats when their engine shut down on the pad in 1965, and the *Soyuz 18* cosmonauts never reached earth orbit but flew their spacecraft to a landing in southeast Siberia. The 1-man Mercury capsule had a rocket-boosted escape system to carry the spacecraft cabin away from its rocket engines and fuel tanks; the 2-man Gemini had ejection seats to fire astronauts from the cabin like jet pilots from disabled aircraft; and the 3-man Apollo carried a large solid-fuel rocket motor that could pull the 18 000-kg spacecraft away from its tower of engines sec after trouble hit the engines or fuel tanks. Only remaining provision for the Shuttle was a pair of ejection seats for pilot and copilot of the first 4 orbital test flights; the seats would be removed when 5 more crew members were added for subsequent tests and for operational flights. "There is no way to install 7 ejection seats in the shuttle," the article noted. (*W Post*, 7 May 76, A-3)

- Bradford Johnston was appointed NASA's Associate Administrator for Applications, succeeding Charles W. Mathews who retired 27 Feb., NASA announced. Johnston's appointment would be effective 7 June. Now a private management consultant in Wis., he received a B.A. in economics from Wabash College and an M.B.A. from Harvard Business School. His NASA responsibilities would include planning and directing agency programs to identify and demonstrate useful applications of engineering and science techniques. (NASA Release 76-84; Hq announcement 10 May 76)

- The Smithsonian Institution awarded its Langley gold medal for aeronautics to James E. Webb, former head of the National Aeronautics and Space Administration, and the late Grover Loening. Webb was cited for management skills in leading the U.S. to "pre-eminence in space flight research and development." Loening [see 29 Feb.] was honored as "a pioneer aeronautical inventor" who developed the Loening amphibian plane and the design of the strut-braced monoplane. (*W Post*, 8 May 76, E-3)

10 May: NASA announced that a task team headed by Marshall Space Flight Center, and including representatives from Kennedy Space Center and Johnson Space Center, would assist the Air Force's Space and Missile Systems Organization (SAMSO) in evaluating industry proposals for validating the development of the Space Shuttle interim upper stage (IUS). The IUS, an expendable solid-fuel rocket stage capable of launching one or more spacecraft, would be carried into low earth orbit in the bay of the Space Shuttle orbiter; after deployment, it would be fired to carry the spacecraft into orbits of from 800 to 35 000 km, or to earth-escape trajectories for interplanetary missions. The basic IUS developed to meet DOD objectives would be altered as needed to meet NASA-unique requirements (planetary missions, or economical delivery of several smaller satellites). MSFC would be responsible for establishing the NASA-unique

requirements; its task team would be responsible for planning and coordinating NASA's IUS activities; KSC would be responsible for IUS launch operations, and JSC for IUS-orbiter integration and flight operations. The Air Force was expected to award a contract for IUS development by Sept. 1976. (MSFC Release 76-89)
- NASA announced selection of McDonnell Douglas Corp. and the Singer Company's Simulations Products Division for parallel negotiations, leading to award of a contract to one of the companies for maintenance, modification, and operational support of Johnson Space Center's simulator complex to be used in training flight crews for the Space Shuttle. The 2-yr contract beginning 1 July would provide for optional additional periods of 24 mo and 6 mo respectively. The training complex, consisting initially of a Shuttle procedures simulator and a crew procedures evaluation simulator, would have added to it an orbiter aeroflight simulator and a Shuttle mission simulator. General Electric and Computer Sciences Corp. also submitted proposals. (NASA Release 76-87; JSC Release 76-30)
- Sen. John Glenn (D-O.), first American to orbit the earth in *Mercury 6*, said he would be available as a running mate for Jimmy Carter in the national elections but would not pursue the vice-presidential nomination. Glenn said Carter was clearly front runner in the Ohio primary scheduled for 8 June, and that his "ship is already in." (*NYT*, 11 May 76, 13)
- The European Space Agency announced plans for a NASA presentation on current status of the Space Transportation System to be held in Paris 12 to 14 May, for about 250 representatives of governments, national institutes and agencies, and industry of the ESA member states. Opened by Roy Gibson, director general of ESA, the program would be introduced by Arnold W. Frutkin, NASA's Assistant Administrator for International Affairs. Other NASA speakers would be John F. Yardley, Associate Administrator, Office of Space Flight; Chester M. Lee, director of space transportation system operations; and Harold E. Gartrell, deputy manager of the Shuttle payload integration office. The Space Transportation System would include the U.S. Space Shuttle and the ESA Spacelab, as well as a new upper propulsion stage and a U.S.-developed tracking and data system. ESA representatives scheduled to present the Spacelab portion of the 3-day program would include Bernard Deloffre, director of the Spacelab program for ESA; Heinz Stoewer, Spacelab project manager; Jan J. Burger, Spacelab payload adviser; and Jacques Collet of ESA's planning directorate. (ESA release 10 May 76)

10-12 May: The Senate considered and passed H.R. 13172 authorizing NASA $16 800 000 for increased pay costs for the "interim period" 1 July through 30 Sept. 1976. The authorization—part of the budget for fiscal year 1976—supplemented NASA's research and program management request, and was $3 186 000 less than the original request of $19 986 000. (NASA Ofc of Budget Ops, Chron Hist FY 76, 16 June 76)

11 May: President Ford signed legislation reestablishing the post of White House science adviser, a job abolished by President Nixon 3 yr previously. The signing took place in the White House rose garden, at a ceremony attended by leading scientists. President Ford had asked for legislation to set up the position and the White House Office of Science and Technology on the recommendation of Vice President Rockefeller and at the urging of major scientific groups; the job had previously existed under executive order. Originally established by President Roosevelt during World War II, the position lapsed until 1957 when President Eisenhower reactivated it after the launch of Sputnik. President Ford was expected to make a early choice of science adviser, who would direct the Office of Science and Technology Policy and would also be a member of the Domestic Council and adviser to the National Security Council. (*W Post*, 12 May 76, A-21)

- Communications Satellite Corp., the international telecommunications consortium, had begun contract negotiations with U.S. manufacturers on building a new generation of comsats with a capacity of about 12 000 simultaneous telephone calls plus television, to meet growing demand for international communications expected by the early 1980s, the annual meeting of ComSat stockholders was told. Joseph V. Charyk, president of ComSat, noted that the 2 new Intelsat IV-A satellites scheduled for launch this year had a capacity of about 6250 simultaneous telephone calls, plus television; the first such satellite—*Intelsat I*, the 1965 *Early Bird*—could carry either 200 telephone calls or one TV channel. Charyk said negotiations had been held with Hughes Aircraft (maker of Intelsat IV), TRW Inc., and a division of Ford Motor Co. for construction of the new satellites. However, he said, ComSat's share of future international satellite operations would depend on the outcome of its appeal from an FCC order of last Dec. calling for sharp reductions in rates. Had the FCC rates been in effect for all of 1975, ComSat profits would have been $1.60 a share rather than $4.62 a share, he said. (*W Post*, 12 May 76, D-9)

- The Air Force awarded a $1 million supplement of a previous cost-plus-incentive-fee contract to General Electric's Aircraft Engine Group at Cincinnati, O., a labor-surplus area, for extended flight-test spare parts support for B-1 aircraft engines, the Department of Defense announced. (DOD Release 213-76)

- Clocking the speed of plasma at 50 km per sec in the active regions of the sun was "among the most interesting results" of investigations carried out aboard the USSR space station *Salyut 4,* Dr. Konstantin Feoktistov wrote in the *Bulletin of the USSR Academy of Sciences.* Feoktistov, one of the cosmonauts on the flight of *Voskhod 1* in Oct. 1964, was quoted in a Tass broadcast as saying that data from the two manned expeditions to *Salyut 4* were still being processed, but that the missions had established man's ability to "work well in conditions of weightlessness" for more than 2 mo. Citing the stay of Pyotr Klimuk and Vitaly Sevastyanov aboard the station from 25 May to 25 July last year, and describing the use of exercises and pressurized suits to maintain health during that time, Feoktistov said the mission results offered hope that orbital stations and piloted space flights would be further developed. (FBIS, Tass in English, 11 May 76)

12 May: NASA announced a first-time use of satellite relay of medical data from a moving ambulance to a hospital. Scientists and engineers at the National Space Technology Laboratories (NSTL) at Bay St. Louis, Miss., worked with General Electric's Science Services Laboratory to develop a special portable transmitter and antenna that could continuously transmit voice and medical data—including electrocardiograms—from a moving ambulance to the satellite and down to a hospital receiving station. During demonstrations of the system on a highway near Bay St. Louis, communications from the ambulance had been received as far away as N. Mex. The NSTL system, using the data-collection system on *Goes 3*, was similar to a telemedicine system being demonstrated by Johnson Space Center at the Papago Indian Reservation in N. Mex., and use of the *Ats 6* for medical communications in Alaska. Use of an inexpensive receiver at the medical center could make remote health care economically feasible; the system might eventually lead to development of a special medical satellite to relay emergency data from remote hospitals, ships, offshore oil platforms, and other remote locations to major medical centers for consultation. (NASA Release 76-86)

- A joint U.S.-British astronomy project to study the remnant of an exploded star failed when a rocket-motor malfunction kept the x-ray telescope aboard from acquiring its target, a supernova remnant called Puppis A some 10 trillion km from earth. An English Skylark sounding rocket fired from the Woomera Rocket Range in Australia carried a NASA-designed flight telescope, fabricated and assembled at Marshall Space Flight Center, to obtain information on evolution of stars and the formation of neutron stars. Although the telescope assembly, as well as the detectors and electronics supplied by the United Kingdom, worked perfectly, a hole burned through one side of the rocket's aft end produced more spin than the despin device could offset, said Richard Hoover of MSFC, principal U.S. investigator for the project. "Even with the residual spin, we scanned a portion of the sky in the area of the prime target and acquired data on the diffuse x-ray background [a secondary objective] . . . but we couldn't lock on Puppis A in order to get a high-resolution map as desired." The payload was recovered by parachute about 160 km downrange in excellent condition, Hoover said. (NASA Releases 76-62, 76-92; MSFC Release 76-54; *MFSC Star,* 17 Mar 76, 3)

- NASA announced plans for a 2-day course, "Technology Exchange Between the Textile Industry and Government," to be held at Clemson University, Clemson, S.C., to acquaint industry executives with new developments in textile research stemming from government-sponsored programs. Co-sponsored by NASA's Technology Utilization Office, the College of Industrial Management and Textile Science at Clemson, and the Economic Development Administration, the course would offer discussions by experts from industry, government, and the academic community on a wide range of subjects including new fiber developments, fire-retardant materials, and innovations in textile manufacture. Course themes would include industry-government cooperation, new needs and opportunities for cooperation, industrial developments adopted by the government, industry-government information systems presently available, and consumers' choices at retail. (NASA Release 76–88)

13 May: Ames Research Center announced award of a $939 000 contract to Battelle Memorial Institute of Columbus, O., for aid in administering an early-warning aviation safety system for the Federal Aviation Administration. Under the 2-yr contract, Battelle would implement a reporting system to obtain from persons in the national aviation system information on potential threats to flight safety; the reports would be processed to preserve anonymity of the informants and to make the data quickly usable to avoid or reduce aircraft accidents. An aviation-safety reporting system instituted by FAA in May 1975 had met with reluctance on the part of the public to report directly to a regulatory agency. NASA, invited to act as a third party, agreed to act as collection point for safety reports to encourage participation by pilots, controllers, and others using the nation's airways. (NASA Release 76–52; ARC Release 76–34)

- Successful testing of a solar receiver developed with heat-transfer technology used by Rocketdyne Division, Rockwell International Corp., for the U.S. space program was "a major step forward in development of solar electricity generating systems," said Dr. Jack Silverman, director of energy systems for Rocketdyne. Using a large field of mirrors individually focusing the sun's rays to a central receiver on a high tower, where the concentrated heat served to boil water into superheated steam, the system generated the steam at 1366 K and 738 newtons per cm^2. The central receiver was exposed to high heatflows approaching those encountered in rocket engines, and many times higher than those in conventional steam boilers. The development program was sponsored by the Energy Research and Development Administration; Rocketdyne was also under contract to McDonnell Douglas and ERDA to develop a similar receiver and thermal storage subsystems for a 10 000-kw solar electricity-generating pilot plant to be in operation by the end of the decade. (Rockwell Release RD–9)

13—15 May: NASA launched *Comstar 1 A-1*, first of a series of three Comsat General Corp. satellites planned to provide 14 400 two-way high-quality voice circuits in a telephone-communications network serving Hawaii, Alaska, Puerto Rico, and contiguous U.S. An Atlas–Centaur fired from ETR at 6:28 pm EDT 13 May put the Hughes Aircraft-built comsat in an elliptical transfer orbit; firing of the apogee motor at 6:42 pm 15 May would put the Comstar on station over the equator at 128°W, south of San Francisco, at just over 35 000-km altitude, by 4 June. The spin-stabilized cylinder 6.1 m high and 2.4 m in diameter, weighing about 816 kg in orbit, with 14 000 solar cells mounted on the cylinder surface, would carry 24 radio repeaters each capable of handling 1200 one-way voice channels in the 4- to 6-ghz range, using a technique of cross polarization that would double satellite capacity by more efficient use of the frequency spectrum. Comsat General Corp. would own and operate the satellites and associated earth facilities, leasing to American Telephone and Telegraph Co. and to GTE Satellite Corp., a subsidiary of General Telephone and Electronics Corp. The $43-million satellite was called *Comstar D-1* after achieving orbit. (NASA Release 76-75; *NYT*, 13 May 76, 51; Comsat General Release CG 76-115; MOR M-491-201-76-01 [postlaunch] 27 May 76)

14 May: Dryden Flight Research Center announced it would begin a series of YF-17 test flights in mid-May to aid designers of future highly maneuverable aircraft. The YF-17, built by Northrop Corp. for the Air Force's lightweight fighter program, is a twin-engine aircraft incorporating many innovations to give it high performance and maneuverability. Objectives of the NASA research program would be to collect in-flight data on pressure around the engine nozzles and afterbody, for comparison with wind-tunnel data to improve prediction techniques for future fighter-aircraft design; and to continue NASA studies on improved maneuvering for fighter aircraft in the areas of buffet, stability and control, acceleration, and handling. Under a $255 000 contract funded jointly by NASA and the U.S. Navy, the YF-17 would be flown for about 8 wk by Gary Krier, DFRC's YF-17 project pilot. (DFRC Release 7-76)

15 May: A U.S.-European Space Shuttle project starting in 1980 would recruit 30 more astronauts—including women and participants from the USSR—for the planned 200 flights in the project, U.S spokesmen said at a Paris news conference held at the end of a 3-day meeting between NASA and European Space Agency officials. John F. Yardley, NASA Associate Administrator for Manned Space Flight, told newsmen that NASA was already training more than 30 astronauts for this program. Each flight, carrying 3 U.S. astronauts and up to 4 Europeans on space-science missions lasting between a week and a month, would cost between $18 million and $21 million; European scientists had expressed concern that the flights would cost so much that less money would be left for scientific experiments than they had hoped. Arnold W. Frutkin, NASA's Assistant Administrator for International Affairs, when asked if the Soviets might be involved in the project, said, "We have made it clear in informal discussions that this would be available to them." Those using the Shuttle would be expected to foot the bill, officials emphasized. (*NYT*, 15 May 76, 2; *W Post*, 15 May 76, A-15; *W Star*, 16 May 76, A-3)

17 May: SPAR II, a Black Brant VC sounding rocket carrying materials-processing experiments, was launched from the White Sands Missile Range in the second of a series of 15 planned during the next 5 yr in the SPAR (space-processing applications rocket) program managed by Marshall Space Flight Center. The 225.7-kg payload consisted of 10 experiments, 6 similar to those carried on SPAR I launched 11 Dec. 1975; the rocket reached an altitude of 188 km, providing about 5 min of near-weightlessness during the coast phase, and the payload was recovered by parachute and delivered to the investigators for analysis. MSFC employees Carolyn Griner and Dr. Mary Helen Johnston were principal investigators for 2 experiments, both on dendrite remelting and macrosegregation. Other experimenters were Robert B. Pond (Johns Hopkins University), 2 experiments on solidification of lead antimony eutectic; Dr. James W. Patten (Battelle-Northwest Research Institute), 2 experiments on producing closed-cell metal foams; Dr. S.H. Gelles (Battelle

Memorial Institute), agglomeration in immiscible liquids; Dr. H. Ahlborn (University of Hamburg), behavior of aluminum alloys under zero gravity; and Dr. Louis Raymond (Aerospace Corp.), 2 experiments in casting thoria dispersion-strengthened composites at zero gravity. (MSFC Release 76-83, 76-94)

- Payload responsibilities for the first 6 flights of the Space Shuttle had been allocated by NASA, according to *Aviation Wk and Space Technology* magazine. The first 6 orbital missions scheduled for 1979-1980 would be verifications of the overall Shuttle system; the first flight would carry no payload except a 45 000-kg data-analysis system, but the other 5 flight-test missions would carry scientifically meaningful payloads. NASA had assigned the second and fifth flight tests to its Office of Applications; the fourth and sixth, to the Office of Space Science; and the third, to the Office of Aeronautics and Space Technology. The Office of Applications would be involved with space-processing hardware, and the Office of Space Science would oversee the use of existing research instruments; the OAST had been considering a space "workbench" on which various instruments could be deployed for 3- to 9-mo periods. The orbital flight-test missions would begin with a duration of only a few orbits, which should increase to about 7 days by the sixth flight mission; the first 4 flights would land at Dryden Flight Research Center, and the latter 2 at Kennedy Space Center. (*AvWk*, 17 May 76, 21)

- Astronaut Paul J. Weitz, Captain, USN, would retire from military service 1 June but would remain in his present job with the space agency as a civilian, NASA announced. Weitz, pilot on *Skylab 2* (the first manned mission 25 May to 22 June 1973), was part of the crew who saved the mission by erecting a "parasol" shade to reduce overheating caused by damage to *Skylab 1* during its launch, and by unjamming a solar-power wing to provide enough electrical power for their mission and 2 follow-on missions of 59 and 84 days respectively. Retiring after 22 yr Navy service, Weitz was currently working on payloads and flight-crew documentation for the Space Shuttle. (JSC Release 76-33)

- International golfing great Arnold Palmer was scheduled to pilot a Learjet 36 "business aircraft" from Denver, Colo., for an assault on the speed-around-the-world record that had not been surpassed for 10 yr. A Learjet 24 had circled the earth in 1966 in 65 hr 38 min 39 sec elapsed time, at an average speed of better than 563 kph (50 hr 20 min actual flying time) including 17 intermediary stops. The National Aeronautic Association had sanctioned the new attempt, designating journalist Robert Serling as official onboard observer. Two additional pilots would accompany Palmer and Serling on their 2-day flight, scheduled to cover nearly 37 000 km, under the auspices of the Aviation Space Writers Association. Nine stops were planned for the new flight: Logan Airport at Boston; LeBourget at Paris; Mehrabad Airport at Tehran, Iran; Bandanaraike International Airport at Colombo, Sri Lanka; Kemayoran Airport, Jakarta, Indonesia; Manila International Airport; Wake Island; Honolulu International Airport; and Arapahoe County Airport, Denver. Besides NAA sponsorship, the flight was sanctioned by the American

Revolution Bicentennial Association and the Aviation Historical Foundation. (NAA newsletter, May 1976, 1)

18 May: Ames Research Center had contracted with Goodyear Aerospace Corp. of Akron, O., for studies of 2 lighter-than-air vehicle concepts for civilian use, NASA announced. New requirements for transporting heavy loads in power-plant construction, transferring ship cargoes to shore points, and providing quiet energy-saving intercity transportation had revived interest in lighter-than-air vehicles, formerly the only means of nonstop rapid travel across the world's oceans. Use of airships for military missions also had come under consideration; principal potential use was transport of ship cargoes over a beach to shore points. NASA, in conjunction with the U.S. Navy, was also studying military applications for conventional airships that would use the great endurance potential of the airship in activities such as antisubmarine warfare and sea control.

One of the concepts under study, a feeder airliner 60 m long carrying 80 passengers, would be used as a short-haul transport system, landing and taking off vertically and cruising at 160 knots. The other concept, a vehicle to transport large heavy payloads over comparatively short distances, would combine features of large dirigibles and helicopter rotors to provide lifting capacity far beyond that of either type of vehicle alone; the dirigible buoyance would lift the empty weight of the vehicle, and the total lifting capacity of the rotor system would lift and support the payload. The heavy lifter would be most likely to have immediate application, NASA said, because of the need for transport of heavy power-generating equipment or other outsize industrial equipment to a remote destination not served by any other heavy transport systems. Increased engineering knowledge and better understanding of weather phenomena, as well as substituting inert helium for the volatile hydrogen used in German airships of the 1920s, would make a modern airship safe. (NASA Release 76-93)

- The European Space Agency announced completion of the launcher integration site for its Ariane launch vehicle on schedule. The site—a building near Paris measuring 105 m long, 50 m wide, and 33 m high—constructed on its own land by Aerospatiale, the contractor called "the industrial architect of the Ariane programme," was for development tests and assembly, integration, and acceptance of the complete launcher before its dispatch to the launch site. Four Ariane flight tests were scheduled for 1979 and 1980. (ESA release 18 May 76)
- Marshall Space Flight Center and Dryden Flight Research Center officials signed an agreement to conduct jointly a comprehensive program of tests on the parachute recovery system for the Space Shuttle's solid-fuel rocket booster. The program, to begin early in 1977, would consist of drogue-parachute tests and main-parachute tests, using single parachutes, and test deployment of the 3-parachute cluster (actual flight configuration) to be used in recovery of the solid rockets, largest ever flown. The parachutes, some 36.5 m in diameter, would be the largest used in the space program; Apollo spacecraft parachutes were about 24.5 m in diameter. The Dryden center would provide the B-52 aircraft

for the test drops, as well as the flight and maintenance crews, and would perform the tests over the National Parachute Test Range about an hour's flight from Edwards AFB in Calif. MSFC engineers would evaluate the test data to determine the adequacy of the system. (NASA Release 76-94; DFRC Release 6-76; MSFC Release 76-87)

19 May: Arnold Palmer, piloting the Learjet "200 Yankee" named in honor of the U.S. Bicentennial, landed at Arapahoe County airport near Denver after a record-setting flight around the world that took 57 hr 25 min 42 sec and covered 37 000 km. Palmer's flight was 28 hr 43 min 19 sec faster than the longstanding record set by Arthur Godfrey and Dick Merrill in 1966. Accompanied by the official observer and timer for the National Aeronautic Assn., Robert J. Serling, and by copilots James E. Bir and L.L. Purkey, Palmer averaged better than 770 kph on the 2.5-day flight that made 9 stops in 7 countries. In meeting officials at each of the stops, Palmer presented Bicentennial flags and bronze replicas of the Declaration of Independence; at Wake Is., he left a silver plaque commemorating the U.S. Marine Corps defense of the island in World War II. In addition to the speed-around-the-world record, NAA planned to claim for the Palmer flight 18 additional records for speed over recognized course: e.g., Boston to Paris, Paris to Tehran, etc. Since he took up flying in the 1950s, champion golfer Palmer had logged about 4500 hr as a pilot. (NAA *News,* July 76, 1)

- Ground-based equipment incorporating computers that would work with components aboard a Shuttle orbiter to guide it to a safe landing had been shipped from the contractor, Cutler Hammer's AIL Division, NASA announced. The equipment, called a microwave scanning-beam landing system (MSBLS), would be installed on a runway at Dryden Research Center, where initial flight tests of the orbiter would begin in mid-1977; a second set would be installed on a newly constructed runway at Kennedy Space Center, where initial Shuttle orbital missions would be launched in 1979. Both locations would be equipped for approach from either direction, and each system would be fully redundant, including a comprehensive monitoring system with automatic switchover and an uninterruptible power supply. As the Shuttle orbiter would descend in a steep glide moderated to a soft touchdown, onboard computers would direct the vehicle through commands to the control surfaces and must know the vehicle's precise position at every instant; standard instrument-landing electronics could not provide this information, so that the type of beam created by the MSBLS would be needed.

  The MSBLS system would cover a total field of positions through all possible paths the orbiter would take, instead of providing a single straight path for the orbiter to follow; its scanning beam would sweep a wide flat course across the landing sector, and pulses from a ground transmitter would be coded to identify the exact angle at which the beam pointed at each instant of the sweep. A receiver in the descending orbiter would pick up the pulses and decode them to determine the track; the onboard computer would accurately compare the orbiter's location with the desired location and automatically correct any discrepancy. The

MSBLS would provide an unprecedented degree of position-guidance accuracy. (NASA Release 76-97; JSC Release 76-35)
- Isaac T. Gillam IV, NASA Hq Program Manager of Small Launch Vehicles, was designated Director of Space Shuttle Operations at Dryden Flight Research Center, effective 23 May. In his new position, Gillam would be responsible for development of test support facilities, institutional support of test operations, and flight and industrial safety for test operations in support of Shuttle carrier-aircraft testing and orbital approach and landing tests conducted at DFRC. Before coming to NASA Hq in 1963, Gillam served in the U.S. Air Force as pilot, missile launch crew commander, and ROTC instructor; he had done graduate work at Tenn. State Univ. while serving as assistant professor of military science. From 1963 to 1966, he was resources management specialist at NASA Hq and was then appointed Assistant Delta Program Manager in the Launch Vehicles Directorate. Named Delta Program Manager in Sept. 1968, he became Program Manager of Small Launch Vehicles (including Delta and Scout) in June 1973. He received NASA's Distinguished Service Medal for his work in the launch-vehicle program. (NASA Release 76-96)
- The Air Force announced award of a $2 067 113 cost-plus-fixed-fee contract to Teledyne McCormick Selph of Hollister, Calif., for unsymmetrical dimethyl hydrazine, a chemical used in Titan missiles and NASA–USAF space boosters. The contracting activity was the San Antonio Logistics Center at Kelly AFB, Texas. (DOD Release 223-76)
- In an address to the American Institute of Industrial Engineers, meeting in St. Louis, W.F. Rockwell, Jr., chairman of Rockwell International Corp., warned that the U.S. in its search for energy independence should not "go driving up the wrong street" but should make use of nuclear power. Claiming that solar power, nuclear fusion, or other exotic power sources were solutions for the distant future, Rockwell described nuclear energy as "the bridge that will allow us to make a smooth transition to those future energy sources" and as being "safe, clean, abundant and economical." Rockwell noted the 6- to 10-yr period needed between the concept and the start-up of a power plant, adding that today's Americans have roughly 36 mo to "get it all together if we're going to have enough power in 1990." (Rockwell Release R-20)
- Dr. Wernher von Braun was admitted to hospital in Alexandria, Va., 14 May and was reported in fair condition yesterday, the *Washington Post* reported. Hospital spokesmen refused comment on von Braun's illness, citing requests from his doctor and family. Von Braun, who left the government in 1972 to become a vice president of Fairchild Industries, Inc., in nearby Germantown, Md., underwent an operation for cancer last year at Johns Hopkins Hospital in Baltimore, a hospital official there said. The German-born scientist, now 61, had led development of the U.S. space program in the early 1960s. (*W Post*, 19 May 76, B-4)

20 May: The U.S. Navy launched an experimental ocean-surveillance satellite—part of the Whitecloud system developed by the Naval Research Laboratory—on 30 April from Vandenberg AFB aboard an Atlas launch vehicle, reported *Defense/Space Business Daily*. The satellite was in a near-circular orbit of 1122-km apogee, 1100-km perigee, inclined 63.5°, with a period of 107.5 min. (*SBD*, 20 May 76, 21)

- NASA Hq conducted a news conference on the Viking mission to Mars, with Dr. Noel Hinners, Associate Administrator for Space Science; Robert Kraemer, director of planetary programs; Walter Jakobowski, Viking program manager; James Martin, Jr., Viking project manager; and Dr. Gerald S. Soffen, Viking project scientist, to review the mission and answer questions. Dr. Anthony Calio, NASA's Deputy Associate Administrator for Space Science, opened the session by recalling the work done on Viking over the past 7 yr that would culminate in Viking's reaching Mars within the next few weeks. The U.S. for the first time would be operating 4 spacecraft simultaneously, 2 in orbit and 2 on the planet's surface. Kraemer mentioned Galileo's sighting of Mars in 1609, the detection of "canali" on the planet's surface by Schiaparelli and others in the 1800s, and Percival Lowell's founding of the Lowell Observatory in 1894 followed by his publication of reports on "Mars and Its Canals" and "Mars as the Abode of Life." Edgar Rice Burroughs, author of the Tarzan stories and "grandfather of science fiction writers," began writing stories about civilization on Mars that had kept the public "waiting ever since to get down to the surface and see what is there." Although, as Kraemer pointed out, 4 Mariner missions had shown that civilization did not actually exist on Mars, nothing had ruled out the existence of life, and that was a purpose of the Viking mission.

  Jakobowski reviewed the history of Voyager, Surveyor, and Viking, especially the 1971 Mariner flyby that revealed Mars as a dynamic planet. Martin, Viking project manager at Langley Research Center, said the 2 Viking spacecraft had been flying for a long time and had turned in "exceptionally good" performance; this would be the first mission to use optical navigation to confirm radio-tracking data, and the first set of results had shown the spacecraft to be "right on course." Martin added that—as an expression of confidence in his team—a quiet period had been declared the coming week when most of the flight team would be on vacation; "the team needs that time off to get ready for a pretty active and hectic summer," he said. A.T. (Tom) Young, Viking mission director at LaRC, showed slides of mission operations and possible landing sites, reviewing the reasons for site selection and what NASA hoped to learn in both scientific and engineering areas, concluding with a summary of expected activities during the first 20 days on Mars. Viking-mission audio from JPL would be available starting about 15 June.

  Dr. Soffen noted that the scientific questions about Mars covered more than the presence of life there, which was only one of 13 investigations to be conducted by Viking, and suggested that the press representatives "ought to prepare yourselves from an educational point of view about all the experiments other than just the biology." The first question raised by the press was on the indicators of a safe landing; Martin replied that there would be 5—telemetry from a footpad switch; startup of the lander computer, accompanied by turnoff of the descent-engine heaters; drop on equipment-power bus with shutoff of entry equipment; and switch in data rate from the orbiter-to-lander 4-kilobit rate to a 16-kb rate for sending pictures to the orbiter. A question about turning on the backup lander computer got a detailed response involving detectors and switches in a complex procedure that occur when "the second computer hollers for help," as Young put it. Other questions concerned dust storms, camera resolution, and chances of success; Soffen concluded by noting "if we knew the answers, we wouldn't have to do this mission." (Text, 20 May 76; NASA Releases 76-90, 76-98)

- NASA announced delivery of experiment hardware for the first of three High Energy Astronomy Observatories (HEAO) to the prime contractor, TRW Systems of Redondo Beach, Calif. The hardware consisted of 4 experiments to survey and map x-ray sources during the 6-mo mission of HEAO-A in 1977. A Naval Research Laboratory experiment (Dr. Herbert Friedman, principal investigator) would use a large-area survey instrument to locate x-ray sources and obtain data for studying the physics and evolution of energy sources. An experiment built by Goddard Space Flight Center with assistance from the Calif. Institute of Technology would measure emissions and absorptions of diffuse x-rays and correlate results with radio and visible light ray emission; principal investigators were GSFC's Dr. Elihu Boldt and Caltech's Dr. Gordon Garmire. A third experiment, for the Smithsonian Astrophysical Observatory and MIT, was a scanning modulation collimator to determine precise celestial positions of selected x-ray sources and investigate their size and structure; principal investigators were SAO's Dr. Herbert Gursky and MIT's Dr. Hale Bradt. The fourth experiment, for the University of Calif. at San Diego and MIT, would determine the position, spectrum, time variations, intensity, and other properties of hard x-rays and low-energy gamma rays; principal investigators were UCSD's Dr. Laurence Peterson and MIT's Dr. Walter Lewin. Data gathered by the HEAO missions could lead to new theories about energy production and high-density nuclear matter. Marshall Space Flight Center, which would manage HEAO for the Office of Space Science, directed TRW in designing and building the observatories and integrating and testing the overall system, including the experiments. (NASA Release 76-100; MSFC Release 76-86)

- NASA announced signature of an agreement between the Universities Space Research Association (USRA) and the Langley Research Center to provide university participation in the Long Duration Exposure Facility (LDEF) space research program managed by LaRC, which would use the Space Shuttle to launch and retrieve an earth-orbiting multipurpose-experiment carrier starting late in 1979. USRA was established in 1969 through the National Academy of Sciences as a consortium of 50 universities in the U.S., through which the academic community could work with NASA on scientific and technological developments in the space program; its headquarters was at the University of Virginia in Charlottesville. The agreement with LaRC called for USRA to solicit, select, and implement university experiments for each LDEF mission, with funds provided by the sponsoring organizations. NASA would also seek participants in the LDEF program through a formal announcement of opportunity inviting private companies, research and engineering institutes, and other government agencies to provide space experiments for LDEF missions. (NASA Release 76-99)

22 May: The Air Force's *P76-5* satellite was successfully launched by NASA from the Western Test Range on a Scout vehicle at 0041 am PDT. The 72.6-kg spacecraft, designed to evaluate propagation effects of disturbed plasmas on radar and communications systems, was placed in a near-circular sun-synchronous orbit with apogee at 935 km and perigee at 863 km, inclination of 99.6°, and period of 105.9 min. The launch was requested by the Air Force Nov. 75 and agreed to by NASA 3 Dec. 1975, under a 1962 memorandum of understanding on joint NASA-DOD use of the Scout vehicle, and costs of the launch would be borne by the USAF. Mission objectives would be achieved, according to a postlaunch report. (MOR M-490-602-76-01, 21 Apr 76, [prelaunch] 5 May 76, [postlaunch] 24 Sept 76)

- NASA announced issuance of a patent to engineers of Scott Aviation for a new type of breathing apparatus invented under NASA contract, suitable for use as emergency equipment for firemen [see 17 May]. The apparatus included an alarm system working on the compressed air delivered to the wearer through a face mask, warning by a whistle audible only to the wearer if the air pressure fell. The equipment was also lower in weight and bulk, more comfortable, highly visible, and easier to operate. The weight of the apparatus, fitting over the face and around the head, was supported from the hips. Scott Aviation had planned to deliver 50 sets of the equipment to the Boston fire department in July. (*NYT*, 22 May 76, 31)

- Two engineers working for Lockheed Aircraft Corp. had reported that windmills could supply up to a fifth of U.S. energy needs by 1995, the *Washington Post* said. The engineers, Michael Dusey and Ugo Coty, reporting on a year-long investigation funded by the Energy Research and Development Administration, said that large wind-turbine generators could save 2 billion barrels of oil a year; construction of 54 000 generators with rotor blades more than 160 m end-to-end could furnish 1 trillion kw hrs of electricity annually. Other benefits would include reduction of air pollution and creation of a new industry that would employ thousands, they said. The Lockheed researchers put no price tag on their findings, but an earlier study by the University of Hawaii had said such windmills could cost up to $50 000; the price would be regained within 7 yr and the windmills could remain in operation up to 50 yr, according to the 1974 report by Hawaii's Donald Grace. A professor from the Univ. of Mass.—William Heronemus—had proposed some years ago a network of 957 windmills each about 113 m high just to meet the needs of the state of Vermont. (*W Post*, 22 May 76, F-1)

24 May: Commercial supersonic transatlantic transport service began with the arrival of Concorde aircraft from London and Paris at Dulles airport near Washington, D.C. The Anglo-French plane had been barred from New York's John F. Kennedy airport under a ruling by the Port Authority of N.Y. and N.J. after Transportation Secretary William T. Coleman,

British Airways Concorde rolling out at Dulles Airport near Washington, D.C., after first touchdown 24 May 1981. (Photo courtesy British Airways)

Jr., granted limited scheduled flights into the U.S. for a trial period not to exceed 16 mo [see 5 Jan.–4 Feb.]. The Concordes arrived at Dulles 2 min apart, British Airways arriving at 11:54 am EDT and Air France at 11:56, taxiing into a nose-to-nose position in front of the main terminal building. About 8000 persons were on hand for the landing, many of them leaving their cars and walking in on the limited-access highway because of the traffic backup; of the crowd, airport officials estimated that only about 20 were there to protest the Concorde landings. The French flight carried 80 passengers, the British flight 75; both Air France and British Airways had limited passenger payloads to 80 on Concorde flights until they gained more experience with the aircraft's fuel consumption.

The two carriers had selected Monday for the inauguration of North Atlantic service because neither would ordinarily offer a Washington-bound flight that day. Air France planned 3-per-wk flights, leaving Paris on Sunday, Wednesday, and Friday and returning to Paris on Monday, Thursday, and Saturday. British Airways would offer 2 trips per wk, leaving London's Heathrow on Thursday and Saturday and returning on Friday and Sunday. One-way fare London to Washington would be $633.60; Paris–Washington, $827. The difference would arise from fluctuations in currency exchange. The trip took 3 hr 50 min for the Air France plane, spending 2 hr 53 min at supersonic speeds on the way. The British plane timed on its return flight took 3 hr 57 min. The Federal Aviation Administration had instrumented the Dulles airport with 8 permanent and 5 portable recorder units to measure the perceived noise in decibels (pndb) for a data base on all airport noise and especially Concorde noise. (*NYT*, 25 May 76, 1; *W Post*, 25 May 76, A-1, *Av Wk*, 24 May 76, 27; 31 May 76, 22)

24–26 May: A "new view of Jupiter," as the *New York Times* called it, was the subject of a 3-day Ames Research Center symposium on Pioneer discoveries. Scientists from NASA, the Smithsonian Institution, and 15 universities were resource persons for 4 news briefings per day, each scheduled for an hr or more, and 5 science workshops, meeting in the morning and afternoon each day, with a wrapup session the afternoon of the third day. Topics discussed were Jupiter's atmosphere and weather; magnetosphere and radiation belts; origins, characteristics, and composition of the planet and its moons; origin of life on the planet; and *Pioneer 11* targeting toward Saturn upon its departure from the vicinity of Jupiter.

Pioneer findings about Jupiter—characterized by low density and high mass, 318 times that of earth, containing three quarters of the planetary material in the solar system—included discovery that Jupiter's Great Red Spot, three times the size of earth, was not the only such spot; several smaller ones detected in the atmospheric flow around Jupiter moved in and out of the area occupied by the Great Red Spot without losing their own identity. Pioneer measurements of Jupiter properties— internal heat, composition of its moons—permitted the first detailed calculations of the process that formed the planet. Its history was apparently that of a failed star: it went from a gas cloud to a body heated red-hot by gravitational contraction, then to a 4.5–billion-yr cooling, still in process, that made it resemble a "white dwarf." The four planet-size inner moons—Jupiter's "Galilean" satellites—proved scientifically the most interesting; they exhibited a regular density gradient and composition that constituted the best direct evidence for Jupiter's history. The protostar phases deduced for both Jupiter and Saturn indicated that the outer solar system underwent 2 different periods of relatively high temperature. Jupiter had retained for 4 billion yr its primordial heat, which could be neither radiated nor conducted away because of the planet's composition. A "new metal"—liquid metallic hydrogen, simplest of the alkaline metals—identified only inside Jupiter is thought to constitute the planet's interior, beginning about of a quarter of the distance from the surface to the center. As the substance has low thermal conductivity and is opaque to radiated energy, the heat cannot be dissipated, and the temperature at Jupiter's center is about 30 000°, 6 times that of the surface of the sun. As Jupiter's liquid regions are too hot for life, living organisms would have to exist in its gaseous atmosphere. Between the temperatures for freezing and boiling water, Jupiter's atmospheric pressure increases with depth from 5 times earth's atmospheric pressure to 10 times.

Pioneer data suggest slow atmosphere-turnover times of months or years that would permit survival of atmosphere-borne life, said Dr. Cyril Ponnamperuma of the Univ. of Md. Jupiter was 1000 times as large a region for forming complex organic molecules as earth, and a wide variety of environments would be available. The origin of life on a planet would depend less on huge time periods than on how often nature mixed the proper ingredients in the proper environment, Dr. Ponnamperuma said; enough billions or trillions of attempts could produce a highly complex molecule able to replicate itself. The fact that Jupiter's atmosphere might provide hospitable sites for life raised the problem of contamination of that atmosphere by earth organisms riding on atmosphere probes launched by the U.S. and other countries.

Pioneer also found that Jupiter acted as a giant vacuum cleaner, sucking small dust particles from a vast region of space (comparatively little dust was present in the Asteroid Belt), and 170 times as many meteoroids were striking Jupiter's atmosphere as struck the earth in areas of similar size. Concentrations of these particles at Jupiter would be the result of the planet's enormous gravity field. Jupiter's magnetosphere—a million times the volume of earth's—was found to leak high-energy electrons; Jupiter was the major source of such electrons in the solar system, and the sun and Jupiter were the only known important sources of high-energy particles in the solar system. The energy of these particles was constantly increased by recirculation through the planet's radiation belts, said Dr. James Van Allen of the Univ. of Iowa. Earth-orbiting satellites and other spacecraft had detected Jupiter electrons as far away as Mercury, said Dr. John Simpson of the Univ. of Chicago; an input of more than 100 billion watts would be needed to maintain the observed energies. Jupiter was also known to be the strongest radio source in the sky, except for the sun; the massive radio bursts were caused by the moon Io moving through the Jovian magnetic field, said Dr. David Morrison of the Univ. of Hawaii. Scientific detective work revealed that the bursts occurred every 21 hr—half of Io's orbital period—and one of 3 places on Jupiter's visible surface had to face the earth, which Pioneer data proved to be true. (NASA Releases 76-80, 76-91; *NYT*, 31 May 76, 14)

25 May: The Environmental Protection Agency, at a Washington briefing for science writers, described its techniques of airborne remote sensing—combined use of aerial photography and heat-sensing instruments—to detect 98% of water-pollution discharge points in the U.S. John Moran, EPA director of monitoring, said the agency had long used aerial photos to map the extent of oil spills; other devices in EPA's "growing arsenal" included a laser system to measure ground contours and determine whether strip-mined land had been properly restored; an airborne laser firing energy pulses toward the ground to define particle layers in the atmosphere below the aircraft and thus measure the air-inversion ceiling during urban air-pollution alerts; and a system for laser fluorosensing that would monitor water pollution by measuring surface oils, dissolved organic matter, and even algae, by sensing responses to an ultraviolet beam. An earlier EPA report described use of techniques or devices developed for other purposes by NASA or DOD, to detect smoke drift in the atmosphere, oil spills, runoff from illegal stock feedlots, and sick vegetation. (*W Star*, 26 May 76, A-11)

26 May: NASA announced plans for a 3-mo demonstration this summer of using *Ats 6*—known as the "teacher in the sky"—to help some of the world's poorest people boost food production, improve health and nutrition, expand family planning, and raise income levels, through applications of remote sensing, space communications, and high-resolution aerial photography. The program, a joint effort of NASA and the Agency for International Development, would consist of both filmed and live portions broadcast to as many as 30 countries in Asia, Africa, and Latin America. AID would provide $3 million to finance the demonstration. At a UN conference in Nairobi 6 May on trade and development, Secretary of State Kissinger emphasized the use of satellite technology to improve cooperation between industrialized and developing nations; depending upon the response to the broadcasts, AID would be prepared to propose to Congress a long-range follow-up technology program. *Ats–6* was completing a year-long program of instructional television for India; upon completion of this mission 31 July, NASA would move the satellite to a location over the Western Hemisphere where it would beam audio or audiovisual presentations in black and white or in color to special receivers on the ground. AID officials and NASA specialists would set up one transmitter-receiver unit in the capital of each participating country, and up to 5 receivers in outlying locations; each receiver terminal would be equipped with color-television monitors. The live portion of the demonstration would feature 2-way discussions between U.S. personalities and representatives of the recipient countries; a local program would be developed in the host country to demonstrate the communications abilities of the ATS. NASA would be responsible for moving the equipment from country to country and for operation of the satellite and associated equipment. (NASA Release 76-101)

- A new device developed by the USAF Rocket Propulsion Laboratory at Edwards AFB, Calif., would permit an annual "physical examination" of the motors of intercontinental ballistic missiles while the missiles were still in the silo. The practice of removing the motors for inspection had resulted in high costs for handling and transportation as well as high risk of damage to the motors. The AF Space and Missile Systems Organization (SAMSO) at Los Angeles asked AFRPL to find a compact, safe, and economical means of inspection, and acoustical holography was selected. The system when fully developed would be able to scan the outside of an intact missile top to bottom, probing the propellant inside its motors. The device worked much like sonar, transmitting high-frequency sound waves and registering reflected vibrations; the reflected waves, processed electronically, would reproduce any flaws in the propellant on a cathode-ray tube for instant viewing or photographing, and the information would be stored in a computer for later evaluation. Future uses of acoustical holography might be made in medical science, improving on current x-ray techniques. (AF Release OIP 174.75)

26–27 May: John H. Disher, Director of Advanced Programs, NASA Office of Manned Space Flight, was keynote speaker at a space industrialization symposium at Marshall Space Flight Center sponsored jointly by MSFC and the Alabama section of the American Institute of Aeronautics and Astronautics. Defining space industrialization as "the use of space to produce a salable/profitable product or a service which companies as a business expense or citizens through their taxes are willing to pay for," Disher reviewed uses of space for communications, weather prediction, navigation and mapping, and earth-resources survey; he then described the advantages of space environment—earth overview, zero gravity, vacuum, heat and waste disposal, and uninterrupted solar energy—for future uses, concluding with a prediction of a growing tourist industry in space. The symposium concluded 26 May at the annual banquet of the Ala. AIAA with an address on "Space Industrialization as a Concept" by Dr. Krafft Ehricke, science advisor for North American Space Division, Rockwell International Corp. Sessions of the symposium dealt with space habitation, space transportation, space processing of materials, and space power. The symposium was organized to disseminate recent information on expanded space operations after the advent of the Space Shuttle to all segments of the scientific and technical community; participants included NASA centers, other government agencies, industry and universities. (NASA Release 76-81; MSFC Release 76-64, 76-79; Disher text)

28 May: Marshall Space Flight Center announced a new approach to issuance of 2 parallel contracts, to be awarded for studies of a space industrialization effort that would run from 1980 to 2010. One of the study contracts would be awarded to an aerospace firm, the other to a research or "think-tank" firm, to obtain different viewpoints. The proposed work would consist of 2 phases, each lasting about 8 mo, and each contract would cost about $200 000. Replies to today's requests for proposals were due 29 June, and contracts were to be awarded this fall. Whereas the space program had so far emphasized scientific and other exploratory activities, space industrialization would emphasize production of goods and services for economic benefit. First phase of the study would define goals, establish the rationale for a space program on the basis of terrestrial alternatives, and set up a time frame with options and budgetary considerations. The second phase would study future mission opportunities and the optimum role of man in the program, responsiveness of the concepts to national and industrial needs and goals, and technological implications as against costs. (MSFC Release 76-98; NASA Release 76-107)

30 May: Third Century America, the U.S. Bicentennial exposition on science and technology, opened to the public with 30 000 persons attending; 32 000 employes of the Kennedy Space Center had attended a preview the day before. Built within 4 mo, after President Ford in Feb. called on NASA to create a Bicentennial fair at Cape Canaveral, the exposition occupied 30 acres and 15 geodesic domes clustered around the mammoth Vehicle Assembly Building, so large that the U.N. Secretariat could be mounted on wheels and rolled easily through its doors. "The message from NASA, 16 other federal agencies, and a conglomeration of defense contractors," said the *Washington Post,* was that America had spent her technology tax dollars well. The show was put together with $3 million in grants from the Dept. of Commerce, a $500 000 loan from NASA's budget, and an $800 000 advance from the KSC visitor center account. Before it opened, the show was expected to draw 800 000 to 2 million persons; in July, average daily attendance was about 5000, and expectations were reduced to 500 000 before the fair closed 7 Sept.

Many Fla. attractions reported a drop in attendance this summer. (*W Post*, 25 July 76, G-5)

During May: Rockwell International Corp. awarded contracts for the Air Force B-1 strategic bomber program. On 18 May, Rockwell awarded a $2.4-million contract to Hamilton-Standard Division, United Technologies Corp., for air conditioning and pressurization equipment and air-recirculation loops for the fourth B-1 in the 4-plane prototype program. The firm had supplied similar equipment for 3 earlier B-1s. On 24 May, Rockwell awarded a $1.8-million subcontract to Martin Marietta Aerospace for tail assemblies, consisting of 2 horizontal stabilizers and a single vertical stabilizer, for the fourth B-1 prototype aircraft. Martin Marietta had supplied stabilizers for the other 3 B-1s. Rockwell, system contractor on the B-1 program, had developed the B-1 to modernize the strategic bomber force; its advantage over previous heavy bombers was the design ability to avoid enemy defense by flying at nearly sonic speed at treetop height to avoid radar detection. The first 2 prototypes were in flight test, and the third was scheduled to fly later in 1976. (Rockwell Releases LA-3, LA-4)

• *Spaceflight*, publication of the British Interplanetary Society, announced that Soviet authorities after a 2-yr delay had released a photograph of Capella—the brightest star in the constellation Auriga—obtained by the "Orion 2" observatory aboard *Soyuz 13*. Capella is a yellow giant star 150 times as luminous as the sun and 47 light-years away. Soviet scientists believed they had found a "new association of stars" in the several dozen hot stars of extremely low radiation discovered around Capella; they claimed this was the first time in the history of astronomy that a telescope had obtained a spectrogram of a planetary nebula—a huge gaseous formation with a high-temperature star at the center—and had obtained pictures of a known star with a gaseous envelope rich in

Third Century America, sole Bicentennial exposition paid for by the U.S. government, opened to the public 30 May 1976 at Kennedy Space Center. (NASA 76-H-445)

silicon. The Orion 2 equipment could make precision study of stellar objects in the ultraviolet spectrum and provide spectrograms of stars of 13th magnitude "to extend the limits set by the American Skylab, whose crews took pictures of stars down to 7.5th magnitude," the Russians said. (*Spaceflight*, May 1976, 177)

- The National Aeronautic Association reported that Karl Striedieck and L. Roy McMaster were jointly claiming a world soaring record for out-and-return distance of 1299 km, for flights they each made on 17 March along the Allegheny Mountains. Striedieck, an Air National Guard pilot from Port Matilda, Pa., flew in a Schleicher AS–W17 sailplane from Lock Haven, Pa., to Mendota, Va., and back, on course for 11 hr averaging more than 117 kph, claiming by this flight the world out-and-return record for the fifth time. McMaster, an accountant from Elmira Heights, N.Y., flew the same course in a Schempp–Hirth Standard Cirrus sailplane, landing about an hour later than Striedieck. If approved by NAA and the Federation Aeronautique Internationale, the world-record claims would give McMaster a U.S. National Standard Class record for out-and-return distance; Striedieck's AS–W17 would not qualify under standard-class rules. (NAA Newsletter, May 1976, 3)

June 1976

1 June: Recently published criticism by columnist John Keats of "a modest NASA effort to provide meals for elderly people" not only "managed to misunderstand just about every aspect of what we are doing" but also "put down the elderly and [raised] my ire in the process," astronaut Joseph P. Kerwin wrote in a letter to the *New York Times.* Kerwin pointed out that the NASA program was a response to the Texas Governor's Committee on Aging, which had asked the agency for help in developing good-tasting easy-to-prepare and easy-to-deliver meals for people not reached by programs such as Meals on Wheels or group meals sponsored by Congress in various city centers. Kerwin said NASA's engineers "know a little about packaging and shelf life," and "we care because we have relatives who are old, and because we'll be old ourselves soon—if we're lucky." The technology being used to do a job for the old was good, Kerwin concluded, "but it's the caring of which I am most proud." (*NYT,* 1 June 76, 34)

- During its 3 most active yr, NASA had awarded only 19% of its procurement funding to concerns in the industrialized Northeast, according to a Library of Congress study commissioned by Rep. Michael J. Harrington (D-Mass.), whereas the Sunbelt states—ranging from Maryland to Texas—received 33% of the total awards. Of $8.7 billion spent by NASA in procurement contracts for 1968, 1971, and 1975, the Northeast received $1.7 billion and southern regions received $2.9 billion, the report said. "While the aggregate populations in these combined regions are roughly comparable," Harrington said, "the NASA procurement contracts run almost 2 to 1 against us . . . NASA's contract award procedures now join that long and growing list of federally funded activities which discriminate against the industrialized Northeast states." (*NYT,* 2 June 76, 40)

- NASA announced an agreement with the Indian Space Research Organization (ISRO) to add a solar-energy experiment to the cooperative satellite instructional television experiment (SITE) now under way using *Ats 6,* the applications technology satellite. In May, ISRO was sent 2 solar arrays capable of producing 260 watt-hr of power a day under Indian sunlight conditions; the arrays would provide electricity to run 2 of the SITE TV receivers during the 4 hr each day that Indian programming would be broadcast to some 5000 villages using *Ats 6.* When India's National Committee for Space Research studied power alternatives for the SITE terminals in 1969, solar cells were used almost exclusively in spacecraft and were considered too expensive for use on the ground; since that time, demand had increased sharply, the price had come down, and the sharp rise in petroleum prices had made the price of solar power more attractive. Although current cost of a solar-power system for a SITE terminal was estimated as $311 per yr (26% higher than the $247 cost of kerosene generators), research and production should bring the cost down to about $139 a yr by 1979, while the price of kerosene-generated electricity could well be higher. Also, arrays from solar cells could be produced in India and other developing countries by using existing technology, which would add to the advantages of solar power for use throughout the world.

Solar power had proved desirable for use in areas not having central power stations; the solar-power systems were simple to install, had no moving parts, and needed only cleaning of array surfaces and maintenance of electrolyte levels as operational maintenance. NASA's Lewis Research Center had demonstrated a solar array as a power source for an *Ats 6* ground station last year; the India project would be one of many conducted by LeRC to support the U.S. Energy Research and Development Administration (ERDA) in demonstrating terrestrial applications of solar-cell-generated electricity. An ISRO engineer would visit LeRC for training in assembly and operation of the solar arrays. (NASA Release 76-95)

2 June: NASA announced selection of IBM Corp., Gaithersburg, Md., for award of a $24-million 44-mo cost-plus-award-fee contract to supply a Space Shuttle data-processing complex for the mission control center at Johnson Space Center. The complex would consist of 3 computers and peripheral equipment for support of the Shuttle program; work to be performed would include design, fabrication, delivery, installation, and checkout of the computer complex and associated software. Control Data Corp. of Minneapolis also submitted a proposal. (NASA Release 76-104; JSC Release 76-37)

- The USSR would probably put men on the moon within 10 yr to do more ambitious exploration than that done by U.S. astronauts, according to a report on Soviet space programs 1971-1975 by the Science Policy Research Division of the Library of Congress's Congressional Research Division. Dr. Charles S. Sheldon, II, chief of the Science Policy Research Division, stated in the report that the USSR did not abandon lunar plans when *Apollo 11* got to the moon in July 1969, but that the Soviet program had been plagued with "hardware and systems . . . quite inadequate by our standards." In a summary of Dr. Sheldon's report, the *Washington Star* noted his statement that the "race for the moon" was closer than many believed at the time; "the Soviets probably wanted to send the first manned flight around the moon by November 1967, when a test failed." Although a Soviet Apollo would probably not appear for the next 3 yr, Dr. Sheldon said, "within the decade there will probably be a Soviet landing on the moon that will be a generation beyond the Apollo flights . . . ." Describing the extensive Soviet military uses of space in addition to scientific and economic purposes, the report noted "the seriousness and steadiness with which the Russians are adding to their space facilities and their space operations, building versatility and experience in depth." The report did not mention the absence of this element in the U.S., the *Star* noted. The report would be published later in 1976 by the Senate Committee on Aeronautical and Space Sciences, for which it was prepared. (*W Star*, 2 June 76, A-3)

3 June: NASA announced plans to roll out the first of 2 rotor systems research aircraft (RSRA) 7 June at the Sikorsky Aircraft Division in Conn. The RSRA, capable of conventional cruise flight as well as typical helicopter performance, was the product of a joint NASA-Army research program to reduce rotor noise and vibration—with attendant high maintenance costs—and improve performance and cruise speed of future civil and military helicopters. The RSRA was the first helicopter designed with research capability in mind; devices in the main rotor-support structure would permit accurate measurement of rotor forces in flight, and force-measurement systems were also incorporated in the wing, tail rotor, and auxiliary engine to permit measurement and control of the lift and drag of the rotor system. The Army, world's leading user of helicopters, would make the RSRA available to the U.S. helicopter industry for rotor-systems tests; most of the technology would also be applicable to civil vehicles. NASA's goal would be to meet predictions of increased usage in the civil helicopter industry to enable the U.S. to compete favorably in the growing world market. (NASA Release 76-105)

4 June: Kennedy Space Center announced award of a $40 214 contract extension of 12 mo to Ky. State Univ. for continuing its studies of the effects of oxygen atmospheres on animals. The animal under study was the vinegar fly—Drosophila melanogaster—which was grown and tested under 5%, 20%, and 60% oxygen concentrations (normal oxygen content of earth's atmosphere is 21%). Studies over 2 yr showed that, at the 5% and 20% oxygen levels, survival rates of the initial generation of the flies were the same. Flies in a 60% oxygen environment had a survival rate like that of the flies in the 20% environment, for up to 20 days; after 20 days, flies in the 60% atmosphere died. Tissue studies indicated that exposure to a 60% oxygen environment resulted in physical changes, including accelerated aging and problems in the nervous system. The contract extension would permit studies of the mortality, fertility, and gene frequencies of the flies over 10 or more generations. The data would aid in establishing proper oxygen content in manned spacecraft atmospheres during future manned missions, and in selecting humans to participate in manned missions and in deep-sea dives. (KSC Release 198-76)

- A fully metallic replica of the Spacelab—called a "hard" mockup—had begun assembly near Bremen, West Germany, at the plant of ERNO, a subsidiary of VFW-Fokker and prime contractor for the Spacelab. The role of the replica was to ensure that all nonelectrical elements of the complex craft would be compatible; 4 additional Spacelabs would be constructed, including the flight unit, only one of which would be constructed. The next model would be the "high fidelity" model, to be built after final design details are agreed on later in 1976; it would be used for crew training. The next—Engineering Mockup 1—would be sent to the U.S. in 1978 for fitting into the Shuttle. The flight unit would be scheduled for shipment to Kennedy Space Center a year later; finally, Engineering Mockup 2 would be built and kept in Europe to help members of the European Space Agency (ESA) prepare missions over the 10-yr operational period. The single flight version, designed to make at least 50 flights, would carry a wide range of scientific and technological experiments for which the experimenters would be expected to pay a fee. The Europeans hoped to modify this policy in view of their nearly $400 million investment in the project. (*NYT*, 4 June 76, A-15)

7 June: NASA announced selection of Computer Sciences Corp. of Falls Church, Va., for negotiation of a 2-yr $6 million cost-plus-award-fee contract to provide technical support services at the National Space Technology Laboratories, Bay St. Louis, Miss. The contractor would provide support services for NASA's on-site launch-vehicle and rocket-engine static testing and certification programs, and for NASA and other resident agencies working in space and earth-environment programs. (NASA Release 76-111)

8 June: The Communications Research Centre (CRC) of the Canadian government's Communications Department in Ottawa announced successful demonstration of a new satellite-aided search-and-rescue concept that could reduce time, fuel-dollar, and other costs associated with conventional ways of finding downed aircraft. The demonstration was funded by the Canadian Department of National Defence. The project, begun in May 1975, used the Radio Amateur Satellite Corporation's *Oscar 6* satellite with simulated distress signals to show that crash sites in Canada—and elsewhere in the world—could be pinpointed with accuracies as good as 1.6 km, generally within 3 km, in as little as 15 to 20 min after the spacecraft first heard the signal. The conventional emergency locator transmitter (ELT) mandatory for aircraft in Canada and the U.S. would automatically signal on crash impact, as designed, providing a signal on the international distress frequency of 121.5 hz for at least 100 hr for search-and-rescue aircraft to home into. The search range until now had been within 50 km of a crash site, flying crisscross patterns involving many planes and dozens of costly as well as risky flying hours.

The satellite system would depend on two things: highly precise orbit predictions (knowledge of the satellite's exact position at any instant) and sophisticated computer processing of distress signals relayed to a ground station. The system works by measuring the Doppler shift in the ELT signal frequency as the satellite at about 1100 km altitude passes over the crash site; locations of about 60 "crashes" simulated by transmitters as far away as Winnipeg had been fixed by processing *Oscar 6* signals with increasing degrees of accuracy. An operational system would include 3 satellites with a design lifetime of 7 to 10 yr, with total spacecraft and launch costs about $30 million. When the satellite nearest to a crash site appeared over the visible horizon, it would alert ground stations that it had received an alarm; 15 min later, at the conclusion of its pass, an immediate fix accurate within 112 km would be possible, and a position fixing the site within 3 km would be delivered within 2 to 15 min depending on the capacity of the computer. A $3 million annual cost over 10 yr would be only a small part of what Canada was now spending on aerial searches. (CDC release 8 June 76)

- Marshall Space Flight Center announced it had used aerial thermal-scanning techniques to locate high heat-loss areas as part of its energy resources management program to reduce energy consumption agency-wide. The Lewis Research Center had made thermal scanning flights for all NASA centers, and MSFC had been scanned twice, once in October 1975 and again in February 1976, using a C-47 aircraft with onboard scanners that recorded on digital tape the average temperature of each area covered. Fed into a computer, the digital tape produced printouts and live mosaic maps projected on a television monitor to show areas where excessive steam-line losses occurred and to identify buildings losing excessive heat through roof structures. The MSFC facilities office validated the data through visual inspection of steam lines and roof insulation, and issued work requests for repair and replacement of insulation in 88 steam-line locations. Longer range action was planned to repair and replace deteriorated roofing and replace sections of the steam lines. (MSFC Release 76-104)

9 June: NASA launched *Marisat 2*, second in a series of ComSat maritime communications satellites, at 8:09 pm EDT from the Eastern Test Range on a Delta vehicle, into a transfer orbit with apogee of 36 924 km, perigee of 185 km, and 26° inclination. The apogee boost motor fired 11 June would put the spacecraft into a 36 000-km-altitude synchronous orbit at 176°E over the Pacific. Marisat was designed to transmit voice, data, facsimile, and telex messages to and from ships at sea, through special stations in Conn. and Calif. interconnected with existing domestic land networks; initially, the system would be used by the U.S. Navy, until its requirements terminated in the late 1970s. The cylindrical satellite, weighing 655 kg at liftoff and 317 kg in orbit, measured 2.1 m in diameter and was 3.6 m long. Like *Marisat 1*, which had been operating over the Atlantic since its launch 19 February, it carried 3 UHF channels for government use, activated or deactivated by ground command, and 2 4-hz channels operating in the L- and C-bands respectively to carry ship-to-shore and shore-to-ship signals; the full ship-to-shore capacity would always be available independent of government use of the UHF channels, which would use separate receiving facilities. (NASA Release 76-83; MOR 492-205-76-02 [prelaunch] 7 June 76, [postlaunch] 14 Oct 76)

- Johnson Space Center announced "the most successful balloon flight of its type ever conducted" when a football-field-sized balloon carried a scientific instruments package at a 40-km altitude across the central Texas night sky for 12 hr to gather information on far distant giant and super-giant stars. The 589-kg package known as BUSS—for balloon-borne ultraviolet stellar spectrometer—gathered data on 16 separate stars, including Arcturus, a giant star in constellation Bootes; super-giant Antares (alpha Scorpii); Vega (alpha Lyrae); and Spica (alpha Virginis), super-hot star and one of the brightest observed during the flight. Of special interest was super-giant Deneb in constellation Cygnus, 1400 light years distant, whose light recorded by BUSS was emitted before the Anglo-Saxons settled in England; Deneb was so large that, if the sun were located in its center, the earth would orbit the sun entirely within the star's outer limits. The BUSS payload was the culmination of a 3-yr international collaboration between JSC and the Space Research Laboratory of Utrecht in the Netherlands. Dr. Yoji Kondo of JSC was U.S. coprincipal investigator, with Dr. Thomas H. Morgan and Dr. Jerry L. Modisette of Houston Baptist University; the Dutch team was led by Dr. Cees de Jager. Purpose of the balloon flights in a series of star studies that began in 1971 was to evaluate the experiment systems for use on the Space Shuttle; the BUSS package was designed to obtain data on spectral variations of a variety of stars and aid scientists in determining their structure and evolution. Previous flights had obtained some information but only a fiftieth of the spectral range covered by the BUSS payload. The package was landed by a parachute about 26 m in diameter, similar to those used on the Apollo command module; the payload, chute, and balloon were recovered between Abilene and Ft. Worth by NASA engineers who returned the material to Houston. The data would be analyzed by the U.S. team and by the Dutch scientists, who had returned to Holland, in preparation for another balloon flight scheduled for the fall of 1976. (JSC Release 76-39)

- Marshall Space Flight Center announced selection of 11 industrial firms to negotiate for award of contracts totaling $200 000 for marketable solar-heating and cooling subsystems—solar collectors and control systems—to be tested and evaluated for inclusion in complete solar-heating systems that would be installed in residences and commercial buildings throughout the U.S. Data gathered from these installations would be used in a national solar-energy development program administered by the Energy Research and Development Administration (ERDA), for which MSFC is managing the heating and cooling demonstration. Of the 11 companies selected, 8 were small businesses. The companies were: Northrup, Inc., Hutchins, Tex.; Rho Sigma, Van Nuys, Calif.; Solar Energy Products Co., Avon Lake, O.; Solar Energy Systems, Inc., Pennsauken, N.J.; Solargenics, Inc., Chatsworth, Calif.; Solaron Corp., Commerce City, Colo.; Sunworks, Inc., Guilford, Conn.; Ying Manufacturing Corp., Gardena, Calif.; Honeywell, Inc., Minneapolis, Minn.; Libbey-Owens-Ford Co., Toledo, O.; and Martin Marietta Corp., Denver, Colo. (NASA Release 76-109; MSFC Release 76-106)

- Lewis Research Center announced the first of a number of solar-cell demonstrations it would conduct as part of the national photovoltaic conversion program directed by the Energy Research and Development Administration: a solar-powered refrigerator, standard-size camper's model, fitted with 3 panels of photovoltaic cells that convert sunlight into electrical energy. Designed for use in remote locations, the solar cells would run the refrigerator during daylight hours and charge conventional automobile batteries located under the unit; the batteries would power the refrigerator during darkness and on overcast days. With the cooperation of the Interior Department's National Park Service, the refrigerator was being used for perishable foods at a trail-construction camp at Isle Royale National Park in Michigan, a roadless area near northern

Lake Superior accessible only by boat or floatplane and typical of places that have no regular electric power. The solar-cell arrays, although relatively expensive, cost less to use in remote locations than the fuel and transportation would cost for alternative power sources. (NASA Release 76-110)

- Dr. James C. Fletcher, NASA Administrator, named the Ames Research Center to lead a program to strengthen helicopter research and development and serve as a focal point for industry participation and program management. A special helicopter management advisory group had made a presentation to Fletcher 28 May reviewing research needs and stressing the need for improvements in helicopters if the U.S. industry were to get a fair share of the market. Overall direction of the program would come from ARC; Lewis Research Center would conduct research on propulsion, and Langley Research Center would do research in structures and materials, avionics, and noise. ARC would also conduct research in aeromechanics, including technology integration and large-scale testing and simulation. (NASA Release 76-112)

- Successful completion of the *Da Vinci II* scientific balloon flight by 4 crew members was announced by the Energy Research and Development Administration. Launched west of St. Louis on 8 June, the balloon had flown over the city during the day and moved eastward across the Mississippi River toward evening; it had flown across Ill. and landed near Griffin, Ind., a distance of more than 240 km in 24 hr. Purpose of the flight—a joint project of ERDA, the National Oceanic and Atmospheric Administration, and the Environmental Protection Agency—was to follow industrial and urban air pollution across the surrounding countryside and to record its changing concentration and chemistry. The crew reported excellent scientific data, especially on concentrations of sulfur dioxide. The crew, its 3m-square gondola, and more than 900 kg of scientific equipment landed safely and in good condition. (ERDA Release 76-170)

- An article in Moscow's *Krasnaya Zvezda* reviewed the increasing use of satellites for military communications, especially in the U.S. Space communications had become available in Britain, France, the Federal Republic of Germany, Canada, Brazil, and Norway, as well as other countries, and though basically designed for commercial purposes "there is no doubt" that most of the channels were used by the military, the article said. As an example, the national system of space communications in Iran was said to be developed to provide guidance for combat aircraft. Not satisfied with the potential of commercial systems, the Pentagon and the NATO command had established communications satellites exclusively for military purposes, the article said, quoting *U.S. News and World Report* as saying that since 1960 "the Pentagon has sent 83 communications satellites . . . into orbit but has not got what it wanted. Of all these satellites only 13 still continue in orbit, and not all 13 are functioning normally." The Soviet magazine also pointed out that 6 launches of DSCS-2 satellites had been planned, but 2 had been lost during launch and 2 more had broken down in orbit; the 2 currently operating were over the Pacific and the Atlantic at about 36 000 km, and another 6 had been ordered, the first pair due for launch in spring 1977. As press statements showed that two thirds of the military communications with foreign countries were going through commercial systems, "foreign observers" had concluded that the military systems had proven less reliable than Intelsat; the Pentagon had explained that requirements on military satellites were considerably higher, as they had to work with small mobile stations in widely scattered locations, needed increased protection against jamming, and must be durable to withstand possible combat effects. The Russian commentator said the U.S. had begun developing a third generation of military satellites—the DSCS-3—with launch scheduled for 1981, and that these new spacecraft would be powered by radioisotope thermoelectric generators with longer working life and greater capacity. Elimination of bulky solar panels would also make the new craft more difficult to detect in orbit. After describing the Fleetsatcom and Afsatcom systems, the article pointed out that "the creation of a manned multipurpose transport spacecraft" [the U.S. Space Shuttle] would have a major impact on communications-satellite systems; such spacecraft would be able to launch large mass and capacity comsats, eliminating "complex operations and devices [needed to prepare] the satellite for work following its injection into orbit," and would simplify technical servicing and repair of the comsats while in orbit. Although many technical problems remained, space facilities had become the basis of military communications abroad, the article concluded. (FBIS, *Krasnaya Zvezda*, 26 May 76, 3)

10 June: Johnson Space Center announced delivery of the first of 2 training aircraft to Ellington AFB, Tex., for JSC's use in training Space Shuttle crews. The aircraft (STA) was a modified Grumman twin-engine Gulfstream II jet designed to simulate flight characteristics of the Space Shuttle orbiter, using thrust-reverser engines and direct lift control to vary the jet plane's dynamics to resemble those of the orbiter. After JSC personnel conducted a receiving inspection on the plane, it would be returned to Grumman at Bethpage, N.Y., for installation of an electric aileron-trim system; this modification should take about a week. The second STA, scheduled for delivery in late July, would remain at Bethpage to ensure that the plane could duplicate the various landing flight modes of the orbiter during the flight tests. Remaining tests would verify STA's ability to match orbiter trajectory during the period from 10–11 km altitude through touchdown and would check out recent engineering changes. (NASA Release 76-108; JSC Release 76-38)

11 June: A 2-wk experiment to observe the Gulf Stream about 580 km east of NASA's Wallops Flight Center used a combination of spacecraft, aircraft, and water craft to check the ability of remote sensors to measure the magnitude and boundaries of the Gulf Stream from space. Gulf Stream measurements would aid in planning more effective use of coastal waters, as in cooling nuclear power plants, or in constructing offshore drilling rigs or airports located on manmade islands. Scientists also wanted more data about the Gulf Stream, which carries heat energy that influences the global balance of energy in the atmosphere and the ocean, affecting weather and climate as well as coastal water movement, besides carrying nutrients important to the fishing industry. Traditional methods using closely spaced ship stations or buoys to obtain data had proved inadequate for studying fast-changing or large-scale phenomena. Satellite measurements used in the experiment included infrared signatures that identified the Gulf Stream boundary from space photographs. A precision altimeter on NASA's *Geos 3* (geodynamic experimental ocean satellite) measured surface deviations in the ocean within 20 cm, from which velocity of a current could be calculated; the experiment also used the National Oceanic and Atmospheric Administration's *Noaa 4* weather satellite. A C-54 research plane from Wallops carried instruments on 3 flights to measure the current by observing wave interactions; surface measurements were taken by the research vessel *Advance II* from Cape Fear Technical Institute at Wilmington, N.C., manned by ocean scientists from N.C. State University at Raleigh. The project was part of a scientific research task directed by NASA Hq's Office of Applications. (NASA Release 76-115; WFC Release 76-7)

- Detente, to be meaningful to U.S. scientists, must include progress toward more intercommunication and openness, said Dr. Bruce Murray of Cal-Tech and Merton E. Davies of RAND Corp. in an article in *Science* magazine. Noting that for nearly 2 decades the space program of the U.S. and the USSR were bound together through rivalry, competition, and—most recently—cooperation, the authors found it appropriate upon completion of the Apollo-Soyuz mission to review relationships between the 2 countries and focus on areas of possible common interest. The interrelationship with the Soviet space effort had stimulated both societies and had influenced the character of individual programs, the authors said. Future joint activities could take three forms: data exchange, cooperative experiments, and joint operations. Examples cited were exchanges of weather-satellite pictures, lunar soil and rock samples, and limited data on Mars obtained during simultaneous missions in 1971; the U.S. biological experiments carried on *Cosmos 782*, launched and returned to earth in 1975; and the joint operation of Apollo-Soyuz, which demonstrated the practicality of such missions despite differences in language, institutions, technology, and style. Joint scientific progress in space would depend on broader social objectives and activity, supported by popular enthusiasm for intellectual and generally human adventure, the authors said. (*Science*, 11 June 76, 1067)
- The board of governors of the International Telecommunications Satellite Organization (INTELSAT) announced that, at its recent meeting in The Hague 19—27 May, it adopted a 10-m class Standard B ground station to supplement the Standard A station long the foundation of INTELSAT's global system. Currently, 114 Standard A ground stations with 30-m antennas were operating throughout the world, as well as 29 smaller stations used for domestic service, telemetry, tracking and control, monitoring, and limited telecommunications between 2 or 3 destinations. Broadest use of the Standard B station would be in developing countries

that had limited telecommunications requirements. Under study for some years, the Standard B station would use single-channel-per-carrier (SCPC) equipment that allowed routing of traffic on a circuit-by-circuit basis instead of in large groups or bundles, minimizing the capacity loss associated with smaller antennas; it also worked with voice activation, which would be mandatory, using satellite capacity for telephone service only when activated by voice signals. Adoption of the second standard for international service would mean a 40% drop in rates paid for use of the satellites, plus ability to route traffic to any ground stations equipped to receive the user's signals, with no restriction on the total amount of traffic. Previously, the drain on satellite power limited small ground stations to providing service to only 1 or 2 other points. (INTELSAT Release 76-17-I)

11—30 June: The scheduled Fourth of July landing of *Viking 1* on the planet Mars was briefly complicated by 2 problems announced in the news media 11 June: a mechanical problem with a leaky valve in the Viking's fuel tank, and a threatened strike by about 200 technicians at a desert tracking station at Goldstone, Calif., one of 3 in the world that relayed signals to and from the Viking. Neither proved insuperable; by 18 June, John D. Goodlette, chief Viking engineer, was describing the mission as "shooting right down the pickle barrel." "If we do nothing to the spacecraft from here on out," he added, "we'd miss our target by less than 60 miles [95 km]." *Washington Post* staff writer Thomas O'Toole explained that the target was a spot about 9600 km from Mars, where the onboard braking engine would slow the spacecraft from about 14 200 kph to about 9600 kph, putting it into an elliptical orbit to survey possible landing sites.

Viking 1 would scan and map the surface for 2 wk; its cameras would photograph the prime landing site near an ancient rift valley, and other instruments would measure surface temperatures and search the Mars atmosphere for signs of water. The new orbit would bring Viking over the planned landing site once each 24.6-hr Mars day (called a sol). The 2 wk of photographs and measurements were needed so that scientists at Jet Propulsion Laboratory, where the flight had been directed since launch in April 1975, could choose the safest and most interesting spot for their 4-ton bird to alight. The landing attempt would be the first U.S. try at putting a spacecraft down on another planet; the USSR had made 5 unmanned attempts, 3 on Venus and 2 on Mars. The 2 Venera spacecraft had survived on the surface long enough to send back 1 picture each; the third crashed. One Mars spacecraft had crashed, and the other was blown over after it landed by winds clocked at more than 300 kph.

After Viking's braking engine was fired, the main job of flight directors would be to navigate it into a path where its orbit would be lowered again, with apogee reduced from about 19 000 km to 12 500 km. The firing was successful 19 June, and the Viking neared the end of a 690-million-km voyage that began 10 mo ago. William J. O'Neill of JPL said the precision of the flight could be likened to "shooting a basketball in Los Angeles and putting it through a hoop in Madison Square Garden in New York City." Viking was following a northeasterly path around Mars that would bring it within 1450 km of the planet's surface; flight directors planned a second maneuver to bring it still closer over an area 20°N of the equator, called Chryse, where Viking would attempt to land 4 July, selected as a good place to search for signs of life or fossilized life that thrived when a river flowed there a billion years ago. Harold S. Masursky of the U.S. Geological Survey, chief geologist for Viking, said that Chryse resembled California's Death Valley or the water-worn dry valleys of Nevada or Wyoming, where ancient rivers overflowed their banks millions of years ago to deposit alluvial soils on the valley floors. Rock and soil in the Chryse valley appeared red from millions of years of iron oxide deposits brought down by the river overflow; Masursky and his colleagues hoped that the red soils would contain organic molecules, signs of a once-existent life on Mars.

A photograph taken 19 June and released by JPL on Sunday, 20 June, showed the "awesome" Valles Marineris from about 360 000 km up— parallel canyons south of the Mars equator that, if extended together on earth, would reach from Calif. to Penna. The smaller was about 3 km deep, 64 km wide, and 645 km long; the larger was 6 km deep, 96 km wide, and more than 3200 km long. (Arizona's Grand Canyon, largest on earth, is about 2 km deep, 21 km wide at its greatest breadth, and only 350 km long.) The northern hemisphere of Mars appeared covered with haze, which scientists said was probably from water vapor misting out of the cold surface as the sun warmed it during the longer summer days.

Successful firing of the braking engine at 1:25 pm EDT Monday, 21 June, put *Viking 1* into a lower orbit so that its cameras could begin to photograph steep slopes, craters, boulder fields, and other potential obstacles. On 27 June, Viking project manager James S. Martin announced that the 4 July landing was off; pictures of the primary landing area showed "too many unknowns," and the site appeared too hazardous to risk landing without investigating other sites. As the *Viking 1* cameras could not distinguish objects "smaller than a football field," objects as yet invisible could endanger the landing. However, as there was no assurance of any hazard-free area roughly 20 km by 100 km anywhere on Mars, some scientists had wanted to go ahead with the planned landing. Previous closeups taken by *Mariner 9* were not as detailed as those from Viking, and dangerous details of the primary landing site became more obvious with photos from *Viking 1* coming in more clearly.

The *New York Times* reported that the first alternative site was an area called Chryse Phoenicia, a basin about 30 km northwest of the original site; if that site appeared too rough, scientists would look at Tritonis Lacus, about 6400 km east of Chryse, although if this happened, the landing might be delayed until as late as the first wk of August. Project Manager Martin suggested delaying the *Viking 1* landing, if it were not down by 25 July, until the *Viking 2* (now 8 million km away) arrived in Mars orbit; possibly 2 Viking spacecraft might be orbiting simultaneously. JPL said flight tracking and control facilities there were insufficient to handle full operations of both spacecraft simultaneously, and a number of preorbital maneuvers for *Viking 2* would be required starting about 27 July. John Noble Wilford said in the *New York Times* that the earliest possible *Viking 1* landing would be 9 July, and other alternatives would take longer. Scientists, he said, were "dismayed by the rugged Martian terrain"—the primary landing area being heavily pocked with craters, steep escarpments, and a sprinkling of knoblike features probably remnants of erosion like that in the Grand Canyon.

On 28 June, *Viking 1* photographed a site about 300 km northwest of Chryse—the "northwest territory"—after pictures of the prime site showed it to be far more rugged than expected; the scientists wanted a landing as close as possible to the original site because they thought it offered the best chance of finding signs of existing or fossil life. The alternate site would be somewhat flatter and safer, but less interesting scientifically, O'Toole said in the *Post* 29 June. Photographs released 29 June at JPL showed a plateau in the Chryse valley originally picked as the *Viking 1* landing site, where a curving and rugged coastline at the plateau's base could have been formed only by an ocean-like surf. Pictures taken from an altitude of about 1450 km showed a crater with lava-flow traces more than 30 km wide, "a truly enormous scale" according to chief geologist Masursky. "The biggest lava flooding . . . on earth are the basalt floods on Iceland, which are no bigger than a few kilometers across."

Temperature changes mapped by Viking were also dramatic: instruments on Viking registered differences of more than 60°C (140°F) between day and night in the same parts of the planet. Dr. Hugh H. Kieffer, atmospheric scientist on the Viking team, said that the atmosphere on Mars was almost perfectly transparent to sunlight, as little water was present to absorb it; "whatever sunlight reaches the planet gets right to the surface, which then radiates the heat back to the atmosphere to warm it up." An instrument aboard Viking could tell when day broke on the Mars surface although it had no way of "seeing"; it measured water vapor in the atmosphere, and registered its presence when the sun warmed the permafrost that "literally pops [water] into the atmosphere when that sun hits the surface first thing in the morning," according to Dr. Barney Farmer of JPL. Besides surveying possibilities for its own landing, *Viking 1* was photographing potential landing sites for *Viking 2* to give scientists an idea whether they might have to shift sites for that spacecraft also. (*W Post*, 12 June 76, A-12; 19 June, A-3; 20 June, A-1; 21 June, A-9; 22 June, A-17; 28 June, A-3; 29 June, C-4; 30 June, A-27; *NYT*, 21 June, 10; 28 June, 17; *Av Wk*, 28 June 76, 14)

13 June: Photographs from U.S. remote-sensing satellites had revealed the existence of water, oil, uranium, and other minerals in the Egyptian Sahara and the Sinai peninsula, the *Washington Post* reported. Egyptian scientist Ahmed Abdel Hady, head of an Egyptian-American scientific team in charge of a remote-sensing project that had been receiving data from satellites since 1972, said in an AP interview that the satellites revealed enough water in the Sinai desert "to turn most of it green." (Most of Sinai was still occupied by the Israelis, who seized it in 1967.)

The Egyptian government, working in cooperation with Okla. State Univ. and the Univ. of Mich., had an annual budget of $1.2 million for study of the untapped resources, Abdel Hady said.

Although the satellite data had shown that highlands in northern Sinai and the coastal strip of Wadi el Arish, totaling more than 770 km^2, concealed a huge water potential, he declined to go into details because "I don't want to make it difficult for Egypt when it negotiates the next Israeli pullout." However, he described photographs of an area about 1350 km^2 in Sinai that showed three different areas rich in petroleum and natural gas: the Gulf of Suez, with oil reserves already proven; the Mediterranean offshore area in northern Sinai, not yet explored; and a large area in southern Sinai marked by sedimentary rocks carrying natural gas. Uranium was also detected in northwest and southern Sinai, and west central Sinai contained huge amounts of silica that could lead to a glass industry, Abdel Hady said. Besides the desert potential, he added, the images also showed significant food-growing potential in a previously unnoted fertile area measuring about 965 km^2 near the Nile basin; the images were also used to study the Qattara Depression in the desert west of the Nile, site of a huge proposed hydroelectric project. (*W Post*, 13 June 76, G-2)

14 June: Marshall Space Flight Center—lead NASA center for Spacelab development—announced that the full-scale model of the Spacelab module developed by the European Space Agency for orbital research and applications missions in the Shuttle orbiter had been shipped from Germany and was on display at the U.S. Bicentennial exposition on science and technology at Kennedy Space Center, open until 7 Sept. The Spacelab access tunnel and pallet section manufactured at Huntsville for MSFC had been shipped to KSC for assembly with the module. The pressurized module would be used by U.S. and European scientists working in space in a shirtsleeve environment; the unpressurized pallet would contain unmanned equipment such as telescopes, antennas, and instruments requiring direct exposure to space; and the access tunnel would be used by the crew to transfer from the orbiter to the module and return. Spacelab was designed to be fully reusable for an operational lifetime of 50 missions or 5 yr, whichever is reached first; nominal mission duration would be 7 days, but could be extended as long as 30 days. Spacelab was managed jointly by ESA, NASA Hq, and MSFC. (MSFC Release 76-110)

- The Air Force announced that its third prototype B-1 bomber successfully completed its maiden flight, having left Air Force Plant 42 at Palmdale, Calif., at 3:37 pm PDT and landed at nearby Edwards AFB after a flight of 2 hr 9 min. Since its rollout 11 May, the prototype built for USAF by Rockwell International's B-1 Division had undergone fuel and propulsion-system checkout, subsystem tests, low- and high-speed taxi tests, and a complete review of flight readiness. The B-1 was the first large swing-wing aircraft to complete a complex 8-mo series of structural integrity testing so early in its design life. The third B-1's primary flight-test objective was to acquire data to verify predicted design loads in test missions at both subsonic and supersonic speeds. Highest altitude reached during the maiden flight was about 3 km, with top speed at .45 mach. Crew members were Doug Benefield of Rockwell, pilot; AF Lt. Col. Ed McDowell, copilot; and Rockwell flight-test engineer Jack Baldwin, who was riding in the B-1 for the first time. (USAF release 14 June 76)

15 June: Secretary of Transportation William T. Coleman, Jr., proposed a $1 billion program to muffle noisy jet engines and replace older planes with quieter new ones, the *Washington Post* reported. The proposed program, part of a comprehensive aviation-noise policy submitted to the Office of Management and Budget, would be paid for by a surcharge on airline tickets—probably 2%—that would go into an escrow fund to be used by the nation's airlines over a 6- to 8-yr period. The "retrofit" program—most controversial part of the noise-abatement policy—would consist of modifying older 2- and 3-engine jets (such as the McDonnell-Douglas DC-9 and Boeing 727) by surrounding them with sound-absorbent materials to make them quieter. The fund could be used as part of the cost of replacing older 4-engine jets (the Boeing 707 and the McDonnell-Douglas DC-8) with newer, quieter planes. Although the Air Transport Association (ATA) representing most U.S. airlines had long favored replacement of the 4-engine planes, the *Post* said, it had vigorously opposed retrofit of 3-engine planes because the cost would outweigh the benefits. Retrofit had been advocated by the Environmental Protection Agency, the Federal Aviation Administration, and many citizens' groups to reduce noise levels around airports and take advantage of the useful life left in many of the older planes; however, the *Post* noted that opposition to retrofit in the administration had been based on expense, its possible contribution to inflation, and the fact that newer quieter planes would eventually solve the problem anyway. The *Post* quoted "no better than 50-50" chances that Congress would go along with providing public financing to help privately owned airlines make their planes quieter, according to a Capitol Hill source. Under Secretary Coleman's proposal, the 8% surcharge on airfares now going into a trust fund for improving terminal and navigational facilities would be reduced by 2%, which would go into the retrofit fund, so that no actual increase in ticket prices would occur. (*W Post*, 15 June 76, A-3)

- Marshall Space Flight Center announced development of a solar-simulator test facility, believed to be the largest of its kind in existence, to compare the efficiency of solar-energy collectors in MSFC's program of assisting the Energy Research and Development Administration in its solar energy program. Designed to provide variable levels of energy similar to ground-level solar radiation, the simulator contained a lamp array with 405 tungsten halogen 300-watt lamps paired with the same number of Fresnel lenses to provide an illuminated area for solar collectors up to 1.2 by 2.4 meters. After a 6-wk checkout of the lamp array, power equipment, and solar-collector fluid-flow systems, the simulator would serve to test solar-energy collectors that used either air or liquid as the heat-transfer medium. (MSFC Release 76-111)
- The Energy Research and Development Administration announced selection of 2 firms, one in Mass. and the other in Calif., to study potential

wind-power use by utilities to generate electricity in their own localities. The JBF Scientific Corp., a small research company in Wilmington, Mass., would study potential wind-power use by the Cambridge Light Co. and the New Bedford Gas and Edison Light Co. in New England; the utilities would provide information about their power needs and other data. The Aerospace Corp. of El Segundo, Calif., would assess potential wind-power use by the Hawaiian Electric Co., which serves most of that state; the Univ. of Hawaii would also participate in the study. Besides examining the utilities and their power needs, the contractors would study other ways of generating electricity (such as coal and nuclear plants), comparing them with wind energy to see if conversion would be practical and economical. The regional studies would last 1 yr each and would cost about $500 000 overall. The resulting data would supply a model for utilities throughout the country to assess the use of wind energy in their own regions. (ERDA Release 76-180)

16 June: Western Union announced that it had signed a 7-yr contract with the Corporation for Public Broadcasting (CPB) for transmission services via the Westar satellite. Duration of the contract was a possible 15 yr. Contingent upon approval by the Federal Communications Commission, Western Union would begin 31 December 1978 to provide CPB with service on 3 fulltime fully protected transponders for $800 000 per transponder per yr, or $2.4 million annually, permitting CPB to offer national distribution of programs to 165 public-television stations through its manager, the Public Broadcasting Service (PBS). The contract would give CPB an option on a fourth Westar transponder and establish rates for both scheduled and unscheduled service, and anticipated satellite distribution of radio programs to about 170 public radio stations through National Public Radio (NPR). Use of a 3- or 4-channel satellite system would serve many of the Congress's policy goals for public broadcasting, CPB officials said: autonomy of local public stations, increased flexibility in programming, and lower cost of expanding public broadcasting coverage. (Western Union release, 16 June 76)

- Dr. John L. McLucas, head of the Federal Aviation Administration, predicted that the U.S. would produce a second-generation supersonic transport, probably in cooperation with the Europeans, as a followup to the Concorde program. Speaking at a luncheon of the Air Force Association and in a subsequent interview with the *New York Times*, Dr. McLucas said that Congress probably would enact legislation to sharply reduce aircraft noise over the next 6 to 10 yr by modifying and replacing present aircraft. Biggest question would be how much financial help the airlines would get to carry out the plan: McLucas predicted that the airlines would have to settle for loan guarantees rather than outright subsidy. Calling the aviation industry "an endangered species," he said it was threatened by low profits, lack of capital for new equipment, and absence of a national aviation policy; he asked for more cooperation between industry and government, and exploration of joint projects with European partners. "A joint effort to advance . . . the design of the next-generation SST would be in keeping with this new spirit of international cooperation," he added. Dr. McLucas was secretary of the Air Force until President Ford transferred him to head the FAA. (*NYT*, 17 June 76, 49)

17 June: NASA studies in the year past had shown that 5 of 6 types of products considered for space processing could be expected to become profitable, Dr. William R. Lucas, director of Marshall Space Flight Center, told the subcommittee on aerospace technology and national needs of the Senate's Committee on Aeronautical and Space Sciences. NASA's space-processing program had deliberately pursued a policy of performing experiments on available missions and sponsoring formal experiments by members of the scientific community, Lucas said; the experiment program had been "reasonably representative" of the scientific potential that scientists foresaw for materials processing in space. Although available mission capabilities had severely restricted what could be done, the program had included demonstrations of crystal growth, solidification of homogeneous mixtures of immersible materials, processing biological materials for medical application, and other phenomena relating to physical metallurgy and fluid physics. Potentials identified with glasses and ceramics had not yet undergone experimentation.

Citing the success of experiments in electrophoretic separation and crystal growth on the Skylab and Apollo-Soyuz missions, Lucas emphasized the necessity of developing novel techniques for virtually every experiment; experience so far had justified the early interest in materials processing in space, although many lines of investigation had not been completed because of the program's operational constraints. Wide response to the program both in the U.S. and abroad was evidence of the scientific community's interest, he said. Next steps in development would be to finalize payloads for early Shuttle missions, then proceed with design and development; "We expect to proceed with research . . . both through the SPAR [rocket] project and through an expanded program of ground-based research," he added. "We shall need to reemphasize the ground-based program . . . to replenish the fund of new ideas in space processing and build a base of technical support with the breadth that Shuttle experiment operations will need." (MSFC Release 76-112)

18 June: NASA launched the Gravitational Probe-A (GP-A) from Wallops Flight Center at 7:41 am EDT on a Scout vehicle into a trajectory with peak altitude of 889 km and total flight time of 6960 sec. The 102.5-kg probe was designed to measure the gravitational red shift predicted by the equivalence principle laid down by Einstein in 1907 as the cornerstone of the general theory of relativity: GP-A would directly determine the effect of gravitation on time by comparing the rate of a rocket-borne clock to an identical clock on earth. During the flight, the probe clock would be in a weaker gravitational field than the identical clock on earth; as the probe clock rose through increasingly weaker gravity to its maximum altitude, it would appear to run increasingly faster, and its rate would slow as it returned to the stronger gravity at lower altitudes. The experiment required not only extremely accurate telemetry throughout the probe's ascent and descent, but also an extremely accurate set of clocks; those used were atomic-hydrogen masers of extraordinary stability, according to Dr. R.F.C. Vessot of the Smithsonian Astrophysical Observatory, principal investigator. ("Maser" is an acronym for microwave amplification by stimulated emission of radiation.) The clocks were expected to provide measurement accuracy within 5 thousandths of 1% (5×10^{-5}) of the predicted effect. The experiment and its support systems performed normally, but Dr. Vessot said that 1 to 3 mo of data reduction would be needed to determine whether the scientific objective had been met. (NASA Release 76-106; MOR S-879-76-01 [prelaunch] 14 June 76, [postlaunch] 23 June 76; MSFC Release 76-113)

- A 2-mo study this summer at the Ames Research Center would bring together a specially chosen panel of a dozen scientists—"authorities on every aspect of space exploration"—to make an in-depth examination of problems anticipated in the colonization of space. Sponsored by NASA, the group would also consider whether satellite power stations for earth could be built by space colonists. In a report by James Barron for the *Washington Star*, Princeton physicist Gerard K. O'Neill detailed his ideas for a kilometer-plus-wide cylinder with living quarters for 10 000 colonists on its outer rim, the length to be determined by the number of colonists desired; O'Neill emphasized that the colony could be ready by 1990 using existing technology and materials mined from outer space. O'Neill last year took his theories to a NASA-sponsored study group at Stanford University that included 28 physicists, engineers, and social scientists; the panel recommended that the U.S. move toward the colonization of space, saying that it "would represent 'cashing in' on the scientific information returned to us by Apollo." (*W Star*, 18 June 76, A-1)

- The USSR *Venera 10* spacecraft, launched a year ago and in orbit around the planet Venus for the past 8 months, was continuing its research work although its companion, *Venera 9*, had ceased to function, a Tass broadcast announced. Venus was passing behind the sun "from the point of view of an observer on earth," and a radio beam sent to earth by *Venera 10* had passed only 1.5 million km from the sun's surface. Analysis of *Venera 10* signals had shown that streams of near-solar plasma were very heterogeneous and subject to rapid changes in time; also studied were the possibilities of receiving information and controlling spacecraft depending on conditions of radio beams passing near the sun. (FBIS, Tass in English, 18 June 76)

18-30 June: Johnson Space Center announced its plan to use a space-environment simulation chamber to dry records and documents damaged in Houston's flood 15 June. The first batch, scheduled to be dried 19 June, consisted of medical records from Methodist Hospital and valuable irreplaceable books from the Contemporary Arts Museum, as well as museum records. The material would be put inside the chamber on shelves that would be heated to 48°C, and the chamber would be pumped to a vacuum; the process would take from 48 to 72 hr. The technique was pioneered by McDonnell Douglas Aircraft Corp. for the USAF when the latter's records were water-damaged in a fire. A later announcement said the drying was successful and that JSC was in process of drying material from St. Joseph's Hospital and the Univ. of Houston's law library. (JSC Release 76-42; JSC release 30 June 76)

20 June: The USSR launched *Intercosmos 15* on 19 June, a Tass broadcast announced, as an example of "socialist countries' cooperation" in space research. The spacecraft orbital parameters were 521-km apogee, 487-km perigee, 74° inclination, and 94.6-min period. *Intercosmos 15* was designed to "test new systems and units," including a telemetry system developed and manufactured with the participation of specialists from Hungary, Poland, Czechoslovakia, and the German Democratic Republic, as well as the USSR. Satellite data would be transmitted to receiving stations in Hungary, Czechoslovakia, and the GDR. (FBIS, Tass in English, 20 June 76)

- An official Indian government press release broadcast from Bombay said that 2 Centaur rockets and 2 M-1008 rockets would be fired from the equatorial rocket-launching station at Thumba on 28 or 29 June, and 1 July respectively. (FBIS, Samachar News Agency in English, 20 June 76)

21 June: The U.S. Senate by unanimous vote of 88 ratified the multination convention on registration of objects launched into outer space. All 37 countries now members of the United Nations Committee on the Peaceful Uses of Outer Space—which drafted the convention—were expected to ratify it; France and Bulgaria had already ratified, and 24 nations including the USSR had signed it. Registration would be with the secretary general of the UN, reporting was mandatory, and the information to be provided "as soon as practicable" included the name of the launching state (or states), appropriate designator of the space object (or its registration number), date and territory or location of the launch, basic orbital parameters (apogee, perigee, orbital period, inclination), and general functions of the space object. The convention embraced objects launched into earth orbit or into space transit—lunar or deep-space probes—but did not require reporting on objects in brief transit through outer space, such as sounding rockets or ballistic-missile test vehicles. The convention would supersede the voluntary system that had been in operation since 1962; the U.S. State Department had told the Senate Foreign Relations Committee that, although nearly all countries launching objects into outer space had respected the voluntary system, there had never been agreement on the kinds of information to be supplied and the information conveyed to the secretary general had never been uniform.

The committee's only question regarding cooperation was the People's Republic of China, which had not reported on its outer space launchings; Senator Clifford P. Case (R-N.J.) had asked Secretary of State Henry A. Kissinger to find out if mainland China would register under the new convention. NASA had told the committee that the U.S. would not incur substantial costs from the convention's requirement that participating states with space tracking facilities help in identifying a space object, when requested by another participant on the grounds that the object to be identified was "hazardous or deleterious." NASA was already tracking nearly 5000 space objects, some as small as a dinner plate, and knew which nation owned most of them. (NASA Ofc Legis. Affairs, report for 21 June 76; *Av Wk,* 28 June 76, 63)

- President Ford announced his intention to nominate Alan M. Lovelace, NASA's Associate Administrator for Aeronautics and Space Technology, to succeed George M. Low as Deputy Administrator. Lovelace had occupied his position since September 1974. Before coming to NASA, he had been director of research and development for the Air Force Systems Command at Wright-Patterson AFB, O., 1967-1972; director of science and technology for Air Force R&D, assigned to Andrews AFB; and acting Deputy Assistant Secretary for Research and Development for the Air Force in October 1973. The nomination required Senate confirmation, but no opposition was expected; if the Senate did not act by the end of the week, before its 4 July recess, confirmation might be delayed until after 19 July. (Pres Doc, 21 June 76, 1072; *Av Wk,* 28 June 76, 21; NASA Release 76-118)
- NASA would buy the Department of Defense's 2 Space Shuttle orbiters if the administration would give NASA an extra $835 million or so for that purpose, *Aerospace Daily* reported after a meeting between George M. Low, departing NASA Deputy Administrator, and Dr. Malcolm R. Currie, DOD's chief of research and engineering. The 2 agencies reportedly agreed that the nation needed 5 orbiters—2 more than NASA had planned to buy—and that NASA should buy them. NASA Administrator James C. Fletcher and Defense Secretary Donald H. Rumsfeld were reported ready to meet to "firm up a proposal" to the Office of Management and Budget and to President Ford; OMB might contest the need for 5 orbiters, *Aerospace Daily* said. (*Aero Daily,* 21 June 76, 22)

22 June: Tass announced the launch of the *Salyut 5* orbital scientific station "under the program of studying the outer space." Parameters were 260-km apogee, 219-km perigee, 51.6° inclination, and 89-min period. Control posts inside the Soviet Union as well as vessels of the USSR Science Academy "sailing in the waters of the Atlantic Ocean" were tracking the flight, receiving telemetry, and controlling the station, the announcement said. Although the official announcement did not mention manned flights as part of the *Salyut 5* program, the *Washington Post* reported that a Western specialist in Moscow had predicted the new station would be manned "sooner or later." The *Washington Star* noted that 2 visiting cosmonauts—Pyotr Klimuk and Vitaly Sevastyanov, who lived aboard *Salyut 4* for 63 days last summer—speaking at an international meeting in Philadelphia 2 wk ago said the new station would be able to accommodate 6 cosmonauts for as long as 3 mo, and would have 2 entry ports, so that 2 spacecraft could dock with it at the same time. The 2-mo flight of Klimuk and Sevastyanov set a Soviet record for time spent in space; however, the 84-day flight of 3 Americans (Gerald P. Carr, Edward G. Gibson, and William R. Pogue) aboard *Skylab 4* had set the overall endurance record in early 1974.

A USSR manned mission would be the first launching of cosmonauts since the joint Apollo-Soyuz flight with the U.S. in July 1975; the last USSR orbital station, *Salyut 4,* was launched 26 December 1974 and was still performing experiments, according to official announcements. The *Star* said that the Soviets were directing their resources toward perfecting permanent stations in contrast to the U.S. approach of developing a reusable space shuttle. (FBIS, Tass in English, 22 June 76; *W Post*, 23 June 76, A-11; *W Star*, 23 June 76, A-15)

- Switzerland and the U.K. became the first 2 countries to sign a protocol authorizing the European Space Agency to "undertake the exploitation phase" of its meteorological satellite Meteosat. The agreement, signed in Paris, would remain open until 30 September for signature by the other participants (Belgium, Denmark, France, Germany, Italy, and Sweden) and for other countries to participate. Meteosat, being developed for ESA by the Cosmos consortium with Aerospatiale (France) as prime contractor, was one of four current ESA applications programs; it covered the design, development, construction, launch, management, and control of a preoperational satellite. ESA's responsibility for checking out Meteosat in orbit for the first 6 mo after launch (scheduled for the third quarter of 1977) would be extended by the protocol to include the subsequent 2.5-yr "exploitation phase" of the satellite to be carried out with meteorological authorities in the participating countries. Other ESA applications programs were OTS (telecommunications), MAROTS (maritime communications), and Aerosat (air traffic control). (ESA release 23 June 76)

23 June: Marshall Space Flight Center announced award of a $209 368 contract to a small business firm, Northern Research and Engineering Corp. of Cambridge, Mass., for designing, assembling, and demonstrating a portable self-contained firefighting module conceived as a joint effort of NASA's Technology Utilization Office and the U.S. Coast Guard. The module would be a lightweight unit complete with pump, hose, firefighter suits, and other equipment that could be transported by helicopter to the deck of any medium-sized boat in the area of a shipboard or dock fire; it would be able to pump about 5700 liters of sea water per min for an uninterrupted 3 hr. (MSFC Release 76-116)

- The USAF Arnold Engineering Development Center operated in Tenn. by the Air Force Systems Command announced that it was using infrared equipment, designed in Europe to locate tumors in the human body, for studying heat patterns on the surface of the Space Shuttle manned orbiter vehicle under simulated flight conditions. Infrared cameras were known to detect a tumor because its heat-radiation characteristics differed from those of normal tissue. In the USAF tests, a model heated in a wind tunnel's supersonic airflow produced infrared radiation received by a camera and transferred to a detector that could record infrared radiation from 7000 points in the field-of-view.

Earlier Shuttle tests had shown the need for a protective system when the vehicle plunged into the dense atmosphere and for data on expected temperature levels, important to the selection of structural materials and to the design of internal parts. Ceramic-like tiles had been used in a scanning test to see the effect of a slight misalignment on the overall heat distribution on the orbiter's underside; the resulting data had demonstrated advantages over more traditional methods. Standard procedures required the application of surface coatings—either a paint that changed from solid to liquid at a specified temperature, or compounds whose ultraviolet radiation changed with temperature—but such materials could create a surface roughness that would alter the results. Also the melting-paint technique required washing and repainting the model after each test. The infrared camera could transmit its findings to a computer for analysis within minutes, and could provide a permanent color record if required. Use of mechanical heat-measuring devices was limited by the number of sensors that could be installed, and the time required to do so. (Although the infrared system would minimize the need for mechanical instrumentation, tests calling for simultaneous temperature data from all portions of the model surface would still require installation of sensors.) About 90% of the system used at AEDC was off-the-shelf hardware; computer programs and methods of interconnecting components were devised by ARO, Inc., the center's operating contractor. (OIP 114.76)

25 June: NASA announced selection of 8 scientists as a team to develop experiments for a proposed unmanned lunar mission in 1980, under study at the Jet Propulsion Laboratory. The mission would use a low-cost instrumented polar-orbiting spacecraft with a smaller companion subsatellite, launched together from Cape Canaveral on a single Delta vehicle, to orbit the moon for a yr and examine nearly the entire lunar surface from pole to pole, on both near and far sides, to expand knowledge gained from the small areas visited by previous U.S. and Soviet missions to the whole of the moon. This first U.S. lunar mission since *Apollo 17* in 1972 would also be the first global survey of a body other than earth.

The 8 experiment areas would provide measurements for scientists to determine the moon's gravity, magnetism, and heat flow, and the chemical and mineral composition of the moon's surface. Advanced remote sensing—gamma-ray, x-ray, and reflection spectroscopy—would be used in 3 of the experiments to create chemical maps of the lunar surface; a fourth experiment, spectrostereo imaging, would provide photographic data to match these maps with individual features of the lunar landscape. Magnetometer and electron reflector experiments would map electrical and magnetic properties of the lunar surface and subsurface; the remaining 2 experiments—heat flow and gravity field—would provide information on the deep interior of the moon. The gravity field would be mapped by tracking the polar-orbiting spacecraft circling at an altitude of 100 km above the lunar surface; the subsatellite, in a higher orbit up to 5000 km, would track the orbiter when it was hidden from earth by the moon. Mascons (mass concentrations) on the near side of the moon had been discovered by analyzing similar tracking data during the early lunar-orbiter flights in the 1960s; the proposed experiment would determine whether such mascons existed on the far side of the moon, besides providing information on the lunar interior.

Principal investigators named were: gamma-ray spectrometry, Dr. James R. Arnold, Univ. of Calif. at San Diego; spectrostereo imaging,

Merton E. Davies, RAND Corp.; electron reflection, Dr. Robert P. Lin, Univ. of Calif. at Berkeley; infrared spectrometry, Dr. Thomas B. McCord, MIT; microwave radiometry, Dr. Duane O. Muhleman, Cal-Tech; gravity and altimetry, Dr. Roger J. Phillips, JPL; magnetometry, Dr. Christopher T. Russell, Univ. of Calif. at Los Angeles; x-ray spectrometry, Dr. Jacob L. Trombka, Goddard Space Flight Center. Each of the 8 principal investigators would be accompanied by groups of co-investigators; the 71-member science team would be aided by JPL staff members. JPL scientist in charge of the study was Dr. T.V. Johnson. (NASA Release 76-119)

- NASA announced it would use starlight that left the surface of a nearby star 200 yr ago to signal the lighting of a giant 200-candle Bicentennial birthday cake at the new Superdome stadium in New Orleans just before midnight on 3 July. Photons of light that left Gamma Boötes just before midnight 3 July 1776—when the signers of the Declaration of Independence were officially witnessing the birth of the U.S.—would arrive on earth to activate a 1000-cycle tone at the Goddard Space Flight Center. The tone, transmitted to New Orleans on commercial telephone lines, would signal the lighting of the cake. GSFC's orbiting astronomical observatory *Copernicus*, passing over Hawaii in its 740-km orbit, would receive the photons on its star detector and transmit a digital signal to the NASA ground station in Hawaii; this signal, retransmitted to the West Coast and via microwave to GSFC, would activate a relay connecting the tone to New Orleans, where it would be heard on the stadium's public address system for 1 min before the candles would be lighted. Astronaut Story Musgrave would participate in the Superdome activities. (NASA Release 76-121)

- Kennedy Space Center announced a broad program of research projects to use the KSC spaceport and its resources to obtain information on thunderstorms and the hazards they offered to launch operations. Meteorological and electronics scientists from the U.S. and abroad converged on the center for sessions of observing the thunderstorms, whose incidence is high in Fla. during the summer months. More than 70 scientists, many of them graduate students, would participate in experiments under the program; 19 principal investigators would represent organizations such as the Naval Research Laboratory, the National Oceanic and Atmospheric Administration's environmental research laboratory, the Atlantic Science Corporation, Mass. Institute of Technology, N. Mex. Institute of Mining and Technology, S. Dak. School of Mines and Technology, and the Universities of Miami, Arizona, Florida, New York at Albany, and Manchester (England). NASA's Johnson Space Center and Goddard Space Flight center also were participating.

Experiments would investigate precipitation formation, lightning behavior, air motions in thunderstorm clouds, and intracloud electrical discharges, among the numerous projects beginning this June and continuing through the summers of 1977 and 1978. Experimenters would provide their own instruments for the experiments, but much KSC equipment and material would be available, such as the instrumented aircraft used in previous lightning studies; weather radars; the detection and ranging system and electric-field measuring system used to monitor buildup of electrical charges that might interfere with launches; a weather-information network display; a satellite imagery acquisition system, with equipment for processing weather pictures, and an automatic picture-transmission (APT) recording system with timing and camera devices. An earlier program in 1948 studied dynamics of thunderstorms; the current program would emphasize electrical characteristics. (KSC Release 279-76)

- NASA announced that one of its C-54 aircraft based at Wallops Flight Center had photographed several Long Island beaches at the request of the Environmental Protection Agency. The beaches, recently polluted by waterborne trash, were flown over and photographed 23 June with 2 T-11 aerial cameras—one for natural color and one for infrared—and the films were developed at WFC and delivered to EPA's Region II Surveillance and Analysis Division in N.J. EPA would interpret the photographs to obtain information about the extent and expected duration of the problem; NASA would also determine whether recent Landsat imagery of the area would be useful in EPA's analysis. (NASA Release 76-120)
- Dr. John F. Clark, scheduled to retire 1 July as Director of Goddard Space Flight Center, had been selected to serve after that date at NASA Hq as special assistant to the Associate Administrator, according to an announcement by Dr. Robert S. Cooper, GSFC deputy director and Dr. Clark's successor. Dr. Clark agreed to stay at GSFC through 14 August to review actions needed to finalize design activities associated with a stability problem in the Delta second-stage engine, Dr. Cooper said, and would also serve as adviser on the Delta's readiness for the Palapa and Itos launches later in 1976. (GSFC announcement 1997)
- The European Space Agency announced that Bernard Deloffre, who began serving as director of ESA's Spacelab program 1 March 1975, had tendered his resignation 22 June to the chairman of the ESA council, through the Director General. A forthcoming meeting of the council would discuss a replacement. (ESA release 25 June 76)

26 June: The USAF launched an early-warning satellite from Cape Canaveral at 11 am EDT on a Titan IIIC. The satellite—product of the TRW/Aerojet Defense Support Program—was reported to be a replacement for a satellite launched 14 Dec. 1975 that malfunctioned. Orbital parameters were: apogee, 35 600 km; perigee, 35 600 km; inclination, 0.5°; period, 1433 min. DOD, which had 10 launches in 1975, had launched 7 so far in 1976. (*SBD*, 29 June 76, 6-F)

28 June: Dr. William H. Pickering, who retired as director of the Jet Propulsion Laboratory at CalTech 31 March, was named to receive a 1975 National Medal of Science, reported *Aviation Week & Space Technology* magazine. The medal is the U.S. government's highest award for distinguished achievement in science and engineering. (*Av Wk*, 28 June 76, 11)

30 June: In a foreword to Volume 2, Program Implementation, of the Energy Research and Development Administration's massive report, *A National Plan for Energy Research, Development & Demonstration: Creating*

Energy Choices for the Future, ERDA Administrator Robert C. Seamans, Jr., called attention to the 30% increase in funding for ERDA in the 1977 budget, whose purpose was to speed the achievement of long-term energy independence, encourage cost-sharing with private industry (to avoid federal funding of projects "more appropriately the responsibility of the private sector"), and support commercial demonstration of synthetic fuel production through loan guarantees. The plans contained in the more than 400 pages of Vol. 2, he said, represented the agency's considered judgment as to which were reasonable to pursue based on current information, but left room for frequent reassessment in view of technical development, economic situations, and environmental and social conditions. (Text, ERDA 76-1)

- Bernard Deloffre, director of the European Space Agency's Spacelab program since 1 March 1975, resigned his position. ESA announced that the Director General, Roy Gibson, would personally manage the program with the assistance of Prof. Massimo Trella, ESA technical inspector, until appointment of a new Spacelab director. (ESA newsletter Aug 76, 3)

During June: Prompted by the crash of an Overseas National Airways DC-10 at Kennedy International Airport in Nov. 1975, when a General Electric CF6 engine on the plane disintegrated after a massive encounter with seagulls, Webster B. Todd, Jr., chairman of the National Transportation Safety Board, wrote to Federal Aviation Administrator John McLucas that the GE engine was the only one in operation that had disintegrated completely on several occasions after bird ingestion. Todd asked that the CF6 engine be modified and that bird patrols be set up where known problems existed at airports served by CF6-powered aircraft.

FAA bird-ingestion tests allowed certification of engines that could absorb the impact of only 10 birds weighing 0.67 kg each (1.5 lb); Todd said that the FAA's birds were too light and that they should have been fired into the engine simultaneously instead of one at a time. The gulls that caused the ONA crash weighed more than 2.2 kg each (5 lb). Although GE immediately protested the letter, only the CF6 had been implicated in a massive bird-strike explosion, whereas Pratt & Whitney and Rolls Royce engines in similar occurrences did not explode.

GE said it was planning to use aluminum honeycomb to replace the phenolic-resin shroud on the abradable seal around the compressor-booster stages; fan blades detached by bird ingestion had scraped off the resin, which accumulated in the combustion chamber and mixed with the fuel, creating overpressure and ultimately an explosion. Aircraft using the CF6 included the USAF's C-5A Galaxy, 190 DC-10s, 6 B-747s, and the A300 Airbus. ONA had decided on CF6 engines for a recent order of 3 DC-10s. (*Interavia,* June 76, 502)

- The 50th anniversary of scheduled airline service in the U.S. had brought little or no traffic growth in a declining economy, the Air Transport Association reported in its yearly review for 1975. When airline passenger travel began in 1926, no more than 6000 persons were carried between a handful of markets. In 1975, the nation's scheduled airlines carried more than 205 million passengers over a network of 58 000 city-pairs (pairs of cities between which a scheduled service carries people, mail, or freight on more than 13 000 daily flights). The passenger total in 1975 was 2 million less than for 1974, although miles flown remained constant; increased traffic during the last quarter of 1975 had brightened the outlook for the Bicentennial year, ATA said. Freight service in 1975 totaled more than 4.7 billion ton-miles, down 2.5% from 1974. Airline safety improved, with only 3 fatal accidents in 1975 (fewest recorded since records began in 1949) and 124 fatalities, lowest number since 1957. Scheduled airline travel was statistically 10 times safer than by private automobile; in the early 1930s, it was considered 8 times as dangerous. (ATA report, 1976)

- *Symphonie 2*, the Franco-German telecommunications satellite that drifted out of its normal geostationary orbit late last month, was gradually repositioned by using its cold-gas stabilization system, *Aviation Week and Space Technology* magazine reported. French and German space technicians determined that the malfunction was caused by a still unidentified defect in one of 4 valves feeding the exhaust nozzles of the satellite's hot-gas stabilization system. Transmissions normally handled by *Symphonie 2* were shifted to *Symphonie 1* while the repositioning went on; both satellites were reported working normally again. (*Av Wk*, 28 June 76, 11)

- The USAF announced that use of a new tool made of a graphite-epoxy composite material in fabricating 3 YF-16 fuselages of the same material had exceeded design requirements and saved both time and money. Materials and processes used to make the 227-kg 18.5-m^2 composite tool replaced a multiple-step method used to construct large, complex, highly contoured aircraft fuselage skins or structural shells of graphite-epoxy materials; the new concept eliminated 2 steps in fabrication, saving up to 50% of total tool cost. Steel curing tools used for smaller, less complex components were extremely expensive and could damage a contoured part because of incompatible thermal expansion; parts made with the new tool had mechanical properties and dimensional tolerances that exceeded those made with conventional tools. Whereas 23.5 man-hours per $0.09m^2$ were needed to make the previous tool, only 9.7 man-hours were needed for the new tools, for a total saving of 2660 man-hours. Elimination of materials used in the multiple-step procedure meant additional savings. (OIP Release 096.76)

- Articles in 2 successive issues of *Aviation Week & Space Technology* magazine reviewed progress in changing to the international metric system of units, and concluded that industry would look to the U.S. government for the spur required to convert, and the funds to defray expenses of converting. The Department of Defense, which was preparing a directive that would "fall short of . . . a substantive push to the industry" by stating that DOD's aim was to keep pace with industry, was said to be reluctant to take the lead—and be presented with a bill for the costs—in converting to metrics for a major new weapons system. The magazine had surveyed a dozen firms in the aerospace industry, none of which had plans to "metricate on their own without a requirement by their government or commercial customers." Most industry spokesmen said metrication was inevitable, although a mixed system with metric-English capabilities would probably be in effect for the period during which tools and supplies would be used up or replaced with the new measurements. (*AvWk*, 7 June 76, 44; 14 June 76, 100)

- *Interavia* magazine printed a photograph of Dora, a solar generator being manufactured by AEG-Telefunken under contract to the German Institute for Space Research to provide power for 1980s communications satellites. Dora was designed as a "variably deployable array," meaning that it could begin its service life partially deployed; unfurled gradually at intervals, it could maintain a constant output over the years. It could also be folded down and brought back to earth at the end of a Spacelab mission. It consisted of two 22- by 2.8-m array panels carrying 46 000 solar cells. (*Interavia*, June 26, 504)
- The government of Nigeria awarded a $150-million contract to TCOM Corp., a subsidiary of Westinghouse Electric, for a balloon-borne telecommunications and broadcast system to be installed by the end of 1979, and for training of Nigerian personnel. Ten tethered balloons supporting transmitting and receiving equipment at altitudes of 3 to 5 km would provide an expanded telephone, television, and radio service for the whole of Nigeria. (*Interavia*, June 1976, 573)

July 1976

1 July: In Washington, the Smithsonian Institution's new National Air and Space Museum was opened with a speech by President Ford, in observance of the U.S. Bicentennial. The President and Vice President Rockefeller were escorted through the $41-million museum by museum director Michael J. Collins, who had piloted the *Apollo 11* spacecraft. At a signal given 18 min previously by the *Viking 1* spacecraft in orbit around Mars, a 3-m metal arm was activated to cut a red, white, and blue ribbon to officially open the building. "That's the most expensive scissors in history," the President commented as he began his dedication. The museum was "a perfect birthday present from the American people to themselves." During the next century, he said, the best of the American adventure lay ahead; it could find out how to harness and preserve the forces of nature, explore the "uncharted frontier" of the oceans, turn space into a partner for control of pollution and improvement of worldwide communications, draw more energy from the earth and sun, develop new agricultural technologies, and conquer cancer and heart disease. The President noted that just 100 yr ago Alexander Graham Bell first publicly demonstrated his telephone. Progress, he said, can be measured "not only by the extent of our knowledge but by increasing awareness of all that remains to be discovered."

Described by Smithsonian Institution Secretary S. Dillon Ripley as "a chic hangar" and by *Washington Post* architectural writer Wolf von Eckardt as "a work of art" with "a beautiful, natural quality" and "a modest timelessness about it," the museum occupied 3 blocks on the Mall, stretching more than 200 m and standing about 26 m high. The first 2 floors contained more than 10 acres of exhibit space; the third floor contained offices, a library, and a cafeteria. The design called for a structural-steel frame covered with thin slabs of the same Tennessee pink marble used for the National Gallery of Art directly across the Mall; the building was divided into 7 bays—4 marble boxes connected by 3 bronze-glass bays facing the Mall. The center entrance bay, "Milestones of Flight," contained the only permanent displays: the Wright brothers' Kitty Hawk flyer, Lindbergh's *Spirit of St. Louis*, the Bell X-1 (first supersonic plane), and the North American X-15. The problem, said the *New York Times*, was to design a building that "couldn't possibly hold" all it was meant to display; the Saturn V rocket measured 4 times as high as the building, and a Boeing 747 fuselage was longer than the building's width. "Space Hall" in the east bay contained the full-size rockets formerly on display near the Smithsonian "castle"; a pit had been dug and the floor lowered to accommodate the 22-m missiles, and the 30-ton backup Skylab had been cut out so that visitors could walk through the astronauts' living area. The west bay was for "Air Transportation"; more than 100 spacecraft and 64 aircraft, nearly 10 times as many as were previously on display, would be exhibited in the new building.

Chartered by Congress in 1945, the new museum had been quartered in temporary structures until after the Vietnam war. All the stories on the opening noted that it was constructed on time and within its budget. The Smithsonian had begun its aeronautical collection 100 yr previously with a group of Chinese kites presented for the 1876 Philadelphia centennial exposition; its interest in aeronautics had begun even earlier. Joseph Henry, first Secretary of the Smithsonian, had sponsored a balloon ascension from the Mall in 1861 to encourage President Lincoln to use balloons for military observation. In 1916, the Smithsonian had begun its 29-year association with Dr. Robert H. Goddard, father of the liquid-fuel rocket.

The museum expected 7 million visitors a year; opening date had been changed from 4 July to accommodate the Bicentennial crowds expected in D.C., and the museum lobby on Independence Ave. had been open to the public since 2 Feb. (*CSM*, 14 May 76; *W Star*, 25 June 76, C-1; 1 July 76, A-1; *W Post*, 27 June 76, E-1; 1 July 76, B-1; 2 July 76, A-1; *NYT*, 2 July 76, B12; 4 July 76, 22; *C Trib*, 2 July 76, 4-16; Newport News (Va) *Daily Press*, 25 Apr 76)

2-12 July: By "unanimous" vote among nearly 30 Viking project scientists, *Viking 1* would land 17 July in a site called the "Northwest Territory" on the Plain of Chryse that would be safer than the original choice, said Project Manager James S. Martin, Jr. A mosaic of more than 40 photos released by JPL showed the plains area more hospitable than the Valley of Chryse, which turned out to be a network of dry riverbeds too steep and too rough to risk a landing. The scientists hoped that the new choice of site would feature sediments, deposited by ancient Martian rivers, that might contain fossils or other traces of life. Being at a lower altitude, the site could harbor water in a liquid form; "snow, ice, and even liquid water are possible on the site we picked," said Dr. James C. Fletcher, Administrator of NASA, in announcing the new plan. (*W Post*, 2 July 76, A-2; *W Star*, 2 July 76, A-8)

A *NY Times* editorial mentioned "super technology" in the taking and transmitting of Mars photographs that led to selection of an alterna-

Mall view of the National Air and Space Museum, opened to the public 1 July 1976. (NASM Photo)

tive landing site, "an extraordinarily impressive example" of the potential of "robot explorers of the solar system." (*NYT,* 3 July 76, 20)

Argon gas detected by *Viking 1* in the Mars atmosphere suggested that Mars had experienced major volcanic periods that might still be in progress. Carbon dioxide, the only gas previously identified in the Mars atmosphere, had obscured the presence of argon, which had a similar molecular weight. As Viking instruments had detected a reduction in atmospheric temperature approaching the south polar cap, colder than had been forecast because of pure carbon dioxide, the explanation was a dilution by another gas that would conform to the low temperatures, identified as argon. Only about 1% of earth's atmosphere was argon, but Viking had revealed the Mars atmosphere to be as much as 8% argon above the south polar cap. (*W Post,* 3 July 76, A-2)

The new landing space chosen for *Viking 1* in the Chryse region was about 80 km north of an island formed millions of yr ago by flood waters rushing from the canyon areas near the Martian equator. (*W Post,* 4 July 76, A-4)

New data received from *Viking 1* had provided more questions than answers, such as the reason for unexpected readings of Martian temperature and atmospheric moisture. The Mars south pole proved colder than earlier believed, below the condensation point of carbon dioxide, and had as much as 40 to 50 microns of water in the atmosphere, more than expected. The newly selected landing site would contain sufficient moisture to provide ground fog, frost, or even snow. (*W Star,* 4 July 76, A-12)

The Viking project demonstrated that man had not lost his "primal urge to explore," said a *NY Times* editorial 4 July, even with little immediate prospect of economic gain or space colonization. People might worry about the expenditures and priorities of such endeavors, said Dr. Bruce C. Murray, Director of JPL, "but very few say the product itself, the discoveries, is unworthy or immoral." On the Bicentennial Day it was proper to remember that the U.S. was a product of the urge to explore. Whatever it learned about Mars, *Viking 1* proved that there was robust life on earth. (*NYT,* 4 July 76, 4-1)

Failure to find life on Mars would mean a cut in funding for the space program, NASA biologist Dr. Harold Klein predicted. Although he estimated odds at 50 to 1 against Viking's finding life on Mars, he noted that the issue would not be settled because the experiments had built-in limitations. Whether or not life was detected on Mars, most scientists believed life existed elsewhere in the universe, the *NY Times* added. (*NYT,* 6 July 76, 15)

JPL released the best photographs to date of the "Martian Grand Canyon," a gorge 10 times the size of Ariz.'s Grand Canyon but relatively small by Martian standards. The Capri Chasm was as much as 32 km wide and nearly 2 km deep, compared to Ariz.'s which was no deeper than 1.6 km, no more than 21 km wide, and about 350 km end to end. The length of the larger canyon of the Valles Marineris on Mars had been estimated at more than 3200 km. Rockslides detected in the

Viking 1 photographs indicated either massive quakes on Mars, meteor impacts, or effects of winds reaching speeds over 480 kph, JPL scientists said. (*W Post*, 7 July 76, A-10)

The *Viking 1* landing on Mars was put off until at least 20 July after radar observations indicated the alternative landing site was rougher than previous Viking photographs had shown. The spacecraft's orbit would be shifted to permit examination of an area west of the previously selected landing point. (*W Post*, 8 July 76, A-2)

An altered orbit put *Viking 1* over an area called Plateau of the Moon, third to be scouted in the search for a smooth landing site. Radar echoes, which had revealed dunes or boulders on the Plains of Chryse invisible to the Viking cameras, showed the plateau to be "twice as smooth as the plains." (*W Post*, 7 July 76, A-2)

A JPL spokesman said preliminary pictures of another landing site proposed for *Viking 1* showed Martian terrain as rugged as that in 2 spots previously rejected. *Viking 1* was scheduled to take more photographs of a region called West-Northwest, beyond the area that controllers last wk said was too rough for a landing. (*W Star*, 12 July 76, A-6)

6 July–24 Aug.: A 2-man spacecraft, *Soyuz 21*, was launched from the Baykonur cosmodrome at 3:09 pm Moscow time (8:09 am EST) to carry out "joint experiments" with the USSR's *Salyut 5* station in orbit, launched 22 June. Col. Boris Volynov, 41, who flew on *Soyuz 5* during the first Soviet linkup in 1969, was commander, accompanied by flight engineer Lt. Col. Vitaly Zholobov, 39, 35th Russian to fly in space. *Soyuz 21* orbital elements were: apogee, 253 km; perigee, 193 km; inclination, 51.6°; period, 88.7 min. The docking with *Salyut 5* occurred 7 July and the 2 spacecraft would remain linked for 48 days, the crew returning safely 24 Aug. On board *Salyut 5*, the crew performed experiments on the development and behavior of fish, melting of metal and growing of crystals in weightlessness, and environmental studies. In April 1975 a Soyuz crew had failed to link with the *Salyut 4* station, reportedly because of a rocket malfunction, but a second crew was successful in May and spent 63 days aboard. Although Soviet authorities offered few details on mission plans, the *Salyut 5* had 2 docking ports, making it possible for 2 Soyuz craft to be docked at once. U.S. sources speculated that the mission might try to break the record of 84 days in space set by the third crew of Skylab in 1974.

Tass reported 10 July that Volynov and Zholobov were using an exercise device that allowed them to "run" in the weightlessness of outer space; a report from the cosmonauts recorded on that date described a mass-meter installed on *Salyut 5* to measure the crew's weight in space through a vibration mechanism. On 28 July a report from the flight-control center said the cosmonauts were checking whether their training on the running track and other activities producing "microshocks" would affect the formation of crystals aboard the station. (FBIS, Tass in English, 6-30 July 76; *W Post*, 7 July, A-22; *W Star*, 7 July 76, A-2; *NYT*, 8 July 76, 23; *Spacewarn*, SPX-273, 20 July 76, 2; *Spaceflight*, Jan 77, 36-37)

8 July: A "Big Bird" reconnaissance satellite, built by Lockheed, was launched by the USAF from Vandenberg AFB, Calif., on a Titan IIID. A previous Big Bird launched 4 Dec. 1975 had remained in orbit until 1 Apr. 1976. (*Av Wk,* 19 July 76, 243)

- NASA launched *Palapa 1*, first of a series of Indonesian comsats, into a synchronous transfer orbit from the Eastern Test Range at 7:31 pm EDT on a Delta vehicle. Orbital elements were 36 504-km apogee, 231-km perigee, and 24.6° inclination. At the 10th apogee, the boost motor was fired (4:30 pm EDT on 11 July) and the spacecraft was maneuvered into a position above the equator at about 83° E. Satellite status was reported satisfactory. The satellite's name came from a 14th-century prime minister, Gaja Mada, who vowed not to eat a popular delicacy called palapa until Indonesia was united. The 8 July launch date was selected to ensure that the satellite would be operational on 17 Aug., when Indonesia celebrated its 31st anniversary of independence. Produced by Hughes Aircraft, the Palapa satellites would be identical to the Telesat Canada and Westar satellites except for antenna modifications; each satellite could relay 12 color TV channels or up to 4000 telephone circuits. A second Palapa was scheduled for launch in 1977. (NASA Release 76-117; MOR M-492-208-76-01, [prelaunch] 24 June 76, [postlaunch] 14 Oct 76)

- Marshall Space Flight Center announced that 3 different solar collectors procured from industry in an ERDA-sponsored program on technical management and support of solar-energy development were undergoing tests of efficiency versus cost. The collectors, after exposure to weathering, were put into an active situation in which the heat-transfer medium (usually water) passed through them, and measurements were made to determine their efficiency. At the same time, 4 low-cost collectors developed by MSFC were being tested to evaluate the thermal performance of black-nickel and black-chrome absorber coatings that could reduce the amount of collector area required, and thereby lower the cost of a specific application. (MSFC Release 76-121)

- Discrepancies in forecasting the size of the 1976 Soviet grain crop underlined the difficulty of accurately predicting worldwide harvests and food needs, the *Washington Post* reported. The U.S. Dept. of Agriculture estimated 22 June that the USSR would produce 190 million tons of grain; other sources objected that the estimate was several millions of tons below the figure set by the Central Intelligence Agency and between 4 and 10 million tons higher than estimates by 2 private firms (Cargill Investor Services of Chicago, and Schnittker Associates of Washington). The size of the Soviet crop was a crucial statistic, the *Post* noted, because—if low enough—it could send world grain prices on a rapid rise; the price of grain would affect the U.S. economy, triggering higher food prices and adding to domestic inflation. The world had had no excess food stocks since 1972, so that faraway droughts or political developments affecting food trade could make U.S. grain prices fluctuate wildly. Government and industry had invested millions of dollars to get information from satellites, computer banks, private weather consultants, translations of foreign broadcasts, and tipsters, the *Post* said. Banks, shipping companies, railroads, and other enterprises in agricultural trade or investment had increased their research in crop data to be ready for price changes or sudden shifts in markets. Despite the flood of information, global agricultural predictions remained inexact.

A satellite equipped for crop surveys swept across the Soviet Union last year and "helped government analysts assess a major Russian crop disaster," the *Post* noted. The information, fed into complicated computer models that collated soil moisture and temperature data with other inputs, produced a projection of Soviet grain production; however, the computer models of the Dept. of Agriculture and the CIA last year produced differing estimates. (*W Post*, 8 July 76, A-2)

- NASA issued a call for Space Shuttle astronaut candidates—15 pilots and 15 mission specialists—to be selected by Dec. 1977. Applications would be accepted through June 1977, and candidates would report to Johnson Space Center in July 1978 for 2 yr of training. Appointment as an astronaut would depend on satisfactory completion of training. Minority and women candidates were encouraged to apply. NASA spokesman Bob Gordon said the agency expected no problems if a woman were selected; the Space Shuttle was "designed to accommodate women astronauts...the waste management system is the only problem really and that has been designed for both males and females." Since the astronaut program began in 1959, 73 pilots and scientists had been selected; 31 persons were available as Space Shuttle crew, 28 astronauts assigned to JSC and 3 holding government positions in Washington, D.C. Pilot astronauts would control the Shuttle during launch, orbit, and landing, and would maintain vehicle systems; mission-specialist astronauts would coordinate orbiter operations in flight planning, use of consumables, and management of the payload. (NASA Release 76-122; JSC Release 76-44; DOD Release 310-76; *NYT*, 8 July 76, 12)

9 July: Marshall Space Flight Center announced selection of 3 firms to negotiate contracts for development of solar heating and cooling systems for residential and commercial use, under an agreement with the Energy Research and Development Administration to install and demonstrate such systems in a substantial number of residential and commercial structures in a wide range of geographic locations [see 1 July]. Firms selected were AiResearch Manufacturing Co. of Calif., General Electric Co. Space Div., and Honeywell Inc. Each of the contractors would deliver prototype solar heating or heating-and-cooling systems, at an estimated cost for the 3 contracts of about $14 million over 3 yrs. (MSFC Release 76-125)

- NASA announced that Dr. Noel J. Hinners, Associate Administrator for Space Science, had selected Drs. Charles R. Chappell and James L. Burch of Marshall Space Flight Center as members of the flight team for the proposed Electrodynamics Explorer (EE) satellite mission, consisting of 2 satellites in polar orbit to study physical processes in the atmosphere, ionosphere, and magnetosphere from high and low orbit. Dr. Chappell's proposal for the mission was a cold-plasma instrument, a retarding ion-mass spectrometer; Dr. Burch proposed a high-altitude instrument for measuring medium-energy plasma. The 2 scientists participated in the project study phase during the past yr, and would work on the flight team with the EE study office at Goddard Space Flight Center in preparing a project plan for the dual-satellite mission. (NASA Release 76-123)

9-18 July: NASA announced plans to launch a 4-stage Javelin sounding rocket from Wallops Flight Center carrying a 76-kg payload that would release red, white, and blue chemical clouds visible along the East Coast from Charleston, S.C., to Boston and as far inland as Cleveland, O. Observation of the rocket's trails would enable the measuring of winds and magnetic and electrical fields in space. The rocket, launched at 9:23 pm local time 18 July from Wallops, fell into the Atlantic before releasing its chemicals. According to a spokesman at the center, "We really won't know what happened for several days, after we've studied ground trackings, but ground observation indicates there was a mechanical failure of some sort." (NASA Release 76-124; WSC Release 76-9; *W Post*, 18 July 76, B-2)

9-10 July: NASA Administrator Dr. James C. Fletcher, carrying greetings from President Ford, was among foreign dignitaries attending the celebration in Yugoslavia of the 120th anniversary of the birth of inventor Nikola Tesla. Born in 1856 in Croatia (then part of Austria-Hungary), Tesla emigrated to the U.S. in 1884, became naturalized, and worked for a time at the Edison company in N.J. His concept of a rotating magnetic field led to many improvements in the fields of radio and electricity: the Tesla coil, the Tesla reduction motor and system of alternating-current transmission, generators of high-frequency current, a transformer to increase oscillating currents to high potentials, and systems for wireless communication and wireless transmission of electric power. Tesla designed the power system at Niagara Falls, N.Y., and lived to be 93.

At ceremonies in the town of Smiljan, Tesla's birthplace, President Josip Broz Tito of Yugoslavia praised the "orientation" of Tesla and his compatriots in the U.S. in their support for Yugoslavia during World War II. Later, at a luncheon, Tito reviewed Tesla's achievements and urged the Yugoslavs to create conditions that would induce talented youths to remain and not leave their country to go abroad. At an informal meeting Dr. Fletcher presented Tito with photographs of Yugoslavia taken by a U.S. satellite, and received from the celebration committee a "golden plaque" bearing Tesla's picture to be presented to President Ford. President Ford's message at the ceremonies had stated that Tesla's work "stands as a link between our two countries in the history of notable accomplishments in science and engineering." (FBIS, Tanjug Domestic Service, 9-10 July 76, 11-16)

11 July: Tass announced that the yachts *Rodina* and *Rossiya* had left the port of Vladivostok in an attempt to repeat the route sailed by Russian navigators 250 yr ago on 2 expeditions to Kamchatka under the leadership of Vitus Bering (1681-1741). The 7-man crews of the ships would begin by visiting Okhotsk, the port where Bering built the ship that went to Kamchatka, then proceed to Kamchatka about 3700 km away. In 1977 and 1978, the expedition would cruise to the Bering Strait and go ashore in North America where Bering's ship *St. Peter* visited 17 July 1741. (FBIS, Tass in English, 11 July 76)

12 July: First flight tests of a unique oblique-wing aircraft model designed for future supersonic use would begin early in Aug. at the Dryden Flight Research Center, NASA announced. The remotely piloted research vehicle (RPRV) to be tested had a wingspan of 6.7 m, weighed 400 kg, and was powered by a 4-cylinder 90-hp engine; its wing angle relative to the fuselage could be varied during flight, and angles up to 45° would be tested. The RPRV would be flown by radio control from a ground cockpit, with a TV camera in the aircraft nose giving the ground controller a pilot's-eye view. Use of the oblique wing was a proposal of Dr. Robert T. Jones, senior scientist at Ames Research Center, to alleviate sonic boom and increase energy effectiveness of supersonic aircraft; positioned at right angles to the fuselage in slower flight, the wing would allow landing and takeoff with minimum power and less noise, but rotated with respect to the fuselage it would provide the high speed possible with the swept-wing design. Studies indicated possible fuel savings even at speeds up to 1600 kph. (DFRC Release 13-76)

13 July: Marshall Space Flight Center announced that MSFC engineers had developed an optical processing system that could use earth pictures taken by satellites such as Landsat or Seasat to survey highway traffic. When the New Orleans, La., planning commission asked the Center if space technology could provide an automated way to conduct thorough, reliable, and inexpensive traffic surveys, MSFC responded with a technique used several years ago of scanning phototransparencies with a laser beam; the new laser scanner would sense differences in intensity of the filtered beam, recognize vehicle sizes and shapes, and store size and location data for later use. Data from an aerial traffic survey using the system could be in the hands of users within hours. The new method must start with aircraft photographs, said MSFC engineer Joseph H. Kerr, who patented the rapid survey system; present Landsat satellites passed over a target only once in 18 days, but the project gave NASA a chance to refine the method for use on satellites in the 1980s. Whereas processing of satellite data might now take months, Kerr said his system could produce data 100 times faster. (NASA Release 76-128)

• Solar power would charge batteries in 2 electric vehicles used by the National Capital Parks, Dept. of the Interior, during this summer's Bicentennial Festival of American Folklife in Washington, D.C., NASA announced. The golfcart-like vehicles, each equipped with 6 standard traction batteries, would get energy from photovoltaic cells capable of providing 1.7 kwh of electricity during peak sunlight; the array was prepared by Lewis Research Center as one of several demonstrations of solar energy it was conducting as manager of ERDA's photovoltaic test and demonstration project. Cells were provided by NASA's Jet Propulsion Laboratory, to which ERDA assigned responsibility for the low-cost silicon solar-array project. The solar-powered vehicles would be used from July through Sept. at the festival, one for transportation of festival workers, the other fitted with a vacuum cleaner for refuse pickup. (NASA Release 76-129;.ERDA Release 76-218)

14 July: Kennedy Space Center announced award of a $2 250 000 contract to the Beech Aircraft Corp., Boulder, Co., for development of a Space Shuttle orbiter fuel-cell servicing system. An orbiter would carry 3 fuel cells using cryogenically stored hydrogen and oxygen to provide electrical power; each cell would connect to a separate power bus, but only 2 cells would be used during minimum power load. Peak or average power load would use all 3 cells. The contract called for Beech to design, fabricate, test, and deliver a complete cryogenic remotely controlled unit that would provide liquid hydrogen and liquid oxygen for the fuel cells, delivery to be completed by 1 Dec. 1977. (KSC Release 333-76)

15 July: NASA announced appointment of Robert A. Newman as Assistant Administrator for Public Affairs, effective 26 July. Newman was vice-president of community affairs for TRW, Inc., and president of the TRW Foundation at Cleveland, O. In his new NASA post, he would be responsible for all information services of the space agency except for technical publications. He had received bachelor's degrees in journalism and sociology from the Univ. of Mo., served in the USAF during the Korean War, and had been a member of the board of directors of the Air Force Academy Foundation. (NASA Release 76-131; Fletcher anno. 15 July 76)

16 July: NASA announced selection of Hamilton Standard Division, United Aircraft Corp., for negotiations leading to award of contract for development and production of space suits to be used by men and women during Space Shuttle flights. The basic cost-plus-award fee contract would cost about $15 million through Sept. 1980. The suit system, consisting of a basic suit and built-in support system including breathing-atmosphere and cooling components, would be adjustable instead of custom-made for each astronaut (as in earlier programs) and would come in small, medium, and large sizes to fit candidates varying in height from 1.5 to 2 m. Hamilton Standard would provide hardware and spares needed for 7 suits and support equipment, as well as training, manpower, and equipment needed to support the program at NASA field centers. Johnson Space Center would have technical direction of the contract. (JSC Release 76-46)

- NASA announced that its *Nimbus 6* weather satellite would be monitoring the cross-country progress of a 1914 Dodge participating in the Bicentennial world antique auto race that began 15 June in Istanbul, Turkey, and wound through 11 countries in Europe. Having crossed the Atlantic by ship, the four entrants—all American—were scheduled to rally 16 July at Times Sq., New York City, to begin the final leg of the race—New York to San Francisco—that would recreate the first leg of the Great Race of 1908. The tracking device carried on the Dodge belonging to Ed Schuler of Morrison, Ill., had been checked out last May on a run from Ill. to the Goddard Space Flight Center in Md. The device, a rugged self-powered unit normally carried on balloons, buoys, and ice packs, is one of more than 800 such instruments assigned to experimenters since the launch of *Nimbus 6* in mid-1975. Valued at about $1800, the device would be returned to GSFC after the race for checkout and future use on another platform. A similar device used in 1971 to track a light plane that flew around the global poles had previously been carried around the world on a balloon that came down in Central Africa, where a French member of the World Health Organization found it hanging in a tree and returned it to NASA; after the airplane trip, the unit served another year on a buoy before being retired from service. A 1914 Model T Ford originally scheduled to carry the tracking device [see 15 Feb.] was withdrawn from the race by its owner. (NASA Release 76-126)

17-19 July: Project scientists said the high proportion of argon in the Mars atmosphere might damage the life-detection instruments on *Viking 1*. In March 1974, a spectrometer pump on a Mars-bound spacecraft failed to work properly and Russian scientists concluded the problem was caused by argon, an element reluctant to form chemical compounds; the Russian pump apparently could not handle the high argon content of the Mars atmosphere. *Viking 1's* gas-chromatograph mass spectrometer contained filter material to screen out carbon dioxide and carbon monoxide in the search for organic compounds in Mars air and soil samples, which would be heated in the instrument to produce gas for analysis by the chromatograph; the high argon content might interfere with the readings. (*NYT*, 17 July 76, 24)

JPL flight directors were checking out the instruments on *Viking 1* for its Mars landing scheduled for 20 July. Immediately upon landing, the spacecraft would measure air temperature and pressure and wind speed and direction, and would detect any interior motions of the planet such as moving lava or shaking of the crust. It was also scheduled to take 2 photographs, one downward to record the terrain of the landing site and the other a panorama shot to show the general surface of Mars. Life-detection experiments would not begin until about a wk after the landing. (*W Post*, 18 July 76, A-10)

Commands from JPL to turn on the $153-million instrument package to be landed by *Viking 1* on the Plains of Chryse took between 17 and 18 min to reach the spacecraft nearly 346 million km away from earth. First spacecraft responses to the signals indicated the instruments and navigational equipment were in working condition. Landing was scheduled for 8 am EDT on 20 July; *Viking 1* would separate early in the morning into 2 components, 2300-kg orbiter and 590-kg lander. The orbiter, powered by purple solar cells on panels, would remain in orbit to serve as a radio relay for the lander and to photograph nearly half the Martian surface over the next 12 mo. The lander, powered by an atomic battery, would coast for about 3 hr, slowed by a parachute and braking engines, and land on 3 aluminum legs at about 8 kph, more gently than most parachutists land on earth. (*W Post*, 19 July 76, A-3)

Odds were heavily against existence of any form of life on Mars, said a *Washington Post* editorial, and against *Viking 1's* finding evidence of it even if it were there. The Viking flights were "in some ways more remarkable an undertaking" than the Apollo flights because the spacecraft was on its own with no humans to correct errors and a sizable time lapse between sending and receipt of radioed instructions. The mission, said the *Post*, was "already a stunning accomplishment." (*W Post*, 20 July 76, A-16)

The Viking spacecraft was "a breed apart," the *Wash. Post* reported, as different from previous spacecraft "as a Cadillac from a Pinto." Almost nothing on the lander had flown in space before; almost everything had to be built super-small to save weight. As the first U.S. spacecraft to land on another planet, the Viking had to be designed to withstand chemical and heat sterilization to avoid contaminating Mars with earth organisms. The tape recorder, for instance, had to be made of bronze coated with phosphor, because no plastic tape could have survived the heat. (*W Post*, 20 July 76, A-2)

19 July: President Ford proclaimed 20 July as "Space Exploration Day" to coincide with the date 7 yr ago when U.S. astronaut Neil Armstrong became the first man to set foot on the moon. NASA's *Viking 1* lander was expected to touch down on the surface of Mars about 8 am 20 July, reported nationally on network television; transmission of the first picture from the lander should be completed before 9 am. In his proclamation, the President said, "We begin our third century with . . . the most ambitious of all deep space explorations . . . Wherever we reach, we will have come in peace for all mankind." In an accompanying statement to NASA employees, Administrator James C. Fletcher said "the President has in effect commended all of you who work (for) the national space agency." (Text; NASA Release 76-133; Fletcher anno, 20 July 76)

First panoramic view of Mars surface from Viking 1, 20 July 1976: upper half of picture is left half of view, lower photo joins it on the right. Upper view shows housing for Viking sampler arm; lower view shows (at right) high-gain antenna for direct communication with earth. (NASA 76-H-557)

- Marshall Space Flight Center announced that about 80 scientists from universities, medical facilities, and industry, serving as the Universities Space Research Association, would meet at Huntsville 26-30 July to evaluate experiments proposed for the first Spacelab mission. NASA's Space Science Steering Committttee would use the evaluations in selecting experiments to be carried on *Spacelab 1*, scheduled for launch in the 3rd quarter of 1980. The 10 categories in which experiments were proposed were: astrophysics, upper atmosphere, space plasma, biomedicine, atmospheric observations, biology, technology/behavior, communications/navigation, drop dynamics/atmospheric cloud physics, and technology experiments. MSFC was assigned management responsibility for defining, integrating, and operating the payloads for the first 3 Spacelab missions. (MSFC Release 76-130)

20 July: The *Viking 1* lander touched down safely on a plain at the western edge of the Chryse region of Mars at 7:53 am EDT, 11 mo to the day after its journey of more than 700 million km through space began at KSC Launch Complex 41. Landing was within 17 sec of the predicted time; velocity at impact, predicted to be 2.4 ± 0.9 kps, was actually 2.49 kps. Within minutes of the landing, the Viking began relaying a photograph of one of its footpads on the nearby surface up to the Viking orbiter for transmission to earth; a panoramic view of the landing site followed. Distance from earth to Mars at landing time was more than 322 million km, and radio waves took 19 additional minutes to travel back to earth to signal touchdown. Spacecraft controllers at the Jet Propulsion Laboratory broke into applause and cheers when the data arrived showing a successful landing; telemetry showed the descent curve fitting preplanned values as the lander's aeroshell, supersonic parachute, and terminal descent engines braked the descent to the surface of Mars. The site at Chryse Planitia was chosen for the landing after Viking orbiter data and the giant telescope at Arecibo, P.R., showed that the primary landing site and the first alternative were too hazardous. NASA Administrator Dr. James C. Fletcher, who was at JPL for the landing, said that the intensive search for a safe landing site "really paid off."

The Mars landing came on the 7th anniversary of man's first landing on the moon 20 July 1969 during the *Apollo 11* mission. President Ford, who telephoned JPL with congratulations after the landing, asked Viking Project Manager James S. Martin, Jr., of LaRC if there were plans for a Viking 3. Martin replied that there were not, but that his team was "ready to take on Viking 3 plus Viking 4, 5, and 6." The *Viking 2* was expected to go into orbit around Mars about 7 Aug. (Mission Status Bulletin 35, 20 July 76; *Spaceport News*, 23 July 76, 1)

- The Tass news agency reported the soft landing of "the automatic interplanetary station" *Viking 1* on the surface of Mars, adding that the first pictures of the surface showed "a foot-pad of the craft on . . . dust-covered terrain strewn with small sharp stones . . . According to the preliminary assessments, the ground consists of basalt rocks and volcanic lava. It is believed to be comparatively soft." Tass noted the postponement of the Viking landing, originally scheduled for 4 July, and the search for bacterial life on Mars which was "one of the aims of this exploration." (FBIS, Tass in English, 20 July 76)

- Rocketdyne Division of Rockwell International Corp. announced it had been awarded a $10.8-million contract by Lewis Research Center for 8 Atlas MA-5 propulsion systems, with an option for 2 additional systems. The contract would run through August 1978. The engines would be used to launch high-energy astronomical observatories (HEAO), or commercial comsats for the Communications Satellite Corporation on a reimbursable basis. Originally developed by Rocketdyne for the USAF's first intercontinental ballistic missile, the Atlas had flown space missions since 1958 when it launched *Echo*, first U.S. comsat. Atlas had powered more than 400 payloads with launch reliability of more than 99%. (Rockwell Release RD-18)

- "Eventually there will be an American SST program," Dr. John L. McLucas wrote in a letter to the *Phila. Inquirer*. Dr. McLucas, Administrator of the Federal Aviation Administration, agreed with an *Inquirer* editorial that stated, "Concorde hasn't made it yet," but pointed out that costs of research and development and other expenses of producing a new aircraft had virtually halted the introduction of new aircraft designs "suitable to the major airlines." New transport aircraft would offer advantages such as greatly reduced noise, a high level of safety, and 30% improvement in fuel efficiency; however, continued development of new planes and their purchase by the carriers "is moving beyond the financial scope of private investors singly or collectively," Dr. McLucas wrote. Some combination of government and private initiative would be needed to finance the next generation of air transports; "it's more likely to be the result of a cooperative American-European effort," Dr. McLucas concluded. (*P Inq*, 20 July 76, 1-E)

- India had used deuterium from the U.S. to detonate its first atomic explosion, said acting Assistant Secretary of State Myron Kratzer at an "unprecedented" hearing by the Nuclear Regulatory Commission on licensing of future sales of reactor fuel to India. The State Dept. said that stopping the sale of the reactor fuel would constitute "severe social and economic hardship" to 80 million Indians dependent on power generated by a plant that had been operating since 1969. Adrian Fisher, "former high official in the U.S. Arms Control and Disarmament Agency," said India had broken its word to Canada and the U.S. by using nuclear materials to create an atomic device in 1974; India said the device was "peaceful" and did not contravene U.S. bans on use of the material to build weapons. Herbert Scoville, another former disarmament official, said India's nuclear generating stations were producing enough plutonium to build 10 to 20 bombs per year, and that India now had a 4-yr stockpile. Scoville and Fisher said India should be required to sell back its plutonium and accept further restrictions on use of nuclear material. A strong U.S. stand on nuclear materials to India was the only hope— "perhaps a vain hope," Scoville said—of stopping the spread of nuclear weapons. Rep. Clarence D. Long (D–Md.) said that Iran, Egypt, South Africa, and Argentina were countries that would be encouraged to develop nuclear weapons if the U.S. did not impose sanctions on India and restrict the flow of nuclear material. The commission was examining a large pile of documents on the history of the U.S. nuclear relationship to India; some petitioners (e.g., the Sierra Club) had claimed that evidence as far back as 1963 showed India preparing to develop a nuclear explosive device for political purposes. (*W Star*, 21 July 76, A-8)

21 July: The *Marshall Star* reported on a Bicentennial celebration at MSFC's Michoud Assembly Facility in New Orleans, La., that included planting of a "moon" tree to commemorate the 7th anniversary of the first Apollo landing. Michoud was the place where giant Saturn moon rockets were fabricated. Participating in the program were Lt. Gov. James Fitzmorris of La.; MSFC Director Dr. William R. Lucas; and Robert Littlefield, manager of the Michoud facility. The tree was a pine grown from seed carried to the moon and back in 1971 on the *Apollo 14* mission as part of research into effects of weightlessness on seed germination. The U.S. Dept. of Agriculture had planted the returned seeds, and the seedlings had been set out in various public areas over the U.S. such as in Philadelphia's Washington Square. As the tree was planted in front of the Michoud administration building, a time capsule was deposited to be opened during the nation's Tricentennial in 2076. The capsule contained documents, tapes, and pictures relating to the space program. (*Marshall Star*, 21 July 76, 4)

- Marshall Space Flight Center announced award to Martin Marietta's Denver division of a cost-plus-incentive-fee/award-fee contract for $9 282 667 to deliver parachute decelerator subsystems for recovery of 12 solid-fuel rocket boosters from 6 Space Shuttle flights. The contract, effective through Dec. 1980, covers design, development, manufacture, and refurbishment of the subsystems. Work was authorized to start 6 July at Martin's plant in Colorado and at Pioneer Parachute Co. of Manchester, Conn., subcontractor. (MSFC Release 76-135)
- An enormous radiotelescope "the size of the earth" to be used in studying the process of star formation in earth's galaxy had been constructed by linking installations in Washington, D.C., California, Australia, and the Soviet Union, the Naval Research Laboratory announced. Four large radiotelescopes distributed around the world had joined in using very long-baseline interferometry (VLBI) to record microwave signals received at each antenna on videotape, and comparing the recordings. The resultant huge antenna would be able to resolve a source less than 2 ten-thousandths of a second of arc, comparable to the thickness of a human hair at a 161-km distance, or a man's footprint on the moon. An instrument this size would be needed to study primordial clouds in the constellation Aquila containing water-vapor molecules that emit intense microwave radiation through maser amplification, analogous to the operation of optical lasers. These water-vapor masers seem to be associated with star formation taking place in giant clouds in the galaxy; sizes of the individual masing sources, which may number more than 100 in a single cloud, range from half to 3 times the distance between the sun and the earth. The masing radiation offered a way to study the clouds that

obscure radiation at other frequencies and to estimate the physical conditions that surround star formation. Installations in the experiment were NRL's 26-m antenna at its Maryland Point, Md., observatory; NASA's Deep Space Network 64-m antenna at Tidbinbilla, Australia; the Crimean Astrophysical Observatory's 22-m antenna at Semeiz, USSR; and the Calif. Institute of Technology's 39-m antenna at Owens Valley, Calif. The maximum spacing achievable is the earth's diameter (12 600 km), a distance approximating the separation between the Md. and Australian antennas (12 091 km). (NRL Release 45-7-76B/HF; NYT, 20 July 76, 13)

22 July: Images from 2 Landsats circling the earth 912 km up had helped the state of Md. monitor strip-mining damage and inventory its surface coal-mining activities, NASA announced. The state did not know the location or extent of orphan mines—those abandoned before 1967—and began a mapping program to determine the extent of the disturbed land and the procedures and expense that would be required to reclaim it. A 1967 law required private companies to operate under state-approved reclamation plans; acreage disturbed before 1967 would be reclaimed by the state. The Landsat images showed from 25 to 30% more acreage affected by the mines than anyone had realized. Arthur T. Anderson of Goddard Space Flight Center's information transfer laboratory compared Landsat images taken over a 3-yr period ending in 1974 to identify and measure effects of strip mining in Western Md. with an accuracy better than that obtainable from aerial photography or field-inspection techniques. Other states using Landsat imagery for reclamation projects were Ohio, Tenn., W. Va., Ky., Pa., S.C., Fla., and the Great Plains states. (NASA Release 76-134)

• President Ford would name Dr. H. Guyford Stever, director of the National Science Foundation for the past 4 yr, to head the recently reestablished White House Office of Science and Technology, the *New York Times* reported. The post had been abolished by President Nixon 3 yr ago "amid reports that he did not like the advice he was getting from it, particularly concerning antiballistic missiles and supersonic planes," the *Times* said. *Aviation Week & Space Technology* magazine had reported earlier that 4 conservative Republican senators had warned President Ford that Stever's nomination would be an "affront" to Congress; the NSF had been a target of some congressmen because of funding projects that Congress considered questionable, *Av Wk* said. Other candidates for the position had shied away because of a possible change in administration after the forthcoming elections. Sen. Frank E. Moss (D-Utah) urged quick action on the nomination; the Senate Appropriations Committee had just cut $1 million from a $3.3-million budget estimate for the Office of Science and Technology. (*NYT*, 22 July 76, 37; *Av Wk*, 5 July 76, 13)

21-31 July: The *Wash. Star* "Portfolio" column reported that television was "barely there" as the *Viking 1* spacecraft touched down on Mars and began to relay "those sharp, stark pictures to earth." Hardly had reporters began asking explanations from scientists "who looked stunned with pleasure at the accomplishment," when "it was commercial time . . . and the 9 a.m. programs." Viewers were "left wondering whether there was really an event of such magnitude being televised or more science fiction," the *Star* commented. (*W Star*, 21 July 76, B-1)

Commentary on the Viking landing on Mars: The *Chicago Tribune* hailed the mission as "an epoch-making success . . . The achievement is the greater because the world has known from the beginning what we were attempting to do. There would have been no way to hide a failure." (*C Trib*, 21 July 76, 1-7)

The *Phila. Inquirer* called the landing "another giant leap for mankind" that "underscores the sophistication of unmanned spacecraft . . . Risks can be taken . . . that could not be justified with persons aboard . . . Its performance in the first hours on Mars speaks more eloquently than words." (*P Inq*, 21 July 76, 5-A)

The *NY Times* said the TV networks were criticized for their coverage; Dr. J. Richard Keefe, former NASA scientist now at the Univ. of Louisville, said the coverage was "downright disgusting . . . Talk about being blase about space exploration, this was just incredible . . . I think the population has become apathetic about the whole space program. It's kind of sad, I think." Spokesmen for the network told the *Times* they had given 20 to 25 min of coverage, showing the 2 pictures transmitted by *Viking 1* and released by JPL where the transmission was received. NBC said its New York City switchboard had about 30 calls asking for more; ABC reported 25, and CBS reported 8 or 10. ABC said the number of calls was significant "for the time of day." "What they may not have understood was that we showed whatever we had," the spokesman said. "That was it." (*NYT*, 21 July 76, 12)

The *Wall Street Journal* said in an editorial that it could not "speculate with any confidence on why yesterday's *Viking I* landing on Mars was important. And yet we are confident that it was." The TV networks that "spent tedious hours broadcasting the dullest political convention on record" a wk ago "were preoccupied with game shows" while Viking was sending back its first pictures of the surface of another planet. However, those who said the $1 billion cost of the project was wasted unless Viking's cameras pick up " something interesting, like a local resident," would be missing the point, *WSJ* added. The project demonstrated application of the most advanced technology to a single objective, advanced managerial skills as "a triumph of human organization," and left mankind with the psychic effects of the visit to another planet. " . . . After each leap deeper into space, nothing is quite the same again." (*WSJ*, 21 July 76, 22)

A *Washington Star* editorial said that *Viking 1* may have "ended generations of fanciful imaginings" about life on Mars, and asked if it was worth 7 yr of effort and $1 billion "to get a rather sobering view of what our imaginations told us was a fascinating planet." Although it was much too early to predict the "ultimate scientific harvest," the *Star* said "we have no doubt that the ultimate payoff will be more than adequate in scientific if not in fictional terms." (*W Star*, 22 July 76, A-14)

The first color photograph of Mars, showing a dramatic blue sky against a rust-red soil surface, was in error, Viking project scientists announced. Astronomer Carl Sagan said that someone had given the wrong weighting to the color filters in reconstructing the electronic image on earth, and that corrected photos would show a pinker sky. The seismometer on the Viking lander had apparently jammed and would impair detection of "marsquakes" unless corrected; Lou Kingsland, deputy mission director, said the proper command might not have been issued or the problem might be in the electrical circuits. (*W Star*, 22 July 76, A-1)

Having successfully set down the *Viking 1* lander in a safe spot, Project Manager James S. Martin, Jr., announced that *Viking 2* would be sent to a region likely to be more scientifically interesting even with less assurance of safe touchdown. Cydonia, a region halfway between the Mars equator and north pole, considered because of a greater likelihood of water, required for the presence of life as we know it. Viking project scientists said they were eager to survey a site as different physically from the desert Chryse site as possible. *Viking 2*, still more than 1.6 million km from Mars, would arrive in orbit there early in Aug. and land about 4 Sept. if all went well. Debate on the second landing site focused on possible hazards in the northern region, out of range of the radar signals that led to choice of the first site. Astronomer Carl Sagan said the first landing was "reasonably lucky," with chances of success calculated at only 60%, and the northern site would be less promising. However, he and the other scientists voted overwhelmingly for Cydonia because, as Sagan said, "it is not only the probability of success but the significance of success" that merited consideration. (*W Star*, 22 July 76, A-6)

Asked about the role of luck in the successful *Viking 1* landing on Mars, Project Manager James S. Martin, Jr., replied, "I don't plan on luck . . . . I believe that most of what you call luck you make yourself. It's people doing that extra job . . . " Martin's management approach during his 8 yr with the Viking project had been to "get into the details of everything . . . . I've never known any other way to be successful . . . . It inspires and motivates the next level of management to get involved." (*NYT*, 22 July 76, 24)

The "most electrifying new information" received from Mars since the *Viking 1* landing was that the Martian atmosphere contained 3% nitrogen, an element essential to life as we know it, said the *NY Times*. Until this week, the absence of nitrogen on Mars had been the most compelling argument against existence of life "in any form" on Mars; now this argument had been removed. (*NYT*, 22 July 76, 30)

JPL scientists were still hedging on the color of the Martian sky, after announcing that photographs released earlier showing a blue sky might have been processed wrongly. Astronomer Carl Sagan noted that the "boos given to (the) announcement about a pink sky reflect our wish for Mars to be just like the earth." Dr. Thomas Mutch, leader of the Viking imaging team, said computer verification of the filtered signals for color reconstruction could take another wk or two. (*W Star*, 23 July 76, A-1)

Viking Mission Director A. Thomas Young told a news conference that the problem with the lander's soil-sampling arm was probably an error in the computer command that failed to drop a locking pin near the arm's "wrist." A new set of commands to extend the arm further and free the pin would be sent 24 or 25 July. The arm would scoop up soil and drop it in a hopper on the spacecraft that would distribute it to 3 biology experiments aimed at detecting signs of life on the planet. Other scoops would be used for chemical analysis. (*NYT*, 24 July 76, 1)

Besides the trouble with its soil sampler, the *Viking 1* lander was having radio problems, said Project Manager Martin. One of two receivers getting direct communications from earth was not working properly, and the radio for relaying data from the lander to its orbiter was operating in a low-power mode. Also, the instrument for detecting "marsquakes" had not been released from its "caged" position on the lander despite repeated commands. A special scientific team was checking to see if all the problems were related, perhaps in the computer signals to the electrical circuitry. (*W Star*, 23 July 76, A-3)

Mars was a very much richer and larger planet than the airless and waterless moon, said a *NY Times* editorial, with a firm surface and virtually every essential required for life. Barring the catastrophe of a nuclear war, Mars could be sustaining a human population "by the time of this nation's Tricentennial," the paper said. (*NYT*, 25 July 76, A-16)

"I really do not know what the fuss is all about—landing on Mars, I mean. I landed there 40 years ago," said Buster Crabbe, who played both Flash Gordon and Buck Rogers in movies of the 1930s. Crabbe, who won the 400-m freestyle swimming gold medal in the 1932 Olympic games, was hired by Paramount Pictures to do a Tarzan film, and went on to star in space pictures. "People still remember old Flash," the 68-yr-old Crabbe said. "Here I am still around, no ill effects from my space travel . . . . I do feel sorry for our space people though, having to go through all the things we did years ago. I don't think they know what's in store for them. Just wait till they run into the clay men and the hawk men and the shark men." (*NYT*, 25 July 76, E17; Tucson *Ariz. Daily Star*, 30 July 76, A-19)

Viking 1 did not relay any picture of life-sized Martians to earth because it landed in the wrong place, charged columnist Art Buchwald in the *Wash. Post*. Buchwald reported his friend, a science-fiction buff, as saying the scientists were interested only in finding "a smooth place to set down the camera." If the Martians sent a camera to earth, the smoothest place to land it would be the Sahara desert. "If you lived on Mars you wouldn't pitch a tent on some barren spot where nothing was happening," argued the friend. "If Viking had landed next to one of their canals, the Martians would have sailed their tall ships up to it and given us the greatest show ever seen on television." (*W Post*, 25 July 76, H-1)

A series of simple commands from earth freed the mechanical arm on the *Viking 1* spacecraft reported to have jammed 22 July. The extendable arm would be able to dig into the Martian soil as planned, and the second Viking spacecraft would be able to land in Cydonia now that the Chryse area could be explored. The Viking was commanded to extend its arm twice as far as previously and to rotate it several times to free a locking pin inserted to protect the arm from vibration during its flight in space and landing on Mars. The command was timed to precede transmission of a spacecraft camera image taken soon after receipt of the command; the boom would not be in the picture if it had not responded to the order to extend itself in front of the camera eye. When the picture came in a half hour later, the mechanical arm was in clear view, and a surface shot clearly showed the locking pin had fallen free. Flight directors had not succeeded in uncaging a seismometer, composed of 3 delicate balancing beams to detect movements in the planet's crust, that had been wired together for protection from vibrations. Spacecraft computer signals to electrify and burn the wire were not received, and a second command bypassing the computer apparently failed also. Detection of marsquakes might have to depend on *Viking 2*, scheduled to land in the Cydonia basin about 4 Sept. (*W Post*, 26 July 76, A-1)

The atmosphere of Mars was once 10 times thicker and richer in gas than it is today, and "we could breathe that atmosphere" if it were richer in oxygen, said Dr. Tobias Owen of the State Univ. of N.Y., a member of the project science team at JPL, where the first samplings by the *Viking 1* mass spectrometer were analyzed. Mars's atmosphere was now 95% carbon dioxide, 2 to 3% nitrogen, 1 to 2% argon, 0.3% oxygen; earth's atmosphere is 78% nitrogen, 21% oxygen, 0.9% argon-40, and 0.03% carbon dioxide. "Mars is an intermediate planet between the earth and the moon," said Dr. Owen. "We do not see a hugely abundant biology, but that doesn't say there's none at all." Color pictures of the Martian surface near the Viking lander revealed a soil even redder than in the original picture received last week. Rocks strewn over the desert sands exhibited a red coating laid down years ago when Mars's atmosphere was richer in oxygen and water, "through oxidation of the iron and hydration of the minerals," said Viking geologist Alan S. Binder. "We propose to test the hypothesis by cracking open one of the red rocks . . . to see if there is any rust inside." (*W Post*, 27 July 76, A-1; *W Star*, 27 July 76, A-3; *C Trib*, 27 July 76, 5)

Viking 1's mechanical arm scooped its first sample of the Mars surface 28 July, leaving a sharp-edged trench clearly visible in a picture transmitted afterward by the unmanned lander. The 3-m arm, a tube about 38 mm in diameter, ended in a sharp-edged scoop that could dig into the surface with a force of about 13 kg if necessary. When the cover snapped shut on the sample, the arm retracted and swiveled to sift the sample into a funnel feeding into the "biology box." This container, about the size of a large milk carton, encased the 3 experiments designed to detect life. Two other experiments, also located under funnels, were the gas chromatograph mass spectrometer to detect organic molecules and an x-ray fluorescence spectrometer to identify inorganic materials, Sensors inside the funnels would tell whether the samples were sufficient.

The 3 biology experiments were the "pyrolytic release" experiment of Dr. Norman Horowitz, to look for microorganisms that used the sun as a source of energy like photosynthesis on earth to build organic compounds, through exposure of the sample to radioactive carbon dioxide under artificial sunlight; a "labeled release" experiment by Dr. Gilbert Levin of Biospherics, Inc., to look for signs of microbial metabolism by tracing radioactive organic compounds injected into the Mars sample to see if they were consumed, releasing gaseous wastes that could be detected; and a "gas exchange" experiment by Vance Oyama of NASA's Ames Research Center to provide a rich variety of foods in a liquid form—"chicken soup"—to the Mars sample and look for changes in gas composition inside the experiment chamber that would indicate the food was being used.

The biology-box samples would incubate for from 5 to 12 days before results were announced, and a positive finding would be confirmed by subsequent control experiments. The science team anticipated a long wait for any kind of answer, as well as the possibility that the answers might be contradictory. Dr. Gerald Soffen, chief scientist for Viking, pointed out that results would be difficult to interpret: "you have to put yourself on the level of a microbe," he added. The microbes might drown in the liquid, ingest the wrong food, or burn up in the warm experiment chambers. (*W Star*, 28 July, A-1, A-5)

One of the five soil-testing experiments on *Viking 1* did not receive enough soil to actuate it, project officials announced, and two attempts to start the gas chromatograph mass spectrometer working on a sample apparently failed. Project Manager Martin said the flight engineers were not yet able to determine what went wrong, because the sampling arm apparently worked as planned but the sample never got into the instrument. The motor driving a stirring rod to brush particles through a strainer did not come on, and scientists speculated the sample might still be sitting on the strainer. A photograph of the trench dug to obtain the sample had shown that the soil was surprisingly cohesive, "like wet sand," according to Princeton Univ. scientist Dr. Robert B. Hargraves. After delivering soil to the experiments in the biology box, the mechanical arm made 2 attempts to dig and deliver a sample to the mass spectrometer but the instrument did not signal receipt of the sample either time. It was not clear, Martin said, whether the arm lost the soil on the swing back or was off the mark when it deposited the soil, or the soil in the hopper never filtered down to the instrument. The arm responded properly to the following command and delivered a spoonful of soil to the x-ray spectrometer with no difficulty. (*NYT*, 29 July 76, 1; *W Post*, 29 July 76, A-3; *W Star*, 29 July 76, A-3; *WSJ*, 29 July 76, 1)

Viking project scientists debated the makeup of the Mars soil sample that might have been too thick to filter into an instrument opening. Dr. Ronald Scott said that, although the material was somewhat cohesive, "certainly something must have gotten through." Deputy Mission Director Kingsland said the most likely causes of the apparent malfunction were failure of the mechanical scoop to obtain a full sample, or failure of the "level full" indicator in the instrument to signal properly. Dr. Klaus Biemann of MIT, chief of the gas chromatograph experiment, said the arm might have come up empty from the trench after scooping for the biology experiments, and that a repeat attempt next week would take both possible explanations of the problem into account. Kingsland said the material seemed to have a consistency "something like wet clay." Dr. Scott emphasized that this did not indicate water in the soil. The other instruments seemed to be processing their samples normally, and results from the inorganic-chemistry sampling in the x-ray spectrometer should provide the first assay of Martian soil as early as tomorrow. (NYT, 30 July 76, A-22; W Star, 30 July 76, A-5; C Trib, 31 July 76, 2)

The first experimental results from Mars, returned 30 July by *Viking 1's* lander, showed that the surface consisted of iron, calcium, silicon, titanium, and aluminum, in amounts that would be determined within a few days. The entire surface around the lander was covered with a very thin coating of vivid orange-red iron oxides that gave the plant its red or rusty appearance. "Mars is a painted desert," said Dr. Gerald Soffen, one of the project scientists. Dr. Priestly Toulmin, leader of the inorganic-chemistry team, noted that Mars soil up to now could be analyzed only by long-distance methods such as light reflections from the surface; Viking had "picked up a piece of Mars and put it into an analytical instrument." Dr. Benton Clark, also of the inorganic-chemistry team, said that Viking had not detected trace elements like vanadium or molybdenum considered essential for plant growth on earth; this did not mean the elements were not present, nor had elements like arsenic been detected that would make earth soil sterile. The results "do not rule out the possibility of some form of life," Dr. Clark added. Results from the 3 biology experiments would be transmitted to earth within a few days. (NYT, 31 July 76, A-1)

22-24 July: NASA launched the second in a series of Comstar satellites at 6:04 pm EDT 22 July from Complex 36 at Cape Canaveral, Fla., on an Atlas-Centaur AC-40 into a transfer orbit with 42 171-km apogee, 2930-km perigee, and 21.8° inclination. Like its predecessor launched 13 May, Comstar-B (called *Comstar D-2* in orbit) was owned and would be operated by Comsat General Corp. under lease to American Telephone and Telegraph Co. as part of a three-satellite domestic communications system. *Comstar D-2* went into orbit lacking FCC approval for its operational use; AT&T engineering director Robert Latter said details for use of the system had not been worked out. On 24 July at 6:12 pm EDT, Comsat General commanded firing of the apogee kick motor to put the Comstar into station at 95°W, above the equator for coverage of the southwestern U.S., at approximately 35 793-km altitude. A third Comstar would be launched in 1978 to allow for growth of service, and a fourth Comstar would be built as a spare. (NASA Release 76-127; Comsat General Release CG-76-117; NASA MOR M-491-201-76-02; [prelaunch] 20 July 76, [postlaunch] 13 Dec 76; D/SD 26 July 76, 127)

23 July: Proponents of the B-1 strategic bomber, "vigorously opposed for years by people inside and outside of government," seemed likely to win their struggle to get approval for the $22-billion program, said *Science* magazine. A House-Senate conference committee voted in mid-June to spend $960.5 million in procurement funds for the first 3 B-1 planes, although the Senate had passed an amendment delaying spending of the money until a new administration should take office.

This year's fight to procure the B-1 was the culmination of a 15-yr effort that began with the shooting down of a U-2 plane over the Soviet Union in 1960; the USAF projected the building of a low-flying manned bomber, the ideas converging in 1969 with AMSA (advanced manned strategic aircraft), a project opposed by then Secretary of Defense Robert McNamara. In 1969, when Melvin Laird became President Nixon's Secretary of Defense, AMSA studies were concluded and the final design "metamorphosed into the B-1," Rockwell Intl. Corp. becoming prime contractor. At the same time, other Pentagon planners were backing a strategic armed cruise missile decoy system (SCAD), assuming that Congress would not approve both programs. At the urging of the USAF, Secretary Laird canceled the SCAD program in July 1973, but the chairman of the Senate Armed Services Committee (Sen. Thomas McIntyre, D-N.H.) ordered it reinstated.

Both the bomber and the cruise missile had encountered development problems and increased expenses; internal Rockwell documents from Jan. 1974 showed company concern about "competitive threats in the form of the standoff missile and . . . the launch aircraft" that other companies might seek to build. A massive public-relations campaign was mounted to protect the B-1; ultimately, *Science* said, the B-1 would win but the cruise missile would also have its day, quoting John B. Walsh, Deputy Director of Defense Research and Engineering (DDR&E) at the Pentagon, as saying "you need both bombers and cruise missiles." (*Science*, 23 July 76, 303)

- Marshall Space Flight Center announced that Barbara S. Askins, a chemist in MSFC's Astronomy Branch, had applied for patent on an improved process, autoradiographic film intensification, that would improve photographic images after films or plates had been developed. Exposing silver or other imaging material on a developed film or plate to a radioactive environment would convert the material to a radioactive compound; placing the radioactivated material in contact with a receiver emulsion on an unexposed film or plate would reproduce the original image with increased contrast and density. The hazards and complexity of handling radioactive material had limited the use of such a process, but the new process did not require special training, expensive equipment, or extraordinary safety precautions, as it used an isotopic organo-sulfur 35 compound and could be adapted easily to an ordinary photographic laboratory. Askins said the process would be used in astronomy and other scientific research where low light levels were encountered. (MSFC Release 76-13)

- The U.S. had the smallest Air Force since the beginning of the Korean War, General William J. Evans told a luncheon of the National Security

Industrial Association in Los Angeles. Gen. Evans, commander of the Air Force Systems Command, said that the Air Force, in an effort to reverse this trend, was in the midst of the greatest modernization period of its short history, citing "the effective combination" of "land-based terminals and spaceborne satellites to provide attack warning, weather data, communication links, and positioning and navigation for American forces wherever they may be." The U.S. was building a modern Air Force "with inefficient equipment that would be at home in a museum," he said, calling on industry for more investment in modern equipment and advanced technology. Contractor testing had not been realistic, and performance of contractor products had failed to meet operational requirements; the government presence in contractor plants, to which industry objected, could be reduced with better designs, streamlined operations with better visibility, and measures such as greater use of computers to reduce inefficiencies. (AFSC Release OIP 152.76)

- The organizational structure of a large earth-orbiting space station was the topic of a doctoral study at Fla. State Univ. for Dr. James Ragusa of Kennedy Space Center's Science and Applications Office. The study was an offshoot of work on future space projects (with Gene McCoy of KSC's Shuttle Payload Integration Office) before Dr. Ragusa left to work on his Ph.D. The topic was one about which little had been written, and was timely in that the structure should be included in plans for such a mission. *Spacelab*, with a 7-man crew, needed no complex organizational structure; a space colony with a projected population of thousands would require a government rather than an organizational structure. A space base housing 50 to 100 persons would bring together people of varying backgrounds for extended periods of time—the station being designed with a 10-yr operating life—and large-group behavior would be a factor in considering the best type of organization. Ragusa investigated the organizational structures used on submarines and destroyers, interviewing the submarine captains and taking cruises on the nuclear sub *Nathanael Greene* and the research sub *Benjamin Franklin*.

  Among those interviewed for the study were Dr. Wernher von Braun; Arthur Clarke, author of *2001* and father of the communications satellite; several astronauts; and Gene Roddenberry, creator of the TV series "Star Trek." Ragusa found that the organization on Roddenberry's *U.S.S. Enterprise* was modeled after that used on 18th century English ships, which in turn was based on that used on Phoenician ships 2500 years ago, in which the ship was considered an extension of the country itself and abided by the laws and traditions of the homeland; the captain was king, and his word was law. The structure best fitting the needs of a space base was the "total matrix model" with four levels of hierarchy, best for efficient and orderly management of the crew because of its adaptability. Of 8 models, the "Star Trek" structure ranked fifth because of its lack of flexibility and efficiency. (*Spaceport News*, 23 July 76, 5)

- NASA announced that the U.S. Government Printing Office had published the 3-volume English edition of the joint U.S.-USSR *Foundations of Space Biology and Medicine,* summarizing biological and medical results of the first 15 yr of space flight. The work was produced by a joint editorial board on space biology and medicine formed in Oct. 1965 by NASA and the Soviet Academy of Sciences; the USSR had published a Russian-language edition in Moscow. The text consisted of 45 chapters, 19 authored by U.S. scientists, 20 by USSR scientists, and 6 by teams of authors from both nations. Volume I was called "Space as a Habitat"; Volume II, "Ecological and Physiological Foundations of Space Biology and Medicine," responses of man, plants, and animals to space flight; Volume III, "Space Medicine and Biotechnology," technology and procedures needed to sustain life and permit living creatures to function in space. (NASA Release 76-135)

23-24 July: The scientific balloon *Da Vinci III*, launched west of St. Louis, Mo., at 7:25 am CDT on 23 July, landed safely at about 9 am EDT on 24 July about 80 km east of Lexington, Ky. Purpose of the flight was to follow industrial and urban air pollution into the surrounding countryside, recording the changes in concentration and chemistry of the pollutants. Early findings indicated that high levels of ozone and sulfur dioxide—2 major air pollutants—persisted throughout the flight. The 4-member crew of *Da Vinci III* included 3 who manned both the *Da Vinci I* flight in Nov. 1974 and the *Da Vinci II* flight in June 1976: Dr. Rudolf J. Engelmann of the National Oceanic and Atmospheric Administration; pilot and project originator Ms. Vera Simons; and pilot Jimmie Craig of the U.S. Naval Weapons Center. The new member of the crew was Preston B. Herrington of Sandia Laboratories, Energy Research and Development Administration, which would collect and analyze the flight data as well as conduct all flight operations. Da Vinci was a joint project of ERDA, NOAA, and the Environmental Protection Agency. (ERDA Releases 76-211, 76-245)

25 July: Blueprints for future Mars explorations, triggered by the *Viking 1* landing on Mars, included not only the dispatch of roving robot vehicles to the surface of Mars but also collection and return of Mars samples by a "sailing ship"—a spacecraft carrying an onionskin-thin sail of aluminum-coated plastic measuring more than 185 m^2 and powered by the pressure of solar electromagnetic radiation. Space flight would become possible with less reliance on heavy and expensive rockets, said John Noble Wilford in the *New York Times;* a space-sailer could be launched from an earth-orbiting Shuttle, deploy landers to collect and return samples, and return to earth with larger amounts of Martian samples than possible by other methods. JPL's plans for future missions might include a fleet of the sailers (to be called Yankee Clippers) "flying the routes from earth orbit to the vicinity of Mars, hauling supplies and portable habitats, and finally explorers, to a Mars base." (*NYT*, 25 July 76, 44)

- Rep. Les Aspin (D-Wis.) charged that the U.S. Navy planned to spend $187.6 million to repair F-14 Tomcat jet-fighter engines that it planned to replace at an additional cost of up to $1.9 billion. ". . . The Navy hopes to pull a double whammy on the taxpayers by replacing the repaired engine with a brand new one," Aspin said. The Navy conceded that replacing engines in the jet fighters would cost $1.6 to $1.9 billion, but claimed the planned repairs to current engines would cost only $86 million. The Navy said it wanted new engines for installation beginning in 1981. (*W Post,* 26 July 76, A-1)

25–31 July: An almost equal division of opinion in the U.S. Senate on whether to go ahead with production of the B-1 bomber was occupying politicians and commentators in Washington. The *New York Times Magazine* story headed "Will It Bomb?" asked if the B-1 were really needed and concluded that it was not, admitting that the project had a "seemingly irreversible momentum" and already had "a backhanded commitment to production." Election-year oratory was calling for a decision to be postponed until a new administration took office. A lobby group called Campaign to Stop the B-1 Bomber, claiming to represent 26 national organizations, sent a telegram to Democratic presidential nominee Jimmy Carter asking him to send back Sen. Walter F. Mondale (D-Minn.), vice presidential nominee, who was meeting with Carter in Ga., for a "crucial" vote in which Mondale's presence was "essential." The vote to postpone in the Senate Appropriations Committee was 15 to 14; the House of Representatives had voted twice, in April and again in June, to proceed with the B-1, but on both votes the nonvoters could have changed the outcome. (The April vote in the House was 210 to 177, with 46 not voting; the June vote was 207 to 186, with 37 not voting.) A spokesman for Rockwell Intl. Corp., which would be prime contractor for the B-1, warned that postponement would increase costs by at least $500 million, cost 3000 workers their jobs, and shelve plans to hire another 7700 workers. Defense Secretary Donald F. Rumsfeld said delay in producing the B-1 would be "unsound from a cost standpoint, from a management standpoint," and Air Force Secretary Thomas C. Reed said delay would add "half a million dollars to the cost of the program for no purpose."

A *New York Times* story 30 July said that a breakfast meeting with Secy. Reed for a group of Senate aides had been paid for by the Air Force Association, a private group with 155 000 members—largely retired Air Force personnel—and chartered as a veterans organization not required to register as a lobbyist. The Air Force Legislative Liaison Office, which arranged the breakfast, seemed to have been "concerned that it might be engaged in a form of lobbying of doubtful legality," said the *NYT,* pointing out that a 1948 law "specifically provides that no funds appropriated by Congress can be used by a Government agency directly or indirectly 'to influence in any manner' the vote of a member of Congress." (*NYT* magazine, 25 July 76, 7; *W Post,* 26 July 76, A-13; *C Trib,* 29 July 76, 10; *NYT,* 30 July 76, A6; *Av Wk,* 26 July 76, 23)

26 July: Johnson Space Center announced delivery and acceptance of the Sperry Rand computer system, Univac 110/46, that would drive the Shuttle mission simulators. Dr. Bruce B. Johnson, technical manager of the contract, said that JSC had tested the computer complex round the clock for 30 days before its acceptance, and that the complex had demonstrated reliability "far in excess of contractual requirements."

Completed 3 wk ahead of schedule, the Univac 1100/46 contained 10 processors, making it one of the most powerful digital computers built to date, with 900 000 36-bit words in the main memory and 2 billion characters in mass storage capacity. The computer would simulate a wide range of mission situations—prelaunch, ascent, aborts, orbit rendezvous and docking, entry, and landing—in the process of training flight crews and ground personnel in all phases of the Shuttle program. Total cost of the computer complex was $7 934 876. (JSC Release 76-47)

- The U.S. Air Force announced production of a new material with electrical properties suitable for solid-state laser applications: anisotropic yttrium aluminate doped with neodymium ($Nd:YAlO_3$), a substance difficult to make into high-quality single crystals. The new material would replace neodymium YAG in solid-state lasers because it loses less heat than conventional lasing materials, requires no external component to polarize the laser light, and stores more energy for electro-optical switched-laser operations. The USAF sponsored the research at a pilot plant of Lambda-Airtron Division of Litton at Morris Plains, N.J. which produced 40 rods of the new material; the longest, about 100 mm long, was valued at $4000. The high cost resulted from the cost of the raw materials and the iridium crucibles used in production, as well as the large amounts of power needed. The Air Force Materials Laboratory at Wright-Patterson AFB, O., would lend the tested rods to qualified government agencies and contractors. (AFSC Release OIP 118.76)

27 July: The *Viking 1* orbiter photographed a tiny moon of Mars called Phobos at a distance of just over 8000 km, providing a picture so clear that more than 100 craters were visible on the surface only 22.5 km across. At least 6 craters measured nearly 2 km across, meaning that meteorites the size of small towns had crashed into Phobos during the last 3 or 4 billion yr. The face of Phobos in the picture released by JPL was the same one photographed in 1971 and 1972 *Mariner 9*, first spacecraft to orbit Mars. "This means that Phobos always shows its same side to Mars," said Dr. Thomas C. Duxbury of JPL, as earth's moon always shows the same side because of the pull of earth's gravity. Another *Viking 1* picture showed small channels in the Argyre basin of Mars's southern highland that were the best evidence yet for heavy rainfall on the planet in past ages. Present atmospheric pressure on Mars was only about 7.7 millibars, 1/200th of that on earth and insufficient for liquid water to form or rain to fall. (*W Post*, 28 July 76, A-8)

- Johnson Space Center announced that NASA had awarded the Singer Co. Simulations Products Division, Binghamton, N.Y., a $6.5-million 2-yr contract for maintenance, modification, and operational support of JSC's simulation complex at Houston to be used in training flight crews for the Space Shuttle program. Initially, the simulation complex would consist of the Shuttle procedures simulator and the crew procedures evaluation simulator; later, an orbiter aeroflight simulator would be added, followed by the Shuttle mission simulator. In addition to systems and hardware engineering, configuration control, installation and tests of modifications, and development, drafting, and illustration of software, the contract

- would require maintenance, servicing, and operational support of the equipment as well as documentation and logistics support. Extensions of 24 and 6 mo would be optional. McDonnell Douglas Corp. also negotiated for this contract (see 10 May). (JSC Release 76-48)
- NASA announced the appointment of Harold E. Pryor as Deputy Administrative Assistant for Technology Utilization in the NASA Hq Office of Industry Affairs. Pryor had been with NASA since 1964, serving as director of the management systems office, the staff operations division in the Office of Procurement, and the NASA/DOD Contract Administration Services Office in the Defense Supply Agency Hq. He was also executive assistant to the Director, Manned Space Flight Management Operations, until July 1973, when he became director of the Scientific and Technical Information Office. Pryor earned bachelor's degrees in naval science and tactics and in aeronautical engineering; he served in the U.S. Navy during World War II. He would succeed Clare F. Farley, recently retired from NASA. (NASA anno. 27 July 76)
- Tass reported from Moscow the launch of *Intercosmos 16*, carrying scientific apparatus developed in the German Democratic Republic, Czechoslovakia, and Sweden, as well as in the USSR. Orbital parameters were said to be 523-km apogee, 465-km perigee, 94.4-min period, and 50.6° inclination. The probe was to study ultraviolet and x-ray radiation from the sun and the effects of this radiation on upper-atmosphere structure. Observatories in Bulgaria, Hungary, the GDR, and Czechoslovakia were making ground observations of the sun simultaneously with measurements taken aboard the satellite. (FBIS, Tass in English, 27 July 76)
- Reporting on the opening of the new National Air and Space Museum in Washington, D.C., *Pravda* mentioned the life-size model of the docked Soyuz and Apollo spacecraft, saying that it "arouses the greatest interest among the museum's visitors" and adding, "The USSR and the United States must continue cooperation." The news story, headlined "First Anniversary," quoted both Dr. James Fletcher, NASA Administrator, and his deputy Dr. George M. Low on the technical and political importance of the ASTP mission. (FBIS, quoting Tass in *Pravda*, 16 July 76, 5)
- U.S. intelligence sources reported the third failure of the USSR this year in testing an antisatellite system, when a Russian Hunter satellite failed to stay in orbit 21 July. The Hunter's mission was to destroy the *Cosmos 839* launched into orbit 12 days earlier; U.S. analysts did not know what was causing the Soviet problem. Soviet efforts to develop an antisatellite system dated back about 10 yr, with only 5 successful launches out of more than 20 attempts before testing stopped in 1971. All 3 attempts since testing resumed in Feb. 1976 had been failures, the U.S. intelligence sources said. The satellite-destroyer was reported to be about 8 m long and to weigh about 2.5 tons at launch; equipped with 5 main rocket engines, the interceptor was said to be able to close on a target at about 4 km per sec and to get as close as 30 m to the target before exploding on ground command. The Associated Press report noted that the U.S. had not tried to develop a similar system since the 1960s;

Dr. Malcolm Currie, DOD Director of Defense Research and Engineering, said last winter that "we are very much concerned about satellite vulnerability." (*W Post*, 27 July 76, A-2; *C Trib*, 27 July 76, 2)

27-28 July: NASA's Flight Research Center announced that Edwards AFB, Calif., was the scene of 3 air records set by pilots in the SR-71. Maj. Adolphus (Pat) Bledsoe, Jr., pilot, with Maj. John Fuller as reconnaissance system manager, flew the SR-71 at 3400 kph, surpassing the previous "absolute" (on a straight course) speed record of 2981 kph as well as a world-class speed record of 2920 kph set by Soviet pilots, both in Oct. 1967. Next day, Capt. Robert C. Helt, pilot, with Maj. Larry A. Elliott as recon officer, flew the SR-71 to an altitude of 26 km, surpassing the previous record of 24.4 km (in horizontal flight) set in 1965 by USAF Col. R.L. Stephens in a Lockheed YF-12A interceptor (prototype sister ship of SR-71). On the same day, Capt. Elden Joersz, pilot, with Maj. George T. Morgan, Jr., as recon officer, flew the SR-71 at a speed of 3530 kph to surpass the previous record of 3331 kph over a closed course set by Col. Stephens in 1965 in the YF-12A. The National Aeronautic Association verified the speed and altitude measurements, but the records would be unofficial until accredited by the Paris-based Fédération Aeronautique Internationale. (FRC issuance 27-28 July 76; *W Star*, 28 July 76, 5)

28 July: Kennedy Space Center announced that Mae Walterhouse, coordinator of the Federal Women's Program at KSC and recently elected national president of Federally Employed Women, Inc. (FEW), would participate in the 1976-1977 Career Development Program at NASA Hq. Ms. Walterhouse would report to Washington 7 Sept. for a yr of experience in the Office of Resources Management. The NASA program would assist the centers in developing potential supervisors and managers at all levels throughout the agency, foster understanding of Hq functions on the part of center personnel, and provide training in specific disciplines or functional areas. Ms. Walterhouse, who received a national award this summer for service to the FEW organization, planned to return to KSC and join the Administration Directorate after the yr at NASA Hq. (KSC Release 376-76)

- The Department of Defense authorized development and testing of 2 prototypes of an improved STOL (short takeoff and landing) aircraft for the Marine Corps by McDonnell Douglas Corp. at a cost of about $400 million, the *Wall Street Journal* announced. The prototype, called the AV8B Harrier, would be an adaptation of the existing AV8A Harrier built by Britain's Hawker Siddeley Aircraft Ltd. The Marine Corps had purchased 110 earlier Harriers from the British firm; if McDonnell Douglas could develop a stronger version, the Corps would buy "several hundred" from that company. The new versions were expected to cost about $5 milllion each. The Hawker Siddeley group had been authorized to build an improved Harrier for the Royal Navy; the Pentagon said it expected "mutual cooperation in the procurement of supplies and the exchange of information" as the 2 programs proceeded. (*WSJ*, 28 July 76, 6)

- DOD announced plans to sell West Germany 500 Sidewinder missiles and support equipment valued at $43 million. The announcement was one of several made by the Pentagon on foreign weapons sales, since law required that Congress must be notified and have 20 days in which to disapprove such sales. DOD also planned to provide $26 million in pilot training under a Northrop Corp. contract to upgrade the Saudi Arabian air force, through the sale of F5E jet fighters and construction of support and training facilities. (*WSJ*, 29 July 76, 3)
- Under a 3-nation agreement announced by Britain, West Germany, and Italy, the Tornado MRCA (multi-role combat aircraft) dubbed "the spearhead of Europe's coming offensive against the United States for equality in the skies" would go into production and would be operational by 1980. The agreement provided for construction at a total cost of $10 billion of 809 swing-wing "wonder weapon" planes that could fly just under the speed of sound at treetop level, and twice the speed of sound at higher altitudes; it was designed to replace 2 of the most profitable planes ever produced by U.S. makers, the Phantom used by Britain's RAF and the F104 Starfighters used by West Germany and Italy. The Tornado carried terrain-following radar that could fly it under enemy radar, and could vary its wingspan from 8.5 to 14m. Panavia (consisting of British Aircraft Corp., Messerschmitt-Boelkow-Blohm, and Aeritalia) would build the Tornado airframe and Turbo-Union (Rolls Royce, Motoren-und-Turbinen Union, and Fiat) would build the engine. The U.S. press noted that, for the first time, American aerospace firms would be facing a competitor as formidable as themselves. (*W Post*, 15 Aug 76, A-14; *W Star*, 18 Aug 76, A-4)

29 July: NASA launched ITOS-H, fifth operational spacecraft of the second-generation improved Tiros operational satellite (ITOS) series, from the Western Test Range at 1:07 pm EDT (10:07 am PDT, local time) on a 2-stage Delta vehicle into a synchronous polar orbit with 1523.6-km apogee, 1512.5-km perigee, 102.1° inclination, and 116.2-min period. After successful injection into orbit, the satellite was redesignated *Noaa 5*; upon completion of in-orbit checkout, spacecraft control would be transferred to the National Oceanic and Atmospheric Administration (NOAA) as part of that agency's National Operational Meteorological Satellite System (NOMSS). Of 4 previous Noaa polar-orbiting spacecraft, 2 still provided limited information but had deteriorated after 2.5 and 1.5 yr in orbit respectively because of harsh environmental conditions in space.

ITOS spacecraft were designed to provide complete global coverage of earth's cloudcover and atmospheric structure on a daily basis, using both daytime and nighttime instrumentation. The 345-kg boxlike spacecraft measured roughly $1 \times 1 \times 1.25$ m and carried 3 winglike panels covered with solar cells, as well as earth-orientation devices and 4 communications antennas. Sensor systems included 2 vertical-temperature profile radiometers, 2 very high-resolution radiometers, and 2 solar-proton monitoring systems, as well as a scanning radiometer system to provide both stored picture coverage for transmission on demand and direct transmission of images to receiving stations within range. Visible-channel resolution was 3.7 km, infrared resolution 7.4 km. (NASA Release 76-130; MOR E-601-76-17 [prelaunch] 28 July 76, [postlaunch] 30 July 76, 1 Oct 76; NOAA Releases 76-143, 76-152)

- U.S. airline traffic increased more than 12% in the first half of 1976, the Air Transport Association reported; increasing numbers of passengers plus increased fares and cost-cutting would produce better profits for the airlines. Analysts said Americans were flying in unprecedented numbers because they felt the recession was over; even "ailing international carriers" like Pan American World Airways and Trans World Airlines were trimming their operating losses and might show some profit for the year, the *Wall Street Journal* reported. (*WSJ*, 29 July 76, 15)

30 July: NASA announced that its Viking Undergraduate Intern Program, permitting U.S. college students to participate in the Viking mission to Mars, was under way at the Jet Propulsion Laboratory in Pasadena, Calif. Fifty-eight students from about 600 applicants had been selected to spend 30 days during the summer working with Viking scientists in a number of scientific areas at JPL, where more than 72 science teams were conducting a detailed examination of the planet, including a search for life. The program was the idea of Prof. Thomas Mutch of Brown Univ., leader of the Viking imaging team, who was assisted by a teammate, Prof. Carl Sagan of Cornell Univ.; the 2 professors reviewed each of the applicants' qualifications and selected the 58 students for the summer's work. The program, designed to directly involve undergraduates with a strong interest in planetary science, was funded by the Alfred P. Sloan Foundation and by NASA's Planetary Geology Office. (NASA Release 76-139)

- A "tethered" satellite—suspended by a cable from the Space Shuttle orbiter's cargo bay—could serve to deploy and control materials used in constructing a space station, or to transfer articles from one manned vehicle to another, said NASA scientists at the Marshall Space Flight Center. The orbiter could also "troll" a tethered satellite through a low-altitude earth orbit to explore the atmospheric region between 80 and 120 km above earth's surface; previous exploration vehicles—sounding rockets and low-altitude satellites—could not remain in that region long enough for extensive studies before being forced back to earth by atmospheric drag and gravity. The Smithsonian Astrophysical Observatory would aid NASA in system-dynamics studies for the tethered satellite system. MSFC would manage program development, integrate experiments, and support orbital operations. If the tethered system proved feasible, MSFC would begin spacecraft system design, and the first mission could be scheduled as early as 1980. (NASA Release 76-138)

- NASA announced it had awarded a $300 000 grant through its Office of Aeronautics and Space Technology to the Univ. of Wash., Seattle, Division of Ceramic Engineering for research into the nature and properties of ceramic materials. The grant would help educate engineering design students in the search for new materials capable of sustaining high temperatures and exhibiting physical properties needed in an increasing number of applications. Engineering contributions to the field of ceramics had led to improvements in fiberglass, optical communications, nuclear fuels, catalytic converters, self-cleaning ovens and ceramic-top ranges, eyeglasses that darken in the sun, and human implants not susceptible to reaction with body mechanisms. Principal focus of the Univ. of Wash. program would be development of gas-turbine engine components that could withstand high temperatures and severe mechanical stress, and economical production of such components. (NASA Release 76-137)

- NASA announced selection of a team consisting of General Electric Co. and Hamilton Standard Division of United Technology Corp. for negotiating with NASA and ERDA a $7-million contract to design, build, and test by 1978 a 1.5-Mw wind-turbine electricity-generating system, biggest in history. The experimental system, a windmill with 2 horizontally rotating fiberglass rotor blades about 61 m long mounted on a 45-m tower, could produce annually enough energy to supply more than 500 homes at a site with average winds of 29 kph. Located at a utility-company site, the NASA-ERDA system would supply electricity to the local utility for public use, to determine the economics and operating characteristics of large wind turbines coupled to conventional power plants. Utility companies had proposed 17 sites across the U.S.; ERDA would measure their wind characteristics over the coming year and would choose the site late in 1976. Lewis Research Center would manage the project for ERDA. The new system would be bigger than the 1.25-Mw system, 53 m in diameter, built near Rutland, Vt., in the 1940s; that project could not compete economically with the then low cost of fossil fuels, and had been abandoned. (NASA Release 76-136)

- Officials of the Professional Air Traffic Controllers Organization (PATCO) threatened a slowdown at major U.S. airports in the form of strict adherence to rules on spacing of aircraft, the *Chicago Tribune* reported. The union, in a salary dispute with the U.S. Civil Service Commission, would use the slowdown to protest failure to raise flight controllers' pay. Federal Aviation Administration rules would permit a separation between airplanes of from 4.8 to 8 km; if controllers spaced flights the full 8 km apart, schedules could be delayed by as much as 6 hr at busy airports such as Chicago's O'Hare. The tactics, although inconveniencing travelers, would not affect air safety, said PATCO president John Leyden. FAA had issued a request that controllers ignore the slowdown. (*C Trib*, 30 July 76, 5)

31 July: Johnson Space Center employees Herbert A. Zook and Richard W. High received a patent this week for invention of a "capture cell," a device to capture meteoroids traveling through space and preserve them for scientific analysis. The cube-shaped cell had an open side covered by a thin membrane; a meteoroid striking the membrane would enter the cell and shatter, but the remnants and any vapor created by the heat of impact would remain inside the cell, which was constructed of a material like polyethylene. Large arrays of the cells would be exposed on a spacecraft for a long time, up to a year, to allow time for a number of strikes to occur. Interest in the composition of meteoroids springs from the possible information on larger bodies such as comets or asteroids on which they originated. (*NYT,* 31 July 76, 27)

During July: Scientists at JPL completed a list of "purple pigeons"—bright birds of the future—as unmanned planetary missions considered feasible for the period 1980-1990. In hopes that *Viking 1's* successful landing on Mars might prompt wider support for such missions, JPL proposed 7 advanced missions that would probably be reduced to 3 or 4 to be forwarded to NASA Hq. The list includes: return to Mars with rover vehicles to extend exploration of the Martian surface; a lander for one of Jupiter's moons, accompanied by a satellite around the planet; a similar mission to Saturn and its moon Titan; asteroid rendezvous and near-range photographic mission; a radar-mapper satellite for Venus; flight to Halley's Comet with a solar-sailer satellite (propelled by solar radiation); and establishment of an automated station on the moon for extended study of the lunar environment, probably by an international scientific group. (*Av Wk,* 26 July 76, 16)

- Marshall Space Flight Center announced a number of activities, centered on the Space Shuttle, including testing, procurement, and completion of facilities.

—During July, MSFC awarded 14 contracts each of which totaled more than $50 000. (MSFC Release 76-155)

—MSFC completed preparation of a 2.8% scale model of the Space Shuttle solid-fuel rocket booster (SRB) that would be tested in Ames Research Center wind tunnels (the 2 supersonic facilities and the 4.2m transonic facility). MSFC had used a 0.5% scale model in its smaller trisonic (over mach 3) wind tunnel to develop preliminary environmental data; the smaller tunnel was cheaper to operate, and a smaller model was cheaper to alter. Data from the ARC tests would be use in SRB component vibration and acoustic design and in establishing qualification-test criteria, and finally for generation of data on the acoustic environment for reentry of an actual full-scale SRB. (MSFC Release 76-120)

—Completing the first phase of acoustic testing that used scale models of the Space Shuttle with models of the Kennedy Space Center launch pad, MSFC announced that it had begun another test program using a scale model of the Vandenberg AFB, Calif., launch pad and a Shuttle model having working main engines and solid-fuel rocket boosters to the same scale. The Vandenberg launch pad differed from that at KSC, but MSFC said that much of the information obtained since Aug. 1974 in 150 test firings to evaluate different acoustic methods for the KSC launch pad would be useful in suppressing noise at the Vandenberg site. The tests were made to determine the best way to control noise and ignition overpressure when Shuttle engines and boosters were ignited at launch; uncontrolled pressures and acoustics could damage the Shuttle or its payload. Tests were scheduled for completion in Dec. 1976. (MSFC Release 76-117)

—MSFC announced award of a $247 363 contract to Rockwell's Space Div. for a 12-mo study of Shuttle booster and external tank options that could reduce cost per flight and increase payload weight capability. (MSFC Release 76-118)

—MSFC announced that the first flight-configuration nozzle for the Shuttle main engine under development by Rockwell's Rocketdyne Div. had been completed and successfully proof-tested. Completion of the nozzle represented a significant design and manufacturing achievement; the nozzle consisted of 1080 precisely formed tapering tubes brazed to a shell and stiffened by bands. It was about 3m tall and 2.7m in diameter, weighing about 454 kg. Attached to the main combustion chamber of the main engine, the nozzle would allow exhaust gases to expand to obtain maximum thrust, contributing to highly efficient burn. (MSFC Release 76-128)

—MSFC announced completion of tests conducted with KSC to determine the effects of salty and brackish water on materials used in elements of the Space Shuttle. Recovery of 2 Shuttle boosters after each launch would require towing the boosters in the salty Atlantic Ocean and into the brackish waters of a Fla. river. The studies used an integrated test bed constructed from the same materials and covered with the same sealants and paints planned for use on flight rockets, carrying samples of electrical and electronic equipment and hydraulic systems being considered for Shuttle use. The test bed, shipped to KSC, was put into the ocean for 7 days and in the Banana River for 3 days, at a depth that closely simulated actual booster-recovery conditions. Returned to MSFC, the test bed had been subjected to study by engineers and technicians to discover the effects of the ocean and river environments, and to look for corrosion or marine biological growth. After this study, the refurbished test bed would be used again for tests in the ocean by KSC personnel. (MSFC Release 76-131)

—MSFC announced it had requested bids from industry on development of range-safety receivers for the Space Shuttle. Each Shuttle's 2 booster solid rockets would carry 2 receivers and the external tank 1, a total of 5 for each launch. The receivers would be part of a safety system that would let the range safety officer destroy the vehicle in case of a malfunction at launch. NASA planned to reuse the receivers up to 20 times, recovering and returning the equipment to the manufacturer for testing before it was shipped to the assembly contractor for reuse. (MSFC Release 76-132)

—MSFC announced that assembly of the first Space Shuttle external tank had begun at the Michoud Assembly Facility near New Orleans. Assembly should be completed and the tank delivered to the Natl. Space Technology Laboratories at Bay St. Louis, Miss., in mid-1977 for use in the test program for the main propulsion engine. The tests would use the tank, three Shuttle main engines, and other components in actual firings to verify the design and operation of the main propulsion system. The program would consist of 15 test firings, 11 at or near the full 8-min duration. (MSFC Release 76-137)

—MSFC announced preparations for the arrival of the first Space Shuttle orbiter in mid-July for a series of tests at MSFC's dynamic test facility. The orbiter would arrive at the Redstone airfield in 1978 atop a modified Boeing 747 and be off-loaded by KSC personnel. On the ground, the orbiter would become the responsibility of MSFC personnel, who would move it to the test facility and mate it with the rest of the Space Shuttle elements for vibration tests. This would be the first time all elements of the Shuttle were united. Tests were scheduled to run from spring to winter of 1978. (MSFC Release 76-141)

—MSFC announced completion of tests on minitank 5, a small aluminum tank coated with a spray insulation that would be used to protect the Space Shuttle's external tank and its highly volatile liquid hydrogen contents from engine exhaust or aerodynamic heat. This was the fifth of 13 such tanks to be tested at MSFC. The tests were to determine the effectiveness of Martin Marietta's foam insulation against stress encountered during Shuttle launch and space flight. Besides acoustic and vacuum tests, the minitanks would undergo pressure, boiloff, and holding tests with liquid hydrogen; the hold test would determine the effect of a 7-hr idle period on a full tank's insulation system. The test period of 7 hr was estimated to be the longest time a fueled Shuttle would have to wait on a launch pad before liftoff. (MSFC Release 76-142)

—MSFC announced completion of the only totally new structure built exclusively for testing the Space Shuttle external tank: the pneumatic test facility at Michoud Assembly Facility near New Orleans, which would accommodate the tank measuring 47 m long (more than half the length of a football field) and 8.4 m in diameter. Empty, the tank would weigh about 34 000 kg. The pneumatic facility would be used for proof tests and leak tests of the external tank's liquid hydrogen tank; the proof test would pressurize the tank with gaseous nitrogen and apply a series of external loads with 9 hydraulic cylinders. The leak test would verify that no out-of-specification leaks existed. Martin Marietta Aerospace was prime contractor at Michoud for external tank production. (MSFC Release 76-143)

- Johnson Space Center announced it had awarded to Klate Holt Co. of Houston a 1-yr $1.5-million extension of contract for custodial support services at JSC. The cost-plus-award-fee contract was originally awarded in July 1975. The contractor had provided custodial services in 82 buildings at JSC and to buildings occupied by JSC at nearby Ellington AFB. (JSC Release 76-45)

- The *Newsreview*, published by the Air Force Systems Command, reported that a long-winged U-2 aircraft on loan to NASA had successfully completed tests in the "icy grip" of the main chamber at McKinley Climatic Laboratory during which it underwent exposure to temperatures as low as $-57°C$. The laboratory, operated by the USAF Armament Development and Test Center at Eglin AFB, Fla., conducted the tests at NASA's request to isolate flight-control malfunctions occurring in the cold temperatures of high altitude. The U-2, a single-seat single-engine jet designed to operate at altitudes above 21 km, had been used by NASA as an earth-resources survey tool, making observations in astronomy, high-altitude atmospheric physics, and geophysics in addition to supporting general earth-resources programs. (AFSC *Newsreview*, July 76, 6)

- Gene Sivertson of Langley Research Center's Space Systems Div. had developed a new idea for search and rescue operations using passive reflectors with imaging or side-looking radar to locate persons in emergency situations. The passive reflectors could be rigid, inflatable, or erectable structures that would show up on a radar system as bright spots easily distinguishable from clutter signals of the terrain. The reflectors could be carried as standard equipment by aircraft, ships and small boats, earth vehicles, and individuals; different packages could be tailored to the user's needs. The inflatable concept would use a balloon containing a reflector structure of aluminized mylar, with a small helium or hydrogen gas generator in the canister holding the mylar-saran balloon. Deployed in an emergency situation, the reflector would mark a distress site and provide a radar target. The concept had been field-tested in Mich. and Fla. with excellent test films as a result. Sivertson hoped to have the imaging radar on satellites with continuous global monitoring, and to test the system on the Space Shuttle in the 1980s. (*Langley Researcher*, 9 July 76, 3)
- Dr. John A. O'Keefe, geophysicist at Goddard Space Flight Center, published a book, *Tektites and Their Origin*, which noted that a piece of the moon could be purchased for a small amount. Tektites, the glass pebbles that fall from the sky, are ejecta from lunar volcanoes and not debris of meteorite impacts on the moon or the earth as previously suggested. Lunar missions that discovered tektite-like glass in the dust at some sites, the study of microscopic tektites from the bottom of earth's seas, and the volcanic appearance of some layered tektites caused the change in ideas. (*Goddard News*, July 76, 3)
- Calling the Landsat photographs "some of the most fascinating, most valuable photographs ever taken," *Readers Digest* magazine said the satellites "will have enormous effect on our lives and on the lives of our children," as they revealed "things never before known about the earth and man's activities on it, vastly improving our ability to make the planet more habitable." The article quoted NASA Administrator James C. Fletcher: "If I had to pick one Space Age development most likely to save the world, I would pick the Landsats and the satellites that will evolve from them later in this decade." The article described the process by which Landsat images were collected, processed, and used for a number of purposes: detecting oil and mineral deposits, finding sources of fresh water, assessing crops, forcecasting and minimizing damage from natural disasters, and monitoring population growth and pollution, among others. The Landsats, despite their importance, were still "in their infancy," the article said, looking toward the day when "few activities will be undertaken on...the globe without first consulting these electronic oracles in space." (*RD*, July 1976, 13)
- An unusually high loss of total blood hemoglobin after the first 16 days of flight was one of the medical findings from the *Salyut 4-Soyuz 18* mission reported by Cosmonauts Vitaly I. Sevastyanov and Lt. Col. Pyotr Klimuk to the 19th annual meeting of the Committee on Space Research (COSPAR) of the International Council of Scientific Unions. If valid, the Soviet data would represent a "significant departure" from earlier data

on body mechanisms governing red cells in zero gravity, said *Aviation Week and Space Technology* magazine. Klimuk told the Phila. meeting that hemoglobin had decreased 16% in his blood and 25% in Sevastyanov's by the 16th day of the flight; although past data showed hemoglobin loss averaging 1% daily, Sevastyanov must have had a mechanism that "destroyed more cells than normal" to achieve the 25% deficiency. U.S. Skylab missions reported that crewmen normally recovered their ability to produce red blood cells with more time in orbit; the cosmonauts did not discuss their recovery from the hemoglobin deficiency, but reported that nothing medical had ben discovered during their 63-day flight to prevent increased duration of Soviet space flight. (*Av Wk*, 5 July 76, 49)

- New federal policies deriving from the Occupational Health and Safety Act and the National Environmental Policy Act would affect about 30% of the USAF's Aerospace Medical Division investment in biotechnology research and development, said an article in *Aviation Week* magazine. "USAF has yet to appreciate...how important these two laws are going to be. The day is rapidly coming when even major weapons systems will never get through [a Defense System Acquisition Review council] without having to come to grips with...the occupational safety of the people that have to work around the system and the impact of...noise, electromagnetic radiation effects and toxic effluents," said Col. George C. Mohr, director of AMD research and development.

    Among items of special interest were the rocket fuels—unsymmetrical dimethylhydrazine and monomethylhydrazine—and the hydrochloric-acid fallout expected from Shuttle motor exhaust during launches at Vandenberg AFB; effects of laser exposure on eyes and skin; effects of multiple stress, including that caused by locating or operating equipment; and development of hardware, such as a lightweight 1-kg helmet to replace the standard 2-kg helmet that caused "high moment" effects on the wearer during sustained high-gravity maneuvers in the new generation of fighter aircraft. Another environmental-safety hardware item was a new 9-layer 25-mm windshield for planes to overcome the bird-strike problem, in place of the standard 2-layer 8.5-mm-thick windshield.

    A major problem, Col. Mohr said, has been that biotechnology standards and criteria often were not considered until after the system concept had been finalized; another problem is the DOD policy of tri-service planning, an attempt to ensure that the services do not duplicate research. Use of mathematical modeling in biotechnology had tended to "drive aerospace medicine research toward an engineering discipline," the article said; the data output is usable by design engineers, while new legislation prevents use of humans in hazardous testing, and cost of laboratory animals has soared. (*Av Wk*, 19 July 76, 219)

- "To prove that enthusiastic but untrained high school and college-age students can build a 5-kw solar generator," a project called Sunfire managed by JPL's Space Exploration Post would build such a generator for inhabitants of Pitcairn Is. in the South Pacific about 8000 km east of Australia. In 1974, Howard Broyles of JPL became interested in using solar energy to supply power at low cost, and discovered that the Pitcairn residents could use the power; he interested a physics teacher at the local high school in the actual construction of a solar electric generator, and work began as a class project. When the site of construction became unavailable at the end of the school year, the project, under the name Seep (solar energy experimentation project), was offered to the JPL Space Exploration Post and approved by Dr. William H. Pickering, JPL Director. Sunfire—backward acronym for Energy for Remote Islands From the Sun—would be a 5-kw generator converting solar rays into electricity by using a parabolic mirror to focus the rays on a boiler that makes steam, turning a double set of turbines that run alternators to produce the power. The Pitcairn residents, fewer in number "than JPL's Building 180 has employes," are mostly descendants of the crew of HMS *Bounty*, which mutinied in 1790 and found refuge on the small remote island. (*LabOratory*, 1976-4, 8)

August 1976

1 August: Scientists at JPL, where the *Viking 1* mission was being directed, were startled at responses from 2 of 3 instruments reporting on the first Mars surface samplings. The labeled-release instrument, designed to count radioactive molecules released by a process feeding tagged nutrients to a soil sample, had counted 4 times as many radioactive releases as would have resulted from a similar earth sample: 4537 per min for the first 9 hr 20 min it counted, compared to about 100 counts per min from Calif. soil sampled in a prelaunch test, and no more than 750 per min from soil gathered in the dry valleys of Antarctica. The rapidity and magnitude of the response were "quite surprising," said Dr. Gilbert Levin of Biospherics, Inc., which built the instrument.

Another Viking instrument, the gas-exchange experiment that dampened a surface sample with nutrients and monitored the resultant release of gas, detected an increase in oxygen released by the sample 15 times greater than was explainable by the Mars atmosphere or by oxidation of minerals similar to earth soils. The same instrument 24 hr later reported an increase in oxygen release 30% greater than that detected the first day. Scientists said the results might be attributed to superoxides, produced by intense ultraviolet solar radiation on the Mars surface, reacting with the extemely iron-rich Martian soil, not shielded by atmosphere as earth's surface is. Further information would be needed to reach firm conclusions, said Dr. Harold Klein of Ames Research Center, chief Viking biologist. (*W Post*, 1 Aug 76, A-1; *NYT*, 1 Aug 76, 4-8)

- Having completed a yr of service to India's Satellite Instructional Television Experiment (SITE), *Ats 6* began a 4-mo journey back to an orbital location over the Western Hemisphere to take part in experiments using direct broadcasting for education and health care. During its journey, it would be used by NASA in a project with the U.S. Agency for Intl. Development (AID) to demonstrate the potential of direct broadcasting to officials in more than 24 developing countries.

The first set of demonstrations—known as AIDSAT, for AID space-age technology—would begin between 1 and 26 Aug. for 11 developing countries and one international conference; the program would consist of 3 films created especially for the purpose, one on communications technology for national development, one on use of satellites for natural resources monitoring, and one on use of satellites for disaster prediction and relief. After these films the host country would transmit a 30-min program originated in the terminal in each nation's capital; then a 2-way discussion would be shown in which U.S. officials conversant with space technology and U.S. assistance would talk with representatives of the host country. President Ford would present an initial greeting, and astronaut Owen K. Garriott would be moderator of the first 5 programs.

Of the 12 programs planned, those to Thailand, Pakistan, and Bangladesh were broadcast first, and subsequent ones were scheduled 9 Aug. to the United Arab Emirates, 10 Aug. to Oman, 16 Aug. to Jordan, 17 Aug. to Kenya, and 18 Aug. to Yemen. On 23 Aug. the program would address the Conference on Applied Science and Technology in the Arab World at Rabat, Morocco; programs to Libya, Sudan, and Morocco would be broadcast during the remainder of August. A second group of demonstrations would begin in late Sept., and 15 more countries in Africa, Central and South America, and the Caribbean had been invited to participate. (NASA Release 76-140)

- By the yr 2000, mankind would have chosen between global cooperation and mutual destruction, said biochemist Isaac Asimov, professor at the Boston Univ. School of Medicine and one of the most prolific science-fiction writers. In an article copyrighted by the Phila. *Bulletin,* Asimov set forth his predictions which envisioned a single world power: a global community would grow in which more and more of the world's activities would be under control of multinational organizations, approaching a 21st-century global government, unless the nations were willing to settle for mutual suicide. The U.S. population would be less than now predicted, Asimov said, about 265 million, with government policies keeping growth to a minimum. The world population would have reached about 6 billion, a 50% increase from that of 1976. The U.S. would be searching for ways to use its food supplies to encourage a saner population policy; it would not be able to hoard its food for profitable sales because the welfare of the U.S. would depend on a strong world economy and "as non-desperate a world population as possible." The U.S. would be more nearly a vegetarian nation, both because of the higher yield per acre of grain and other plant food than if used for animal raising and because of the adverse effect on world opinion if the U.S. continued its wasteful eating habits while others starved.

  World opinion would have more power to affect national policy in the yr 2000; improved electronic communications would bring peoples closer and make them effectively part of a global community. Petroleum fuels would diminish and no longer serve as the major energy source; coal would become more important, as would windpower and solar and geothermal energy. By the yr 2000 the world would embark on new technological advance, whatever the apparent benefit, with more caution and foresight. A space colony would be under construction, to serve both as a human habitat and as a solar-power station. As the number of such colonies increased, population expansion might again be possible, plus sufficient energy to satisfy a hungry world. The yr 2000 would probably be a dark time, Asimov said, but would offer hope to all those alive for a brighter future for their children. (*C Trib,* 1 Aug 76, 23)

- Tass, the official Soviet press agency, said that permanent monitoring of earth from space would soon be a possibility because of advances in manned space missions, citing a prgress report on the cosmonauts (Boris

Volynov and Vitaly Zholobov) who had been orbiting the earth in the *Salyut 5* space laboratory since 7 July. Major tasks of the cosmonauts, Tass said, were to survey Soviet territory below the 52nd parallel and compile detailed maps; analyze geological formations for possible gas, oil, and ore deposits; and study seismic activity or storms and forest fires. A previous report described a photographic search for minerals by the two cosmonauts, who photographed the southern Ukraine, Moldavia, the Altai territory of Soviet Central Asia, and the Caspian lowlands. Tass did not say whether a permanent monitoring service would be automatic or man-operated, or if the USSR planned to inaugurate such a service. (*LA Times*, 31 July 76, 1 Aug 76)

2 August: An early burst of activity recorded by an instrument aboard *Viking 1* on the surface of Mars had begun to slow down, said biologists at JPL who had been startled by the high rate of reaction indicated by the labeled-release experiment. The instrument used a sample of Martian surface, fed with a nutrient of amino acids, sugar, and vitamins, plus a radioactive-carbon tracer that would produce radioactive carbon dioxide if the sample used the nutrient to grow or metabolize. A geiger counter in the instrument had counted as many as 4500 radioactive molecules per minute being released from the sampler over 8 hr, going as high as 8000 per min over the period of a day. Dr. Harold P. Klein of Ames Research Center, head of the Viking biology team, said the origins of the activity were not clear, and scientists were not sure whether "something is metabolizing" or not. It would take a wk to reach any conclusion from these results or those of the other two biology experiments, team members said. (*NYT*, 1 Aug 76, 1; *W Post*, 1 Aug 76, 1; 2 Aug 76, C-5; *W Star*, 2 Aug 76, A-1; *WSJ*, 2 Aug 76, 1)

3 August: Cloud-brightness images from orbiting satellites would enable the National Oceanic and Atmospheric Administration (NOAA) to predict the difference between relatively "dry" hurricanes and destructively wet ones such as Agnes (1972), Fifi (1974), and Eloise (1975), said Dr. Neil Frank of the Dept. of Commerce's National Hurricane Center. For years, said Dr. Frank, meteorologists had known that some hurricanes were wetter than others, with those having highest potential rainfall also highest in damage over land. A new technique would estimate hurricane-rainfall potential from cloud images calibrated with computer models of river-basin flooding; the technique would be "a predictive tool" for inclusion in hurricane warnings. Heavy rains from Fifi in 1974 caused one of the worst natural disasters of the Western Hemisphere when heavy rains brought unexpected flash flooding that killed thousands in and around Honduras. Besides killing 118 people, Agnes caused an estimated $2.1 billion in property damage from torrential rains and flooding; Eloise caused an estimated $200 million in property damage, again mainly from rain-caused flooding.

The satellite technique was used retrospectively to estimate the rainfall from 7 hurricanes for which radar and rain-gauge measurements were available; results disclosed little or no relationship between the volume or intensity of a storm and its rainfall. However, the rainfall potential calculated by the new procedure for the 7 hurricanes in the study agreed well with actual experience. Although estimates obtained through the new technique were relative, not absolute, the NOAA scientists viewed the procedure as a valuable tool for developing nations, an inexpensive system to monitor rainfall and improve the meteorological basis of agricultural and water-management planning. (NOAA Release 76-158)

- The rate of production of radioactive gas from a Mars sample monitored by an instrument on *Viking 1* had "plateaued," said Dr. Gilbert V. Levin, who designed the instrument, and the slowdown in activity was puzzling from both a biological and a chemical standpoint. A biological response would generally evolve gas for a longer period, whereas the count—if recording a merely chemical reaction—"took place at a very rapid rate initially and then, uncharacteristically, slowed down and took a long time to plateau." A special team of scientists from several fields was convened to consider all possible nonbiological explanations for the unusual readings from *Viking 1*, and various laboratories around the U.S. were being enlisted to help test the theories. (*NYT*, 3 Aug 76, 18; *W Star*, 3 Aug 76, A-7; *WSJ*, 2 Aug 76, 1)

- Edwin E. (Buzz) Aldrin, Jr., second person to walk on the moon, told an audience at Orange, Calif., that he had become an alcoholic several years before the 1969 *Apollo 11* moon mission and stopped drinking only 2 days before the flight. He had resumed drinking shortly after the mission. Aldrin said that, as a West Point cadet, he had been "caught in the alcohol trap": having decided to give up alcohol when he was sent to Korea as a combat officer, he resumed drinking when he perceived that his military image was measured by "who could drink the most." Aldrin spoke at a hospital where he stayed during a month-long recovery program in the summer of 1975; he had not mentioned his alcoholism in his 1973 book "Return to Earth" although he did discuss his post-Apollo psychiatric treatment, confirmed by the Air Force in 1972. (*W Star*, 3 Aug 76, A-2; *NYT*, 3 Aug 76; *W Post Parade*, 26 Sept. 76, 6)

- NASA announced that a French team of experimenters, using very high-resolution sensors on the *Oso 8*, had observed an oscillation in the sun's atmosphere every 14 min very like seismic activity on the earth, during which the atmosphere moves up and down as much as 1300 km. No one had expected "this huge movement of gas which might well involve the entire solar atmosphere," said Dr. Roger Thomas of GSFC, one of the *Oso 8* project scientists. "However, it may prove possible to use the waves to learn more about the sun's interior, in a manner analogous to using seismic activity to study the structure of the earth or moon." Dr. Roger Bonnet of the Centre National de la Récherche Scientifique in Paris, principal investigator, agreed that the discovery could be one of the most important results from the French instrumentation. *Oso 8* was launched 21 June 1975 carrying 8 onboard instruments for solar research—the most sophisticated and ambitious ever flown, according to NASA—and preliminary results indicated the spacecraft's mission could be considered successful. (NASA Release 76-141)

- Lockheed Aircraft Corp. reported a 24.2% decline of earnings in the 2nd quarter, and 10.1% decline in the first half of 1976, blaming the drop on substantial writeoffs and costs related to the TriStar L-1011 and reduced levels of production. The company had hoped to apportion startup costs of tooling and production over sales of an estimated 300 TriStars; only 13 planes were delivered in the first 6 mo of 1976 and only 3 more were scheduled for delivery during 1976, with firm orders for 22 more. However, 3 of the remaining order were subject to approval by the Japanese government, which was investigating Lockheed bribes paid in Japan to promote its sales there. Robert W. Haack, Lockheed chairman, said that the company was in an "overall stronger position" this year despite the lower earnings, pointing to debt reduction, a new bank-lending agreement, and contracts signed during the second quarter of 1976, which included a $625-million contract with Saudi Arabia for an air-traffic control system and a $697-million order from Canada for 18 of its F3 Orion aircraft. (*NYT*, 4 Aug 76, 41; *WSJ*, 4 Aug 76, 9)

- INTELSAT—the International Telecommunications Satellite Organization—announced award of an $8840 contract to the Univ. of Oulu, Finland, for measurements of low-elevation-angle scintillation. The 1-yr contract covered investigation of the slow fluctuation in received-signal strength from a satellite to earth stations in humid localities, at elevation angles of 5° above the horizon. Water vapor in the lower atmosphere could cause fluctuations in signal strength, intensified where the look-angle from station to satellite is as low as 5°, causing the signal to pass through more of earth's atmosphere than if the elevation angle were greater. (INTELSAT Release 76-21-M)

4 August: Johnson Space Center announced signing of a supplemental agreement with Lockheed Electronics Co. of Houston for $2.57 million, for additional scientific and technical support at the Slidell Computer Complex of the Earth Resources Laboratory in La. The supplemental agreement brought the total value of the contract to $7.1 million. (JSC Release 76-49)

- MSFC reported an unprecedented amount of data distributed to and used by the scientific community as a result of rapport, established under an arrangement known as the Skylab solar workshops, among principal investigators in the Skylab Apollo telescope mount (ATM) project and other scientists. The workshops, supported by NASA, had provided a means of "intense interaction and collaboration" by the participants, MSFC said.

  The first workshop, held in Oct. 1975 on coronal holes, was attended by 60 scientists representing 19 universities and other institutions; a second workshop in Feb. and the last this wk would conclude the series. The workshops met near the High Altitude Observatory at Boulder, Colo., under fairly isolated conditions; the scientists were together most of the time, passing on information, explaining discoveries, and talking shop from breakfast until late at night. Jack Zirker of the Univ. of Hawaii's Institute of Astronomy, director of the first workshop series, said the workshop procedure was "an ideal way to deal with the topic" because participants collaborated closely between meetings, planning and carrying out projects that otherwise might have languished.

The next workshop in the series, on solar flares, was scheduled to begin with a first meeting in the last week of Oct. 1976. Discussions at the first meeting would define the total effort; participants would return home to work on individual pieces of the problem. A second meeting 4 or 5 mo later would be a midterm status review and working meeting, with participants contributing information relevant to the total project. At the final workshop, participants would sum up their findings and identify unsolved problems, preparatory to publishing their results. The workshop would produce a monograph summarizing the state of the art at that point. Scientists taking part in the solar workshop series said the system could be adapted readily to research in other scientific fields and was highly productive. (MSFC Release 76-145)

6 August: The jumbo jet—"a revolutionary class of elephantine airliner" carrying twice as many passengers as the biggest narrowbodied aircraft—was a jumbo bust, said the *Wall Street Journal* reporting on declining business in the aircraft industry, especially on a 3-way battle among U.S. manufacturers for British Airways orders. The government-owned British airline would buy 6 widebody jet liners valued at about $165 million with possible options on 3 more, to replace an aging fleet of Boeing 707s and Vickers VC10s on its long-distance routes, and there was no British-made aircraft of the size British Airways needed.

McDonnell Douglas had been advertising in London papers "the difference to Britain" of $537 million and 13,000 jobs if the DC-10 were chosen, because it would use the Rolls Royce engine now used in Lockheed's slightly smaller TriStars. Lockheed, also competing for the order, said the engines on its L-1011 TriStars had already provided more than $750 million of business for Britain, and buying a long-range version of the TriStar for overseas routes would be simply an extension of a proven relationship. The Boeing Company had also been competing for the order, although its 747 was said to be too big for the British Airways' long-distance routes: it would carry about 400 passengers, compared to 260 for the DC-10 and 240 for the L-1011. Boeing, however, had been talking with Aerospatiale of France on a smaller version of the latter's widebody twin-engine jet, to be marketed in the U.S. with a revised Boeing-built wing in return for French participation in a bigger version of Boeing's 737; the French were reported wary of Boeing and leaning toward an agreement with McDonnell Douglas on a modification of the French 150-seat Mercure.

The *WSJ* reported 10 Aug. that U.S. manufacturers' sales of widebody aircraft had fallen "billions of dollars behind original projections." McDonnell was producing DC-10s at only 20% of capacity; Boeing had cut output of 747s from 84 a yr to 18; and Lockheed's production of L-1011s was "even lower." U.S. airlines, for which the widebodies were originally designed, had not placed a single new order for passenger jumbos since 1973, and planned to buy "only a handful" before 1980. The jumbo was industry's answer to the problem of carrying the passenger load predicted for 2 decades without clogging runways, skyways, and airports; the traffic growth had been curtailed by economic recession and the oil embargo that increased fuel prices unexpectedly. Also, the rise in nonstop service between medium-size cities whose traffic formerly was funnelled through major airports had cut into the traffic of the jumbo jets. Makers of the widebodies were far from breaking even on their design and development costs, which had run over $4 billion; about 1200 planes would have to be sold to cover that investment, the *WSJ* said, and so far the 3 makers reported only about 700 firm orders and deliveries, with another 80 options that might or might not materialize. (*WSJ*, 6 Aug 76, 12; 10 Aug 76, 1)

- NASA announced award of a $7-million contract to General Electric Co. and the Hamilton Standard Div. of United Technologies Corp. for design and construction of the world's largest windmill to be located at one of 17 possible utility-company sites where it could produce energy for as many as 500 homes a yr [see 30 July]. Sponsored by the Energy Research and Development Administration and NASA, the project would be directed by LeRC's wind-power office. Richard L. Puthoff, program manager of LeRC's wind-power office, said that wind as an energy source could provide 5 to 10% of U.S. needs at maximum efficiency and usage. The 100-kw system built in 1975 at LeRC's Plum Brook test area had been used as an experimental apparatus in developing the large windmill. (*NYT*, 8 Aug 76, 20; *WSJ*, 6 Aug 76, 6)

- INTELSAT (International Telecommunications Satellite Organization) announced that its board of governors, meeting in Washington 21-28 July, had approved the loan of a cesium ion thruster to NASA under a 1-yr no-cost arrangement, for use in research on inert gases (argon and xenon) as propellants for auxiliary propulsion in spacecraft and ground-based applications. Like the nickel-hydrogen battery for the Navy's navigation-technology satellite to be launched later this year and an echo canceller to be licensed for use in satellite communications, the INTELSAT thruster was developed under a series of contracts managed by COMSAT Laboratories at Clarksburg, Md. It had exhibited higher electrical and propellant efficiency than other thrusters of comparable size. (INTELSAT Release 76-22-M)

- The board of governors of INTELSAT—the International Telecommunications Satellite Organization—announced award of a 14-mo $2.9 million contract to the Société Des Télécommunications Internationales du Cameroun for additional tracking, telemetry, command, and monitoring services to be provided by a new antenna collocated at Zamengoe, Cameroon, with existing standard communications and antennas. The new facilities, to be operational by October 1977, would augment service as the global satellite system became more complex, providing redundant service especially in the Atlantic and Indian Ocean areas. The contract included an option for extended service for an additional 24 mo. The Cameroon facilities had monitored *Intelsat IV-A* satellites since 1 April 76. (INTELSAT Release 76-23-I)

8 August: A new political-economic battle between the U.S. and Britain was expected as bilateral talks were scheduled in London for Sept. on renegotiation of the "Bermuda Agreement" of 1946 regulating airline operations between the U.S. and Britain, in the Caribbean, and in and out of Hong Kong. Britain had announced 22 June it would withdraw from the agreement. The pact (which originally settled wrangling between wartime allies over peacetime air commerce, and served as model for more than 60 bilateral air-travel agreements between the U.S. and other nations) specified the routes the two countries' airlines could fly and prevented competition on prices. At the time, U.S. carriers had the advantage of new equipment to fly the North Atlantic, and Britain and other war-poor European countries were looking to tourism to bolster their economies.

Britain's objectives in renegotiation would be an equal share of traffic instead of the present 65 to 35% split in favor of the U.S., as well as a cutback in the number of North Atlantic flights. Constantine Menges, director of the international division of the U.S. Civil Aeronautics Board, said the British were interested in equality of results rather than equality of competition, and warned that encouraging airlines to equalize their shares of the market would endanger consumer interests by allowing the airlines to neglect measures that would increase their markets. Political factors would govern any decision: most non-American airlines (including Britain's) were state-owned and symbols of national prestige. On many international air routes, competing airlines pooled operations and shared revenues and expenses. This cartel philosophy had not applied to North Atlantic routes because of the Bermuda principles; however, Britain now wanted to reduce the number of flights (now involving 28 airlines) that had became destructively competitive, with an estimated equivalent of 8 jumbo jets flying empty between the U.S. and Britain every day. The British view was that restricting the number of flights would enable each nation's airlines to fly more profitably, eventually benefitting the consumer by easing the pressure for higher fares.

On the other hand, U.S. legislation that created the CAB in 1938 had exempted international routes from regulation by subjecting CAB decisions affecting them to presidential review, leaving room for political influences. Whereas the U.S. airlines would be sympathetic to restrictions on flight numbers, the U.S. ideological commitment would uphold free enterprise and unrestricted competition. However, renegotiation might offer the U.S. some advantages: changes on other transatlantic routes where it was now at a disadvantage against European national carriers, increasing the use of charter groups to fill empty seats on scheduled flights, and securing greater equality in so-called fifth and sixth freedoms that permitted European carriers to lure transatlantic passengers with cheap stopovers in and transport to and from other cities besides the basic destinations. Dr. Menges suggested that the U.S. might best negotiate with the European Common Market as a whole, instead of with any one nation; the British rejected this idea as an affront to national sovereignty. (*NYT*, 8 Aug 76, 3-1)

9–25 August: At 15:04 GMT (18:04 Moscow time) 9 Aug., *Luna 24*, an automatic space probe, was launched toward the moon "from the orbit of an artificial earth satellite" as the official USSR news agency Tass described it. A course correction 11 Aug. permitted the probe to reach a preset point in near-moon space; a braking operation 14 Aug. put the probe into a circular selenocentric orbit 115 km from the moon's surface, with 120° inclination and 1 hr 59 min period. (*NYT,* 11 Aug 76, 37; FBIS, Moscow Tass in English, 9 Aug 76, 14 Aug 76; *Spacewarn* bulletin, 17 Aug 76, 1; *P Inq,* 10 Aug 76, 6; LC *S&T Alert* #3407, 31 Aug 76)

At 6:36 GMT (9:36 Moscow time) 18 Aug. the probe landed on the moon in the Sea of Crises (Mare Crisium) at 12°45′N and 62°12′E, about 40 km from a northern isthmus between the moon's 2 seas, where the oldest rocks (about 4.5 billion yr) on the moon might be sampled. The area differed from the Sea of Felicity—site of landings by *Luna 16* (1970) and *Luna 20* (1972)—not only in age but also in the presence of mascons, gravitational anomalies caused by mass concentrations of extremely dense materials. (The U.S. *Lunar Orbiter 5* had discovered 5 such concentrations on the moon in 1968.)

The USSR's 8th successful unmanned landing on the moon in 6 yr was expected to make up for the failure of *Luna 23,* which sustained damage when it landed on the moon 8 Nov. 1975 and was unable to carry out its mission. The U.S. had not probed the moon since 1972, date of the last of 6 manned lunar landings. *Luna 24* weighed 1880 kg upon landing; early reports contained no description of the probe, but photographs of it appeared in Oct. in *Aviation Wk and Space Technology* magazine, with comparisons of the earlier probes. (FBIS, Moscow Tass in English, 18 Aug 76, 13 Oct 76; UPI, *Denver Post,* 19 Aug 76, 14-F; *NYT* editorial, 26 Aug 76; *Science News,* 28 Aug 76; *Av Wk,* 11 Oct 76, 53)

Tass announced 19 Aug. that, after being on the moon 22 hr 49 min, *Luna 24* had blasted off the moon's surface at 5:25 GMT (8:25 am Moscow time) and was on its way back to earth with a sample of moon soil obtained by a device that drilled to a depth of about 2m and put the core sample into a hermetically sealed container in the return module. The bulk of the spacecraft remained on the moon's surface; the *New York Times* noted there was no mention of a robot vehicle like the Lunokhods landed previously. Lunokhod 1 (on *Luna 17*) had spent 10 mo on the moon in 1970-1971, traveling about 10 km; Lunokhod 2 (on *Luna 21*) roamed the Lemonnier crater in the Sea of Serenity for 6 mo in 1973. (FBIS, Moscow Tass in English, 19 Aug 76; *NYT,* 20 Aug 76, A-19; UPI, *Denver Post,* 20 Aug 76, 10; *LA Times,* 20 Aug 76, 3)

On 22 Aug. the *Luna 24* return craft parachuted to earth in a "forest site" 2400 km northeast of Moscow, soft-landing at 17:55 GMT (20:55 Moscow time). The landing craft and sample were flown to Moscow from the site. Research on the comparatively large moon sample would proceed at the Geochemistry and Analytical Chemistry Institute of the USSR Academy of Sciences in Moscow, where lunar samples returned by *Luna 16* and *Lunar 20* had been studied. *Pravda* was quoted as stating that the 1970 mission of *Luna 16* brought back particles of lunar iron that "does not rust"; learning how to manufacture such a metal under earth conditions "would repay all the expenditures for space study," the newspaper said. Metallurgy stations might be established in space or on the moon to make nonrusting iron in commercial quantities, a process requiring a powerful vacuum like that in space which would be difficult to establish on earth. The gray-brown lunar material was said to resemble the *Luna 16* sample, also taken from a lunar "sea" area. Mineralogists studying the top of the 30-cm-wide column of material said they expected to find the same Mendeleyev elements existing on the moon as on the earth, but in different proportions because of fewer volatile materials. Scientists would search the lunar sample for any sign of water vapor, regarded as important "not only for space physics but also for the future exploration of space," Tass said.

The *Luna 24* landing craft, still on the moon, would continue to function and flight controllers were still communicating with it. (FBIS, Moscow Tass Intl Serv in Russian, 21 Aug 76; Tass in English, 25 Aug 76; *CSM*, 24 Aug 76, 1; *B Sun*, 24 Aug 76, 5; *SBD*, 31 Aug 76, 223)

9 August: McDonnell Douglas Corp. announced award of a $63.9-million contract to Hughes Aircraft Co. for research, test, and development of radar for the Navy F-18 strike fighter, including 3 renewal options for additional production beyond the development phase. Current Navy and Marine Corps plans called for procurement of 800 F-18s in addition to the test-flight planes; major assembly was to be completed late in 1977 and early 1978, with first flight scheduled for mid-1978, McDonnell said. Hughes Aircraft received Army and Navy contracts totaling $16.5 million a wk later for a location-reporting system and for development of a missile fire-control system for the F-14 fighter plane, the *Wall Street Journal* reported. (*WSJ*, 9 Aug 76, 11; 17 Aug 76, 6)

- More trouble developed for the DOD's "controversial" F-111 and F-14A jet fighter planes, the *Wall Street Journal* reported, as the Air Force announced grounding of 183 General Dynamics F-111s at Nellis AFB, Nev., and Cannon AFB, N.M. The groundings, ordered after inspections showed defects in fan blades, would continue to midmonth and would not affect F-111s stationed elsewhere. A General Accounting Office report on the Grumman F-14A, 14 of which had crashed at a loss of $161.6 million, questioned the plane's ability to cope with attacks by antiship missiles. The GAO said that problems with equipment and inadequate supplies of parts had reduced the readiness of the F-14A to 37.2% during 1975, and that the electronic equipment's reliability was extremely low, some of the major systems exhibiting only 6% to 14% of the desired objective. Cost of the F-14A had risen from $18.2 million each in 1973 to $20.2 million. Designed to defend the Navy against assaults by antiship missiles, the F-14A could engage current Soviet fighter jets, said GAO, but the Navy itself feared the plane's ability to cope with the missiles might be "marginal." (*WSJ*, 9 Aug 76, 11)

- The U.S. Senate, by a vote of 78 to 6, confirmed the nomination of Dr. H. Guyford Stever to be director of the newly created Office of Science and Technology Policy in the White House. Dr. Stever, director of the National Science Foundation, had been under criticism for NSF actions in reviewing applications for research grants; also, conservative senators had objected to social-science textbooks prepared under NSF auspices in the 4 yr that Dr. Stever headed that agency.

The confirmation marked reestablishment of a White House science advisory post after its elimination by President Nixon early in 1973. On 7 Nov. 1957 President Eisenhower had announced appointment of Dr. James R. Killian as special assistant to the President for science and technology, whose prime role was to speed rocket and missile development in the wake of the first satellite a month earlier. Dr. George B. Kistiakowsky, who succeeded Dr. Killian, wrote later that the science advisory committee set up in the White House helped President Eisenhower make major military decisions.

The Natl. Science and Technology Policy, Organization, and Priorities Act of 11 May 1976 would also provide 4 new executive-branch agencies: the Office of Science and Technology Policy, whose director would advise the National Security Council upon request, but would mainly assist the Office of Management and Budget in decisions on funding federally supported R&D and would prepare an annual science and technology report for Congress as a counterpart of the State of the Union Message; the President's Committee on Science and Technology, consisting of 8 to 14 specialists in a wide range of fields who would study for 2 yr and report on the nation's science, engineering, and technology policies, and disband after submitting the report unless the President chose to continue it; the Federal Coordinating Council for Science, Engineering, and Technology, to deal with interagency problems, chaired by the Science Adviser; and the Intergovernmental Science, Engineering, and Technology Panel, consisting of the Science Adviser as chairman, the Director of the NSF, and at least 10 members representing state interests, to define civilian problems at state, regional, and local levels which could be resolved by science, engineering, and technology. Dr. Richard T. Atkinson, Deputy Director of NSF, would act as Director until the end of the present administration in January. (*CR*, 9 Aug 76, S 13954-13963; *NYT*, 6 Aug 76, A1; 10 Aug 76, A3; 15 Aug 76, 4-6; 24 Aug 76, 28)

10 August: By a vote of 82 to 6, the U.S. Senate approved a $104-billion defense appropriations bill that prohibited expenditure of $1 billion voted by the House for the controversial B-1 bomber, until after 1 Feb. 1977, when the next President would be in office. The House had included funds for initial orders for a fleet of 244 supersonic B-1s that would eventually cost more than $22 billion. The bill would go to conference where the House-Senate differences would be settled, after Congress returned 23 Aug. from a week's recess for the Republican convention. The date of the production decision would be important because the current Republican administration proposed to place orders for the first 3 planes in Nov.; the Democratic presidential nominee, Jimmy Carter, opposed B-1 production and wanted a review of prototype test data early next yr. The Senate bill was $3.9 billion less than President Ford's budget proposal and $1.4 billion less than the House voted; however, it was $11.6 billion more than appropriated for FY 1976, which ended 30 June.

(*NYT*, 10 Aug 76, 8; *WSJ*, 10 Aug 76, 3; *W Star*, 10 Aug 76, A-2)
- The U.S. Air Force announced selection of the Boeing Aerospace Corp., a unit of the Boeing Company, to design and produce an interim upperstage (IUS) solid-fuel rocket motor for the Space Shuttle, to lift Shuttle craft from low earth orbit to higher mission altitudes. Boeing, chosen over Martin Marietta, Lockheed, and General Dynamics Corp., still had to negotiate a contract, estimated to be ultimately worth about $300 million. (*NYT*, 10 Aug 76, 45; *WSJ*, 10 Aug 76, 3)
- Aeronutronic Ford Corp. would probably win a $250-million contract to build Intelsat V, the next generation of seven global comsats owned by INTELSAT (the 94-nation Intl. Telecommunications Satellite Organization), said the *New York Times*. Scheduled for launch starting in 1979 to expand message-handling capacity of the INTELSAT network, particularly across the North Atlantic, Intelsat V would be first to use both a spatial separation technique used on Intelsat IV and a cross-polarization technique used on two ComSat domestic satellites leased to American Telephone & Telegraph and the General Telephone and Electronics Corp. INTELSAT's board at its July meeting chose Aeronutronic for final negotiations over Hughes Aircraft, which had build all but one of the preceding generations of Intelsat spacecraft; other competitors eliminated earlier were TRW Systems (which had built one previous generation of the Intelsats) and Lockheed Missiles and Space Co. All the competitors had selected West European or Japanese subcontractors.

    Unlike previous Intelsats, which were cylinders coated with solar cells, Intelsat V's would be boxlike containers with flat solar-cell panels stretching from each side like windmills; weighing nearly a metric ton apiece, the spacecraft would be stabilized by a 3-axis system similar to those used on the Franco-German Symphonie satellites over the Atlantic Ocean, and on the U.S. domestic satellites built and operated by RCA. Each Intelsat V would carry 27 signal-transmitting transponders with capacity of 12 000 two-way telephone conversations, compared to 6000 through Intelsat IVA and 4000 through Intelsat IV; each satellite would also transmit color TV. Intelsat V would replace Intelsat IVA which had replaced the Intelsat IV, seven of which were in orbit above the Atlantic, Pacific, and Indian Oceans. (*NYT*, 10 Aug 76, 37)
- Robert G. Deissler, technical consultant to NASA's Lewis Research Center, received the 1975 Max Jakob Memorial Award at the 16th National Heat Transfer Conference in St. Louis. The award, established in 1961 by the American Society for Mechanical Engineering and the American Institute for Chemical Engineering, commemorated a pioneer in the science of heat transmission, and was given for distinguished service in the field of heat transfer. Deissler, who joined Lewis in 1947 and gained early recognition for work in turbulent flow and heat transfer of variable-property fluids in pipes or tubes, was cited for "outstanding contributions to the theory of turbulence and turbulent transport...for his ability to perceive and derive the fundamental theory required to advance applied research and development in convective heat transfer." Author of about 70 technical papers, Deissler had received an Exceptional Service Award

in 1957 from NACA, NASA's predecessor. (LeRC *News*, 6 Aug 75, 1; Cl *PD*, 10 Aug 76)
- Both the Boeing Co. and Japan's official transport-development corporation announced pending agreement on Japanese participation in a billion-dollar project to develop a new medium-range Boeing jet liner in the 1980s. The new plane, temporarily designated the 7×7 (middle number to be given later), would be a 198-passenger 2-aisle craft, probably with 3 jet engines: one under each wing, and a third on the tail. It would be low-noise and fuel-conservative, with a high-performance bent wing to operate close to the speed of sound. The medium-range jet would fly one-stop trips between N.Y. and San Francisco. The pending agreement was announced separately, Reuters news agency reporting from Tokyo that the Civil Transport Development Corporation representing government and leading aerospace manufacturers had said the parties were ready to proceed; Boeing headquarters in Seattle termed the announcement premature, but verified progress in negotiations. Proposed Japanese share in the venture had ranged from 50 to 30 to 20%, the last amount currently being discussed. Aeritalia, the Italian aerospace-development concern, had agreed to put up 20% and Boeing had favored French, British, and West German participation. (*NYT*, 11 Aug 76, 43)

12 August: NASA announced plans for rollout of the first Space Shuttle orbiter on 17 Sept. at NASA's Rockwell Intl. Space Div. at Palmdale, Calif. The ceremony was scheduled for 9:30 am PDT, and media representatives were alerted. (JSC Release 76-51)
- WFC announced launch of a rocketborne experiment to investigate effects on the upper atmosphere of metal ions from a meteor shower. A 2-stage solid-propellant Nike-Tomahawk was launched at 11:54 am EDT, lifting a 149.3-kg payload to a peak altitude of 147.5 km. The payload included 2 experiments, one for the Univ. of Ill. and the other for the Univ. of Bern, Switzerland, as well as a solar-aspect sensor and a clamshell nosecone provided by WFC. The Illinois experiment also carried an instrument for GSFC. The meteor shower (the Perseids) deposited metallic debris in the atmosphere from about 120 to 90 km; measurements of the relative concentration of the different metals—magnesium, iron, calcium, potassium—could help identify the nature and origin of the meteor. A second objective would be to study the dispersion and concentration of metal ions in the atmosphere, thought to cause unusual propagation of radio and TV signals. The meteor shower was predicted to last for 4 days beginning 11 Aug. (WFC Release 76-12)
- McDonnell Douglas Corp. would join Aerospatiale and Dassault-Breguet of France in a transatlantic consortium to develop and market a new medium-range jetliner for the 1980s, announced France's Transportation Secretary Marcel Cavaillé at Toulouse, capital of the French aerospace industry. Tentatively named Mercure 200, the 160- to 180-seat twin-jet aircraft would have a range of about 3000 km; a modification of the current Mercure plane, it was estimated to cost about $240 million to develop, whereas a totally new plane would have cost more than $1 billion. The new plane, designed to save about 20% of fuel

costs over the current Mercure, would be powered by a CFM 56 engine jointly designed by General Electric and France's government-owned Snecma (Société Nationale d'Etude et de Construction de Moteurs d'Aviation). Aerospatiale—the state-owned National Society of Aerospace Industry—would get 40% of the construction work, which would cushion it from layoffs when construction of the Concorde ended this year; Dassault, the private company that designed the plane, would get 5%, probably the final assembly. McDonnell would get 15%, the rest to be offered to other European manufacturers.

Including McDonnell Douglas in the consortium would give the Mercure 200 an unprecedented entry into the American market, which amounted to half the world market, noted the *New York Times*. Although the new aircraft would compete with Boeing's 727 and 737, which previously had a monopoly on the medium-range market, exploratory talks with Boeing had come to nothing because French officials feared the size of Boeing's operation would relegate the French to little more than a subcontractor. the preliminary agreement with McDonnell Douglas awaited a McDonnell promise not to build a plane competing with the French-German passenger-transport Airbus. (*NYT*, 13 Aug 76, D3; *WSJ*, 13 Aug 76, 4)

- NASA announced it would cosponsor a conference 21 Sept. at the Univ. of Conn. on transfer of biomedical instrument technology. The conference, also sponsored by the New England Research Applications Center, the Conn. Dept of Commerce, the Conn. Product Development Corp., and the Univ. of Conn. Health Center, would display technology developed by NASA for the manned space program to manufacturers and show how the technology could be applied to develop improved medical equipment and techniques. Hundreds of improvements based on NASA technology were already in use by the medical community, ranging from bio-isolation garments to a rechargeable cardiac pacemaker. At the conference, medical experts from NASA Hq, JSC, and NASA's biomedical and technology applications teams would describe the types of technology available to industry and the medical community, and the role NASA could play in helping to commercialize new products. NASA's program of licensing patents would also be explained. (NASA Release 76-144)

13 August: MSFC announced completion of 2 of 8 major facilities for production of Space Shuttle solid-propellant rocket motors at Thiokol Corp.'s Wasatch Div. in Brigham City, Utah. Thiokol was selected as prime contractor by MSFC for development of the solid-fuel rocket motor, main element of the reusable solid-fuel rocket booster being developed by MSFC. The 2 completed units were a nozzle-bearing test facility, used to test a flexible bearing allowing the rocket-motor nozzle to turn up to 8° in any direction for thrust-vector control in steering, and an x-ray facility for inspecting propellant after casting into the motor-case segments. A case-preparation building, where motor cases would be prepared for propellant loading by sandblasting, cleaning, relining, and painting, was reported 95% complete. Three facilities for loading the propellants—casting pit 1, a casting house, and casting-pit covers—were about half completed. A case-refurbishment facility to handle the motor cases after recovery would be under construction shortly. (MSFC Release 76-151)

15 August: In an attempt to predict the weather on the planet Jupiter, Dr. Gareth P. Williams, researcher at NOAA's Fluid Dynamics Laboratory at Princeton Univ., applied a computer model of earth's atmosphere to Jupiter and found that it not only reproduced patterns known to exist there but also provided new explanations of visible features such as the Great Red Spot and horizontal stripes, as well as revealing a 4-yr cycle of heat transfer.

Using a set of equations describing physical processes that produce atmospheric behavior on earth, Dr. Williams fed in conditions known to exist on Jupiter—the planet's enormous size, amount of sunlight it receives, and rate of rotation (which dominated the behavior of the Jovian atmosphere)—and found that the basic behavior of atmosphere was "quite similar." The purpose of atmosphere, he said, was to transfer heat from the equator to the poles; the terrestrial heat cycle took about a month, but the NOAA model revealed a corresponding cycle on Jupiter that took 4 yr. Whereas the earth received its heat from the sun, Jupiter received only 4% as much solar heat as earth, but the infrared measurements showed that the planet was warmer than it should be if its only heat source were the sun. Scientists believe that the planet generates heat of its own in an amount about equal to what it receives from the sun; this extra heat complicated the job of modeling Jupiter's atmosphere because there was no way of telling how it might be distributed, but it also provided a confirmation for the model.

Atmospheric motions on Jupiter tended to perpetuate themselves instead of dying out, as they do on earth because of interaction with the earth's surface. Jupiter has no "surface" as we know it, Dr. Williams noted; "it just gets denser toward the interior and motions gradually dissipate downward...." Circulation patterns producing eddies in the planet's atmosphere seem to be stable; the Great Red Spot had been apparent since the 17th century, when it was first sighted. The researchers hoped that *Pioneer II*, now about halfway between Jupiter and Saturn, might provide a positive test of the Princeton model; "we think Saturn is like Jupiter, but we don't know," said Dr. Williams. "Maybe it has a red spot." (NOAA Release 76-170)

- Because supersonic aircraft are inherently dirtier than those flying slower than sound, the Environmental Protection Agency's emission standards for future generations of supersonic aircraft would allow SSTs to emit 4 times more air pollution than subsonic jets. EPA described the standards, scheduled for issuance 16 Aug., as "the most stringent that can be imposed" by the 1980 effective date; additional requirements for reducing amounts of oxides of nitrogen would take effect 1 Jan. 1984. EPA recommended new ground procedures to reduce duration of SST engine operation—and resulting pollution—while the planes were on the ground.

    New standards for 1980 engines would allow carbon monoxide emission at 58% below the current Concorde level, and hydrocarbons 77%

lower, with nitrogen oxides at about the current level; 1984 levels would be 88% less for carbon monoxide, 94% less for hydrocarbons, and 44% less for nitrogen oxides. The 1984 allowances would still be much higher than those for subsonics. EPA estimated it would cost the aircraft industry about $10 million to meet the new emission levels for a fleet of 70 planes, and that the standards would cost airlines about $5 per hr per plane for increased maintenance. (*W Post,* 15 Aug 76, A-12; *WSJ,* 17 Aug 76, 8)

16 August: MSFC announced that NASA had extended for 8 mo an existing contract with Sperry Rand of Huntsville, Ala., for continued engineering support services to MSFC's Science and Engineering Directorate. The $3,479,844 extension would run through 31 March 1977. (MSFC Release 76-152)

17 August: A large scientific balloon released 13 Aug. in Sicily by German and Italian scientists discharged a package of equipment for gamma-and cosmic-ray experiments and dropped it to earth near Rutland, Mass., said a meteorologist of the Natl. Weather Service at Worcester Airport. The radio-controlled balloon came down in Gardner, Mass., after a separation explosion at 40.2-km altitude. A spokesman at Hanscom AFB in Bedford said the balloon was owned by the Natl. Center for Atmospheric Research in Palestine, Tex. (*NYT,* 18 Aug 76, 13)

- The 2 *Soyuz 21* cosmonauts aboard space station *Salyut 5,* first manned space mission since last July's Apollo-Soyuz Test Project, were experiencing "a state of sensory deprivation, a sort of sensory hunger," according to the government newspaper *Izvestia.* Although Col. Boris Volynov and Lt. Col. Vitaly Zholobov continued to carry out their scientific and medico-biological experiments as they entered the 7th wk of their stay on *Salyut 5*—the halfway mark toward breaking the U.S. endurance record of 84 days in space set by 3 *Skylab 4* astronauts in 1973—and despite their years of training to combat the problem, *Izvestia* said "the organism still reacts in a peculiar way to space conditions." One symptom of "the increase in their need for communication" was that the cosmonauts were asking ground control more and more often for news from earth. On the advice of psychologists, ground control had begun playing music to the cosmonauts, *Izvestia* said. (UPI, in *W Star,* 17 Aug 76, A-4)

18 August: NASA announced appointment of Jerry Hlass, Director of Space Flight Facilities at NASA Hq, as Manager of the Natl. Space Technology Laboratories at Bay St. Louis, Mo., effective 1 Sept. Hlass came to Hq from GSFC, where he was head of the Data Acquisition Facility Section between 1961 and 1963. He received a bachelor's degree in mechanical engineering in 1949 from N.C. State Univ., and a master's in engineering administration from Geo. Washington Univ. in 1971. As manager of NSTL, Hlass would report to the Hq Associate Administrator for Center Operations. (NASA anno. 18 Aug 76; NASA Release 76-147)

- Lockheed Aircraft Corp. won a significant contract for construction of 6 new widebody jets when British Airways announced selection of the new long-range TriStar L-1011 for delivery starting in 1979 to replace

Boeing 707s and VC-10s on routes where the older planes were too small but the Boeing 747 was too large. Selection of Lockheed followed an intense competition with McDonnell Douglas over the past 6 wk, in which Lockheed was plagued by questions of finance following its near-bankruptcy in 1971 and its role in alleged bribery in the sales of military and commercial planes overseas. Lockheed chairman Robert W. Haack said market research indicated a need for 244 aircraft like the new TriStar by 1985, offering the company improved chances for reestablishing itself in the jetliner field. (*NYT,* 19 Aug 76, 51; *WSJ,* 19 Aug 76, 3; *W Post,* 19 Aug 76, C13; *C Trib,* 19 Aug 76, 4-9)

- A federal judge ordered NASA to reinstate with back pay and benefits the federal employees who lost their jobs at Marshall Space Flight Center to private contractors, the *Washington Star* reported. In a suit filed in 1967 in the U.S. District Court in Washington, D.C., by the AFL-CIO American Federation of Government Employees, the union contended that 22 MSFC contracts with private companies resulted in the layoff of about 3000 federal workers. Judge Joseph Waddy agreed that the contracts were illegal and ordered that federal employees who lost jobs as a result of work going to the private companies should be rehired, and that any similar contracts be canceled. The decision upheld Civil Service Commission standards that forbid a private contractor from assuming an employer-employee relationship over federal workers. NASA attorney Sid Masri said as many as 2000 employees could be terminated as a result of the decision, which NASA was sure to appeal. (*W Star,* 18 Aug 76, A-2, B-2; *WSJ,* 19 Aug 76, 17)

- U.S. officials reported that Peru planned to buy 36 late-model fighter-bombers from the USSR, a deal that might jeopardize a $200-million short-term loan being negotiated between Peru and a group of U.S. banks. The Soviet Union was said to have offered Peru good terms for a $250-million purchase of 2 squadrons of Sukhoi-22 planes; the price would be paid over 10 yr at 2% annual interest, with a yr of grace. A spokesman for the U.S. banks said they were unaware of the pending deal and could not comment on the effect it might have on the proposed loan, considered essential to enable Peru to pay for a recently expropriated U.S.-owned mining company valued at $100 million; Peru had offered $12 million in compensation. Peru had a foreign debt of about $3.7 billion, one of the highest of any developing nation. U.S. officials said Peru's motive for seeking the fighter-bombers might have been to apply political and military pressure on neighboring Chile, which was dealing with Bolivia to obtain access to the Pacific through territory seized by Chile from Peru in 1883. (*NYT,* 19 Aug 76, 5)

19 August: The National Science Foundation reported that U.S. spending for research and development (R&D) would reach $38.1 billion in 1976, 8% above the 1975 level. R&D expenditures would account for 2.2% of the gross national product in 1976, down from 2.3% in 1975. Of the total R&D spending, federal agencies would account for $20.1 billion; industry, about $16.6 billion; universities and colleges, about $800 million; and other nonprofit institutions, about $600 million. The figures appeared in a NSF report, *National Patterns of R&D Resources: Funds and Manpower in the United States 1953-1976*, updated annually in a continuing analysis of the nation's scientific and technological resources. (NSF Release PR76-68)

- The space share of federal R&D funding for FY 1977 would increase only slightly, according to a Natl. Science Foundation report, *An Analysis of Federal R&D Funding by Function, Fiscal Years 1969-1977*. The 1977 estimate of $2940 million was considerably less than the 1969 level of $3732 million; the space share—13%—was much lower than the 24% for the earlier year. Manned space flight, always the leading subfunction, remained predominant with the Space Shuttle accounting for more than 40% of all space activities. Space science, with 20% of the total, reflected a decline from 1976 and showed effects of lower funding for NASA lunar and planetary exploration as Viking and outer planet missions moved into the launch stages. Space technology and NASA support activities would increase during FY 1977. (NSF, *Space Science Resources Studies*, 19 Aug 76, 1)

- The European Space Agency announced that its exhibit at the Farnborough International Aerospace Exhibition in Sept. would include a full-scale model of the Spacelab, to be exhibited for the first time in Britain to give the British public a better idea of Europe's space activities and to extend contacts with science and industry. ESA's exhibit would also include full-size models of Meteosat and the OTS comsat, plus a tenth-scale model of the Ariane launcher. British firms were prime contractors for the OTS (orbiting test satellite) scheduled for launch in 1977 and for the maritime satellite Marots to be launched early in 1978. (ESA release 19 Aug 76)

20 August: Noaa 5, launched by NASA 29 July for the National Oceanic and Atmospheric Administration (NOAA) into a sun-synchronous polar orbit at 1505 km, was turned over to NOAA after checkout and would become the primary operational satellite providing weather forecasters with a wide variety of environmental data. *Noaa 5* would replace an earlier version, *Noaa 4*, which would become the backup spacecraft; an older polar orbiter launched in 1973 would be deactivated. NOAA's Natl. Environmental Satellite Service operates 2 satellite systems: one polar-orbiting, the other consisting of 2 geostationary satellites remaining in the same relative position to earth's surface at all times. The polar-orbiting craft would scan any spot on earth twice every 24 hr and return visual and infrared imagery of cloudcover, sea-surface temperature, and other environmental indicators. (NOAA Release 76-199; MOR E-601-76-17 [postlaunch] 1 Oct 76; telcon, NOAA PAO, 5 Oct 76)

24 August: Aerospace contracts valued at about $1 billion were hanging fire as DOD and NASA struggled over which agency would ask Congress for funds (beyond those already appropriated for development and testing) to pay for 2 additional Space Shuttles, said the *New York Times*. NASA's commitment of 1971 was to develop the Shuttle system at a cost of about $6.9 billion, to include 2 Shuttles, the start on a third, and a series of preoperational flights into orbit beginning in 1979. The NASA plan envisioned a total of nearly 600 Shuttle flights in the 12 yr beginning with

1980, with a fleet of 5 Shuttles averaging an annual 60 flights starting in the mid-80s.

DOD had earmarked $1.5 billion for Shuttle expenses by 1981, including $700 million for refurbishment of Vandenberg AFB in Calif. for polar-orbit Shuttle launches and $190 million for facilities at Cape Canaveral, Fla., plus $178 million for development of an interim upper stage (a solid-fuel rocket to be ordered by the Space and Missile Systems Organization of the USAF from Boeing Aerospace, for boosting Shuttle-payload spacecraft into extremely high orbit). DOD had been maintaining for 2 yr that it would not have money for the 2 final Shuttles; Dr. Malcolm Currie, DOD Director of Research and Engineering, told Congress during budget testimony 3 March that paying for the orbiters would exceed what DOD considered "cost effective" amounts for participation in the shuttle program.

DOD's refusal to pay had caused "consternation among planetary scientists," the *NYT* said, because NASA might be forced to cut much of its scientific space program in the 1980s to pay for 2 more Shuttles within a fixed budget. However, the matter should be resolved within a few wk, the article added. (*NYT*, 24 Aug 76, 37)

- The *Soyuz 21* manned mission to space station *Salyut 5* ended abruptly after cosmonauts Vitaly Zholobov and Boris Volynov had been in orbit 48 da. Although the first manned space flight since Apollo-Soyuz failed to break the U.S. 84-day record, no problem was mentioned by the Tass news agency; the first hint of termination came about 10 hr before the landing. The *Soyuz 21* descent module landed by parachute on the Karl Marx collective farm near Tselinograd in Kazakhstan at 9:33 pm Moscow time (1:33 pm EST); general condition of the cosmonauts was "satisfactory," said Tass. Experiments returned to earth for analysis were of 3 kinds: geophysical and astrophysical observations, technological, and biological. Technology experiments used processes potentially useful in constructing a larger space station, such as studies of the behavior of gases and liquids in space and a soldering technique using a chemical reaction to produce the necessary heat. Cytological and genetic experiments should provide additional data on the nature and hazards of solar radiation in prolonged space flight. (*Nature*, 26 Aug 76, 735; *Av Wk*, 30 Aug 76, 23; FBIS, Tass in English, 24 Aug 76; *NYT*, 25 Aug 76, 20; *W Post*, 25 Aug 76, B-9)

25 August: NASA announced the appointment of Carl E. Grant as NASA Director of Personnel, effective 5 Sept. For the past 5 yr, Grant had been personnel director for the Small Business Administration; previously he was associate director for personnel administration at the Smithsonian Institution, and had held personnel positions at the Civil Service Commission, Department of Defense, and U.S. Army. He received a B.S. in economics and business administration from the Univ. of Detroit in 1960. In his new post, Grant would be responsible for personnel administration agencywide. (NASA anno. 25 Aug 76)

26 August: First flight of a new fly-by-wire control system for jet aircraft was scheduled for 27 Aug. at Dryden Flight Research Center, NASA announced. The system would use 3 digital computers for primary control (and a 3-channel analog system for emergencies) to translate pilot signals to aircraft-control surfaces by lightweight wires instead of the conventional control systems, reducing the aerodynamic load as much as 20% and saving up to 10% in production costs. In Phase I of the program, a modified G-8 jet was flown 42 times to establish the feasibility of digital fly-by-wire control, using a single-channel digital computer developed for the flight-control system of the Apollo lunar module. Phase II would use a triplex system that could be modified to use Shuttle software, leading to a digital system with applications both to Shuttle and to other advanced transport systems. Project pilot Gary Krier would fly the checkout tests, scheduled to last through 1978 with about 30 flights planned. (NASA Release 76-150; DFRC Release 74-76)

27 August: NASA's Scientific and Technical Information Office announced publication of the first formal report on early scientific results from the Viking mission to Mars: an 80-p book, "Viking 1 Early Results." The book, a concise description of the Viking mission, its scientific experiments, and its objectives, told the story of the *Viking 1* lander's first 25 days of operation on the surface of Mars and gave "a preliminary and tentative account of early conclusions." The book also contained 50 photographs of Mars taken from the surface and from orbit. (NASA Release 76-151)

• A Spacelab simulation mission known as ASSESS II would be flown jointly next spring by NASA and the European Space Agency, and the payload specialists for the 4-engine jet mission would be named shortly, NASA announced. ASSESS was the acronym for airborne science Spacelab experiment-systems simulation. Carlos C. Hagood, NASA mission manager for *Spacelab 2* and ASSESS, said two U.S. candidates would be named as prime payload specialists, with a third as backup; ESA would also select prime and backup specialists. Launch date of the 10-day mission had been set for 15 May 1977. The mission would be flown in the Galileo II, a Convair 990 operated by Ames Research Center, one of several ARC flying laboratories supporting research activities internationally. The aircraft would fly 6.5 hr each day, and the payload specialists would remain in the plane and work another 6 hr on the ground; their off-duty time would be spent in a van containing living quarters connected to the aircraft.

ASSESS was developed to involve individual principal investigators (PI) in defining and operating the scientific payload; to study the investigator working-group (IWG) concept, providing PIs a means of full and direct visibility into mission management and a direct channel for making recommendations; to evaluate the concept of PI selection of payload specialists; and to validate a concept for training and cross-training PI-selected payload specialists. MSFC was lead center for the ASSESS II mission. Dr. Anthony C. DeLoach was MSFC mission scientist; ESA mission scientist was Dr. John Beckman of the European Space Research and Technology Center at Noordwijk, The Netherlands. ESA mission manager was Johannes de Waard, based in Porz-Wahn, West Germany. NASA candidates for payload specialist would be selected by NASA members of the investigator working group, and European candidates by the ESA members; the selections would be recommeded to the ASSESS program manager, Bernard T. Nolan of NASA Hq Office of Applications. (MSFC Release 76-158)

- A Soviet-American symposium of oceanography experts in the Black Sea town of Yalta ended as scientists agreed on a 2-yr program of joint effort in the Atlantic aimed at compiling a theoretical model of the ocean. The model would aid in more accurate weather forecasting, in predicting climate fluctuations several yr in advance, and in combating environmental pollution. Cooperating in the work would be 10 Soviet and American vessels, earth satellites, and meteorological ground stations. (FBIS, Tass in English, 27 Aug 76)

29 August: French interests had spent as much as $2 million to pressure the public, the news media, and various legislators into granting permanent U.S. landing rights to the controversial Concorde supersonic transport, wrote Robert Walters in the *Washington Post Parade* magazine. By the time a decision would be made next year, the expenditure would probably reach $3 million. The French producer of Concorde—Aerospatiale—signed up influential Americans and engaged as principal in its campaign a firm called DGA International, whose chairman was Charles E. Goodell, former Republican senator from N.Y. The U.S. Dept. of Justice found that DGA had a contract calling for a bonus of $500 000 if commercial Concorde service were authorized; U.S. law prohibited representatives of foreign interests from signing contracts in which financial compensation "is contingent in whole, or in part, upon the success of any political activities." The Dept. of Justice in a civil suit filed in federal court in 1975 charged that the contract was illegal; the case was settled without a trial when the defendants agreed to elimination of bonus clauses and to full labeling of their public-relations and propaganda materials. The lobbying, however, continued unabated, the *Parade* article noted, as did employment of former legislators and public officials on behalf of Concorde. (*W Post, Parade,* 29 Aug 76, 17)

- An article in the *Baltimore Sun* magazine reported that more than 3000 amateur radio operators—"hams"—in 90 countries had pooled their efforts to design, build, and finance the ham satellite series known as OSCAR, for orbiting satellite carrying amateur radio. The first OSCAR rode into space from Vandenberg AFB, Calif., 12 Dec. 1961, sixtieth anniversary of the first transatlantic wireless transmission, as a piggyback payload on the Thor-Agena carrying the USAF *Discoverer 36* into orbit. *Oscar 1* weighed about 4.5 kg and consisted of a battery, a small transistor, and an antenna packed into a metal case; constructed by radio hams working for companies building satellites, it operated on a tenth of a watt power and transmitted the greeting "hi" in Morse code before it stopped functioning 3 wk later. NASA agreed to provide launches as deadweight ballast for government satellites to advance public interest in space science, after the hams pointed out the OSCAR's potential use in educational programs and emergency communications.

In 1959 the hams had formed the Radio Amateur Satellite Corp. based in Washington and affiliated with the American Radio Relay League, national association of ham radio operators. After launch of *Oscar 4*, satellite activity shifted east, where AMSAT members working for GSFC and ComSat were eager to undertake assembly projects. By using volunteer efforts and parts donated as tax-deductible contributions by various companies, the hams could produce a satellite for about $60 000, compared to the commercial cost estimated at $2 million. *Oscar 6*, oldest active ham satellite, was about to celebrate its 4th anniversary; *Oscar 7*, most recent (and largest, at nearly 30 kg), had been in use for nearly 2 yr.

Baltimore engineer Jack Colson, an employee of the Johns Hopkins Applied Physics Laboratory, had received an award as the first amateur to be in touch with hams in all 50 states; he took 8 mo to achieve this, 4 mo for Hawaii alone because the satellite path permitted communications with the islands for only 2 min every 3 wk. Colson had also talked with hams in 43 countries and had bounced signals off the moon; he estimated that his ham contacts over the past 25 yr numbered about 30 000. The 750 000 radio amateurs all over the world came from all walks of life and ranged in age from 8-yr-olds to elderly men and women; about 20 000 amateurs were joining the U.S. ham network every year, the article said. Use of the satellite was free to all hams, whether or not they belonged to an amateur organization; OSCARs had transmitted medical data, weather bulletins, and emergency communications such as search-and-rescue mission information. Although shortwave radio could outdo the satellite on sheer distance, its transmissions—unlike those of the satellite—could be upset by atmosphere conditions or solar emissions. To improve OSCAR's time and distance limitations, AMSAT members were working on a new satellite that could transmit continuously over a greater geographical area for up to 10 hr, because of a higher elliptical orbit. Launch was planned for 1980. (*B Sun Magazine*, 29 Aug 76, 10)

30 August: The Peoples Republic of China launched its sixth satellite, referenced as *China 6*, from its facility at Shuang-Cheng-Tze into a highly elliptical orbit with 2147-km apogee, 193-km perigee, 69.1° inclination, and 108.7-min period. An announcement from the China press agency Hsinhua said the satellite was functioning normally, but gave no details on the spacecraft or its mission. (*NYT*, 31 Aug 76, 7; *SBD*, 8 Dec 76, 203; *SSR*, 31 Aug 76, 31; *SF*, Feb 77, 78)

- Return of the USSR's *Luna 24* to earth with a load of material drilled from nearly 2 m below the surface of the moon might be a dress rehearsal for a round-trip Mars probe to offset world headlines gained by *Viking 1*, said the *Christian Science Monitor*. The question why the Soviets would expend more effort on moon samples when they already had three sets (two from previous probes, *Luna 16* and *Luna 20*, and one from the U.S. Apollo program) would be answered if the *Luna 24* mission were being used to perfect techniques for obtaining Mars samples, a move "which would vault the Soviet Union back into the lead in planetary research," said the *CSM*. The USSR had landed unmanned spacecraft on the surface of Venus and obtained information for a short period, and had sent to the moon in 1973 an unmanned vehicle *Lunokhod 2* that crisscrossed the Lemonnier crater for 6 mo, reporting back data on magnetic fields and laser direction-finding. The *CSM* noted that, although the USSR automated machines worked well, the Soviets still lacked boosters to propel cosmonauts to the moon. (*CSM*, 30 Aug 76)

* The 26-m "Venus dish" at NASA's Goldstone tracking station near Barstow, Calif., normally used to communicate with interplanetary spacecraft, was serving as a research tool to study conversion of solar energy from satellites to electricity for use on earth, Goldstone researchers reported. In the tests, the big antenna represented an energy satellite collecting and converting sun energy into microwaves; more than a km and a half away, a set of receiver panels called a "rectenna" played the part of a ground station. The 7.6-m-high receiver panels contained more than 4500 aluminum T-shaped rectenna elements about 100 mm high, working like a TV antenna to gather and filter the microwave energy, converting it to AC or DC that could be fed directly into a utility. Results had been promising, with collection and conversion of microwave beams to usable electricity at an efficiency of 82%. (NASA PAO, 76-H-685, caption)

* Scientists all over the world had been invited to propose experiments for the second Spacelab mission, NASA announced, asking that they submit by 1 Oct. their letters of intent to propose, with actual proposals due by 3 Dec. Final selection of experiments would be made in Aug. 1977; proposals would be evaluated in the areas of space science (including life sciences), applications, and space technology.

Primary intent of the second Spacelab mission would be to evaluate the laboratory's system and subsystem performance, but space and resources would be available for experiments. Second objective of the

The 26-m "Venus dish" at NASA's Goldstone tracking station near Barstow, Calif., in use with rectenna (in background) for solar-energy conversion experiments. (NASA 76-H-685)

mission would be to demonstrate the broad capabilities of Spacelab for scientific research, especially in astrophysics (astronomy, high-energy astrophysics, and solar physics). Proposals should designate a principal investigator as the point of contact with NASA, to manage the efforts of any other persons involved in the experiments.

As the second mission payload would not include a pressurized module as the design for *Spacelab 1* did, instruments would be mounted on pallets exposed to space with a remote manipulator system available if needed. Experiments would be controlled remotely by persons at the payload-specialist station in the Space Shuttle orbiter's aft flight deck. Power, data distribution, and thermal control would be available from flight subsystems in a pressurized igloo on the forward pallet. Spacelab, a reusable space laboratory, was under construction in Europe by the European Space Agency for NASA. Marshall Space Flight Center was the lead center for Spacelab development, as well as for the Shuttle main engine, external tank, and solid-fuel rocket booster. (NASA Release 76-152)

- A slowdown ordered by the 14 000-member Professional Air Traffic Controllers Organization 25 Aug. had caused disruptions in air travel nationwide, airline officials told the *New York Times*. Eastern Airlines experienced 1-to 2-hr delays in rush-hour traffic at La Guardia Airport, and similar snarls were reported by other carriers at other airports, the most severe at Los Angeles. Union president John F. Leyden had asked members to start handling traffic "by the book" and adhere rigidly to rules that required planes to stay about 5 to 10 km behind aircraft ahead of them, omitting efforts to expedite traffic flow as is normally done. The controllers were protesting delay in completion of a Civil Service Commission study of controller job classification, which held hope of higher pay levels, and a CSC announcement that its investigation had found both undergrading and overgrading in the jobs. (*NYT*, 30 Aug 76, A-9)
- Information from space satellites orbiting more than 900 km above the earth was being regularly evaluated in U.S. attempts to measure crops in the Soviet Union, the Peoples Republic of China, and other countries, officials of the U.S. Dept. of Agriculture said. The information had been incorporated with other data in a project known as the Large-Area Crop Inventory Experiment (LACIE) that had been going on since 1974 and would be completed by mid-1978. The disclosure of crop-watching appeared in a weekly issue of *Foreign Agriculture* published by the department's Foreign Agricultural Service. (*W Star*, 30 Aug 76, A-5)

31 August: NASA selected 3 firms for negotiations leading to award of a single contract for assembly, checkout, launch operations, and refurbishment of the Space Shuttle solid-fueled booster, MSFC announced. The firms were Boeing, McDonnell Douglas, and United Space Boosters Inc. of Sunnyvale, Calif., a wholly owned subsidiary of United Technology Corp. The contract would extend through March 1980 to include the first 6 development flights of the Shuttle, with option for renewal through Feb. 1982. The contractor would be responsible for assembling and checking out the booster (forward and aft skirts, attach structures, nosecone, and various subsystems) with the solid-fueled motor. Two solid-fueled boosters would be used to launch the Shuttle, in conjunction with the orbiter main engines; after launch, the boosters would be jettisoned, lowered into the ocean by parachute, retrieved, and returned to KSC for refurbishment and reuse. Work would be performed at MSFC and KSC except for the solid-fueled motor, which would be returned to Thiokol facilities in Utah for reloading with propellant. (MSFC Release 75-159)

- Johnson Space Center announced selection of Mason-Reguard, Lexington, Ky., for negotiations leading to award of a cost-plus-award-fee contract for protective support services, including security, operation of police and fire departments, safety engineering, and emergency ambulance service. Proposed cost of the services from 1 Oct. 1976 through 30 Sept. 1977 would be $1 837 000, with option for extension over two more 1-yr periods. (JSC Release 76-52)

During August: The U.S. Air Force files for Project Blue Book, a 20-yr investigation of unidentified flying objects, had drawn many visitors since they were opened to the public at the National Archives in Washington, D.C., on 14 July, according to *Science* magazine. The USAF investigation had been closed in 1969 after the government decided that none of the 12 618 cases in the file indicated the existence of extraterrestrial vehicles. Only 701 of the total number remained unexplained. Biggest year for sightings was 1952, with 1501 reported; only 146 were reported for 1969, last year of the project, and only 1 of the objects sighted that year remained unidentified. The national preoccupation with UFOs "currently seems to be at a low ebb," said *Science*, adding that the most promising investigations might be in the behavioral or subjective aspects of UFO sightings. (*Science*, 20 Aug 76, 662)

- "Enticing hints but no firm answers" to the question of life on Mars were the product of an increasing volume of scientific data from the *Viking 1* lander; "practically faultless operation" for more than 5 wk had produced enough information to keep the Viking scientists busy, especially the data from the 3 bioexperiments and a related molecular analysis of Martian soil. Results were "certainly not what would be expected" if earth-type organisms were present, but they were not explainable in terms of "simple" nonbiological chemistry, said *Nature* magazine.

The labeled-release (LR) instrument added a small amount of water containing nutrients "labeled" with carbon 14 to its Mars sample, and incubated it at 15°C, warmer than the surface temperature of Mars. "Totally unexpected" was a massive and rapid release of radioactivity into the gas phase—presumably carbon monoxide or dioxide—and the radioactive count reached 4500 in 10 hr, leveling off at about 8500 in 48 hr. A second dose of radioactive nutrient produced an initial burst followed by a drop in radioactivity, and the graph "almost" flattened out for the remaining 4 days of the experiment; however, scientists were surprised to discover a possible trend to very slowly accelerating release of radioactivity. "Unfortunately," said *Nature*, "the experiment (was) stopped for performance of a control test..." A repetition of this experiment later would extend incubation over wk or mo rather than days; "results certainly do not look like anything one would expect" from an earth sample in the same situation. Most popular explanation for the initial burst of activity was oxidation of the labeled nutrients.

The gas-exchange (GE) experiment used gas chromatography to monitor a Mars sample wetted with a "soup" of nutrients to detect any changes in the instrument's content of gas. First dampening of the sample produced "a remarkable burst of oxygen" that leveled off after several hr but remained stable for days. The experiment would continue "for some time," to see if further changes occurred. Viking scientists were trying to duplicate results of the LR and GE experiments in earth laboratories using simple catalysts and oxidizing agents.

The pyrolytic-release (PR) experiment, most technologically complex, proved "least equivocal when it comes to interpretation" of the results, *Nature* reported. A Mars sample incubated in a "Martian atmosphere," with added water vapor and irradiation from a xenon lamp to simulate solar radiation, was heated to 625°C to drive off and account for unreacted carbon monoxide and dioxide and to pyrolyze any organic compounds trapped as vapor. The vapor trap, heated to 700°C, would release and oxidize any remaining organic compounds. A sterile sample should provide a peak ratio of about 500 to 1; the results from the Mars sample were about 75 to 1, "several times more active" than nonsterilized soil from the dry valleys of Antarctica. Results from the control experiment with heat-sterilized soil would be "crucial" because a high peak ratio would indicate that biological, rather than chemical, processes were going on in the soil that could fix atmospheric carbon monoxide or dioxide in the presence of light. If life were present on Mars, science would expect a high peak ratio from a surface sample strongly illuminated in atmosphere of about 90% carbon dioxide. Unfortunately, the initial run of the organic-chemistry analysis was spoiled by a failure of the surface-sampling arm to deliver a full load to the gas-chromatograph mass spectrometer. A further analysis would be performed on a full load obtained subsequently. Lack of detectable organic compounds in this sample would further complicate interpretation of the 3 biology experiments. (*Av Wk*, 30 Aug 76, 22; *Nature*, 26 Aug 76, 734)

- The 1976 summer Olympic games at Montreal 17 July–1 Aug. were the most widely telecast event in history, INTELSAT announced, having been viewed by more than 1 billion persons according to press estimates. Record use of the global system for satellite coverage included more than 930 transmissions during the games, more than 70 telecasts on some days. During peak periods satellites transmitted as many as 5 programs simultaneously across the Atlantic Ocean. The global television coverage used 2 Intelsats over the Atlantic, 1 over the Pacific, and 1 over the Indian Ocean; the 4 satellites sent 2585 half-channel hours of Olympic and related transmissions, more than twice the number of hours of the 1972 Olympics at Munich. Teleglobe Canada, the Canadian communications authority that arranged for the global services, used earth stations at Lake Cowichan, B.C., and Mill Village, N.S., as well as a transportable station at Montreal; other transmissions went through ComSatCorp's Andover, Me., ground station. (INTELSAT Release 76-24-I)

- The Natl. Science Foundation announced that federal funding for research and development would increase in FY 1977, ahead of expected inflation. The long-term trend of the R&D portion of the budget had been downward, with a steady decline from 1967 to 1971; a slight rise in 1972 preceded another decline to a 1975 low that represented the smallest federal support for R&D for the 10-yr period. The estimated increase for FY 1977 would restore "real performance" to a level close to that of 1972, although about 20% below that of 1967. Estimated shares of the FY 1977 total would be 11% for basic research, 23% for applied research, and 66% for development. (The NSF noted that most major NASA projects had been categorized as development, as they primarily generated outer-space transport technology. Substantial parts of these programs were classified in former years as basic or applied research. The shift in NASA categories resulted in lower shares of federal R&D funding for basic and applied research, and a larger share for development.) About 60% of development funds would be accounted for by DOD, which with NASA and ERDA would account for more than 90%. NASA's $99-million increase put its share of the total 1977 R&D budget at about 15%; DOD accounted for an estimated 48% of the federal R&D total in 1977, and ERDA would account for 14%. HEW would account for about 11%, NSF 3%, and USDA about 2%. The next 4 agencies in size of R&D support—DOT, Dept. of Interior, EPA, and Dept. of Commerce—all showed decreases in their federal R&D funding from the 1976 level. (NSF, *Science Resources Studies*, 10 Aug 76, 1)

September 1976

1 September: NASA launched a Navy Transit Improvement Program (TIP-III) spacecraft on a Scout vehicle at 2:14 pm PDT from the Western Test Range into an elliptical polar transfer orbit with 787-km apogee, 341-km perigee, 90.2° inclination, and 96-min period. Although orbit was achieved, the spacecraft's solar panels failed to deploy and the Navy was reported planning corrective action. The NASA portion of the launch was adjudged successful 21 Sept. (MOR 490-601-76-02 [prelaunch] 17 Aug 76, [postlaunch] 24 Sept 76; SSR, 31 Oct 76, 31; Pres Rpt 76, 99)

- Marshall Space Flight Center announced selection of Science Applications, Inc., of Los Angeles and Rockwell Intl. Corp. for negotiation of contracts to study space industrialization. The study program would lead from Shuttle Spacelab and early space-station experiments to permanent, practical, commercial use of space. The space-industrialization accomplishments so far achieved would be expanded into disciplines other than communications and meteorology; the study program would attempt to define a balance between future opportunities (satellite power systems) and relatively immediate benefits (space processing of materials). Possible areas of activity would include development of new techniques and materials, new development of earth resources, and eventually the movement of people to space for tourism or medical purposes, and the industrialization of the moon. The parallel studies would proceed in two phases, each requiring about 8 mo, and would cost about $200 000 each. (MSFC Release 76-160)

2 September: Something found by *Viking 1* on the surface of Mars behaved as if it were alive, said 6 biologists on the science team at JPL, who nevertheless refused to confirm the presence of life on Mars until results from *Viking 2* were in. *Viking 2* was on its way to a landing scheduled for 3 Sept. on Utopia Planitia (the Utopian plains) above which the air was found by *Viking 1*'s orbiter to contain 5 to 10 times as much water as that over the Chryse area where *Viking 1* landed. A surface sample of Chryse itself dug by a mechanical arm was found to contain a surprising amount of water that boiled off when it was heated by an instrument on the *Viking 1* lander. Four other instruments, 3 built to look for life, received samples of the Mars soil: one to look for photosynthesis, the second to look for signs of metabolism, the third to detect "breathing" or "sweating" characteristic of earth-type life forms. All 3 instruments appeared to detect what they were looking for, but the biologists explained the results as an exotic chemistry not connected with biology.

Having quoted odds of a million to 1 against finding life on Mars, the biologists now said they could never state with 100% certainty that

Viking had found life until they could return a Mars sample to earth, which some were afraid to do: Nobel prizewinner Dr. Joshua K. Lederberg of Stanford University had told associates that Martian life forms brought to earth might threaten the 2 million species of life here by competing in unknown ways for food, water, or air needed to survive. (Thomas O'Toole, *W Post,* 3 Sept 76, A–2)

- Marshall Space Flight Center announced that the Space Div. of Rockwell Intl. Corp. at Palmdale, Calif., had received delivery of three dummy main engines for the Space Shuttle—known officially as flight mass simulators—for mounting on Shuttle Orbiter 101, which would be used for approach and landing tests at Dryden Flight Research Center and for ground vibration tests at MSFC. The simulator engines, resembling the real main engines in size and weight, could be adjusted in weight and center-of-gravity and gimballed to provide various positions for testing. Orbiter 101, after vibration testing, would be returned to Palmdale for replacement of the simulators with flight engines and placed in flight status. (MSFC Release 76–162)

- The National Oceanic and Atmospheric Administration announced that a team of fishermen, engineers, physicists, oceanographers, biologists, and computer specialists had used a satellite to find fish off the coast of Louisiana in a government-industry cooperative program begun by NOAA last year. Water turbidity, measured by Landsat sensors as variations in color, would reflect the distribution of fish; the fishing vessels, working with spotter aircraft, confirmed the presence of fish in most areas predicted by the Landsat data. This technique of locating concentrations of fish would aid in better understanding of coastal fishery ecology and better resources assessment and management, NOAA said. On 19 July, *Landsat 1* had passed over a selected study area, sending multispectral scanner data to Goddard Space Flight Center; the data tapes were hand-carried to NASA's Earth Resources Laboratory in La. for processing. Less than 21 hr after the scan, the spotter pilots and fishing-vessel captains were being called to check their findings with predicted locations. The satellite reports were valid, NOAA said. Installation of an operational system would require 3 to 5 yr to develop special computer programs and facilities. (NOAA Release 76–196; JSC Release 76–50; JSC *Roundup,* 10 Sept 76, 4)

- *Soyuz 21* cosmonauts Boris Volynov and Vitaly Zholobov, who landed safely 24 Aug. from a 48-day mission to Soviet space station *Salyut 5,* returned to Moscow from the Baykonur cosmodrome near Tyuratam where they had been resting after their flight. The two received the title of Hero of the Soviet Union, Volynov for the second time. The news agency Tass reported that *Salyut 5* was flying in a "controlled automatic regime." In a *Pravda* interview, Academician Georgy Petrov emphasized the importance of orbiting stations, which he said would serve in preparing for space flights to other planets as well as in studying the earth from outer space. Petrov predicted the building of space stations with changeable crews of 20 to 30, serving for periods of time up to decades, and later the establishment of "super-large multipurpose orbital complexes with crews consisting of a hundred members and more." A special role would be that of stations for moon study in selenocentric orbits, from which crews would land on the moon's surface in "small expeditionary spacecraft." (FBIS, Tass in English, 1-6 Sept 76)

3 September: Viking 2 apparently touched down at about 6:39 EDT in a field of windswept sand dunes on the Utopian plains of Mars at the edge of its northern polar cap, after a communications breakdown that left JPL scientists without contact during the last 3 hr of the spacecraft's journey of more than 650 million km through space. The stabilization system on the orbiter, which served as a relay link between the lander and the deep space tracking network on earth, lost power 26 sec after separation of the lander at 3:40 pm EDT, and the high-power antenna on the orbiter no longer pointed at the earth. The blackout lasted nearly an hr, while mission personnel tried to reestablish communications. A low-power transmitter on the orbiter, installed to provide limited 2-way communications with earth, finally sent engineering data at 6:59 pm indicating that the lander had touched down. Lander signals were strong for 17 min until the orbiting relay link passed out of range.

First pictures from the lander would be relayed about 3 am EDT on 4 Sept. as the first step in a "recovery plan" to put the experiments and systems back on schedule. All pictures and data taken by the lander were being stored on tape in the orbiter for later playback. Except for the orbiter malfunction, the pictures would have been available within 2 hr of the touchdown.

The landing date of *Viking 1*, originally 4 July, had been postponed for 2 wk in order to ensure a safe site; the spot for *Viking 2* was chosen because indications of frost or fog in the area made it likely to encourage the presence of life, scientists said. Results from *Viking 1* suggested either that chemical processes never observed in earth laboratories occurred on Mars, or that life forms existed on Mars that were unknown on earth. Life search by *Viking 2* would not begin for a wk; the 2 landing craft were about 7400 km apart on the surface of Mars, *Viking 1* at Chryse near the equator (at a latitude comparable to that of Mexico City) and *Viking 2* in a warmer area (at a latitude comparable to that of Montreal) more than 1600 km northeast of the *Viking 1* site. (*W Star*, 4 Sept 76, A-1; *B Sun*, 4 Sept 76, A-1; *NYT*, 4 Sept 76, 1; *W Post*, 4 Sept 76, A-1)

- Newest international organization created to exploit space technology "for the benefit of mankind," the International Maritime Satellite Organization (INMARSAT) was chartered in London after 4 yr of study and negotiation. A 1958 United Nations subgroup called IMCO—the Intergovernmental Maritime Consultative Organization—had been responsible for nautical matters of common concern to seafaring nations. The UN Secretariat first report on space activities and resources in 1972 noted IMCO's interest in space for maritime purposes, particularly distress systems, safety of navigation and position determination, operation of maritime mobile services beyond the scope of existing methods, and improved maritime communications. IMCO proposed a new international maritime satellite system for exchange of telephone, telegraph, and facsimile messages and improvement of navigation. IMCO's Maritime Safety Committee in March 1972 had formed a panel of experts to study and recommend a program of experiments and development work that would be necessary to form a new organization.

The panel examined the financial, legal, technical, and operational problems of creating a new entity, reporting yearly to the safety committee. As the consensus favored formation of a new organization, the IMCO assembly resolved in November 1973 to convene an international conference early in 1975 to set up an international maritime satellite system. IMCO's secretary general was to invite all UN member states and interested intergovernmental and nongovernmental organizations.

The international conference, 23 April to 9 May 1975, was attended by delegates from 45 nations and observers from 15 international agencies and other organizations. The conference set up two working committees, one to discuss relationships between governments and their telecommunications and maritime entities and the distribution of powers between the member states and the INMARSAT council; the second working group would consider procurement and financial policy. The conference concluded that an international intergovernmental organization was needed to administer a worldwide maritime satellite system. It agreed to reconvene early in 1976, appointing an intersessional group to draft recommendations on four points: relationships between governments and designated entities, distribution of powers between assembly and council, type and number of appropriate international instruments, and procurement policies. The first conference concluded with a recommendation that all countries permit operation of ship stations (onboard terminals) in certain radio-frequency bands within harbor limits and other waters within national jurisdictions.

As the first session ended, representatives of 13 Western European countries and the United States agreed on major elements of the organizational arrangements that would form the proposed system. The arrangements envisioned the designation by a government of an entity to assume its responsibilities with INMARSAT, vesting of management in a strong governing body with investors making policy in proportion to their use of the system, and a procurement policy awarding contracts on the basis of price, quality, and most favorable delivery time. These issues had been debated during the organization of INTELSAT. INMARSAT differed, however, because of the maritime interests involved (ship owners, maritime unions, national maritime ministries and regulatory bodies) and the presence of the USSR historically had favored international organizations composed of governments only, having been reluctant to enter "mixed" organizations involving states and private enterprises, such as INTELSAT.

The working group's first session in London in August 1975 included representatives of 37 countries and 8 international organizations. The U.S. and the USSR reached an agreement on the basic roles of governments and operational entities, adopted by the session, which also developed a procurement policy. At the second session in October 1975, 31 countries and 6 international organizations were represented; this session discussed membership rules, investments, and information policy. The third session, held in December 1975 at Noordwijk, The Netherlands, was attended by representatives of 26 countries and 4 international organizations. It set up two committees, one to deal with financial, the other with nonfinancial, matters. The U.S. representative commented on "the cooperative spirit" shown at the three sessions.

The international conference resumed in London in February 1976 with representatives from 47 countries and 16 intergovernmental agencies and international organizations. Texts of two documents—a convention to be signed by governments, and an operating agreement to be signed by signatories (governments, or their designated entities)—were adopted in large measure, with some revisions. Three articles of the convention not decided were the maximum voting power of each council member, the question of permitting reservations to the convention, and the official and working languages to be used. The conference agreed to a third session to resolve these points, setting up a preparatory committee for the formal establishment of INMARSAT.

The third session of the conference, called on Sept. 1, 1976, was attended by delegates from 47 countries and observers from Yugoslavia, as well as delegates from 23 international agencies. The delegates agreed that no reservations would be made to the convention or the operating agreement, and decided to omit the matter of language from the convention. Preconference negotiations had led to a consensus on the ceiling permissible on a council member's vote, setting 25% of total voting participation as an upper limit; this wording was accepted by the conference. Observers noted significance in the languages in which the new convention was printed: English, French, Spanish, and Russian. (Stephen E. Doyle, *INMARSAT . . . Origins and Structure*, 13 April 77)

4–8 September: Soviet news agency Tass reported the successful touchdown of the *Viking 2* lander, adding that no data were coming from the lander "owing to failures in the Viking station's orbiting vehicle" that relayed lander communications to earth. (FBIS, Tass in English, 4 Sept 76)

As the long-awaited picture transmission from the *Viking 2* landing site on Mars finally appeared at JPL, scientists were surprised to see rocks and more rocks in the Utopia region where they had expected steep sand dunes. Boulders were so numerous that driving a vehicle across the Utopia plain would be a challenge, Walter Sullivan wrote in the *NYT*. The scientists had hoped that the damper climate at the *Viking 2* site might give more definite clues to organic life than the ambiguous reports from the *Viking 1* lander experiments. They said they were delighted with the clarity of the photographs and the speed with which they were transmitted once the communications problem was solved. A tilt in the horizon appearing in the pictures meant either that one footpad of the lander was resting on a rock (according to Dr. Thomas A. Mutch, head of the surface-imaging experiment), or that the lander was on a slope (according to Project Manager James S. Martin, Jr.). Martin expressed concern about getting the second lander back on schedule after the difficulties it had encountered on descent. Although communication had been reestablished with the orbiter's main antenna after an unexplained stability problem had pointed its transmitters out of earth's range for about 12 hr, the lander might have suffered from a rocky landing when one shock absorber failed to function. An apparent tear was visible in the rim of the 76.2-cm aluminum dish on the *Viking 2* lander, which Martin said probably resulted from the landing, but did not affect the radio signals to earth. Only direct communication with the lander would show whether damage had occurred to the body and its electronics systems. Dr. Carl Sagan of Cornell Univ. pointed out that both Viking landing sites had been chosen purposely "for their blandness," as free as possible from geological hazard; the rocks visible in the *Viking 2* transmission were evidence of "an enormous exuberance of geological processes."

Meanwhile, two of *Viking 1*'s three biological instruments repeated their samplings of Mars soil from the plain of Chryse, with ambiguous results. The experiment that used nutrients to detect life processes in the soil registered about the same number of radioactive counts (suggesting that the nutrient was being metabolized) as the first trial had. The photosynthesis detector registered only a third as much activity as the first sample, however.

The *NYT* pointed out that the *Viking 2* lander would deploy its instruments in an environment different from that of *Viking 1*. Utopia, nearer the Martian north pole than Chryse, was expected to have a layer of water-containing permafrost underlying the surface; as the Martian summer was at its warmest in the Utopia region, scientists hoped for a better chance of life-identifying experiments at the Utopia site. Exobiologists had noted that, if *Viking 2* results were more suggestive of life, they would feel encouraged; if the results were the same as those from *Viking 1*, this at least would show that the first reports were reliable. (*NYT*, 5 Sept 76, 1, 4–5; *W Star*, 5 Sept 76, A–1; *W Post*, 5 Sept 76, A–1)

An editorial in the *NYT* said the Viking engineers from government and private industry had needed luck to achieve two successful landings in two attempts; luck would have been inadequate without the "superb design and technical foresight" that had anticipated and forestalled many problems. (*NYT*, 6 Sept 76, 14)

Project Manager Martin reported 5 Sept. that what had appeared to be a tear in an antenna was probably dirt, spattered up when *Viking 2* landed. The successful uncaging of the *Viking 2* seismometer met with cheers from the geophysicists on the project who had been marking time since the *Viking 1* instrument failed to unlock. Dr. Nafi Toksoz, a seismologist from MIT, said that lack of data from more than one point on Mars would make triangulation (and precise location of seismic disturbances) impossible, but even one instrument might answer questions about the nature of the planet's interior and the level of tectonic activity. He predicted that Mars would be "more active than the moon and less active than the earth." (*W Star*, 6 Sept 76, A–3)

As *Viking 2* prepared to reach for soil samples, project scientists speculated on the possibility of mobile landers to explore the terrain and collect samples for study on earth. Dr. Elliott C. Morris of the U.S. Geological Survey said he felt like a child with nose pressed against a candy-store window, the "goodies forever beyond his reach." Like the Soviet Lunokhods that moved about on the moon, Mars landers with wheels or tractor treads could be set down in a relatively flat area and sent into mountains, canyons, and craters far more dramatic than similar features on earth. Project Manager Martin noted that such a mission could be launched in 1981 to land on Mars the following year: a backup lander in storage at Martin Marietta's Denver plant could be modified to make it mobile, and a spare orbiter was in storage at JPL, control center for the Viking missions managed by LaRC for NASA. Walter Sullivan of the *NYT* commented that the vivid detail in the panoramas transmitted from the cameras on the two landers had given the viewers the urge to go on over the horizon to see what lay beyond. (*NYT*, 7 Sept 76, 21)

The first weather report from the *Viking 2* lander disclosed that Mars weather was almost the same in the northern latitudes as in the tropics this time of year, except that winds in the northern regions shifted more often. Dr. Seymour L. Hess of Fla. State Univ., leader of the Viking weather team, said the findings were as predicted, the winds changing in patterns similar to those in similar regions on earth. Bitterly cold temperatures on the Utopia plains—warmer than at Chryse, where the nights were longer—were no colder than Arctic or Antarctic temperatures at night on earth. The soil at Utopia appeared red in the *Viking 2* photographs, but not as red as the surface at Chryse. The red color resulted from oxidation of iron in the Martian soil, which had an iron content of 14%, richer in iron than almost any soil on earth. The Viking scientists theorized that the paler soil at Utopia resulted from the presence of water, which would wash out some of the redness by forming hydrates or sulfates with other minerals in the soil. (*W Post*, 8 Sept 76, A-7)

6 September: Aviation Week magazine reported that the State Superior Court of Ariz. had found against aeronautical chartmaker Jeppesen Sanderson, sued by insurance underwriters for failing to include on the map of the Philippines a peak near Manila—Mt. Kamunay, nearly 1050 m high—where four crewmen were killed in July 1971 when a Pan American Boeing 707 flew into the mountain. More than $5.8 million was awarded in damages for value of cargo and aircraft in a ruling called a "landmark in product liability litigation." The suit had been brought originally by three of the four widows, whose damage awards had not been set. (*Av Wk*, 6 Sept 76, 47)

• Leadership of Europe's comsat programs had gravitated toward the U.K.—"more by accident" than by design, said an ESA official—since prime contractorship for those programs had fallen to Britain's Hawker Siddeley Dynamics within the European consortium known as MESH (Matra of France, ERNO of W. Germany, Saab-Scania of Sweden, Aeritalia of Italy, and Fokker-VFW of Holland). Current programs included OTS, the orbital test satellite scheduled for launch in 1977 to evaluate items for an operational European telecommunications system; Marots, to be launched in 1978, an experimental maritime comsat; and ECS, the European comsat that ESA hoped to fly in the early 1980s. *Aviation Week & Space Technology* magazine reported that potential markets worldwide for prime-contractor capability had become so numerous that

MESH had divided marketing responsibilities among its members according to their influence in various regions, HSD being strongest in Arab League areas and in Africa. An HSD official noted that Europe was ahead of the U.S. in three-axis stabilization and in higher frequencies for satellite communications; demonstration of effective project leadership would assure future success in the world market. (*Av Wk*, 6 Sept 76, 96)

6–8 September: A MIG-25 (Foxbat) Soviet mach 3.2 fighter plane made an emergency landing at 1:57 pm (457 GMT) at the Hakodate commercial airport on Hokkaido, northernmost of Japan's three main islands. Pilot of the MIG-25, Soviet Air Force Lt. Victor I. Belenko, initially asked for an interpreter and a canvas cover for the plane because it contained "military secrets." Japanese officials, who said at first that the pilot would be returned to the Soviet Union if his landing was merely an emergency, later said Belenko had asked for political asylum and would be transferred to the U.S. on 8 Sept.

Japan's Natl. Police Agency announced that Belenko was the 15th Soviet national to seek protection in Japan with a view to taking political asylum in some other country; of the previous 14, 11 went to the U.S. and 1 each to Italy, Israel, and West Germany. (FBIS, Hong Kong AFB in English, 6, 7 Sept; Tokyo Kyodo in English, 6, 7 Sept; *Av Wk*, 13 Sept 76, 25; *W Post*, 7 Sept 76, A–1)

7 September: Third Century America, the bicentennial exposition on science and technology that opened in May at KSC, closed its gates with a ceremony that included music, fireworks, and cannon salutes. More than 600 000 visitors had seen displays prepared by government agencies, educational institutions, and industry firms, many of which had representatives on the outdoor stage during the closing exercises. A time capsule dedicated during the ceremony would remain on display at the visitor center with duplicates of the items contained in the capsule: 1976 coins, stamps, and medals; items representative of 1976 technology and lifestyle, including space-related materials; newspapers, letters, and a mail-order catalog. The stainless steel cylinder, enclosed in a stainless steel cube with a lexan plastic top, would be opened in 99 yr (on 4 July 2075) in preparation for the U.S. Tricentennial. (*Spaceport News*, 3 Sept 76, 8; 17 Sept 76, 4)

8 September: NASA announced that the rollout of Orbiter 101 had been set for 17 Sept. at the Rockwell Int'l Space Div. assembly plant at Palmdale, Calif. [see 11 Aug.]. First of NASA's Shuttle spacecraft off the assembly line, OV-101 was not yet scheduled for orbital flight: its first job, beginning in Jan. 1977, would be as a test vehicle. It would be launched from a modified Boeing 747 jetliner on which it was riding piggyback, in a series of approach and landing tests at Dryden Flight Research Center. It would then be ferried early in 1978 to Marshall Space Flight Center for ground vibration tests. Mated in a test stand to the 46-m external tank as if for an actual launch, Orbiter 101 would undergo extended vibration and stress loading equal to that experienced during a launch phase, when all the main engines—OV-101's three main engines and the two solid-fuel boosters—would produce up to 30 million newtons of thrust. After the testing at MSFC, Orbiter 101 would be returned to Calif. to be prepared for space flight; Orbiter 102 would be used in initial orbital flights from Kennedy Space Center, scheduled for 1979.

Under construction since June 1974, the Orbiter's main parts came from various contractors: crew module and aft fuselage from Rockwell at Palmdale, mid fuselage (cargo bay) from General Dynamics in San Diego, wings from Grumman Aerospace in N.Y., and tail assembly from Fairchild Republic in N.Y. Orbiter's three main engines (each providing 211 500 kg of thrust at launch) were constructed by Rockwell's Rocketdyne Div. under contract to MSFC. (NASA Release 76-143; *Marshall Star*, 8 Sept 76, 1)

- President Ford named Shuttle Orbiter 101 the Enterprise, over the objections of NASA officials who preferred the name Constitution and had planned the Orbiter rollout ceremonies for 17 Sept., Constitution Day. NASA Administrator Dr. James C. Fletcher had paid a 45-min visit to the White House to brief the President on the Shuttle program and to discuss the naming. The name Enterprise, illustrious in U.S. naval history, had been given to the first nuclear-powered carrier, to a World War II carrier, and to an American sloop in the Revolutionary War. The name Constitution had met with objections that the Shuttle was considered an international effort in which several countries would participate. *Aviation Week* magazine commented on the "power of an aroused involved public—especially in an election year" in getting nearly 100 000 fans of the TV series Star Trek to sign letters and petitions asking the White House to have the first Orbiter named Enterprise (after the space ship in the series) rather than Constitution, the name favored by NASA officials. *Av Wk* said the officials were concerned about commercialization of the name Enterprise in association with the TV show. The *Washington Star* in an editorial said it was "pathetic" that the public desire for drama in outer space had not been killed by the mundane discoveries on Mars, Venus, and the moon, and predicted that "nothing exciting will happen in the real-life Enterprise," even though the naming incident confirmed a public desire "to associate space with adventure and suspense." (*W Post*, 9 Sept 76, E-9, A-2; *W Star*, 19 Sept 76, E-1; *Av Wk*, 13 Sept 76, 26; 27 Sept 76, 11; *Marshall Star*, 15 Sept 76, 1)
- Lewis Research Center announced award of a $73.6 million contract to General Dynamics Corp.'s Convair Div. for 8 Atlas-Centaur launch vehicles to be used in NASA missions over the next 4 yr. Launches would include Intelsat V comsats, High Energy Astronomical Observatory (HEAO) satellites, FLTSATCOM satellites for a worldwide DOD communications system, and the Pioneer Venus mission to provide details on the Venus atmosphere. The fixed-price incentive contract would run from 3 Sept. 1976 through Sept. 1980; work would be done at the contractor's plant in San Diego, Calif. (NASA Release 76-154)
- The Space Segment Board for Aerosat met at Frascati, Italy, to review proposals for development, production, launch, and 7-yr operation of two satellites [see 22 Jan.]. The space segment of the Aerosat program, conducted jointly by the European Space Agency, the

U.S. Federal Aviation Administration, and the government of Canada, would provide an experimental system of satellite communications between transoceanic aircraft and the ground, leading to guidelines for an operational system to be established by the Intl. Civil Aviation Organization (ICAO). During the program, two satellites would be launched into geostationary orbit over the Atlantic Ocean, separated by about 25° longitude, with a 7-yr lifetime; the first launch would occur in 1979, and the second 8 mo later. The space board, representing ESA, Canada, and Comsat General Corp., reviewed the proposals submitted by General Electric, Radio Corp. of America, and TRW; all were reported to be of "a commendably high technical standard," but the price submitted by GE was below that of the other two. The board therefore authorized negotiation with GE toward award of a fixed-price contract. (Comsat General Release CG 76-120; ESA release 10 Sept 76)

8-17 September: About 70 French officials, mostly from the Centre National d'Etudes Spatiales (CNES), and about 140 Soviet representatives of the Cosmic Research Institute at Moscow and the Intercosmos Council of the Soviet Academy of Sciences met in Leningrad to review current joint programs and to discuss new collaborations, especially a joint investigation (possibly in 1983) of atmospheric and surface characteristics of the planet Venus. Under a new protocol of 1975, the two countries were to work together in 11 specific areas including lunar and planetary studies. France had provided some of the equipment carried on *Venera 9* and *Venera 10* to study the Venusian atmosphere; on the future mission, a Soviet Venus orbiter would drop French-supplied pressure balloons to within 55 km of the planet's surface. Joint projects included space biology experiments supplied by France for *Cosmos 782*, and a French spacecraft called Signe 3 to be launched by the USSR late in 1977 and used with a Prognoz satellite for localizing and studying gamma-ray sources. The meeting marked the tenth anniversary of the signature of an intergovernmental agreement between France and the USSR on space cooperation. (*Av Wk*, 27 Sept 76, 26; FBIS, Tass in English, 9 Sept 76)

9 September: An aviation-policy declaration issued by the Ford administration today went beyond a similar policy issued in 1970, the *New York Times* reported, quoting Transportation Secy. William T. Coleman, Jr., who drafted the policy, as saying the timing of the declaration was "85% coincidental" with current talks between the U.S. and Britain on Britain's plan to withdraw unilaterally from the Bermuda agreement on transatlantic air passenger markets. The new policy emphasized the U.S. view that private carriers operating without subsidy could offer the most efficient service to air travelers. Britain wanted a new Anglo-American agreement to reduce the number of transatlantic flights, giving the government-owned British Airways about half of what remained; it currently had about 34% of the market. Charles Robinson, Deputy Secretary of State, had told a press conference that unilateral withdrawal from the Bermuda agreement was an "improper and illegal" way to force the reduction in flight numbers. The new U.S. statement endorsed the idea of trimming excess passenger capacity [see 8 Aug.] but opposed a rigid division of the market into equal shares, pointing out that two thirds of the passengers originate in the U.S. The statement also called for more flexibility in charter rules, the addition of domestic feeder routes to strictly international carriers such as Pan Am to fill seats between domestic cities, and elimination of cross-subsidization (promotional discounts) so that economy-fare passengers would not subsidize bargain-fare customers. (*NYT,* 9 Sept 76, 5)

- A nickel-zinc battery improved by space technology, installed in a utility van and tested against a regulation lead-acid battery, ran the delivery vehicle nearly twice as far at a constant speed of about 32 kph, Lewis Research Center announced. The improved battery would be able to meet the needs of 95% of the nation's drivers for a full-service urban vehicle and could help to reduce dependence on petroleum. The improvement consisted of an inorganic separator adapted from space battery technology. LeRC planned further studies of the battery's life, performance, and competitive cost, using sample batteries in mail pickup and delivery vans with the cooperation of the U.S. Postal Service. (NASA Release 76-155)

- The European Space Agency (ESA) announced opening of a new ground station near Michelstadt/Odenwald, West Germany, that would include two 15m-dia parabolic antennas that would be operational for the launches of ESA satellites Geos and Meteosat in 1977. Geos, first geostationary scientific satellite, and Meteosat, Europe's first meteorological satellite, would remain in uninterrupted contact with the Odenwald station because of their geostationary orbits. The two antennas represented an innovation in the field of telecommunications, ESA said, because of a special construction: To avoid mutual interference with ground-to-ground links operated by the W. German post office, the special construction resulted in an exceptionally low sidelobe level heretofore unattained, without reducing the station's telecommunications performance. The station would be inaugurated 14 Sept. by Hans Matthöfer, German minister for research and technology, and Roy Gibson, ESA's director general. (ESA release 13 Sept 76)

10 September: Scientists using a NASA S-band radar obtained the first detailed pictures of the surface of Venus, NASA announced. Installed at the Arecibo radar observatory in Puerto Rico under a $3-million NASA contract, the S-band system was 50 times more sensitive than equipment previously available for radar observations of Venus. Optical telescopes had been unable to penetrate the dense cloudcover of Venus, but the Arecibo radar had produced high-quality photograph-like images with a clarity like that of optical photographs of the moon taken from earth. Data were obtained during a series of daily 2-hr observations in the months before and after the close approach of Venus to the earth in late August 1975.

Radar echoes of the signal from a powerful transmitter operating at a 12.6-cm wavelength were measured for strength, precise frequency, and time of arrival by a 330-m radiotelescope at Arecibo and an auxiliary 30-m telescope about 10.5 km distant; the two telescopes constituted an interferometer that enabled mapping of areas with detailed definition and precise location of the echoing regions. Venus features as small as 19 km could be distinguished. In the area about 10.25 million sq km mapped by the radar, scientists had observed a large basin bordered by ejecta suggesting its formation by impacts like those that created the maria on the moon, as well as a very bright area "about the size of Oklahoma" tentatively named Maxwell (for the 19th-century physicist) which appeared to result from processes internal to Venus—possibly a large eruption of lava—with long parallel ridges unlike features on either the earth or the moon. (NASA Release 76-153; *NYT*, 10 Sept 76, A-1)

- Part of a Soviet satellite—the rocket body from *Cosmos 854*, tracked by North American Air Defense Command radar—had fallen in Montana earlier this week, DOD sources reported. The rocket body, which had not been recovered, looked like a meteor as it passed over Washington State and Idaho and fell south of the Canadian border. DOD said debris from Soviet space vehicles had dropped on the U.S. previously, but usually in small pieces. (*W Star*, 10 Sept 76, A-5; *Av Wk*, 13 Sept 76, 30)

- The European Space Agency announced selection of General Electric Co. to produce and launch two satellites for the Aerosat program [see 8 Sept.], saying that the GE bid was "significantly below" that of the other two contenders, the RCA Corp. and TRW Inc. A GE spokesman estimated the value of the contract at $60 million, half to be spent in the U.S. and half to go to a consortium of foreign companies known as the Cosmos Group, which had formed a bidding team with GE. (*NYT*, 11 Sept 76, 33)

- A British Airways Trident 3 collided with an Inex-Adria McDonnell Douglas DC-9 near Zagreb, Yugoslavia, at 11:34 am local time in good weather; the midair collision killed 176 persons. Three major airways converged over Zagreb, and four air-traffic controllers had been arrested and were being held on suspicion of responsibility for the disaster. Crash recorders from both aircraft and cockpit voice recorders were to be used in an initial hearing to show whether the Yugoslav transport, taking German tourists from Split to Cologne, was being controlled in the Serbo-Croat language instead of in English (the official international aviation language). Zagreb control had cleared the DC-9 to climb to the point where the collision occurred; a British Airways official said after hearing a tape that the Trident crew may not have seen the DC-9 or known of its approach. (*Av Wk*, 20 Sept 76, 32)

11 September: The Air Force launched the first Block 5D advanced meteorological satellite from Vandenberg AFB on a Thor-Burner 2 at 1:01 am local time into an orbit with 848-km apogee, 818-km perigee, 98.6° inclination, and 101.5-min period. The RCA-built research satellite was a 450-kg cylinder 6.4m long and 1.7m in diameter. USAF later reported that a power-system failure had apparently occurred because no communications were coming from the *Ams 1*; no cause of the malfunction had been determined, and "chances of correcting it are grim," USAF said. (*SSR*, 31 Oct 76, 31; *SBD*, 21 Sept 76, 24; *Av Wk*, 27 Sept 76, 26)

- The USSR launched the second operational satellite in its Statsionar 1 series, also known as *Raduga* (Rainbow), into a circular orbit at 35 900-km altitude, 0.3° inclination, and 1436-min period, Tass announced. Carrying retransmitting equipment for continuous telephone and telegraph communications, *Raduga* would also transmit color and black-and-white TV programs to Central Asia and Siberia. (UPI in *NYT*, 14 Sept 76, 28; *Sf*, Mar 1977, 117)

12 September: Vandals who damaged the moon rock on display at the National Air and Space Museum in Washington, D.C., were not successful in obtaining a piece of the rock, said the *Washington Post*. The lunar sample, a 40-gram triangular chuck of basalt, had been imbedded in a block of glass so that it might be touched by museum visitors; about 2 cubic mm had been chipped away and a missing corner of the triangle and a scratch over the surface had become visible in the single beam of light that illuminated the sample. NASA had sent experts to photograph the damage and to vacuum the exhibit area for any dust or small chips that might have come off the sample. Electronic security measures were being increased, and a fulltime guard had been stationed near the display, as a result of the incident. (*W Post*, 12 Sept 76, B-3; *Av Wk*, 27 Sept 76, 16)

- As *Viking 2*'s lander was preparing for its scoop-and-analyze sequence on the rocky Utopia site of Mars, the *Viking 1* orbiter fired its engines to shift orbit around the planet in 40° longitude jumps each day for 9 days before taking over the relay of communications from the *Viking 2* lander. *Viking 2*'s orbiter would then begin a polar-orbital scan of Mars to search for traces of organic materials—carbon-based molecules found in all life forms on earth—not yet detected by *Viking 1* experiments. *Viking 1*'s lander would remain in standby mode after its weeks of testing at the Chryse site. Project scientist Dr. Gerald A. Soffen, commenting on the confusing results of the *Viking 1* experiments (which hinted at the existence of life but offered no evidence of organic chemicals), said he had been prepared for the discovery of organic chemicals without life, but the suggestion of Martian life without organics was totally unexpected.

Just after 6 a.m. EDT on 12 Sept., the 3-meter arm of *Viking 2* dug its first sample of Mars soil and dropped it into the hopper leading to three biology instruments. First cycle of the pyrolytic-release experiment began when the soil reached the test chamber, to be incubated in a carbon monoxide-carbon dioxide atmosphere containing a radioactive tracer; after 5 days, the atmosphere would be flushed out and the soil sample heated to vaporize any organic material. A detector would measure the radioactive carbon that had been ingested by any organisms. The gas-exchange detector—looking for signs of photosynthesis in the dark—would reveal the presence of organisms that might exhale gases such as oxygen or carbon dioxide through the night as well as in the daytime. (The first results of the same experiment on *Viking 1* at Chryse had showed the soil was releasing six times as much gas as it would have in the absence of photosynthesis; a second trial had produced a signal a third as strong as the first, but still twice what it would have been if no photosynthesis were going on. Neither the readings nor their differences had been explained to everyone's satisfaction.) The third biological instrument—the labeled-release experiment—would add a nutrient liquid to the test chamber on Tuesday and monitor over 10 days the release of radioactive tracer from any metabolized nutrient.

A second soil sample would be used beginning Monday, 13 Sept., for an organic-chemistry analysis to detect carbon molecules, and for an x-ray fluorescence experiment to learn the composition of inorganic chemicals in the Mars soil.

Scientists believed chances of finding definite traces of life would be better at Utopia, where five times more water had been measured in the air than at Chryse. They also planned to reduce chances of ambiguity in the Utopia readings by conducting the photosynthesis search at night, and by moving a rock to take a sample of Mars soil that was shielded from the killing solar ultraviolet constantly impinging on the planet's surface and that was deeper than the sample taken at Chryse. (*W Star*, 12 Sept 76, A-14; *W Post*, 13 Sept 76, A-5; *C Trib*, 13 Sept 76, 6-16)

13 September: A select committee of the National Academy of Sciences, named 2 yr ago at the request of four federal agencies, recommended a 2-yr delay before banning aerosol sprays containing fluorocarbons. The group advised three "urgent" measures: drafting of legislation to regulate fluorocarbon usage when the need arose; action to require labeling of products containing fluorocarbons, so that consumers could stop using them if they wished to do so; and a campaign by public health agencies to reduce people's overexposure to sun and the resulting malignant melanoma, incidence of which had been increasing by 10 to 15% a year, about a third of the current 8500 yearly cases being fatal.

An international conference on threats to the stratospheric ozone, meeting at the State Univ. of Utah, heard Dr. James G. Anderson of the Univ. of Mich. report findings from a 28 July balloon flight of chlorine and chlorine monoxide at 35- to 42-km altitudes in quantities twice as abundant as predicted by theorists. The results had not been available when the NAS findings were issued. The NAS had said that a selective ban on fluorocarbon spray-can propellants would be necessary within 2 yr. The international conference also heard a report that nitrous oxide in the air was increasing steadily as a result of many factors (including fuel combustion and bacterial digestion of fertilizers), becoming as serious a threat to the stratospheric ozone as the Freons, with a 20% ozone reduction possible by the end of the 20th century. Bags of exhaust gases obtained from a federal test center for auto emissions had shown that the catalytic converters designed to remove hydrocarbons from exhaust had permitted considerable release of nitrous oxides. Dr. R. J. Cicerone of the Univ. of Mich. had estimated that automobile contribution of nitrous oxide to the atmosphere within a yr was about a million tons.

Researchers from the Univ. of Calif. Livermore Laboratory at Lawrence had told a meeting of the American Chemical Society that better methods of analysis were needed to permit accurate predictions of the effects of contaminants on the stratospheric ozone. Supersonic transport operations, for instance, were known to affect the stratosphere but not enough was known about the forty-odd possible chemical reactions at high altitude to say if or how the ozone layer was damaged. William H. Duewer, a chemist on the Livermore team, added that, even if all the chemistry were pinned down, the effect of SST operations could not be determined precisely until meteorological factors were fully accounted for. The next 1 or 2 yr should provide new insights into these stratospheric processes, the team said. (*NYT*, 14 Sept 76, 1; 17 Sept 76, A14; *W Post*, 14 Sept 76, A-1; *Av Wk*, 6 Sept 76, 49)

13-16 September: The soil-collecting arm of the *Viking 2* lander apparently jammed some time between delivering a sample of Mars soil to the biology-experiment hopper on Sunday and delivering the rest of the soil to a second hopper leading to an x-ray analysis chamber. The first indication of trouble came when a routine picture transmission of the sampling sequence failed to show the collector head, although another picture showed that the collector had dug a trench and collected a soil sample. The lander had been programmed to halt the arm motion if something went wrong; Project Manager James S. Martin, Jr., said more than a day might be needed to define the problem. (The arm on the *Viking 1* lander had jammed twice while digging at Chryse, and both times the flight directors at JPL were able to get it moving again.) A command would be sent to the *Viking 2* lander Tuesday morning, 14 Sept., to extend the arm from its arrested position to be photographed so the flight directors could see what went wrong. The halt in operations meant that the gas-chromatograph spectrometer intended to analyze the soil for organic molecules never got a sample for analysis.

The biology experiments, which did receive soil samples, were progressing with their tasks. The gas-exchange experiment had already indicated a slower response to oxygen leaving the soil in its test chamber than was shown by a similar *Viking 1* instrument at Chryse, where a high oxygen count had been considered a possible indication of life activity. Scientists had decided the reading probably resulted from "an exotic chemistry." Dr. Harold P. Klein, chief biologist on the project, said the lower readings at Utopia might simply indicate "less oxidizing substances at the *Viking 2* site." The other two biology instruments were incubating their soil samples; both were transmitting data on background radiation for use as reference points later in the experiments, which would use radioactive counts to determine the presence of microorganisms.

After seeing a picture of the lander's arm, showing that it had rotated 180° instead of 45° as it should have, flight directors diagnosed the problem as failure of a switch. Project Manager Martin said it would be possible to override the switch and get the arm moving again by Friday, 17 Sept., when flight directors would order the arm to continue delivery of a soil sample to the x-ray instrument.

At a news conference 16 Sept., project scientists reported that the first measurements from the labeled-release experiment showed 33% more radioactive gas count from the Utopia sample than had come from the Chryse soil: 10 000 counts per minute (compared to 7500) or 20 times the amount of gas that would be registered in the absence of metabolic activity. The scientists still refused to say the activity was proof of life, noting that, if organisms were picking up all the nutrients in the instrument, it should record up to 15 000 counts per minute. The gas-exchange experiment, which had detected some signs of activity in the Utopia sample but less than *Viking 1* found in the Chryse sample, apparently provided an argument against the "exotic chemistry" theory because of the prolonged increase of carbon dioxide in the test chamber. Dr. Vance I. Oyama of Ames Research Center said of this result, which might be attributed to chemical reaction, "if it persisted and was accompanied by other changes, we can ascribe it to biological changes." No decisive report had come from the pyrolytic-release experiment, considered the least ambiguous of the detectors.

Project geologists meanwhile were scanning the photographs from both Viking orbiters and landers, noting especially the enormous scale of Martian topography. "Everything we see is ten times anything on earth," said Dr. John Guest of the Univ. of London, adding that earth in comparison was almost as smooth as a billiard ball. Most prominent features on Mars were the volcano Olympus Mons whose base on earth would reach from New York City to Montreal, and a canyon long enough to stretch from New York to Salt Lake City. The plateau on which the Mars volcano rested stood more than 9 km high, taller than Everest, earth's highest mountain. Dr. Michael Carr of the U.S. Geological Survey suggested that the outsize features of Mars existed because of the lack of plate movement on the planet. Mars, like earth, appeared divided into two "provinces" geologically, one area heavily cratered and higher, therefore assumed to be older; the other area was lower and smoother, considered to be younger. This division might be typical of all bodies of the inner solar system, as a similar distribution of "provinces" appeared on the moon and on earth. (*W Star*, 13 Sept 76, A-5; 15 Sept 76, A-2; *W Post*, 14 Sept 76, A-12; 16 Sept 76, C-1; 17 Sept 76, A-4; *NYT*, 17 Sept 76, A14; *C Trib*, 18 Sept 76, 1-7)

14 September: Marshall Space Flight Center announced plans for a day-long meeting 16 Sept. at MSFC to review the status of the spinning solid upper stage (SSUS) that would deliver spacecraft from the Space Shuttle in low orbit to their operational synchronous orbits. The SSUS, carrying a satellite of the type now being launched by expendable Atlas-Centaur or Thor-Delta vehicles, would be launched from earth in the cargo bay of the Space Shuttle. In orbit, the SSUS with its payload would be extended from the cargo bay on a cradle with a spin table that would stabilize the SSUS as it ejected from the cradle, before its own engine ignited to send it and its payload into the proper orbit. After reaching the required altitude, the payload's propulsion system would position it in the desired orbital attitude. Industry representatives invited to the status review would hear presentations by NASA and Aerospace Corp. on concepts and current mission models for the SSUS. MSFC planned to issue requests for proposals for Phase B (preliminary design definition) about 1 Oct., with contracts being issued in Feb. Plans were to have the SSUS ready for flight in Dec. 1979. (MSFC Release 76-166; *Av Wk*, 6 Sept 76, 46)

- The General Services Administration goofed in agreeing to an exchange of California property with Rockwell Intl. Corp., said Rep. Jack Brooks

(D-Tex.), chairman of the House Committee on Government Operations. GSA agreed 3 yr ago to accept a $19.5-million piece of property on the former Rockwell AFB at Canoga Park, Calif., in exchange for a surplus $16-million building. No other agency had been found willing to occupy the Rockwell AFB property, and the corporation had claimed $1.8 million worth of special installations at the former base. Rep. Brooks urged GSA to fight the corporation's attempt to get more out of the trade. (*C Trib*, 14 Sept 76, 4-7)

- The price of solar cells had come down, the Energy Research and Development Administration announced. Cells that would have cost about $210 ($21 a watt) 6 mo ago today cost $155 ($15.50 a watt). "This 26% price drop in just 6 mo shows that our research and development program is making significant progress," said Dr. Henry H. Marvin, director of ERDA's division of solar energy. "However, we have a long way to go to meet our 1986 goal of 50¢ a peak watt." At the goal price, Dr. Marvin said, solar cells could compete with electric power from conventional sources; even at the present high cost, solar cells were competitive for special uses, such as recharging batteries at remote microwave stations operated by railroads and highway departments in southwestern states. Manufacture and sale of solar cells for such uses doubled in the past 18 mo, and the price during that period had dropped almost 50%. NASA's Jet Propulsion Laboratory would manage solar-cell purchase contracts for ERDA, the announcement said, to establish technical feasibility of uses ranging from remote power sources to integration with existing power systems. (ERDA Release 76-290)

- Cosmonauts from 8 "socialist member-countries of the Intercosmos program" would join crews from the USSR in space beginning in 1978, the official Soviet news agency Tass announced. Lt. Gen. Vladimir Shatalov, chief of training for the cosmonaut program, said that timing of such flights would depend on the readiness of technology and preparedness of the crews: the crew commander would be a Soviet flyer, and the flight engineer and research engineer would represent another participating country. Crewmembers from Bulgaria, Hungary, East Germany, Cuba, Mongolia, Poland, Romania, and Czechoslovakia would train at the Yuri Gagarin Center near Moscow where Soviet cosmonauts have trained. Intercosmos is an association of Communist nations that had already cooperated in launching 16 unmanned satellites, the *New York Times* noted; the only men in space so far had been either Soviet or American. (FBIS, Moscow Tass in English, 15 Sept 76; *NYT*, 15 Sept 76, 72; *W Post*, 15 Sept 76, A-14)

15 September: The U.S. Air Force launched a Lockheed-built photoreconnaissance satellite into a polar orbit from Vandenberg AFB on a Titan IIIB-Agena D. Initial orbital elements were 342-km apogee, 132-km perigee, 96.4° inclination, and 89.2-min period. The spacecraft was a 3000-kg cylinder 8m long and 1.5m in diameter. The last previous satellite of this type, launched 22 March, had reentered 18 May. (*Av Wk*, 4 Oct 76, 26; USAF anno. 15 Sept 76; *Sf*, March 1977, 117)

- Johnson Space Center announced receipt of delivery on the second of two Space Shuttle training aircraft. The first had been delivered 8 June 1976. The modified Grumman Gulfstream II twin-engine jet, flown to Ellington AFB from the plant in N.Y., would simulate Shuttle Orbiter handling qualities, performance, and flight-control procedures during subsonic flight phase, from 10.6-km altitude to simulated Orbiter touchdown. Crew training for approach and landing tests of the Shuttle Orbiter would begin late in Oct. (JSC Release 76-58)
- Lockheed Missiles and Space Co. officially opened a facility in Sunnyvale, Calif., to manufacture a unique all-silica insulation refined from common sand from which 34 000 "tiles" would be made to cover 70% of the Shuttle Orbiter's surface. The material was said to be so efficient that it could be held with bare hands while red-hot; it could be taken from an oven at 1533K and plunged immediately into cold water without damage. The tiles would be expected to survive such temperatures through 100 flights with only minor maintenance, making the Shuttle a truly reusable space transportation system. Each tile would be milled to match the curvature of the Shuttle surface at the exact point to which it would be attached; the JSC *Roundup* said the task of fitting the tiles to the spacecraft skin would be like assembling a three-dimensional jigsaw puzzle twice the size of a basketball court. No two tiles in a "shipset" would be exactly alike. (JSC *Roundup*, 24 Sept 76, 2)
- Soviet scientists studying the movement of Phobos and Deimos, moons of Mars, had discovered that the speed of revolution of Phobos was increasing, *Izvestia* reported. The USSR's institute of theoretical astronomy had processed nearly 5000 observations of the Martian satellites from various observatories worldwide over the past 100 yr. Phobos was found to revolve at a distance of 9400 km from the center of Mars; Deimos was in an orbit 2.5 times further away. The acceleration of Phobos over the years was believed to be caused by a so-called tidal interaction with Mars, and meant that—"in several tens of millions of years from now"—it would fall on to the planet. (FBIS, 15 Sept 76)

15–27 September: The USSR launched *Soyuz 22* into earth orbit 15 Sept. at 12:48 pm Moscow time (5:48 am EDT) from the Baykonur cosmodrome near Tyuratam. Orbital parameters were: apogee 280 km, perigee 250 km, inclination 65°, period 89.6 min. The spacecraft, carrying cosmonauts Col. Valery Bykovsky and civilian Vladimir Aksenov, was the first manned Soviet craft to carry foreign-made equipment—a "multizonal" photographic instrument made by the "state-owned enterprise" Karl Zeiss Jena of East Germany. In a preflight interview, Bykovsky said the camera would occupy "a whole section of the Soyuz spaceship." Previously, "Intercosmos" instruments developed by other socialist countries had been installed only on unmanned Soviet satellites. The Zeiss MKF-6 photographic unit was designed "to simultaneously photograph the earth's surface in six different wave bands," four visible and two infrared, according to a Tass broadcast. Journalists were permitted to inspect the spacecraft's interior, where the "photographic section" had been installed in the space usually occupied by the docking compartment. *Izvestia* announced that the docking equipment had been removed and that *Soyuz 22* would therefore make no attempt to dock with *Salyut 5*, vacated only 3 wk earlier by the *Soyuz 21* crew. *Soyuz 22* had been constructed as a backup vehicle for last year's Apollo-Soyuz test flight, *Izvestia* said. (FBIS, Tass in English, 15-16 Sept 76; Intl Service in Russian, 15 Sept 76; *W Star*, 15 Sept 76, A-4, 16 Sept 76, A-4; *NYT*, 16 Sept 76, 21; *W Post*, 16 Sept 76, A18; *Av Wk*, 20 Sept 76, 25)

Moscow radio said 16 Sept. that *Soyuz 22* photographed the earth's surface and the upper layers of its atmosphere with the Zeiss multispectral camera built in East Germany for this mission. Pictures of the Soviet Union would be used in agriculture, forestry, and geology research; Anatoly Alexandrov, president of the USSR Academy of Sciences, said the new equipment promised much to the Soviet economy in those fields as well as in mineral prospecting. One unusual task was the study of color effects caused by cosmic-ray particles in light-sensitive cells of the eye. The cosmonauts also took samples of the cabin air to record changes in its composition during the flight.

Tass reported 20 Sept. that *Soyuz 22* had taken pictures of objects that had never been "targets of space photography": the northern regions of the Soviet Union, with simultaneous earth and aerial photography to provide the fullest possible comparative data on surface phenomena and processes.

The cosmonauts were concluding their final photography sessions 22 Sept. in preparation for their return to earth, the Moscow domestic service reported, packing tapes, logbooks, and other materials in the landing module and checking its engine.

After spending 8 days in orbit, the crew of *Soyuz 22* soft-landed 23 Sept. at 10:42 am Moscow time (3:42 am EDT) about 150 km northwest of Tselinograd in Kazakhstan. Cosmonauts Bykovsky and Aksenov were feeling well, Tass reported, and the flight data were being processed and studied.

Soyuz 22's East German-made Zeiss multispectral camera did not survive the mission; it was built into the orbital module, normally jettisoned before Soviet reentry procedures. The flight director, cosmonaut Aleksey Yeliseyev, emphasized that the crew "enjoyed a large measure of independence in controlling the ship and in carrying out various investigations." Vladimir Shatalov, in charge of cosmonaut training, said that Bykovsky (as commander of *Soyuz 22*) oriented and stabilized the ship so that Aksenov (as flight engineer) could position and operate the camera. The crew "carried out all operations . . . by heart, so to say, without consulting their instructions," Tass reported.

Postflight medical checks performed at the Baykonur cosmodrome near Leninsk showed the *Soyuz 22* cosmonauts to be in good health, Tass reported 27 Sept. One of the scientific directors of the mission said that all twelve of the photo cassettes brought back from *Soyuz 22* were in good condition, and that the crew had completed "an immense amount of research" in space during their week-long expedition.

Aviation Week & Space Technology magazine noted that launching *Soyuz 22* on a mission that did not dock with a Salyut space station contradicted an earlier statement by cosmonaut Pyotr Klimuk, commander of the *Soyuz 18* mission to *Salyut 4*, who had told a COSPAR meeting in June that no more Soyuz vehicles would fly missions independent of the Salyut stations. (*Av Wk*, 27 Sept 76, 20; FBIS, Moscow domestic service in Russian, 16 Sept 76, 17 Sept 76, 22 Sept 76, 27 Sept 76; FBIS, Tass in English, 17–20 Sept 76, 23 Sept 76, 27 Sept 76)

17 September: The Enterprise—Space Shuttle Orbiter 101, described by NASA as the flagship of "the new era of space transportation" and by Sen. Barry M. Goldwater (R-Ariz.) as "probably the best investment the U.S. Congress has ever made"—was rolled out of Rockwell Intl.'s assembly facility at Palmdale, Calif., just before 1 pm EDT, to the strains of the theme from the Star Trek TV show and applause from about 2000 spectators. The audience included Sen. John B. Tunney (D-Calif.); Rep. Olin E. Teague (D-Tex.), chairman of the House Committee on Science and Technology; and Rep. William M. Ketchum (R-Calif.), "representing Antelope Valley" where the Orbiter was built. Also on hand were test-flight crewmen Fred W. Haise, Joe H. Engle, Charles G. Fullerton, and Richard H. Truly, the astronaut team who would put the Orbiter through approach and landing tests scheduled for 1979. Special guests were six members of the original Star Trek cast and ST creator Gene Roddenberry [see 8 Sept.].

NASA Administrator James C. Fletcher told those attending that the ceremony was "a proud moment" that would mean "the evolution to man in space—not just astronauts." (Plans called for the Orbiter to serve as an all-purpose space trainer, carrying a crew of two or more plus up to five scientific or technological investigators working in space in a shirtsleeve environment.) Willard F. Rockwell, Jr., board chairman of Rockwell Intl., predicted "one of the most exciting chapters in American history" in the productive use of space that the Shuttle would make possible.

The first Shuttle in orbit was to carry 4500 kg of instruments to measure stress, and the second, a small satellite to be left in orbit; the last of six orbital test flights would carry a payload of about 30 000 kg or as many as six satellites to be left in orbit at altitudes up to 160 km.

Resembling "a space-bound DC-9," according to Thomas O'Toole in the *Washington Post*, the Enterprise was more than 36.5 m long and weighed nearly 70 000 kg, having gained more than 2200 kg over the past 2 yr. Strengthening the wings, landing-gear and payload-bay doors, parts of the fuselage, and the enormous fuel tanks during Shuttle development had added to both the weight and the cost of the vehicle, meaning that its price would be more than the $5.2 billion (in 1971 dollars) estimated by NASA. One change alone, from aluminum to a boron-epoxy composite for landing-gear doors, would cost about $9000 per kg to save 112 kg of weight. Replacing aluminum castings in the fuselage with titanium would cost about $18 000 a kg to save slightly more than 200 kg of weight.

Sen. Goldwater reminded those at the rollout that the first U.S. manned orbital flight (that of John Glenn in *Friendship 7*) had occurred 14 yr previously, in Feb. 1962, and predicted that the Space Shuttle would make manned space flight a "routine" experience. (NASA Release 76-149; *NYT*, 18 Sept 76, 1, 8; *W Post*, 18 Sept 76, A-2; *W Star*, 18 Sept 76, A-1; *C Trib*, 18 Sept 76, 1-7, 2-7; *Av Wk*, 20 Sept 76, 12; 27 Sept 76, 12; KSC *Spaceport News*, 17 Sept 76, 1; JSC *Roundup*, 24 Sept 76, 1)

* NASA announced plans to launch a third maritime satellite, Marisat-C, for Comsat General Corp. from Cape Canaveral on a Delta rocket 14 Oct. *Marisat 1*, launched 19 Feb., was in orbit over the Atlantic Ocean at 15°W; *Marisat 2*, launched 9 June, was in orbit over the mid-Pacific at 176.5°. The two Marisats currently provided communications services to the U.S. Navy as well as fulltime commercial voice and data communications to the maritime industry; Marisat-C, which would be in synchronous orbit over the Indian Ocean, would be used initially only by the Navy, which planned to lease UHF capacity it found to be surplus. Comsat General would reimburse NASA for the cost of launch vehicle, launch, and other administrative expenses, and would supply all ground-station support. (NASA Release 76-156)

18-20 September: The *Viking 2* lander signaled to mission scientists at JPL that its sample-collecting arm was working and had moved toward the x-ray instrument that would analyze the mineral content of the Mars soil. If the collector had delivered its sample and the analysis was proceeding, the signal expected on the next relay would confirm that the arm was available for further sampling. The mission scientists said they planned to instruct the arm to turn over a rock on the surface and take a sample

Enterprise—Space Shuttle Orbiter 101—on public display for the first time at Rockwell's Palmdale, Calif., facility, 17 Sept. 1981. (NASA 76-H-854)

from the underlying soil that had not been exposed to lethal ultraviolet rays. Geologists had characterized the appearance of nearby soil structure as resembling "caliche," a calcium carbonate crust formed on earth surfaces by evaporation of water that deposited mineral salts on the soil particles and cemented them together. The cementing action might increase the difficulty of dislodging a rock with the lander arm, but scientists noted the ease with which the arm had taken the first sample. As the content of that first *Viking 2* sample so closely resembled that from the *Viking 1*, scientists wanted to try for a sample with different properties, especially some trace of carbon-based chemicals that would indicate life processes. The caliche sample would be taken 25 Sept. and the attempt to move a rock would be made 8 Oct., a 3-hr strategy meeting decided 20 Sept. Project Manager James S. Martin, Jr., said the x-ray instrument had never received its sample for analysis, so a sample for it would be dug 3 Oct. An editorial in the *New York Times* asked whether, in the absence of carbon compounds, the mission scientists should consider the possibility of Martian life based on some chemical other than carbon. (Mission Status Bulletin 42; *NYT*, 19 Sept 76, 29; 20 Sept 76, 32; *W Post*, 19 Sept 76, A12; 20 Sept 76, A-5; *W Star*, 19 Sept 76, A-2)

19 September: A NASA study reported that relocation of crew training and mission-control activities from Johnson Space Center in Tex. to Cape Canaveral in Fla. would offer "no management, technical or budgetary advantages" and "would seriously affect a smoothly functioning, highly efficient organization" at JSC, delaying the Space Shuttle program by up to 2 yr. The study, requested in June 1975 by Rep. Don Fuqua (D-Fla.), chairman of a subcommittee on space science and applications of the House Committee on Science and Technology, said that relocation would cost up to $842 million by 1983, and that the funds would not be recoverable. (*NYT*, 19 Sept 76, 24)

20 September: Aviation Week & Space Technology magazine reported that the first hitch in operations of the *Viking 2* lander—trouble with its digging arm—had been solved: the problem turned out to be in a microswitch. Ground control at the Jet Propulsion Laboratory in Calif. had commanded the lander's computer to ignore signals from the malfunctioning switch in future digs. If the corrective commands freed the arm for duty, the experiments needing soil samples would begin receiving them and results would become available within a week or 10 days. *Viking 2*'s lander cameras had returned excellent pictures of the Utopia terrain around the landing site; the meteorological experiment was collecting and returning weather data regularly; and the seismometer (whose parallel on *Viking 1* was never uncaged) was ready to monitor marsquakes, the magazine said. (*Av Wk*, 20 Sept 76, 25, 58-61)

- The council of Intersputnik—the International Organization for Cosmic Telecommunications—ended its fifth meeting in Berlin after 4 days of deliberations. Attending were delegations from Bulgaria, Hungary, the German Democratic Republic, Cuba, Mongolia, Poland, Czechoslovakia, and the USSR; also attending were observers from the Intl. Organization for Radio and Television (OIRT), Intercosmos, and the Standing Commission for Post and Telecommunications (CEMA). The meeting adopted documents on the use of technical installations for transmitting TV, radio, telephone, telegraph, pictures, and data by satellite. (FBIS, ADN Intl. Service in German, 20 Sept 76)

- Evaluation of data and pictures from a dendrite-remelting experiment flown in Dec. 1975 on a space processing applications rocket (SPAR 1) confirmed the belief that space processing could produce better materials than was possible on earth, NASA announced. The experiment, one of nine sent to a 225-km altitude for about 5 min of near weightlessness, included a camera to record solidification of aluminum chloride solution in a transparent vial in the absence of gravity. No fluid motion was detected in the liquid, which solidified uniformly without the formation of crystals (dendrites) that broke off and settled to the bottom of the solution in a control experiment on earth. This experiment was the first in which scientists had viewed the process of solidification under weightless conditions, said the co-investigators, Carolyn Griner and Dr. Mary Helen Johnston of Marshall Space Flight Center. Controlling the crystalline structure of materials would enable scientists to make them uniformly strong to meet specific needs. (NASA Release 76-159; MSFC Release 76-167)

- Nine sites for the world's first solar electric-power plant had been proposed by utility companies and government units in the U.S. in response to a request in July 1976, the Energy Research and Development Administration announced. Sites suggested were in Arizona, California, Florida, Mississippi, Puerto Rico, Rhode Island, and Texas. The agency planned to complete an evaluation and negotiate a contract with the selected proposer early in 1977; the selected proposer would become a partner with ERDA in constructing and operating a plant to generate electric power from high-pressure steam produced by concentrated power from the sun. Power from this pilot plant would be distributed by the utility to its customers; the plant would produce 10 Mw (10 000 kw) under optimum full-sun conditions, enough to supply a community of 10 000 population. Construction of the plant would begin in 1978 and reach completion in 1980, at an estimated cost of $100 million. (ERDA Release 76-294)

20-25 September: The committee on planetary and lunar exploration of the Space Science Board began meeting 20 Sept. at the Calif. Inst. of Technology to consider proposals for future space exploration, including a radar mission to Mars to determine the thickness of the ice at the polar caps and investigate the terracing of the planet as a clue to past climate. Terrace edges in both polar regions had been found to lie along concentric circles centered close to, but not identically with, the poles. Scientists had suggested the terraces were formed when Mars was spinning around a slightly different axis under climate conditions other than those at present. A change in the *Viking 1* orbiter's path to take over communications relay for the *Viking 2* lander would free the *Viking 2* orbiter by 24 Sept. to proceed with a polar-observing mission.

The *Viking 2* lander's pyrolytic-release instrument, which had been directed to look for evidence of photosynthesis activity in a dry Mars soil sample in the dark to avoid ambiguity, had returned an ambiguous reading, said Dr. Norman H. Horowitz of the Calif. Inst. of Technology, who designed the instrument. The count of radioactive molecules released during the analysis was 21 per minute, far less than the 96 per minute counted in the first similar assay by the *Viking 1* lander, but more than the 15 counts per minute the instrument would show in the absence of photosynthesis. The puzzled scientists planned another sampling, this time with a moistened sample under a sun simulator, which had been omitted from the first run to prevent overheating the *Viking 2* instrument, working in a location warmer than the Chryse area tested by *Viking 1*.

Although neither of the Viking landers had turned up more than a suggestion of life, *Viking 2* had confirmed the presence of water on Mars, and more than had been expected, project scientists announced at a JPL news conference 22 Sept. The permanent northern polar cap was found to be composed entirely of frozen water, not of frozen carbon dioxide as had been thought. A thin layer of dry ice covering the polar caps during most of the Martian year had suggested that both caps consisted of frozen carbon dioxide. Two *Viking 2* orbiter measurements that led to the discovery were the unusually high water-vapor content in the atmosphere over the northern pole, and the average-surface-temperature readings on the icecap, which were too warm for frozen carbon dioxide to exist. Dr. Hugh Kieffer of UCLA predicted that a survey of the south polar cap would find it also made entirely of water ice. The new evidence suggested that, when first formed, Mars might have had twice as much water as the earth had at a similar period in its development; even now, scientists described Mars as "a planet-size iceberg." Other Viking instruments had detected traces of krypton and xenon in the Mars atmosphere. This discovery was the first clue that the planet once had a considerably denser atmosphere that could have supported liquid water and even rainfall. Dr. Gerald Soffen, chief Viking scientist, said that the krypton finding was a "major marker" that would permit deductions about the origin of the Mars atmosphere.

On Saturday, 25 Sept., mission scientists at JPL were "relieved" to see photographs transmitted by *Viking 2* showing that the lander had followed an order radioed from earth to dig a trench and dump the soil sample into the organic analyzer to screen it for 2 wk in a search for microorganisms. (Mission Status Bulletin 44; *NYT*, 21 Sept 76, 18; 26 Sept 76, 5; *W Post*, 23 Sept 76, A-2; 24 Sept 76, A-2; 26 Sept 76, A-6; *W Star*, 23 Sept 76, A-3; 26 Sept 76, A-14)

21 September: NASA announced selection of Western Union Telegraph Co. and RCA Global Communications, Inc., for competitive negotiations leading to award of a single contract to provide tracking and data-relay satellite services (TDRSS) to support earth-orbiting spacecraft for 10 yr beginning in 1980. Goddard Space Flight Center would manage procurement of two relay satellites in synchronous earth orbit, plus ground-terminal facilities to be located at White Sands, N.M. The system would support all of NASA's scientific, applications, and manned spacecraft missions in earth orbit up to 5000-km altitude, including Space Shuttle, Spacelab, and the automated spacecraft which the Shuttle would insert into orbit in the 1980s. TDRSS, when fully operational, would provide coverage over 85 to 100% of each orbit as compared to a present average of 15%, and permit closing down many government-owned ground stations and leased communications circuits now needed to support earth-orbiting satellites in the Spaceflight Tracking and Data Network (STDN); portions of the latter system would be retained only for synchronous satellites and those in higher altitude elliptical earth orbits. (NASA Release 76-158)

- Marshall Space Flight Center announced development of a solar concentrator-collector that would use a large plastic Fresnel lens to focus sunlight on a tube containing heat-transport fluid, for use in constructing or converting cooling systems in large buildings or in manufacturing processes requiring heat in the 200-370°C range. A Fresnel lens, consisting of a thin transparent material grooved on one surface, would serve the function of a heavier ordinary curved lens in focusing incoming light in the desired direction. MSFC's lens, thought to be the largest tried in a solar-energy application, was assembled from multiple lens panels and measured 1.8 by 3.6 meters; it had been undergoing tests at Wyle Laboratories in Huntsville since July 1976. It would be moved to MSFC and integrated into a solar test bed for further analysis of performance. Primary advantage of the Fresnel lens would be its adaptability to mass-production techniques, permitting low cost; besides the light weight and space saving of the acrylic plastic, MSFC cited durability and mechanical strength, as well as ease of cleaning, of lenses exposed to weather. The test article, to minimize technology requirements and to interface with existing systems, used off-the-shelf materials and hardware as fully as possible. MSFC findings would be made available to industry after further testing and evaluation. (MSFC Release 76-169)

- MSFC announced selection of two NASA payload specialists and a backup for ASSESS II, a Spacelab simulation project to be flown in May 1977 in a four-engine jet aircraft. ASSESS II, a joint mission of NASA and the European Space Agency (ESA), would use the two U.S. payload specialists and two selected by ESA [see 27 Aug.]. The two selected, both of the Jet Propulsion Laboratory, were Dr. Robert T. Menzies and David S. Biliu; Leon B. Weaver of MSFC was chosen as the backup. Dr. Menzies would be principal investigator (PI) for the laser absorption spectrometer experiment; Biliu was a member of JPL's synthetic-aperture radar research team; Weaver, a veteran of the ASSESS Lear 4 mission of Oct. 1974 and presently assistant mission manager, would represent MSFC's Spacelab Payload Projects Office, and was selected from a group already involved in ASSESS II whose experience would permit training at minimal cost. The ASSESS II payload would include both NASA and ESA experiments in earth resources, monitoring of atmospheric pollution, and infrared astronomy. (MSFC Release 76-170; *Marshall Star*, 22 Sept 76, 4)

- A year-long cooperative experiment using *Ats 6* in a mission called SITE—Satellite Instructional Television Experiment—to broadcast daily programs on agriculture, health and hygiene, family planning, and national integration to an estimated 5 million persons in 2400 remote villages in India never before exposed to television had "met its primary goal," said Prof. Yash Pal, director of the space applications center operated by the Indian Space Research Organization at Ahmadabad. *Ats 6* had relayed the programs, broadcast from a ground station near Ahmadabad, to small ground terminals consisting of conventional TV receivers augmented with small electronic components and inexpensive antennas of chickenwire mesh, all built in India; another 2600 terminals near conventional TV stations had rebroadcast the instructional material to cities and nearby villages.

  Before the programs began, teams of Indian social scientists and engineers had visited more than 6000 villages to select the ones to receive SITE transmissions; an individual in each village had to be identified as TV-set caretaker, and a public site—usually a school—selected as the viewing center, so that "weaker sections of village society" could be assured of equal access to the TV programs. As some appropriate villages lacked electricity, these had to be supplied with power, a valuable side effect of the experiment. Programs for SITE had been developed by India's Ministry of Information and Broadcasting in collaboration with ISRO; program content and format had to be kept simple because of the high rate of adult illiteracy. More than 1200 hr of diversified content were broadcast during the experiment; attendance was particularly high for programs related to animal husbandry and agriculture. Children's programs were received enthusiastically, with resulting improvement in school attendance, and many students brought younger members of the family to view the programs.

  One yr of experimental TV would not "change the face of our villages," said Prof. Pal, but it had created "a cadre of dedicated people and the methodology necessary to sustain an ongoing program." As a follow-on, the Indian government would construct ground transmitters in six cluster areas of villages to include about 40% of those involved in SITE, to resume the educational programs by early 1977. (NASA Release 76-157)

- The DOD did not really want the Space Shuttle, charged Sen. William Proxmire (D-Wis.), and NASA had "conned Congress into buying a pig in the poke" in its efforts to obtain funds for Shuttle development. DOD had shown that it did not want the Shuttle by its actions, said Proxmire, citing DOD's statement that it would "under no circumstances" pay for the fourth and fifth Shuttle orbiters, and the DOD decision to go forward with a new satellite communications system (DSCS-III) "which will not capitalize on the unique advantages of the Shuttle . . . [but] is to be compatible with both the Shuttle and the present Titan III-C." Proxmire charged that the DSCS decision was based not on a 9-mo wait for the Shuttle, but on the fact that "DOD has little confidence in a cost-effective operational Shuttle." Proxmire also cited a General Accounting Office report issued in April that DOD would pay less to launch its satellites on expendable vehicles than on the Shuttle through 1990-1991, assuming that DOD did not plan to recover and reuse any of its satellites. (*SBD*, 24 Sept 76, 105)

- William Allan Patterson, pioneer of many firsts in the airline industry, would receive the Wright Brothers Memorial Trophy at an annual dinner in Dec. sponsored by the Natl. Aeronautic Assn., announced NAA President John P. Henebry. The 77-yr-old Patterson, who retired as board chairman of United Airlines in 1963, had joined Boeing Company in 1929 and in 1934 was elected president of United (created in 1931 by Boeing's merger with 3 other airlines to form a transcontinental system), a position he held for 29 yr. Patterson had declared himself most proud of establishing in 1930 the occupation of air stewardess; he had been instrumental in passage of the Federal Aviation Act in 1958, establishing a single federal agency with full authority for air traffic control, and in obtaining appropriation of federal funds for installation of landing and navigation aids at major airports across the U.S. The trophy, a miniature silver copy of the plane flown by the Wright Brothers at Kitty Hawk in 1903, had been awarded yearly by the NAA for significant public service of enduring value to aviation in the U.S. (NAA release, 21 Sept 76)

22 September: Scientists at the Naval Research Laboratory successfully used a computer system to orient sensors on high-altitude balloons orbiting the earth on observation flights of up to 100 days, NRL announced. The long-duration missions were required to monitor high-energy solar flares, cosmic gamma-ray bursts, and transient x-ray sources. The NRL system, designed for superpressure balloons flying at 40-km altitude, used a microcomputer programmable for as many as 32 observation tasks, which could be modified in flight by ground command. An L-band telemetry link from the balloon permitted receipt and recording of data on position and sidereal time, used by the microcomputer to orient the detector. Data could be relayed through geosynchronous satellites, using the computer to orient a high-gain antenna toward the satellite. Real-time long-duration monitoring by balloons could offer a low-cost and frequently more sensitive alternative to satellite experiments, NRL stated. (NRL Release 44-7-76B)

26 September: Pilot training for the Space Shuttle would begin in October at White Sands Missile Range, the *Washington Post* reported, with instructors completing their training by mid-October. Two of the instructors—David Griggs and Ted Mendenhall—were scheduled to make practice approaches at the range 29 September; chief instructor Al Manson and copilot Ed Rainey made a similar test in August. Aircraft used in the Shuttle training are modified Grumman Gulfstream II jets fitted with Shuttle orbiter controls and instrumentation. (*W Post*, 26 Sept 76, A20)

- Both the United States and the USSR were working on ways to shoot down enemy satellites in earth orbit, reported Henry S. Bradsher in the *Washington Star*. Commenting on a Department of Defense disclosure of "aggressive basic technology research efforts" to protect U.S. satellites from a potential Soviet threat, Bradsher noted that the DOD previously had refused to comment on U.S. work on so-called satellite killers [see 27 July] and had tended to play down any Soviet danger. With the DOD announcement came word that the USSR had conducted 3 satellite-destruction tests this year; it had conducted 5 tests of such systems between 1968 and 1971, but had dropped them, leading Pentagon officials to think the program had been halted in accordance with 1972 agreements that limited strategic defense armament (SALT I) and antiballistic-missile systems. The agreements had been based on each side's ability to monitor the other's compliance by means of reconnaissance satellites; however, recent reconnaissance photographs had been interpreted as showing Soviet launch pads carrying satellite-killer systems.

Secretary of the Navy J. William Middendorf had stated 25 Sept. that the Soviet Union was presumed to be working on guidance systems that could destroy targets such as over-the-horizon naval warfare missiles, greatly heightening the threat to U.S. warships. Asked if the U.S. were developing satellite-killers, Middendorf replied, "We're working in that direction." The Pentagon at first refused to comment on a satellite-destruction program, but released a statement when pressed about Middendorf's answer. (*W Star*, 26 Sept 76, A-1)

27 September: NASA was preparing to issue requests for proposals in Jan. 1977 for the Large Space Telescope, resulting in contractor selection during the next year if Congress approved the fiscal 1978 funding requested, said *Aviation Week & Space Technology* magazine. NASA also hoped to issue an announcement of scientific opportunity to the scientific community in Feb., asking for scientific instruments to be carried on the first mission of the 2.4-meter telescope. (*Av Wk*, 27 Sept 76, 9)

28 September: MSFC announced launchings of two giant helium-filled polyethylene balloons carrying cosmic-ray detectors, last week's flight for a French-Danish team investigating isotopic composition of primary cosmic rays, the upcoming flight to evaluate instrumentation for a planned HEAO launch and for other Space Shuttle experiments. The HEAO flight scheduled for 1979 would survey and map gamma-ray and cosmic-ray flux. The balloon launch site near Sioux Falls, S.D., was chosen because earth magnetic-field lines there would deflect fewer cosmic rays, and the low population density and open terrain facilitated payload recovery. The balloons would reach an altitude of about 41 km, above 99.6% of earth's atmosphere (which fragments cosmic rays), and remain aloft for 40 to 60 hr sending data to the ground. Upon electronic command from the ground, the instrument payloads would separate from the balloons and return to earth by parachute; the balloons would be destroyed by ground command over unpopulated areas. Telemetry tapes and recovered instruments would go to the investigators for analysis. (MSFC Release 76-172)

29 September: NASA announced award of a cost-plus-fixed-fee contract to evaluate a fuel-saving concept applicable to commercial subsonic transport aircraft. The three companies selected were Boeing Co., McDonnell

Douglas Corp., and Lockheed Aircraft Corp.; each company estimated its portion of the contract to be about $1.6 million. The 24-mo contract, to be managed by Langley Research Center, would be for a study of laminar flow control by use of an air-suction system built into aircraft skin surfaces to remove boundary-layer air before it could cause drag and air turbulence that would reduce aircraft speed. (NASA Release 76-161)

- Meetings in Ardashat, Armenia, between Soviet and American scientists on space biology and manned flight safety had concluded, announced the Yerevan domestic radio news service. The meeting heard reports on results of the Apollo-Soyuz joint space mission, and proposed subjects for discussion at another meeting to be held in the U.S. in 1977. The government of the Armenian Republic gave a reception to honor the participants. (FBIS, Yerevan in Armenian, 29 Sept 76)

30 September: Veteran astronaut Thomas P. Stafford, Apollo commander of the U.S.-Russian Apollo-Soyuz mission of July 1975, received the Gen. Thomas D. White Space Trophy for 1975, the Natl. Geographic Society announced. Gen. David C. Jones, USAF Chief of Staff, presented the award, honoring the late Air Force Chief of Staff and given annually to the member of the Air Force, military or civilian, who contributed most significantly in the preceding yr to U.S. progress in aerospace. Maj. Gen. Stafford's citation noted "outstanding contributions . . . by his participation in the Apollo-Soyuz Test Project . . . [He] was instrumental in establishing the necessary rapport and spirit of cooperation vital to the success of the mission. Under his leadership, the Apollo crew laid the basic groundwork for future international technological achievements in space and proved that the United States is the acknowledged leader in space exploration." Attending the ceremony with Stafford's Air Force colleagues and officers of the Natl. Geographic Society were former astronaut Michael Collins, now director of the National Air and Space Museum, and Dr. Robert C. Seamans, formerly Secretary of the Air Force and Deputy Administrator of NASA, now Administrator of the Energy Research and Development Administration. (NGS release, 30 Sept 76)

During September: Ames Research Center's *Astrogram* published an interview with the center's first female aircraft mechanic: a co-op student, Maria-Elena Sanchez, enrolled in a airframe and power mechanics course at the College of San Mateo. She was one of four females in the program, whose total enrollment was about 200. As a mechanic at ARC, Ms. Sanchez was performing general maintenance and upkeep on Lear jets and the Cessna 402; for her work at Ames, she was receiving both monetary compensation and college credit. The college program required a student to attend for about 30 mo a school approved by the Federal Aviation Administration, then pass a written, oral, and practical test to receive a license in Airframe and Power Plant Mechanics. Ms. Sanchez, who held a commercial pilot's license and had been flying since 1971, said she hoped to complete the program and obtain the A&P license within a yr. (*ARC Astrogram*, 9 Sept 76, 1)

- The Marshall Space Flight Center announced 3 Sept. the start of hot-firing tests of the steering system for the Space Shuttle solid-fuel rocket booster (SRB). First series of tests, which would continue into Oct., would confirm design of the thrust-vector control steering system developed by MSFC engineers as part of the center's responsibility for SRB design and development. The system under test would provide power to move the nozzle of the SRB in any direction, steering the Shuttle in the first 120 sec of flight during SRB burn. Data from the first series of tests would be evaluated to refine the system design; after any needed modifications, a second series of tests would certify the system. Thrust-vector control units would be provided in 1978 to Thiokol Corp., MSFC contractor for the SRB solid-fuel rocket motor, for testing under actual firing conditions. (MSFC Release 76-163)

  —MSFC announced it had delivered the first production case segment for the SRB motor to Thiokol's Wasatch Div. in Utah on 27 Sept., beating the schedule by 3 days. All case segments for the first motor were to arrive at Thiokol by the end of 1976. The delivered segment, almost 4 m in diameter and 4.2 m long, weighed more than 5990 kg. Eleven segments would be used in each motor; joined, the segments and nozzle would measure more than 38 m long, over three fourths of the total SRB length. (MSFC Release 76-173)

  —MSFC announced 30 Sept. that personnel of the Rocketdyne Div., Rockwell Intl. Corp., had fired a developmental test engine for the Space Shuttle for 650 sec, longest test to date. The same engine had been fired for 300 sec 2 days earlier. The tests, at the Natl. Space Technology Laboratory in Bay St. Louis, Miss., would produce component- and system-operation data from extended-duration firings at increasing power levels. (MSFC Release 76-174)

- Johnson Space Center announced on 21 Sept. modification of a cost-plus-award-fee contract with IBM, Gaithersburg, Md., to cover software for ground-based computing and data-processing systems at JSC. IBM would develop computer programs for Space Shuttle vehicle management and flight operations, and related scientific and medical operations. Value of the modified contract would be $19 463 000. (JSC Release 76-59)

  —NASA contracted 22 Sept. with Aeronutronic Ford Corp. for support services for the Mission Control Center and other ground-based data systems at JSC, including data hardware and software systems engineering and maintenance, as well as logistics, reliability, and quality assurance. The cost-plus-award-fee contract would cost about $46 550 000 from July 1976 through 30 Sept. 1978. (JSC Release 76-60)

  —JSC announced 24 Sept. selection of Hamilton Standard Div. of United Technologies Corp. for negotiations leading to award of a contract for development and production of a portable oxygen system for Space Shuttle crew and passengers. The cost-plus-fixed-fee contract, costing about $1.9 million, would run from 13 Nov. 1976 to 13 July 1979. The system, designed to meet four special Shuttle needs (emergency oxygen in case of cabin-atmosphere contamination, prebreathing before space-walks, life support during rescue, and emergency oxygen after landing in case of landing-area contamination), would be capable of operating independently or connected to the Shuttle oxygen system. The contract would call for delivery of ten units and 50 recharge kits for NASA use in certification, training, and flight, plus ground-support equipment and manpower; it would contain options for 62 additional units and 310 recharge kits. (JSC Release 76-61)

- A giant Air Force antenna at Hamilton, Mass., that had shared in transatlantic communications with the Jodrell Bank Observatory in England by way of moon reflections would be dismantled and moved to Westford, Mass., for use by the Lincoln Laboratory of MIT in radar studies of earth's upper atmosphere, the AFSC *Newsreview* announced. The big dish, nearly 46 m in diameter, had been used by the USAF Geophysics Laboratory since 1962 for ionospheric and radioastronomy studies, as well as in auroral studies using satellite beacons. Continuing USAF research at Hamilton's Sagamore Hill Observatory would use the facility's 25.6-m antenna and smaller receivers. (AFSC *Newsreview*, Sept 76, 10)

October 1976

1 October: The boost given by the Viking Mars missions, now nearing an end, to those concerned with the space program might persuade Congress of the virtues of possible future projects, *Science* magazine reported. Describing 15 Sept. testimony of Dr. John E. Naugle, acting associate administrator of NASA, before a House subcommittee, the magazine said "some of the juicier projects the agency has in mind for the 1980s" might include a roving-vehicle mission to Mars, shipped aboard a Solar Sailor, "the closest thing the agency has ever proposed to a genuine space ship." The Solar Sailor would be driven by solar pressure—light reflected from the sun, collected by huge lightweight sails—and was referred to as "a reusable interplanetary spacecraft." Other future projects would be a Global Information Services satellite system to combine earth sensing, meteorological and pollution observation, data transmission, and navigation uses, all such information to be available to the public at receiving stations or home terminals; some sort of solar-energy collection system, either on the ground or in space; and moving "a segment of our industrial society into space . . . where there is abundant solar energy and an almost inexhaustible vacuum to act as a sink for thermal and chemical pollution." Naugle and other agency spokesmen had noted that such plans were still in the design stage and the agency was not ready to discuss them in detail; however, *Science* commented, the Viking success might mean that for NASA "perhaps the sky will be the limit, after all." (*Science,* 1 Oct 76, 39)

- NASA announced successful static test firing of one of two Space Shuttle main engines for 650 sec at main-stage operating level (50% rated thrust level) at the Natl. Space Technology Laboratories, Bay St. Louis, Miss., under the direction of MSFC. The firing, on 30 Sept., was the longest test firing of an SSME to date. Previous firings had aimed at testing engine-throttling control, engine start and shutdown sequence, and engine power balance. The current series of tests, conducted by personnel of Rockwell Intl.'s Rocketdyne Div., would obtain component- and system-operation data at increasing levels of power. The NSTL test program combined development and acceptance testing; each Shuttle engine would be fired at NSTL before being certified for flight. (NASA Release 76-162)
- Three products developed by Lewis Research Center had appeared in a list of the 100 most significant products developed during the year, published by *Industrial Research* magazine, said the *Lewis News*. The magazine yearly considers about 1000 entries from industries throughout the U.S., and presents the IR-100 award for the products selected by a panel of nationally known technical judges. The 1976 award brings to 16 the number won by LeRC since it entered the competition 11 yr ago. Products cited were: a continuous-production cyclotron target, used in producing

radioactive isotopes for diagnostic nuclear medicine; a thickness-measuring radar that could measure lake-ice thickness directly beneath an overflying aircraft, used by the Coast Guard to help extend the shipping season on the Great Lakes; and a ceramic thermal-barrier coating that would increase the life of metal parts by protecting them from very high-temperature erosive and corrosive gases, used to prolong the usefulness of turbine blades ten times beyond that of uncoated blades. (Lewis News, 1 Oct 76, 1)

1-21 October: No evidence of organic material in the soil of Mars appeared in *Viking 2* lander's first experiment results, John Noble Wilford reported in the *New York Times*. The experiment—heating a soil sample to vaporize any organic molecules—was one of eight scheduled for the lander. Dr. Klaus Biemann of MIT, leader of the organic chemistry experiment, noted that the *Viking 1* lander had found no trace of organic compounds either. Biology experiments on both landers had returned data indicating some unexplained activity in the soil samples. Gentry Lee, director of scientific analysis for the Viking project, said neither a biological nor a chemical hypothesis was consistent with all the data. (NYT, 1 Oct 76, D13)

A letter to the *NY Times* had emphasized that the absence not of carbon but of "the extremely complex molecules of carbon that are characteristic of earth life" was the puzzling factor in the Viking findings at Mars. (NYT, 2 Oct 76, 24)

Project scientists had offered three possible explanations for the negative results of the attempts to identify organic compounds. First, any organic compounds left on Mars by the solar wind, by meteor fall, or by past life processes could have been destroyed by agents such as ultraviolet radiation, oxygen, or oxidants such as nitrates or metal oxides acting independently or synergistically; "certainly," the project report published in the Oct. issue of *Science* said, "continuous exposure to short-wave ultraviolet light in the presence of oxygen will cause rapid chemical decomposition of most organic compounds." Second, organic compounds might exist in Mars soil in amounts too dilute to be detected by the Viking instruments. Third, organic compounds might be forming on Mars and undergoing rapid destruction. The traces of organic molecules seen in a second *Viking 2* lander chemistry test could well have resulted from contaminants known to have been in the test chamber before the spacecraft left earth, Dr. Biemann said. (NYT, 2 Oct 76, 28)

The *Viking 2* lander pushed at a rock on the surface of Mars but was unable to move it, JPL scientists reported. The shove was preliminary to an attempt to sample soil shielded from solar radiation and more likely to contain organic material. Dr. Louis Kingsland, deputy mission director, announced the lander would be ordered to try again and to turn over a smaller rock if the first choice failed to move. Dr. Priestley Toulmin, director of the inorganic analysis experiments, said the iron-rich topsoil tested by the *Viking 2* lander had a striking resemblance to that tested at the *Viking 1* site; results of the assay were almost a duplicate of those sent to earth 2 mo ago. The similarity lead scientists to think the soil had been affected by ancient weather conditions that swept over large portions of the planet; however, the results could not show whether the material below the surface was likewise uniform around the planet. (*W Star,* 6 Oct 76, A-5; *NYT,* 8 Oct 76, A26)

The third of three *Viking 2* lander tests for dead organic matter in the soil of Mars had shown no sign of fossilized life, Dr. Biemann said at JPL. The last test would use soil exposed by turning over a rock on the Mars surface that scientists said might have been undisturbed for as long as a million yr. Dr. Norman H. Horowitz of Caltech, who supervised experiments on both *Viking 1* and *Viking 2* sites that looked for signs of biological activity, said that the results he had obtained "would be convincing for life on Mars if we had found the dead organic matter." The sample from under the rock would be held for chemical analysis until scientists could examine photographs of the trench where it was dug to make sure the sample was taken from the proper protected place. Another under-rock sample would be dug for biological analysis later, the second *Viking 2* test for synthesis of organic matter to be conducted at the Utopia site. (*W Post,* 12 Oct 76, A-8; *NYT,* 12 Oct 76, 18)

Meanwhile, the seismology instrument on the *Viking 2* lander had recorded no marsquake activity in the 2 wk it had been monitoring the planet's surface, said Dr. Don L. Anderson, seismology team director. The detector, turned up to full sensitivity, could detect a quake as small as a 3 on the Richter scale (enough to cause only slight damage on earth) as far away as 200 km from the lander; Dr. Anderson said it was working well, recording breezes and even picking up vibrations from tape recorders on board the lander. A similar instrument on the *Viking 1* lander never was freed from its packaging. (*W Star,* 15 Oct 76, A-3)

The scoop of soil taken from under a rock on the Mars surface also failed to yield a trace of organic molecules, Dr. Biemann announced. However, mission officials planned to go ahead with biological experiments using a sample from under another rock later in the week; the sample would be tested for signs of life processes—growth, metabolism, and respiration. (*C Trib,* 21 Oct 76, 4-22)

3 October: Dr. William Nordberg, 46, director of applications at GSFC, died of cancer after a 2-yr illness. A native of Austria, he had come to the U.S. in 1953 to work as an atmospheric geophysicist for the Army Signal Corps. As one of the scientists transferred from the Army to the newly formed NASA, he went to GSFC in 1959 as head of the physical measurement section in Satellite Applications Systems Div. and became Director of Applications in 1974. In 1975 he had received both the William T. Pecora Award and NASA's Distinguished Service Medal for outstanding contributions to applications of remote sensing of earth by spacecraft, honoring his work with the Landsat project, in which he had coordinated the work of 300 scientists from 38 countries in demonstrating satellite uses in disciplines such as agriculture, forestry, land use, marine resources and oceanography, mineral and oil exploitation, geology, and environmental impact. (*Goddard News,* Sept 76, 1; *W Post,* 5 Oct 76, C-6)

4 October: The Intl. Telecommunications Satellite Organization (INTELSAT) announced that the Assembly of Parties meeting in Nairobi, Kenya, had confirmed the appointment of Santiago Astrain of Chile, since 1973 the secretary general of INTELSAT, as the first Director General of the international organization. The board of governors had appointed Astrain in July to serve a 6-yr term as Director General, beginning 31 Dec. 1976. The Director General would be responsible for implementing permanent management arrangements and for tasks such as implementing a new high-capacity satellite system to be known as Intelsat V in the late 1970s and early 1980s. About 200 delegates, representing 69 of INTELSAT's 95 member countries, attended the Nairobi meeting. Elected were a chairman, Isaac Omolo Okero of Kenya, and a deputy chairman, Fernando Gaviria of Colombia. Vice chairmen elected to represent the five world regions were Romulo Villar Furtado of Brazil (The Americas); Joachim Jaenicke of the Federal Republic of Germany (W. Europe); Zika Radojlovic of Yugoslavia (E. Europe and N. Asia); and Sribhumi Sukhanetr of Thailand (Asia and Australasia). The meeting also recognized the People's Republic of China as the sole legitimate government of China, and welcomed the PRC to join INTELSAT. (INTELSAT Releases 76-28-I, 76-29-I; *Satellite Pathways*, ComSat, vol. 1 no. 7, Sept-Oct 76)

5 *October:* About 3000 people attended a ceremony dedicating the Space Hall of Fame at Alamogordo, N.M., honoring men and women involved in space exploration. Original inductees, numbering 35, included U.S. astronaut Neil Armstrong and Soviet cosmonaut Yuri Gagarin. Nine of the 35 were from the Soviet Union; eight each from the U.S. and West Germany; three from Austria, two from France, and one each from Great Britain, Hungary, Italy, Romania, and Switzerland. Dr. Frederick Durant of the Smithsonian Institution announced the names of the pioneers, nominated by the Intl. Academy of Astronautics in Paris with final selection by a committee appointed by the governor of N.M. for the Hall of Fame. Lubas Perek, head of the UN's outer space affairs division, told the audience that the choice of persons to be honored "will surely reflect the grand international character of outer space research." (*W Star,* 6 Oct 76, 2; *NYT,* 10 Oct 76, 26; Intl Space Hall of Fame, media release; *Alamogordo Daily News,* 3 Oct 76, dedication edition)

- Responses of Presidential candidates Ford and Carter to questionnaires submitted by U.S. physicists and engineers emphasized their "differing viewpoints," wrote Walter Sullivan in the *NY Times.* One set of 16 questions was from 24 professional engineering societies claiming a combined membership of one million; the other, dealing with three broad issues, was submitted by the American Physical Society's president, Dr. William A. Fowler. To a question regarding overseas sales of nuclear fuel and equipment, President Ford replied that the U.S. must maintain the role of major supplier "for peaceful purposes—so that we can influence others to accept controls to minimize the threat of proliferation." Candidate Carter replied that it was "absolutely essential" to halt such sales "even with safeguards" to prevent use of the materials in producing nuclear weapons. *Physics Today,* publishing the replies to Dr. Fowler's questions in the Oct. issue, noted editorially that both candidates were committed to more support for basic research and "a strong voice for science in the Administration's decision-making." (*NYT*, 5 Oct 76, 33)

- One of the first in a group of 32 commercial solar heating and cooling systems in ERDA's nationwide solar-energy demonstration program went into operation in Lynchburg, Va., NASA announced. Local and state government officials attended an open house at the office of Terrell E. Mosely, Inc., where the system is installed, to view the equipment together with businessmen, engineers, architects, and federal representatives. The 32 pilot projects, selected from 308 proposals submitted to ERDA, include ten office buildings, four schools, three hotels or motels, two fire stations, two factories, one hospital, one laboratory, one library, and some miscellaneous buildings. NASA's Marshall Space Flight Center provided technical management for the installations in support of the ERDA program. (NASA Release 76-163; MSFC Release 76-177)

- The official Manila radio announced an agreement between Domestic Satellite Philippines and the Nippon Electric Co. of Japan for acquisition, installation, and construction of 11 earth-satellite stations in a $26 million satellite communications system, to be operational by 1978. The system would provide long-distance service in the national communications network. (FBIS, Manila FEBC in English, 5 Oct 76)

6 October: A rocketborne experiment to investigate the ionosphere failed when the second stage of the two-stage Nike-Cajun vehicle failed to ignite, Wallops Flight Center announced. Liftoff at WFC occurred at 3:30 pm EDT, and the Cajun with its payload fell into the Atlantic Ocean about 2 min 14 sec after liftoff. The experiment was to determine ion composition of the D and lower E regions of the ionosphere, with special emphasis on the transition region. Planned recovery of the experiment in midair would have spared the instrumentation from impact and exposure to salt water, allowing reflight at substantial cost savings. A committee had been appointed to investigate the cause of the malfunction. (WFC Release 76-14)

- ESA's Science Program Committee announced that it had unanimously approved two new scientific projects at its sixth meeting held 4 and 5 Oct. in Paris. The committee approved participation in the Space Telescope, planned by NASA for launch in 1983, and agreed to finance a plan called GEOSARI (launching a Geos spacecraft into a different orbit on the second flight of the Ariane launcher, scheduled for Dec. 1979). The Space Telescope project depended on a favorable outcome of negotiations with NASA and approval of the project by U.S. authorities in 1977. The committee considered an additional ESA-NASA cooperative project, the Out-of-Ecliptic (OOE) mission to fly two spacecraft around the sun at its north and south poles in a first exploration of the solar system's third dimension; the committee agreed to continue negotiations with NASA and to join in an announcement of opportunities for the mission if necessary before the committee's next meeting, in the spring of 1977. (ESA release 6 Oct 76)

- INTELSAT announced award to the British Aircraft Corp. of a 15-mo fixed-price contract for design, fabrication, test, and delivery of a model onboard processor programmable to perform a variety of tasks, including monitoring and control of satellite electronic functions. A standardized system based on microprocessors, rather than hard-wired logic, for each individual function or system would reduce weight and power consumption and increase satellite reliability and efficiency. Value of the contract was $121 600 in U.S. dollars. (INTELSAT Release 76-27-M)

- The Brazilian Telecommunications Company received bids from international firms to supply space tracking equipment to the Brazilian satellite communications system, according to a broadcast from the Brasilia Domestic Service. The announcement said the first stage of the domestic satellite system, to be concluded in 1979, would consist of 17 receiving and transmitting stations primarily in the Amazon region, and two space satellites; a third would be used for replacement. During a second phase ending in 1981, 19 more stations would be added. The system would consist of 44 stations in all, the last eight to be added later. (FBIS, Brasilia Domestic Service in Portuguese, 6 Oct 76)

7 October: Modifications of the test stand to be used at the Natl. Space Technology Laboratories for static firing of the combined Space Shuttle propulsion system had been finished 2 wk early, MSFC announced. The stand had been built for tests of the Saturn V first stage and converted for use in the Shuttle program by Industrial Contractors, Inc., of Idaho Falls, Ida. In the tests, scheduled to be conducted late in 1977 by workers in Rockwell Intl. Corp.'s Space Division under MFSC direction, the main engines of the Shuttle would be connected to a structure resembling the orbiter; this simulator, using both flight and nonflight hardware, would be attached to an external tank, the Shuttle element containing liquid oxygen and liquid hydrogen for the three main engines. Firing of this combination (known as the Main Propulsion Test Article) would be the first time the three main engines were fired as a system. Engineers would combine acoustic tests with the MPTA firings to obtain data for determining optimum vibration and sound levels; they would also use the data to check tanking procedures. Hardware for the tests would arrive at NSTL next summer. (MSFC Release 76-178)

- The Energy Research and Development Administration announced plans to build a "power tower" in 1978 and put it in operation by 1981. Plans called for a large number of sun-tracking mirrors, called heliostats, to reflect sunlight on to a boiler located at the top of a tower; water in the boiler, superheated by concentrated sunlight, would produce high-pressure steam to drive a turbine. ERDA was seeking a utility company to construct and operate the pilot plant: the utility would provide the site, install a conventional turbine and generator, and distribute the power to its customers over existing lines. ERDA would provide the solar portions (tower, boiler, heliostats, and storage) and charge the utility for the electricity. Sites proposed for the pilot plant were in Ariz., Calif., Fla., Miss., P.R., R.I., and Tex. Three teams—headed by Honeywell Inc., McDonnell Douglas Astronautics Co., and Martin Marietta Corp.—were working on designs for the pilot plant, and Boeing Engineering and Construction Corp. was developing a design for the heliostat only. ERDA would choose the best design and test it at a facility under construction at the Sandia Laboratories of ERDA in N.M. (ERDA Release NF-76-12)

- ESA announced appointment of Michel Bignier, director-general of the French Centre National d'Etudes Spatiales (CNES) from Jan. 1972 to June 1976, as director of the Spacelab program effective during Oct. ESA's director general, Roy Gibson, had been managing the program with the assistance of the Technical Inspector, Prof. Massimo Trella, since the former Spacelab director resigned 3 mo ago. In addition to being responsible for directing the Spacelab program, Bignier as a member of ESA's 9-man directorate would assist in collective consideration of major ESA management problems. (ESA release 7 Oct 76)

8 October: Dr. Maxime A. Faget, director of engineering and development at Johnson Space Center, had been chosen to receive awards from both the American Astronautical Society and the Instrument Society of America, the JSC *Roundup* reported. Dr. Faget, one of the original 35 members of the NASA Space Task Force Group, received from the AAS its Space Flight Award, given for the greatest contribution to the advancement of space flight and space exploration. Dr. Faget had received AAS's Lovelace award in 1971. The ISA had selected him for the Albert F. Sperry Award, for "advancing spacecraft, laboratory and biomedical instrumentation through technical and engineering leadership of manned space flight programs." Dr. Faget in 1969 received NASA's medal for exceptional service and distinguished service medal—the agency's highest award—and had been inducted into the Natl. Space Hall of Fame. (JSC *Roundup*, 8 Oct 76, 1)

- Test pilot William H. Dana had been chosen to receive the AIAA 1976 Haley Space Flight Award, the DFRC *X-Press* announced. Dana would receive the award at the IAF conference banquet 15 Oct. Named for Andrew G. Haley, one of the founders of the Am. Rocket Society, the award is for "outstanding contribution . . . to the advancement of the arts, sciences or technology of astronautics." Dana would be honored for his service as test pilot on the HL-10, M2-F3, and X-24B lifting bodies, and his 16 flights as project pilot of the X-15 rocket-powered research aircraft. (DFRC *X-Press*, 8 Oct 76, 2; LA *Herald-Examiner*, 12 Oct 76, 3; Lancaster, Calif. *Ledger-Gazette*, 7 Oct 76, 10)

- MSFC announced that the Alabama Society of Professional Engineers had chosen the Center to receive the first annual Wonder of Engineering award, for design engineering achievement in producing a study called Ecastar (a project during the summer of 1975, studying energy-conservation measures and their possible effect on society and the environment). Ecastar was one of the projects completed over the past decade in a design engineering program run at MSFC by Auburn Univ. A summer faculty fellowship program sponsored by NASA and the Am. Soc. of Engineering Education had awarded the fellowships to engineering, natural science, and social science faculty members in summer study programs directed by universities at NASA research centers, to enable the participants to design courses at their institutions and to further collaboration between engineering and other disciplines. (MSFC Release 76-183)

- NOAA announced award of a $1 441 116 contract extension to **Management and Technical Services Co.**, Daytona Beach, Fla., for technical support in operating NOAA's geostationary environmental satellite system. The extension would continue the company's work in processing and distributing environmental data. Two GOES satellites at altitudes of 35 800 km were transmitting imagery and other information to receiving stations at Wallops, Va., and Suitland, Md., for communication to NOAA centers in Washington, D.C., Miami, Kansas City, and San Francisco. (NOAA Release 76-227)

10 October: U.S. balloonist Ed Yost, who had ditched his balloon in the Atlantic when loss of helium prevented him from completing the first transatlantic balloon crossing, was rescued by a West German tanker, the *Elisabeth Bolton*, after more than 3 hr of floating about 1200 km west of the Azores. The ship was reported headed for Gibraltar. A rescue plane sent to the scene earlier from the USAF Search and Rescue Center at Ramstein, West Germany, had circled over Yost's position until he was picked up. Yost had stayed aloft just short of 107 hr (exceeding the 1913 record of 87 hr) and had traveled about 4000 km, well beyond the 1914 record of less than 3100 km. He had worked a year and a half on the 2-ton balloon—which had cost him about $100 000 of his own money—and began his journey from the U.S. coast at Milbridge, Me., on 5 Oct. The 57-yr-old balloon manufacturer from Sioux Falls, S.D., said he was "in good spirits because I broke a lot of records . . . but I'm sorry I didn't land on solid ground." (*C Trib*, 11 Oct 76, 3; *NYT*, 12 Oct 76, 5)

- Problems confronting physicists in a major effort to discover "the true nature of the most basic components of matter and the laws that govern them" were the subject of last week's annual meeting of the Am. Physical Society's Division of Particles and Fields, held at the Brookhaven Natl. Laboratory on Long Island, said Walter Sullivan in the *NY Times*. Scientists from the U.S., the Soviet Union, and other countries were considering the construction of a "world machine," a particle accelerator that would dwarf any now in existence, measuring about 48 km in circumference and generating ten thousand billion electron volts. Its cost, estimated as 3 to 6 times that of the largest accelerator in existence, would be met on a global basis with the U.S. and the USSR "playing major roles," Sullivan said. Although the project had been discussed at several international conferences and would not be built until the end of the century, proponents believed that groundwork should begin immediately and had asked the Intl. Union of Pure and Applied Physics to take the lead in organizing the effort. Design decisions would await the results of several machines now projected or being built, that might show how problems should be most effectively handled.

  The most powerful machine now operating (at the Fermi Natl. Accelerator Laboratory in Ill.) had a 6.5-km ring that could boost protons to 500 Gev (one Gev equaling a billion electron volts). A modification now in progress, called the doubler, would increase the potential to 1000 Gev by the late 1970s and would cost between $40 and $50 billion. Fermi's closest rival was a super proton synchrotron at CERN—the European

Center for Nuclear Research at Geneva, Switzerland—also with a 6.5-km ring slightly less powerful spanning the Swiss-French border, deep beneath rolling farmland. According to the CERN newsletter, *The Courier*, the most ambitious project was a complex planned by the USSR called UNK (from its Russian name) consisting of two machines in underground tunnels at Serpukhov, south of Moscow; one tunnel, forming a ring nearly 18 km in circumference, would house a device to accelerate protons to 2000 Gev, while the other tunnel would contain a 20-Gev electron accelerator whose beam would collide with the proton beam. Such collisions, at energies higher than now available, might provide answers to puzzles that baffle the theorists, Sullivan said.

The Brookhaven conference heard plans for Isabelle, a 3.2-km machine for colliding proton beams at energies up to 300 Gev, that would take 3 yr to build and cost $166 million. CERN was drawing up plans for a machine to collide beams of 100-Gev positrons and electrons; energies achieved in such collisions are lower than those from proton accelerators because the positron is less massive than the proton. A powerful machine of this type operating at Stanford Univ., Palo Alto, had been called SPEAR; a larger version known as PEP (positron-electron project), more than 2 km in circumference, would be completed there in 1980 at a cost of $78 million, and would generate beams of 18 Gev. The collision of positrons and electrons was considered a "clean" way to produce exotic particles because the collision would annihilate the primary particles in an extremely intense burst of energy, which would then form itself into new particles.

Among questions discussed at the Brookhaven meeting were the nature of the basic building-blocks of matter such as quarks or maons (particles tinier than quarks, said to have been suggested as a possibility by Mao Tse-Tung when told of the quark concept by a visiting physicist). Recent work had shown "how little we understand about the internal dynamics" of particles, and the use of higher energies would be one way to solve some of the problems, Sullivan suggested. (*NYT*, 10 Oct 76, 7)

10–16 October: More than 1000 of the world's top government and industry space experts assembled in Anaheim, Calif., at the wk-long 27th Congress of the Intl. Astronautical Federation. Theme of the meeting was "The New Era of Space Transportation." Delegates would have the opportunity to hear more than 300 technical papers on topics such as safety in space, the future of space law, and the proposed moon treaty. Also scheduled for discussion were the use of direct-broadcast satellites, space rescue, and the use of solar energy in space. Chairman of the conference was Dr. George E. Mueller, president and chairman of System Development Corp. in Santa Monica and former associate administrator of NASA (1963 to 1969) in charge of manned space flight. Delegates arriving Sunday evening had a choice of Disneyland tour or a wine and cheese reception before the sessions began.

At opening ceremonies on the eve of Columbus Day, Dr. Mueller read a greeting from President Ford and spoke of the Bicentennial and Columbus as symbols of international cooperation. NASA Administrator James C. Fletcher joined with John F. Yardley, associate administrator for manned space flight, in sounding an optimistic note on potential uses of space. Fletcher said that mankind had "entered a new world . . . in less than two decades," pointing to the scheduling of Shuttle flights beginning in 1981 that would make space travel attainable, even economically feasible, for the conduct of previously earthbound activities. Roy Gibson, director general of the European Space Agency, said the member nations of ESA were suffering from an economic downturn and might have trouble maintaining ESA's present funding level, estimated at $550 million a year. However, he maintained that space programs were justifiable on the basis of economic return, and that ESA would recommend no program "unless it makes good economic sense for the users."

Dr. Leonard Jaffe, president of the IAF, told the delegates that astronautics involved all technical fields and affected all human institutions. Dr. Lubos Perek, chief of the United Nations' Outer Space Affairs Division, brought greetings from the UN secretary general and summarized UN activities in space, from identifying objects in orbit to supporting remote sensing and telecommunications—a span of interests, he said, "from Astronomy to Gastronomy."

A panel on space transportation discussed 14 questions, from details of the Shuttle mission to clarification of payload prices and legal liability for aborted payloads. The electronic environment of the session— wireless microphones and wireless translation headsets—illustrated collateral benefits from astronautics, said Rob Weadd of the Los Angeles section, AIAA, in the convention newsletter. On Monday afternoon, officials of the Soviet and U.S. space programs reported jointly and individually on results of the July 1975 Apollo-Soyuz mission. A currentevents session Monday afternoon heard Boris Petrov, chairman of the Soviet Intercosmos Council, call for renewal of the 1972 agreement that led to the Apollo-Soyuz mission, after the agreement expired in 1977. Shigebumi Saito, commissioner of the Space Activities Committee of Japan, said his country hoped to participate in another Spacelab in the 1980s and looked toward cooperation with ESA as a result of the recent visit of ESA director Roy Gibson. Lawrence Morley of Canada's Dept. of Mines and Resources said that Canada was "made for remote sensing" and would be looking for partners in the study of ocean management.

On Tuesday, 12 Oct., the Intl. Institute of Space Law held its 19th colloquium in morning and afternoon sessions, continuing through the remainder of the convention. Since 1967, when an initial treaty of principles was signed, space lawyers had worked out international agreements on liability for space accidents, registration of space objects, and rescue and return of astronauts. Two current problems demanding attention were treaties to govern direct broadcasting from space and the monitoring and sensing of terrestrial phenomena. Also, a draft treaty on moon exploration and resource extraction was under way. Among papers presented, that by Dr. V. S. Vereshchetin of the Moscow Institute of State and Law proposed that the legality of each type of practical use of space should depend on its compliance with principles of state sovereignty and sovereign equality of states. Dr. Stephen Gorove of the Univ. of Miss. law school described the legal regime likely to apply to space colonies, pointing out gaps in existng law and recommending changes. In another paper, Dr. Gorove called the moon treaty a landmark in the development of space law, and cited many areas of general agreement, adding that lack of agreement was equally instructive in revealing problems in the process of international lawmaking. Dr. D. Krstic of Yugoslavia stated that the era of sovereignty was gradually disappearing, and that space activities should be a product of all mankind, not just the strong powers.

On Wednesday, 13 Oct., some of the congress delegates "including a large group of Russian cosmonauts and scientists" took a bus tour and had a briefing on the Space Shuttle program at the Rockwell Intl. plant at Downey, Calif. Dr. Krafft Ehricke, scientific adviser at Rockwell Intl., described studies of Lunetta and Soletta systems that would use huge reflectors on thin plastic assembled in space to illumine areas of earth either by day or by night, lengthening agricultural activities in season or lighting polar regions for increased access to petroleum or mineral deposits in areas where nights might last for 3 mo. Other delegates went in four busloads to the Jet Propulsion Laboratory at Pasadena for an all-day program during which the guests in four groups rotated among exhibitions and lectures on the Viking spacecraft, the space flight operations and spacecraft assembly facilities, and future planetary missions. Speakers included Dr. Bruce Murray, director of JPL, and A. Thomas Young, deputy Viking project manager, as well as John Casani, manager of the Mariner Jupiter-Saturn project.

On Thursday, 14 Oct., IAF delegate R. Gilbert Moore became the first Shuttle customer to hand over a letter of intent to purchase NASA's "getaway special"—a 90-kg 0.14-cu-m package to be carried by the Orbiter on a space-available basis, for $10 000. Moore, general manager of Thiokol Corp.'s Astro-Met plant at Ogden, Ut., said he would buy the ticket for himself, not his company. He had been active for some time in obtaining space rides for student experiments, and said he might open his Shuttle package for students or sublet some of it to recover costs, but had not decided yet what use to make of the package. "It's such a bargain," he said. "You can't get a ride anywhere else for that kind of money." Moore handed his letter of intent to Chester Lee, director of space transportation systems operations in NASA's Office of Space Flight, who was in charge of rounding up Shuttle customers; the letter went by Comsat satellite to Lee's office in Washington, D.C.

Aviation Week & Space Technology magazine reported that the Soviet representatives at the IAF meeting questioned officials about the Shuttle's thermal protection system, its main-engine refurbishment, and its handling characteristics, being "especially interested in the use of cathode-ray tube displays." The briefings were part of the Rockwell tour offered to IAF groups. NASA had approved Soviet inspection of the orbiter, the magazine said, because no technology transfer problems were involved and the Soviets had shown Salyut and Soyuz vehicles to U.S. personnel; however, the Soviets had never discussed or shown future vehicles to U.S. representatives, the article added. Upon questioning the Soviet representatives unofficially about any USSR shuttle plans, NASA was told only that analytical studies were under way.

The *Washington Star* and *Av Wk* reported that Soviet Intercosmos chairman Dr. Boris N. Petrov and Vasily A. Sarychev, another member of the Soviet IAF delegation, were robbed at gunpoint in the lobby of their motel and escaped when a number of people got off an elevator. Petrov and Sarychev did not understand the gunman's demands at first, but finally surrendered $37 before the unidentified robber fled. Although local police were notified, Petrov did not register a formal complaint, and the delegation stayed at the convention two days longer, after which Petrov and other Soviet officials flew to Washington to discuss future joint space missions. Besides Petrov, the IAF delegates included Dr. Vereshchetin, deputy chairman of Intercosmos; V. C. Vachnadze and A. I. Tsarev, Intercosmos members; R. Z. Sagdeyev, director of the USSR Academy of Sciences institute of outer space, and his associate, Y. P. Semenov; and D. S. Chetveryakov, M. V. Sokolov, and V. P. Legostayev.

As the IAF convention closed on Saturday, 16 Oct., Dr. Mueller said the panels and presentations at the meeting had demonstrated again the benefits and feasibility of international cooperation. He said it was up to the members to convince the President and U.S. Congress to provide sufficient funds for the Shuttle over the next 5 yr. (*IAF news 27th Congress*, text; list of abstracts; *LA Times*, 12 Oct 76, 11; 14 Oct 76; *Av Wk*, 18 Oct 76, 25; 25 Oct 76, 22, 23; *W Star*, 20 Oct 76, A-2; *W Post*, 27 Oct 76, A-2)

11 October: Moon rocks—prime celebrities 5 yr ago—had become has-beens, wrote reporter James P. Sterba in the *New York Times*. Astronauts had brought back from the six moon missions about 382 kg of lunar material; Dr. Michael G. Duke, curator of moon rocks at Johnson Space Center, claimed to know the disposition of every gram. Fragments returned from *Apollo 11* went to 137 heads of state and 51 U.S. governors; chunks returned by *Apollo 17* in Dec. 1972 had been dispensed to nations and states. Scientists in 15 nations, 26 states, and the Virgin Is. had been studying samples under NASA-sponsored research grants, and their findings would fill a shelf 1.5 m wide. "We know more about the composition of the moon than we do about the earth," Dr. Duke said, although less than 20% of the lunar samples had been circulated for research and exhibits. Lately the calls for samples to exhibit had dwindled, and the number of scientists studying them had been halved. Security had remained tight because of the efforts of rock collectors to obtain samples; however, the moon rocks had lost ground in public interest compared to the Mars rocks being turned over by *Vikings 1* and *2*, Sterba wrote. (*NYT*, 11 Oct 76, 27)

- Inspection by U.S. Air Force and Japanese technicians of the MiG-25 Foxbat Soviet interceptor plane landed in Japan by a defecting pilot early in September revealed a minimum of innovation and an approach that relied on "brute force," reported *Aviation Week and Space Technology* magazine. Examination of the Foxbat revealed a standard ejection seat, although the pilot had told U.S. interrogators none was carried because of a desire to hold down the weight of the steel-frame aircraft (about 14 000 kg with wings, tail surfaces, and afterburners removed). The number of cockpit instruments was about half those used in the F–4EJ built in Japan under license. None of the avionics in the MiG–25 used solid-state circuitry, relying instead on vacuum tubes; fatigue cracks in the airframe had been repaired by rough welding techniques. Despite the design deficiencies, however, the MiG–25 was "a formidable opponent," *Av Wk* noted. (*Av Wk*, 11 Oct 76, 18)

12 October: NASA announced that LeRC, the U.S. Coast Guard, the U.S. Navy, and the Natl. Weather Service had successfully completed in Alaska a 3–wk demonstration of an all-weather ice-information system developed in response to a Congressional request to see if the Great Lakes could be kept open for shipping all year. Cargoes reshipped because their barges had been iced in or turned back by weather had incurred additional transportation costs estimated at between $30 and $50 million in 1975 alone. The demonstration—performed along the western and northern coasts of Alaska, a region having serious shipping problems caused by thick ice 60% of the time—used a Coast Guard plane equipped with NASA's side-looking airborne radar to obtain daily microwave imagery similar to black-and-white photography, that revealed the type and distribution of ice in any kind of weather regardless of dense cloudcover. The data on coastal shipping were relayed through a Goes satellite by way of Wallops Island to LeRC; after processing, the images were sent back through Canada's CTS satellite to Alaska for the Navy's ice interpreters to use in making navigation charts for vessels moving in offshore ice. The system had been used in the Arctic to aid the Coast Guard icebreaker *Glacier* in operating through ice in zero-visibility conditions. Use of the system had helped to keep the Great Lakes open for shipping for two full seasons for the first time in history, at an estimated saving of hundreds of millions of dollars each year; the 1976–77 shipping would be the third and final year of demonstration. (NASA Release 76–165)

- NASA announced appointment of Dr. James R. Lawson, currently serving as special assistant to the director of ERDA's Office of University Affairs, as NASA Director of University Affairs, effective 8 November. Dr. Lawson, president of Fisk Univ. from 1967 to 1975, received a bachelor's degree in physics from Fisk and M.S. and Ph.D. degrees from the Univ. of Mich. In 1957 he became professor and chairman of the physics department at Fisk, doing research on infrared spectroscopy. A member of Phi Beta Kappa and Sigma Xi, he had received the Rosenwald Fellowship and the Afro-American Natl. Fellowship Award. In his new position, Dr. Lawson would be principal adviser on NASA's relations with universities, including special university programs having agency-wide scope and interest. (NASA anno. 12 Oct 76)

13 October: NASA announced that a data-collection unit called Site Data Acquisition System (SDAS), first in a centralized nationwide network that would monitor the performance of solar heating and cooling systems, had begun operating last wk, transmitting daily the data from a solar heating and cooling system installed at the George A. Towns Elementary School in Atlanta, Ga. The school, built in 1962, operated throughout the year, including the summer months; a large-scale system for heating and cooling, with solar collectors on the roof, had been added to the building last year. The project was conducted jointly by Westinghouse Electric Corp. and the Ga. Inst. of Technology, under a cost-sharing no-fee contract with ERDA. The data-gathering network to which the Towns school SDAS belonged would collect information on climate—temperature, humidity, wind, and available sunlight—and collector inlet-outlet temperature, flow rate, and performance of the energy-storage system. Raw data would be received, processed, and printed out by an IBM facility in Huntsville, Ala., under contract to MSFC.

Total number of units in the program might reach 2000 or more by the end of 1979, NASA estimated. Users of the information would include manufacturers of heating and cooling systems; building and construction firms; architects, municipalities, individuals, and others concerned with building design and construction. The national program was aimed at demonstrating the efficiency of solar energy systems for residential and commercial building, and to stimulate marketing and public acceptance. (NASA Release 76-166; MSFC Release 76-188; *C Trib*, 19 Oct 76, 1-18)

- NASA announced that Robert N. Lindley, director of project management at GSFC, would begin a temporary assignment in Paris 24 Oct. as Deputy Associate Administrator for Space Flight (European Operations), serving as senior NASA adviser to ESA for the Spacelab program. During this assignment, Dr. William C. Schneider, currently Deputy Associate Administrator for Space Flight at NASA Hq, would act as director of project mangement at GSFC. Both officials would return to their original positions at the end of the ESA assignment. (NASA anno 13 Oct 76)

- Rockwell Intl. announced award to the Aerostructures Div. of Avco Corp. of a contract valued at more than $50 million for long lead-time production work on outer wings of the B-1 bomber. Work would cover the first three operational B-1s produced for the U.S. Air Force. The Aerostructures Div. would design and fabricate tools and prepare for actual building of the outer wings. Rockwell Intl. had assembled outer wings for the first three B-1 prototypes, which had already accumulated more than 350 hr of flight tests at Edwards AFB, Calif. Outer wings on each side of the B-1 would constitute a shipset—one of the largest sections of the B-1—weighing more than 13 000 kg, and measuring more than 18 m long and 4.5 m wide. Fuel would be stored inside the wing structure. (Rockwell Release LA-5)

- The European Space Agency announced choice of payloads for its Ariane launcher L02 and L03 qualification flights scheduled between June 1979 and October 1980. Principal passenger on the L02 flight would be the recently announced Geosari satellite, with a lateral passenger, Amsat, a 70-kg radio-amateur space communications satellite proposed by a German organization affiliated with the Intl. Radio Amateur Satellite Organization. Ariane L03 would carry the India comsat Apple (Ariane passenger payload experiment), a three-axis-stabilized spacecraft weighing 616 kg that would continue experiments in the 4- to 6-ghz frequency to be performed by India in the Satellite Telecommunications Experiments Project (STEP). Mounted in the lower central portion of the launcher, Apple would bear the load of a principal-passenger satellite to be chosen by the end of the year from among a second Meteosat prototype;

a flight prototype of Symphonie, Franco-German comsat; an Italian Sirio microwave comsat with a payload different from the first model (see *A & A75*, 10 Mar); and a Canadian CTS spacecraft equipped with a European communications payload. Nominal orbit for the L02 and L03 flights would be 35 800-km apogee, 200-km perigee, and 17.7° inclination. (ESA release 13 Oct 76)

- ESA announced plans to award to Messerchmitt-Bölkow-Blohm of West Germany, as prime contractor for COSMOS (a consortium of European aerospace companies), two contracts worth a total of about $55.38 million (U.S.) for definition of the Exosat satellite (phase B) and for subsequent design and development (phase C and D). Subcontractors from ten member states of ESA would share in satellite development. Exosat, scheduled for launch in 1980–81 by a Delta or Ariane vehicle, would measure position, structure, spectral and temporal characteristics of cosmic x-rays between -0.1 kev and $+50$ kev. Work on the 12-mo phase B contract would begin in Jan. 1977. (ESA release 13 Oct 76)

14 October: NASA launched *Marisat 3*, third and last in a series of Comsat General maritime communications satellites, from ETR at 6:44 pm EDT on a Delta into a synchronous transfer orbit. At seventh apogee 8:11 pm EDT 16 Oct., the apogee boost motor was fired to maneuver the satellite to a position at 36 000 km altitude above the equator at about 182°E; final move to operation location at 73°E above the Indian Ocean would take place in Nov.

The spin-stabilized active-repeater spacecraft, weighing about 700 kg in orbit, was a cylinder with a diameter of about 2 m and an overall length of 3.65 m. Built by Hughes Aircraft Corp., the comsat would act as a relay station to transmit and receive information between ships and submarines at sea and shore stations. It would join *Marisat 1*, operating at 15°W over the Atlantic Ocean since Feb. 1976, and *Marisat 2*, operating at 176.5°E over the Pacific since June. (The *Marisat 3* mission was judged successful 3 Nov. 1976; the *Marisat 2* mission [see 9 June] had been judged successful 6 Oct. 1976.) (NASA Release 76-156; MOR M-492-205-76-03 [prelaunch] 6 Oct 76, [postlaunch] 5 Nov. 76; *W Star*, 15 Oct 76, A-5; MOR M-492-205-76-02, 14 Oct 76)

- Marshall Space Flight Center announced scheduling of further tests in October for prototype cloud-physics experiment hardware to be flown on the Space Shuttle to help researchers understand microphysical processes occurring in the atmosphere. The experiments would develop techniques for accurate prediction and control of weather (increase of snowfall or rain, dissipation of fog, suppression of lightning) to improve man's environment. The Atmospheric Cloud Physics Laboratory (ACPL) being developed by MSFC would use the low or zero gravity during Shuttle/Spacelab flights to perform experiments without contamination by supporting devices—wires, spider webs, blasts of air or electrical fields—used for test objects in earth laboratories. The prototype equipment was being tested in weightless periods provided by parabolic flights in the Johnson Space Center's zero-gravity test aircraft. O. H. Vaughn of the Aerospace Environment Div., Space Science Laboratory at MSFC, had been monitoring performance of the equipment during flights that began in September. (MSFC Release 76-186)

- Ruth Bates Harris, NASA's deputy assistant administrator for community and human relations, announced that she would leave the agency effective 15 Oct. Mrs. Harris said she would return to New York "to attend to pressing personal family needs," but would "spend an number of months continuing to work as a consultant" to NASA, in order to complete some unfinished community and human relations projects. She had been a deputy assistant administrator since Aug. 1974, the first woman to reach that position in NASA, and had earlier been the first director of NASA's Office of Equal Employment Opportunity. Before joining NASA, Mrs. Harris had been director of human relations for the public schools in Montgomery County, Md., and had been director of the Human Relations Commission in Washington, D.C. She had received more than 50 awards and citations for her work in human relations. (NASA Release 76-170)

- The USSR launched a geophysical rocket, *Vertikal 4*, from an undisclosed site "in the medium latitudes" of European Russia to an altitude of 1512 km. Carrying "more than 10 complex and diverse scientific instruments" developed and built by specialists from Bulgaria, East Germany, Czechoslovakia, and the USSR, the rocket was designed to sample a vertical cross section of the neutral upper atmosphere and ionospheric plasma in a 15-min time period. In a Pravda interview, Academician B. N. Petrov noted that composition of the atmosphere and ionosphere changed with height, the lighter gases being at the top. None of the previous Vertikal rockets (launched in 1970, 1971, and 1975) had reached altitudes of more than about 500 km, the oxygen section of the ionosphere. *Vertikal 4* had reached the hydrogen region, or protonosphere, and had measured properties of the plasma affecting radio-wave absorption and investigated effects of shortwave solar radiation on the earth's atmosphere. Results of the investigation would extend knowledge of weather and climate, Petrov said. (FBIS, Tass in English, 14 Oct 76; *Pravda* in Russian, 17 Oct 76; Moscow Domestic Service in Russian, 29 Oct 76)

14-18 October: The USSR launched *Soyuz 23* carrying two cosmonauts, Lt. Col. Vyacheslav Zudov as pilot and Lt. Col. Valery Rozhdestvensky as engineer, from the Baykonur cosmodrome near Tyuratam in Kazakhstan at 8:40 pm Moscow time (1:40 pm EDT) 14 Oct. to link up with the orbiting space station *Salyut 5* and continue research begun there in July by the crew of *Soyuz 21*. The nighttime launch was replayed later on Soviet TV. Initial orbit parameters were 243-km perigee, 275-km apogee, 89.5-min period, and 51.6° inclination.

At what the *NY Times* described as "a restricted press conference before liftoff," the rookie cosmonauts said their mission was to concentrate on practical scientific uses of the Soviet space stations, especially extraterrestrial manufacture of metals, glass, and pharmaceuticals. Rozhdestvensky said that every ruble invested in space exploration had "already been returned tenfold to the national budget in one way or another."

Tass announced at noon Moscow time on 16 Oct. (5 am EDT) that the docking of *Soyuz 23* with *Salyut 5* had been cancelled "because of the off-design regime of [the] approach control system," and the crew had been ordered to return to earth. The *NY Times* noted that this was "the first occasion on which the Russians have made public their problems before the completion of a space flight." Usually no announcement had been forthcoming until the crew had returned to earth. At 8:46 pm Moscow time (1:46 pm EDT)—almost exactly 48 hr after liftoff—the descent module landed 195 km southwest of Tselinograd in Kazakhstan "on the surface of Lake Tengiz" in a heavy snowfall. The search and rescue teams, working under difficult nighttime conditions, recovered the descent module and the cosmonauts from the first splashdown in the 15-yr history of Soviet manned space flight. Announcement of the adverse conditions of the landing was not made until 12 hr afterwards, a delay suggesting some difficulty in locating the capsule. Lt. Gen. Vladimir Shatalov, director of cosmonaut training, said that helicopter and amphibious craft had aided the recovery. Residents of the nearby town of Arkalyk welcomed the returned cosmonauts at the airport and made them honorary citizens. The crew returned to the Baykonur cosmodrome by plane on the morning of 17 Oct.

The splashdown of *Soyuz 23* resulted from chance rather than planning, said the newspaper *Izvestia* on 18 Oct., describing fears at recovery-team headquarters that the landing craft might have gone into swamps near the lake, after a report that the craft had tilted on impact, putting the porthole below the water's surface. The pilot of a plane had reported the capsule's position and a helicopter had towed the capsule to shore amid fog, snow, and broken ice. The official comments were regarded as "an effort to avoid blaming the astronauts for the mission's failure," said the *NY Times*.

The *Soyuz 23* difficulty marked the fourth failure of a Soyuz mission to rendezvous; radar trouble had forced the crews of *Soyuz 3*, *Soyuz 8*, and *Soyuz 15* to return to earth without docking. The *Wash. Post* noted that about one in three Soyuz flights had failed to carry out its mission: *Soyuz 1* and *Soyuz 11* resulted in the deaths of one and three cosmonauts respectively, and two cosmonauts on an unnumbered mission in April 1975 had been forced to abort their flight before reaching earth orbit. (*NYT*, 15 Oct 76, A 13; 16 Oct 76, 6; 17 Oct 76, 29; 18 Oct 76, 32; 19 Oct 76, 17; *W Post*, 15 Oct 76, A-21; 17 Oct 76, A-17; 18 Oct 76, A-16; *C Trib*, 15 Oct 76, 1-5; 17 Oct 76, 1-3; 18 Oct 76, 1-8; *W Star*, 17 Oct 76, A-14; FBIS, Tass in English, 14-17 Oct 76)

15 October: To criticisms of "basic" research, "whose practical applications no one can foresee," the *New York Times* in an editorial said the two Nobel Prizes in medicine for 1976 might have been designed to provide an answer and a rebuke. While Dr. Baruch S. Blumberg in the early 1960s was studying genetic variation in the susceptibility to disease of different people (a "seemingly abstruse topic," the *Times* noted), he found in the blood of an Australian aborigine a strange protein described as Australia Antigen, now usually called Hepatitis B surface antigen. Blood given for transfusions nowadays is regularly tested for this protein to prevent Hepatitis B infection, thus preventing thousands of blood recipients each year from being infected. The other Nobel prizewinner, Dr. D. Carleton Gajdusek, while in New Guinea during the 1950s became interested in the so-called laughing disease, "kuru," that was destroying the Fore, an obscure Stone Age people. His work on that disease revealed a new type of slow-acting virus that could produce brain degeneration, and that might be "involved in multiple sclerosis and other major neurological diseases." The message for financiers and directors of research, said the *Times*, is that "in the long run nothing is more practical than basic studies" and that "basic research yields the richest dividends." (*NYT*, 15 Oct 76)

- Communications Satellite Corporation (ComSat) reported net income for the third quarter of 1976 of $7 607 000, down from last yr's third-quarter net of $11 837 000. The decrease was the result of amounts placed in escrow under an FCC accounting and refund order of 16 June 1976; if the order had not been in effect, the quarter's net would have been $14 603 000. Although operating income dropped as a result of the escrow requirement, the wholly owned Comsat General Corp. for the first time had realized revenues for the entire period from both Marisat and Comstar programs; in the third quarter a yr ago, the subsidiary had reduced ComSat's net operating income by a penny per share. Operating expenses, including taxes, were $30 223 000 for the quarter, up from $24 997 000 for the third quarter of 1975; the increase was due to costs of the launches of *Intelsat IV-A, Marisat,* and *Comstar* satellites. ComSat's gross operating revenues for the third quarter of 1976 totaled $36 260 000, an increase over last yr's third quarter total of $35 116 000; the escrow requirement had prevented the gross revenues from increasing more than $15 million over last yr's third quarter with the beginning of Marisat and Comstar services and continued growth in the number of circuits leased to customers of the Intelsat global communications system. Half circuits leased at the end of Sept. 1976 numbered 4129, a 16% increase over the 3547 leased at the end of Sept. 1975. Money put in escrow under the FCC order represented the difference between customer payments to ComSat under present rates, and the amounts based on lower rates that would be required under an FCC rate decision of 4 Dec. 1975. The rate decision was still under judicial review. (ComSat Release 76–16)

- An enormous explosion in central Siberia that shook the world 68 yr ago might have been caused by the crash of a nuclear-powered spaceship from an alien planet, according to the Soviet news agency Tass, reported in the *Washington Star*. Tass quoted scientist Aleksey Zolotov, who had just returned from a survey of a remote river valley at Tunguska. On 30 June 1908, a blast estimated as up to 2000 times more powerful than the first atomic bomb had shaken measuring devices all over the world and had been heard nearly 1200 km away from Tunguska. Trees had been uprooted as far as 50 km from the site, and ground tremors had thrown to the ground horses that were pulling plows as far away as 380 km. Most scientists had attributed the explosion to the impact of a meteorite or comet with a mass of 10 million tons and measuring mor' than 90 meters across. Zolotov told Tass that his survey team had found higher than normal radioactivity in remnants of trees near the site; also taken were samples of permafrost soil. "Our investigation . . . seems to confirm our assumption that what took place in the Tungus taiga was a nuclear explosion," Zolotov said. "It is from this point of view that we are exploring the possibility of the artificial origin of the Tungus cosmic body." Tass did not give figures for the radiation study, nor did it identify Zolotov beyond saying he was a "noted Soviet scientist . . . who has been studying the Tungus mystery for years." (*W Star*, 15 Oct 76, A-4)

17 October: Two recently published books on "the worst disaster in aviation history—the crash of a Turkish Airlines DC-10 outside Paris that killed at least 346 people in March, 1974—" alleged that the disaster was not only preventable but was predicted in detail years before it happened, the *Washington Post* reported. *Destination Disaster*, by an investigating team from the *London Sunday Times*, aimed at inspiring another congressional investigation into "corporate and governmental interworkings that contributed to the disaster," the *Post* said; *The Last Nine Minutes*, by Moira Johnston, was a more subjective and personal account of the same crash.

Senate and House hearings in the summer of 1974 had revealed that a technically similar accident in June 1972 had exposed an error in the design of the DC-10: a faulty locking system for the door of the large cargo compartment. The 1972 mishap had resulted in a "gentlemen's agreement" between the administrator of the Federal Aviation Administration and the president of the Douglas Division, McDonnell Douglas, that the lock would be fixed but that FAA would not issue a public and legally binding "airworthiness directive" mandating the repair. After the agreement was reached, the director of product engineering for the Convair Division of General Dynamics, subcontractor to McDonnell Douglas for the fuselage (including the door), wrote a long memorandum expressing concern about the door. Neither this memo nor another Convair memo on the question of liability for the cost of modifications ever reached McDonnell Douglas, the authors said, although evidence existed that McDonnell Douglas was fully aware of the problem nevertheless.

Turkish Airlines, which was responsible for maintenance after the DC-10 was delivered, was sharing in liability settlements with McDonnell Douglas, General Dynamics, and the FAA, according to attorneys, who said total liability from the crash would probably set a single-accident record. The *Post* noted that FAA had issued 147 airworthiness directives in 1973; in 1974, year of the crash, the number was 299; in 1975, the number rose to 445.

Private consultant Charles O. Miller, former chief of the Natl. Transportation Safety Board's bureau of aviation safety, who had gone to Paris to assist in investigation of the Turkish Airlines crash, said in a recent interview that the Convair memoranda and other documentation that provided background to the technical decisions had been turned up by the liability lawyers rather than by the crash investigators or by congressional investigators. (*W Post*, 17 Oct 76, A-1, A-12)

18 October: President Ford awarded the National Medal of Science, the highest U.S. award for distinguished scientific achievement, to 15 scientists, one of them now dead:

—John W. Backus, IBM Research Laboratory, San Jose, Calif., for contributions to computer programming

—Manson Benedict, professor emeritus, MIT, for his role in creating the discipline of nuclear engineering and his leadership in developing techniques for uranium-isotope separation

—Hans A. Bethe, professor emeritus of physics, Cornell Univ., for contributions to understanding of the atomic nucleus, the origins of solar heat, and atomic energy

—Shiing-Shen Chern, professor of mathematics, Univ. of Calif. at Berkeley, for work leading to discoveries in geometry and topology

—George B. Dantzig, professor of operations research and computer sciences, Stanford Univ., for invention of linear programming and methods of using mathematical theory in computers

—Hallowell Davis, professor emeritus of physiology, research professor of otolaryngology, Washington Univ., for research leading to advancement in fields ranging from neurology to acoustics and pediatrics

—Paul Gyorgy, late professor emeritus of pediatrics, Univ. of Pa. School of Medicine, for discovery of three vitamins and related research in human nutrition

—Sterling Brown Hendricks, former chief chemist, U.S. Dept. of Agriculture plant industry station, Beltsville, Md., for research in physical and chemical properties of soils

—Joseph O. Hirschfelder, professor of theoretical chemistry, Univ. of Wisc., for contributions to atomic and molecular quantum mechanics

—William H. Pickering, director emeritus, Jet Propulsion Lab., for leadership in exploration of planets and the solar system

—Lewis H. Sarett, senior v.p. for science and technology, Merck & Co., for contributions to chemical synthesis of cortisone and other chemotherapeutic agents

—Frederick E. Terman, v.p. and provost emeritus, Stanford Univ., for his role in creating modern electronics

—Orville A. Vogel, U.S. Dept. of Agriculture, professor emeritus of agronomy and soils, State Univ. of Wash., for agronomic research including development of semidwarf wheats

—E. Bright Wilson, Jr., professor of chemistry, Harvard Univ., for theoretical and experimental contributions to the understanding of molecular structure

—Chien-Hsiung Wu, professor of physics, Columbia Univ., for experiments leading to the understanding of the decay of the radioactive nucleus (*NYT,* 19 Oct 76, 20)

- Sweden's Royal Academy of Sciences announced that three more Americans had won Nobel prizes: Prof. William N. Lipscomb, Harvard Univ., for chemistry, and Prof. Burton Richter of Stanford Univ. and Prof. Samuel C. C. Ting of MIT for physics. Prof. Lipscomb won for his studies of the structures and properties of boranes, and Profs. Richter and Ting shared the physics prize for their independent discoveries of a new type of elementary particle known as psi or J. Americans had won all four of the prizes awarded so far this year; Prof. Milton Friedman of the Univ. of Chicago had won the prize for economics, and the prize for medicine had gone to Dr. Baruch S. Blumberg of the Univ. of Pa. Medical School and Dr. D. Carleton Gajdusek of the Natl. Institute for Neurological Diseases. All the prizes carry an award of $160 000, derived from the estate of Sweden's Alfred Nobel, inventor of dynamite, who established the prizes in 1901. (*NYT,* 19 Oct 76, 1, 34; 24 Oct 76, E-14)

- In separate accounts, *Aviation Week & Space Technology* magazine emphasized problems arising in USSR attempts to reoccupy the *Salyut 5* "military space station." The magazine reported that the *Soyuz 21* cosmonauts were forced to return to earth under emergency circumstances 24 Aug. because of "an acrid odor flowing from the Salyut-5 space station environmental control system." Lt. Col. Vitaly Zholobov and Col. Boris Volynov "tolerated the odor for some time but were unable to find the cause of the problem before the odor became unbearable," the magazine said, noting that the mission was the sixth failure in the last nine USSR attempts to complete manned orbital-station missions. In another story in the same issue, the magazine said that "the Soviets believe they have solved the [*Salyut 5*] environmental control system's odor problem that forced the early return of the Soyuz-21 crew" as demonstrated by the 14 Oct. launch of *Soyuz 23* carrying Lt. Col. Vyacheslav Zudov and Lt. Col. Valery Rozhdestvensky. (*Av Wk,* 18 Oct 76, 13, 25)

- The Energy Research and Development Administration announced that a nuclear explosion reported by the Peoples Republic of China 17 Oct. had been detected by the U.S. Atomic Energy Commission's detection system. The explosion, which occurred underground at 1 a.m. EDT at the Lop Nor test area in western China, was in the low-yield range, ERDA said. (ERDA Release 76-320)

19 October: Physicians at Johnson Space Center were gathering data on female physiological performance and tolerance limits as a basis for establishing criteria in recruiting NASA's first women astronauts. Women employees of the center were asked to volunteer for testing on a treadmill and in a lower body negative-pressure device, to see whether women responded differently from men of comparable age to treadmill exercise and to stresses on the circulatory system induced by decreased pressures on the lower body. On the treadmill test, researchers would monitor the subject's heart rate and blood pressure as they varied the speed and tilt of the moving treadmill. The pressure device, encasing the subject from the waist down, would have the pressure reduced while researchers monitored the reaction of the cardiovascular system to the change in pressure. NASA had issued a call for at least 30 Space Shuttle astronauts (15 pilot candidates, 15 mission specialist candidates) in July 1976, with a closing date of 30 June 1977. Selection would be completed by December 1977. (JSC Release 76-65)

- Two "pilots of note"—Gen. James M. Doolittle and Neil A. Armstrong—had announced plans to establish a Charles A. Lindbergh Memorial Fund to support the work of young scientists, explorers, and conservationists, the *New York Times* reported. Gen. Doolittle, leader of the first World War II air raid on Tokyo, and astronaut Armstrong, first man to walk on the moon, would head a drive to raise $5 million for the fund by 20 May 1977, the fiftieth anniversary of Lindbergh's takeoff on the first solo nonstop flight across the Atlantic Ocean. Income from the endowment, expected to be about $400 000 a yr, would be distributed to Lindbergh Fellows who "combine qualities that made [Lindbergh] a unique human being," said Doolittle at a press conference. He told reporters Lindbergh "would have been happier with a living memorial than one out of bronze and stone." Lindbergh, who died in 1974 at the age of 72, had continued in aviation as a test pilot and airline executive but had branched out into medical technology research and wildlife conservation in his later years, the *NYT* said. The fund's sponsoring committee included representatives of the Explorers Club and the World Wildlife Fund, as well as other areas of interest to Lindbergh: aerospace, ecology, exploration, and science and engineering. Announcement of selections for fellowships would be made yearly on 21 May, anniversary of Lindbergh's arrival in Paris after his 33-hr flight in *Spirit of St. Louis*. (*NYT,* 20 Oct 76, 22)

21 October: Correlations between an astronaut's susceptibility to motion sickness on the ground and his susceptibility in space were the subject of studies by JSC scientists, based on data from U.S. and Soviet manned space flights and on data from tests aboard a zero-gravity training aircraft. Motion sickness had been difficult to study because of the variety of separate sensory inputs to the brain. The three major components of human balance and posture were generally accepted to be visual input, defining the local vertical component; vestibular input from the semicircular canals and otoliths in the inner ear, sensing angular and linear acceleration and the presence or absence of gravity; and muscle sensors, monitoring posture. These interrelated inputs normally function to keep the body balanced; in an earth-normal gravity, balance would be reflexive, and persons would not be conscious of the body-movement patterns that kept them upright. In a moving vehicle, the inputs to the brain might produce contradictory information resulting in a feeling of discomfort; "motion sickness" is not an adaptive response and does not improve the situation, as coughing would do to relieve a throat blockage.

Persons without sight can experience motion sickness, but persons without their vestibular functions intact apparently do not. Although little was known about the interactions among the three sensory systems, motion sickness seemed to be more related to vestibular input. During Skylab missions, in the relatively large areas of living space, much vestibular relearning apparently occurred in the first 2 wk of spaceflight, after which all crewmen became very resistant to motion sickness. No correlation was apparent between an astronaut's susceptibility to motion sickness on the ground and his susceptibiity in space. The studies postulated that the otolith (chalky concretion in the inner ear) was the receptor affording the most direct information on gravity, and therefore was the source of many of the inputs causing disequilibrium or motion sickness. (JSC Release 76-67)

- The B-1 bomber was vital to U.S. security, said Gen. William J. Evans, commander of the Air Force Systems Command, in a speech to the Rockwell Management Club at Los Angeles, Calif. As part of a strategic triad—SLBMs, ICBMs, and bombers—constituting three distinct types of retaliatory weapons, the B-1 would help complicate defense efforts of potential enemies. Although the U.S. now had "a rough equivalence" of strategic strength, Gen. Evans noted that the Soviet Union was engaged in an effort to upset "the present equilibrium" and gain military advantage. Along with deterrent value, the B-1 should be considered a flexible, reusable, appropriate weapon in a conventional tactical role as well as in a nuclear strategic role. A future conflict might start not through "a surprise storm of nuclear missiles," but through "daring but limited acts of provocation" to which the U.S. should be able to respond with at least a show of force or be "perceived as a paper tiger." Gen. Evans congratulated Rockwell and other contributors to the development of the B-1 and urged continuation of the program. (OIP Release 231.76)
- FAA Administrator John L. McLucas announced award to Wilcox Electric Co., Kansas City, Mo., of a $3 720 699 contract for nine Category III instrument-landing systems that would complete a program to provide all-weather landing capability at key airports across the U.S. One Category III system would be installed at the FAA academy at Oklahoma City for training; the other eight would go to Kennedy Airport, NYC; O'Hare, Chicago; Houston Intl., Kansas City Intl., Seattle-Tacoma Intl., Los Angeles Intl., Portland Intl., and Detroit's Metropolitan Wayne County Airport. The new equipment had already been installed at Dulles Intl. near Washington, D.C., Atlanta Intl., San Francisco Intl., and Stapleton Airport at Denver, as well as at the FAA center at Atlantic City, N.J., where it had been used for test and evaluation. Category III equipment would permit landings without visual reference to the ground, under weather conditions with ceiling zero and runway visibility no less than about 200 m. Existing Category II equipment at the sites to be replaced would be moved to other airports, to be named later. (FAA Release 76-97)

22 October: A recently completed manipulator-development facility at Johnson Space Center's Bldg. 9a laboratory would assist NASA engineers and technicians in studying Shuttle payload deployment and retrieval in space, JSC announced. The manipulating system, with a reach of 15.2 meters, was the largest known remotely controlled manipulator system. Sizes of payloads to be carried in the Space Shuttle would range from very small to a maximum length of 18.2 m and diameter of 4.5 m; objective of the facility would be to improve techniques of moving bulky payloads in and out of the orbiter's bay in a weightless environment. The manipulator arm would be operated from inside the orbiter cabin by an astronaut using a pair of hand controllers and viewing the process through a window, aided by a closed-circuit TV system. The laboratory building also housed working models of orbiter components including a full-scale forward cabin section with cargo bay, manipulator station, and a large air-bearing table approximately 25 by 30m on which simulated payloads could be steered on a cushion of air by the remote arm. The air-bearing table would also test the ability of astronauts to move over the smooth interior of the orbiter, using suction-cup shoes instead of cleats locking into grids. An engineer wearing suction-cup shoes and strapped into a cage-like apparatus on the air-bearing table could be familiarized with "walking like a fly" in the day-to-day work environment of the Shuttle. (JSC Release 76-66)

- A new research aircraft combining the precision hover and low downwash of a helicopter with the long range and high speed of a fixed-wing aircraft made its debut at rollout ceremonies in Arlington, Tex., marking "a major milestone in the joint NASA-Army tilt rotor program," according to a NASA announcement. The Bell Helicopter Textron's XV-15 would undergo a 2-yr flight-testing program scheduled by Ames Research Center and the Army's Air Mobility Research and Development Laboratory. The XV-15 was a 12.8-m-long 9.7-m wing-span aircraft with wingtip-mounted engines, transmissions, and 7.6-m propeller rotors that could tilt from the helicopter position for vertical takeoffs and landings and for hovering, to a horizontal position for forward flight at speeds up to more than 482 kph. The tilt rotor would operate with less noise than conventional helicopters or turboprop aircraft of comparable size. The first XV-15 would undergo ground and hover tests, then go to ARC for testing in that center's wind tunnel; the second XV-15 would undergo initial envelope-expansion flight tests at Bell's Arlington, Tex., facility beginning late in 1977. (NASA Release 76-169; *W Star*, 24 Oct 76, A-2)
- MSFC announced completion of a huge self-propelled vehicle called the Straddle Carrier Transporter, fabricated by McDonnell Douglas Astronautics Co. at St. Louis, Mo., and shipped to MSFC for assembly, to be used in moving sections of the Space Shuttle external tank at the Mi-

The XV-15—a tilt-rotor research aircraft combining features of helicopters and conventional airplanes—rolls out at the Arlington, Tex. facility of Bell Helicopter Textron 22 Oct. 1976. (NASA 76-H-804)

choud Assembly Facility in New Orleans. Assembled, the external tank would measure 47.2 m in length and 8.5 m in dia.; empty, it would weigh 33 300 kg, and would carry about 708 440 kg of propellants for launch. The transporter, completely self-supporting, would use electricity from a propane-powered generator for its hydraulic propulsion system, steering, lights, and hoists; it would lift, stabilize, and carry major external tank assemblies over concrete floors and improved roadways. Its five cable hoists included two 10-ton units on a monorail on the aft frame; two 5-ton units on a monorail on the forward frame; and a 2-ton hoist on a centerline rail. The 5- and 10-ton hoists in combination would handle the liquid-hydrogen tank with a volume of 1573 cu m; the smaller hoists would handle the liquid-oxygen tank, with a volume of 552 cu m. (MSFC Release 76-193)

- Dr. Christopher C. Kraft, Jr., director of Johnson Space Center, had received the highest honor given by the government of France to a citizen of another country: the insignia of the Knight of the Legion of Honor, the JSC *Roundup* reported. The award was given for Kraft's "tireless efforts toward better understanding and cooperation between the people of the United States and France." He had also received the Natl. Civil Service League's career-service award for 1976, as one of ten chosen for "excellence in the public service."

  Dr. Robert R. Gilruth, former director of JSC, was one of 35 space pioneers inducted into the Space Hall of Fame 5 Oct. during the dedication of the new facility at Alamogordo, N.M.

  At a JSC ceremony 18 Oct., astronaut Donald K. Slayton received the American Heart Assn.'s 1976 Heart of the Year award for inspiring millions of Americans by "overcoming a heart problem that had grounded him for 10 years, enabling him to resume his distinguished participation in America's space program." (JSC *Roundup,* 22 Oct 76, 1)

24 October: Rockwell Intl. Corp. inspectors testing their inventory of transistors after a government warning to watch for counterfeit semiconductors had found 11 of 11 transistors of one type to be fakes, the *New York Times* reported. Rockwell had concluded that bogus parts had been "unwittingly built into [NASA's] major current project, the Space Shuttle, for which Rockwell is the prime contractor." The *NYT* said that "a rash of discoveries of bogus devices" had plagued the electronics industry for the last year and a half.

The Defense Electronics Supply Center (DESC), a DOD agency that procured almost all electronic devices for defense, found 20 of 60 components taken randomly from stock to be "suspect" and had returned them to the makers for tests to determine authenticity. "Counterfeit" components—devices altered physically to misrepresent their true type or quality, by relabeling, renumbering, or adding false reliability information or dating— would not necessarily be nonfunctional; however, as the purpose of the alteration would be to indicate falsely that a part had passed certain quality tests, and would therefore be much more valuable, the part might fail under strenuous application. Failure of a single component could have consequences far outweighing its 25-cent cost. The transistors tested at Rockwell had been marked to indicate suitability for use in the most rigorous applications; upon disassembly, the parts showed no indication of having undergone any of the extra testing signified by the external markings, which had increased the cost fivefold.

Although one trade organization (Electronics Industries Assn.) set up a task group to combat the problem of counterfeiting, "all sectors of the electronics industry are working against . . . major procedures to alleviate counterfeiting," said the *NYT* report. At issue was the use of distributors, intermediary between producers and end users, who had a "heavy financial interest in maintaining the current system," the report said. Most semiconductor producers sold rejects to junk or surplus dealers for their metal content; the items not disfigured were a very inexpensive source for remarking, and could be introduced into the market as high-quality low-cost items. Virtually all counterfeit components could be traced to the big defense contractors, said the *NYT*. Industry sources contended that the DOD had loose procurement practices, with a purchasing-policy rule that component contracts must go to the lowest bidder. As a DESC spokesman pointed out, "We are simply not allowed to pay a premium price on any contract," even if the low bid is suspiciously below the going rate—"a strong indication that something may be wrong," the *NYT* noted. Accepted methods of the industry made it impossible to guarantee the authenticity of any part that did not come directly from the manufacturer, the report added. Although vendors had predicted that a 1974 DESC edict calling for a strict and immediate upgrading of tests on military-grade components would disrupt the supply, few problems had arisen after the regulations had taken effect. (*NYT*, 24 Oct 76, 3F; *Newsweek*, 25 Oct 76)

25 October: The failure of *Soyuz 23* to rendezvous with *Salyut 5* indicated that the USSR was still encountering the technical and procedural problems that had beset its manned space program for the past 15 yr, said *Aviation Wk & Space Technology* magazine. The failure was the seventh in 11 attempts to complete various space-station missions and the second in a row for Salyut; the unplanned water landing that concluded the failed mission was the first in Soviet space history, but was attributed to "nothing more than bad luck," said *Av Wk*. The rendezvous-system failure was the fourth Soyuz mission plagued by procedural errors or hardware malfunctions associated with docking: *Soyuz 3* failed to dock with *Soyuz 2* in 1968, *Soyuz 8* failed to dock with *Soyuz 7* in 1969, and *Soyuz 15* had been unable to rendezvous and dock with *Salyut 3* in 1974. Because the failure of a system on the *Soyuz 23* ended its mission rather than any problem with *Salyut 5*, Western observers wondered how soon the Soviets would attempt to reman the space station, which *Av Wk* referred to as "military." (*Av Wk*, 25 Oct 76, 23)

- *Viking 2*'s lander dug a soil sample from beneath a rock on Mars and delivered it to the biological instrument package. All three experiments received a part of the sample; initial data from two of them—the labeled-release and the pyrolytic-release experiments—were sent to mission control at JPL. The third (gas-exchange) experiment was in a long incubation period that would continue through solar conjunction—a movement of the bodies which would put the sun between Mars and the earth—preventing communication with the Viking landers and orbiters. Conjunction would begin about 10 Nov. and continue through early Dec.; if the landers were revived successfully after the communications gap, more samples would be collected in mid-Jan. Leonard Clark, surface-sampler engineer, foresaw no problem in reactivating the arm mechanisms, adding that they had worked very well, "just the way they did during ground testing on earth before they left the manufacturer." (*Av Wk*, 1 Nov 76, 15)

26 October: The 2000th major missile launch from Cape Canaveral took place at 4:30 pm EDT when an Army Pershing missile was launched by troops of Battery C, 3d Battalion, 84th Field Artillery, of the U.S. 7th Army from Europe. First launch from "the Cape" had occurred 26 yr previously on 24 July 1950, when *Bumper 8*, a captured German V-2 rocket capped by a WAC Corporal second stage, lifted off the then sparsely settled area. (KSC *Spaceport News*, 12 Nov 76, 1)

27 October: Ames Research Center announced that 18 scientists representing eight institutions would fly on NASA's Galileo II airborne research laboratory on a 3-wk study of atmospheric pollution that would take them nearly pole to pole across the central Pacific Ocean, north to south. Between 28 Oct. and 19 Nov., the scientists in eight teams would measure changes in the upper atmosphere caused by jet aircraft exhaust and would investigate the effect on the upper atmosphere of fluorocarbons and halocarbons from aerosol sprays and other sources. The experiments at many latitudes from far north to far south would gather data in both northern and southern hemispheres to note changes in upper-atmosphere composition and pollutants between altitudes of 10.6 km and 12.1 km. Measuring would begin in Alaska, then proceed at Hawaii, Samoa, Australia, New Zealand, and Antarctica. Observations near Hawaii would be coordinated with similar measurements at about 18-km altitude by NASA's earth-resources survey aircraft, also based at Ames.

The jet-pollutant study was part of a 4-yr cooperative effort between ARC and the Lewis Research Center, which had included experiment packages carried on jets operated by Pan American, United, and Qantas airlines; the Galileo II flights would measure pollution over remote areas not covered by commercial aircraft. The aerosol pollution measurements were part of a broader program by NASA and other agencies to ascertain the constituents of the upper atmosphere in a nonpolluted condition, in order to define changes with time in proportions of ozone, fluorocarbons, water vapor, carbon monoxide, and other compounds, and to determine whether such changes were manmade. Galileo II data from both types of measurements would be coordinated with data collected by NOAA stations in Alaska, Hawaii, and Samoa, as well as by stations of the meteorological services of Australia and New Zealand and the Commonwealth Scientific and Industrial Research Organization. (ARC Release 76-79)

- The Soviet Union's earth satellites had added a new profession—that of printer—to those of meteorologist, cartographer, and communicator, the Tass news agency announced. A Molniya comsat tested in Khabarovsk had transmitted newspapers from Moscow to the Far East more economically and more rapidly than had been possible using the transcontinental cable. The high-resolution capability of the equipment had produced excellent images, the announcement said, adding that "papers received by Far Eastern subscribers will now be absolutely identical to those taken out of their letter boxes by the Moscovites." (FBIS, Tass Intl Service in Russian, 27 Oct 76)

29 October: MSFC announced that it had modified its Saturn facilities to accommodate Space Shuttle work, having completed remodeling of a giant test stand used in the 1960s for static test firing of the Saturn V first stages. The stand would be used in structural tests of the liquid-hydrogen tank portion of the Shuttle external tank; the hydrogen tank measured 29.3 m long and 8.5 m in diameter. Stand modifications included work platforms, changes in the stand structure including pressurization, and an instrumentation and control system. Holddown arms were removed and a flame deflector that channeled rocket exhaust away from the area was taken from under the stand. Liquid-oxygen storage facilities used for Saturn had been converted for liquid-hydrogen storage needed for Shuttle testing, instead of building new storage. The contract with Algernon Blair of Montgomery, Ala., by the Huntsville Div., Corps of Engineers, was for approximately $4 million. (MSFC Release 76-195)

- MSFC announced selection of the Bendix Corp., Teterboro, N.J., for negotiations leading to award of a $7 007 210 cost-plus-award-fee contract for installing, activating, disassembling, and removal of special equipment at MSFC for Space Shuttle structural and dynamic ground tests. The contract would run from 15 Nov. 1976 through Dec. 1979. MSFC responsibilities in developing the Space Shuttle covered three areas: the main engine powering the orbiter, which uses three of these engines; the external tank holding propellants for the main engines during launch and ascent; and the solid-fuel rocket boosters to be jettisoned after burnout and recovered by parachute for future use. MSFC would make structural tests of the external tank and boosters, development tests to prove design concepts of the main engine and external tank, and dynamic tests simulating flight conditions to be encountered by the Shuttle during launch. (MSFC Release 76-196)

- Two Landsats, launched in 1972 and 1975 to monitor earth resources, had been tested for the job of census taker by NASA and the Bureau of the Census, NASA announced. Public Law 521, signed by President Ford 18 Oct. 76, required a complete U.S. census every 5 yr instead of once every 10 yr; the satellites, if their performance in 1980 in regional readouts and verifications should prove satisfactory, would permit significant savings in manpower, the Bureau said. Landsat images, although not detailed enough to count people or houses, could serve to identify geological, agricultural, and societal features, especially residential patterns of growth. Satellite images of areas of Md. and Tex. had been processed in 1975 to identify land cover typical of transition from rural to urban use; conventional census boundaries overlaid on the images enabled a computer to identify urban fringe zones in the test areas, verified by actual census statistics, the Bureau said. (NASA Release 76-176)

- NASA's aviation safety reporting system (ASRS) received nearly 1500 reports in its first 3-mo operating period, which ended in mid-July, NASA announced. As a result of information in the reports, NASA forwarded 130 alert bulletins to the Federal Aviation Administration. Pilots and aircrew members submitted 62% of the reports, and air traffic control personnel 34%, indicating broad support for the program within the aviation community. About 99% of the reports included reporter identification that would permit NASA to follow up on the data if necessary; follow-up was used in more than 150 cases. Twelve of the reports concerned aircraft accidents and were forwarded to the Natl. Transportation Safety Board and the FAA as required; none of the reports contained information relating to a criminal offense. All other reports had the reporters' names removed, as specified in the agreement between NASA and the FAA, before the information was forwarded. Review of the information so far had revealed some "less obvious problems" with the national aviation system, NASA said, and analysis would proceed shortly: problems included equipment malfunction, communications breakdown, flight operations, and personnel workloads. (NASA Release 76-177)
- A mysterious radio signal apparently emanating from the Soviet Union had been so powerful that it had disrupted maritime, aeronautical, telecommunications, and amateur radio operations throughout the world for months, the *New York Times* reported. The U.S. Federal Communications Commission had forwarded four complaints to the USSR Ministry of Post and Telecommunications since 25 Aug., but had not received an answer. Colin Thomas of Leeds, England, worldwide coordinator of interference reports for the Intl. Amateur Radio Union, said that amateurs in Sweden, Norway, West Germany, the U.S., and Australia had complained of the interference; protests by the British Home Office to the USSR had not received a reply, he said. The Intl. Telecommunications Union in Geneva, Switzerland, to which the matter was referred, said it had no power to enforce treaties against interference but tried to mediate such situations. An FCC spokesman said complaints had been received almost daily since early July; direction-finding equipment had confirmed the source of the signals as the eastern side of the Baltic Sea. An extremely wideband signal pulsing 10 times per second was causing the interference, which had generated complaints from every type of shortwave user. What generated the signals, what type of intelligence they might be carrying, and what the purpose was, remained unanswered questions, the *NYT* said. *(NYT,* 30 Oct 76, 5)

During October: Numbers of ham radio operators had received pictures transmitted by the Viking lander from the surface of Mars more than 370 million km away, the Associated Press reported in a story reprinted by the *New York Times.* Retired printing foreman Bob Walton of Des Moines, Ia., said that a third of the 260 000 hams in the U.S. might be capable of receiving the pictures if they spent about $700 on equipment.

Last July the Jet Propulsion Laboratory at Pasadena broadcast on its 20m band that off-duty personnel would relay Mars pictures through their amateur radio station; Walton found a junked TV set, replaced some parts, and began receiving the Mars pictures. The JPL signals, tape-recorded and put through a converter, turned the sound into pictures when played back. Vocal explanations of the pictures could also be taped for playback. Storms in Iowa had prevented Walton from receiving some of the Mars pictures, but ham operators in Tex. and Fla. sent him the ones he had missed, he reported. (*NYT*, 7 Oct 76, 10 C)

Video transmissions of Viking lander pictures from the JPL ham radio station N6V (NASA 1976 Viking) at Pasadena had bypassed ham radio operators in San Diego County nearby, wrote Cliff Smith, science writer for the *San Diego Union*. The shortwave radio signals, good for long-distance communications, had been received by "about 13 000 hams around the world equipped to receive the pictures," but not by the score of ham operators around San Diego. The locals pointed out that special converter devices N6V used for picture transmission and reception had been made by a company in San Diego, and that the local ham organization had donated to N6V the converter used by the station for its Viking transmissions. After a representative of the San Diego Reporter Assn. called the situation to the attention of the N6V broadcasters—all hams employed as technicians at JPL, the Viking control center—N6V corrected the omission by beaming its broadcasts to a repeater station on Otay Mt. that most area hams could receive. Pictures received since the change had included shots of the *Viking 2* mechanical arm trying to turn over a rock; closeups of the Martian moons, Deimos and Phobos; and a weird-looking alien whose image the JPL station transmitted as a joke. (*San Diego Union*, 8 Oct 76)

- MSFC announced development of a new system of vehicle mobility that could greatly extend unmanned explorations on the surface of Mars and other planets. The concept grew out of work on the successful lunar roving vehicle that carried astronauts and equipment on extended exploration of the moon, far from the lunar module's landing site. Scientists had regretted the lack of mobility of the Viking landers on the surface of Mars, because the data on atmosphere and terrain had been limited to the landing site. The new idea, called Elastic Loop Mobility System (ELMS), would use a continuous elastic-loop track in place of the landing pads; the track would distribute lander weight uniformly over a relatively large area, with suspension and drive systems on the spacecraft combined into a single lightweight package. The geometry of the loop would serve to provide "excellent mobility" on soft soils and smooth rides over hard and rough terrain. Tests on models revealed slope-climbing capability, high maneuverability, and power to surmount obstacles and cross crevasses. Combining the ELMS with existing Viking hardware would produce a mobile laboratory that could conduct scientific missions on the Mars surface for 6 mo along traverses up to 150 km.

  Recent reviews of a possible mobile-Viking mission to Mars, using the ELMS concept, considered extending the mission to 2 yr along traverses up to 500 km. The low ground clearance of the Viking spacecraft, together with the low ground-pressure tolerances on soft Martian soil, ruled out the use of wheels for mobility, as well as the use of conventional tracks because of high energy consumption and low operational reliability. Besides use of the MSFC-developed ELMS for planetary exploration, the U.S. Marine Corps had considered the system for a new generation of amphibious landing vehicles for the 1980s, a MSFC spokesman said. (MSFC Release 76-191)

- "Under the guise of lack of funds...a paucity of imagination" had lost the U.S. "a preeminent place in the world of space science and engineering," said a letter in *Aviation Wk & Space Technology* magazine. In the meantime, the USSR "has not turned down any opportunity to use its space science and engineering to advance their well-being for the present or for the future." The letter cited the Grand Tour exploration of the outer planets, during a planetary configuration that "will not occur again until 179 years have passed," and the Large Space Telescope program as lost opportunities for the U.S. to retain technological superiority. (*Av Wk*, 11 Oct 76, 74)

- Johnson Space Center awarded a $96 000 contract to Martin Marietta Aerospace Co. of Denver for a study of equipment needed to construct a solar-power satellite (SPS) in geosynchronous orbit, in or about 1990. The 9-mo contract called for conceptual design and system definition of equipment to support construction of large space systems in orbit, assuming use of the Space Shuttle, and for defining development and maintenance costs of such equipment. (JSC Release 76-62)

—JSC selected Alpha Bldg. Corp. of Houston for negotiations leading to award of a $1 369 000 cost-plus-award-fee contract for construction support services at the Center, including minor construction and other site work—alteration of laboratory systems, facilities, utilities, roads, sewers, walks, etc.—required by space programs and normally for projects estimated at $10 000 or less. The contract would begin 1 December 1976 and end 30 November 1977, with the option for the government to extend the contract for two additional 1-yr periods. (JSC Release 76-63)

November 1976

1 November: NASA would appeal a decision of the U.S. District Court ordering the agency to rehire all Marshall Space Flight Center civil service employees laid off since 1967 as the result of awarding support-services contracts to private business, *Aviation Week & Space Technology* magazine reported. The court had ruled that all such contracts awarded at MSFC since 1967 were illegal. NASA spokesmen estimated that up to 1500 contractor employees now at MSFC could be fired as a result of the ruling, and that 17 000 contractor employees throughout the agency might be affected by the decision, which ordered NASA to rehire the civil service personnel with back pay in amounts depending on the period of unemployment. Had all the employees affected been unemployed for the full 10 yr (the "worst-case situation"), necessary back pay might go as high as $150 million, NASA said, although the actual funds required would be less. An appeals court had already granted NASA a stay until 1 Feb. 1977 before the district court's ruling became effective, so that the agency could ask Congress for supplemental funds to cover the back-pay requirements. (*Av Wk*, 1 Nov 77, 22)

- NASA announced plans to use a new technique for rapid detection of fecal coliform bacteria in water systems, under an agreement with EPA's region 2, to define water quality in coastal areas of the New York Bight along the Atlantic coast. NASA would supply remote data-collecting buoys and a monitor designed to detect coliforms, the accepted indicator of bacterial contamination. Developed at LaRC as a byproduct of early Skylab environmental-control systems technology, the electronic monitoring device could detect human and nonhuman fecal coliforms in a few hours rather than days, permitting health authorities to act promptly upon discovery of large quantities of disease-producing bacteria. Shellfish beds could become infested with pathogenic organisms resulting from ocean dumping of sewage; the sensor could also monitor coliform levels in lakes, public water supplies, and sewage-plant effluent. (NASA Release 76-178)

- The U.S. Air Force Systems Command announced that its space and missile systems organization (SAMSO) had completed tests at MSFC to define effects of sound waves and shock waves on the Space Shuttle during launch, using a 6.4% scale model of the launch pad proposed for Vandenberg AFB. The model launch pad, measuring roughly 13 m long by 6.5 m wide by 2 m high and weighing more than 27 000 kg, underwent engine-induced overpressure and noise environments simulated by Tomahawk solid-fuel rocket motors and high-pressure gaseous hydrogen-oxygen engines. Shock waves from rocket-engine ignition interacted with exhausts on the launch pad to direct overpressure back to the launch vehicle; the flow rate also created an acoustic field causing turbulence in the surrounding atmosphere. The altered environments were modified during testing by altering the configuration to produce the most acceptable design of launch vehicle and payload. (AFSC Release OIP 170.76)

- U.S. nuclear export policy should be conditioned on an international understanding that certain activities are "inherently dangerous," said Victor Gilinsky, commissioner of the U.S. Nuclear Regulatory Commission in a speech at Mass. Institute of Technology. Although the dangers of plutonium—"both a nuclear explosive and the key to a virtually inexhaustible source of energy"—had been regarded as located far in the future, "the future is here," said Gilinsky, citing the accumulation of plutonium-rich spent fuel from civilian power reactors in storage sites around the world, with more and more nations interested in commercial-scale reprocessing. Although the desire for plutonium had been stimulated by the assumption "almost universally held" that its use was a natural, desirable, and indispensable result of using nuclear materials to generate electricity, the assumption had made even more difficult any attempt to restrict the availability of plutonium. Likewise, safeguards based on confusion or misapprehension about the possibilities of misusing plutonium had been nullified by the spread of technological knowhow as a national policy of the U.S. More than 90% of the enriched uranium imported in 1975 into the European Economic Community to produce energy—and eventually plutonium— had been supplied by the U.S., Gilinsky pointed out. Nuclear commerce should be conditioned on a new policy that "unrestricted national development of nuclear power programs is inherently incompatible with a secure world," he concluded, and no distinction should be recognized between military and "so-called 'peaceful'" nuclear explosives. (NRC Release S-14-75)

2 November: Two NASA laboratories had spearheaded "the most ambitious effort so far to detect radio emissions from distant civilizations," Walter Sullivan reported in the *New York Times*. The project, known as SETI (search for extraterrestrial intelligence), would use a specially constructed device called a multichannel spectral analyzer with various radio telescopes to scan simultaneously a million different frequency bands within a range known to scientists as the "waterhole." This part of the spectrum lies between the frequencies emitted by hydrogen atoms adrift in space (1420 mhz) and those from hydroxyl, composed of one hydrogen and one oxygen atom (1662 mhz), which combines with hydrogen to form water; the frequency range had been chosen as "a logical rendezvous for intelligent creatures trying to make contact for the first time." Ames Research Center would have primary responsibility for program management and the Jet Propulsion Laboratory would use its worldwide network of antennas besides providing instruments for the multichannel system, including a data processor capable of handling 6 million channels. NASA officials noted that the search would not depend on deliberate efforts at communication by other civilizations, citing the detectable radio energy emitted by the earth in normal space operations, television transmissions, and so forth. (*NYT,* 2 Nov 76, 16)

- Landsat data analyzed by computer had distinguished accurately between stands of hardwood and pine, offering an improvement in routine forest management, NASA announced. Forestry specialist Darrel Williams at GSFC had achieved an overall accuracy in classifying forest categories of 90%, in comparison with airplane photographs of the same tract, a commercial forest in N.C. owned by the Weyerhaeuser Corp., chosen so that the data could be checked against a known base available continuously. Aerial photos used in forest inventory might be updated only every 3 to 5 yr, because of the expense of airplane flights, whereas Landsat could provide repetitive coverage much more frequently. Use of the Landsat data, besides replacing conventional photographic coverage, could permit continuous land evaluation for insurance and tax purposes and provide evidence of unauthorized cutting, among other benefits. (NASA Release 76-179)

- Britain and France announced their decision to build no more Concorde supersonic jetliners beyond the 16 now under construction or in the air. After a meeting in London with France's transport minister Marcel Cavaillé, British industry minister Gerald Kaufman told a news conference that future projects would concentrate on less exotic subsonic aircraft that would stand a better chance of making money. The Concorde, a project begun by treaty in 1962, had already cost about $1.92 billion for research, design, and development; each plane, with support facilities, would cost about $49.6 million. Nine had been built and sold—five to British Airways, four to Air France—and, of those remaining, three would go to British Airways and two to Air France, while preliminary purchase agreements had been signed with Iran for two others. British Airways had begun supersonic commercial flights in Jan. 1976 with its Concordes between London and Bahrein; Air France had begun similar service at about the same time between Paris and Rio de Janeiro. On 24 May 1976, a trial service began from London and Paris to Dulles Intl. Airport at Washington, D.C. Banned by the New York Port Authority from landing at JFK Intl. Airport, pending results of environmental testing at Dulles, the Concorde was losing money for both its state-owned sponsors; Air France reported losses of from $30 to $32 million in the first 9 mo of Concorde operation. Potential customers for the Concorde had made it clear that they would be interested only in the Europe-to-New York City route, most lucrative of the international flights. Kaufman added that the two countries had not become disenchanted with supersonic transport even though their future plans were to build smaller planes; "the fact that you're going for bread and butter now," he said, "doesn't mean that you'd be disenchanted with caviar later." (*NYT*, 3 Nov 76, 61; *W Post*, 4 Nov 76, A-3; *W Star*, 3 Nov 76, A-5)

- Michel Bignier, Director General of the French CNES from Jan. 1972 to June 1976, assumed his new duties as director of ESA's Spacelab program. (ESA newsletter, Nov 76, 2)

3 November: A Stanford Research Institute study requested by Ames Research Center of ways to detect possible intelligence signals from outer space reported that the most effective and economical technique would be use of a hemispheric antenna more than 3 km in diameter, orbiting the earth opposite the moon but shielded from earth's radio emissions. The orbiting antenna would occupy one of the points in the moon's orbit where the gravity field of earth and moon would balance one another. Two other strategies assessed in the study were an "orchard" of parabolic antennas (similar to that suggested in 1971 for Project Cyclops, a joint ARC-Stanford proposal to use massed 100m antennas for signal-seeking and for radio astronomy), and a group of antennas in craters on the far side of the moon shielded from earth interference to sweep the entire range of the heavens during each month. Choice of the best strategy would depend on estimates of how far the system would need to look to find another technological civilization; Walter Sullivan noted in the *NY Times* "a strong suspicion that the nearest possible civilization may be 500 light years or more away." (*NYT*, 3 Nov 76, 70)

- Dr. Mary Helen Johnston, metallurgist at MSFC, was principal guest speaker at the Federal Women's Day program at Wallops Flight Center. Dr. Johnston, who said she wanted to be one of the first American women chosen for space flight, participated in 1974 as part of an all-women crew of experimenters in a 5-day tryout of the general-purpose laboratory at MSFC, a cylindrical mockup of the Spacelab being built in Europe for Space Shuttle flights. Using the neutral-buoyancy simulator, a working area for handling experiment packages in a zero-gravity environment, the crew developed techniques for use on Spacelab. As principal investigator for three scientific experiments and co-investigator on another, Dr. Johnston had planned the work in hopes of going on orbital missions in the 1980s. Also on the program was Audrey Rowe Colom, director of women's activities for the Corporation for Public Broadcasting and current president of the Natl. Women's Political Caucus. Besides the speakers, the program included skits by WFC employees and a costumed narration of "Two Centuries of American Women on Parade." (WFC Release 76-15)

- Dr. Harrison H. (Jack) Schmitt, astronaut on *Apollo 17*, was the victorious Republican candidate for the U.S. Senate from New Mexico, defeating Democrat Joseph M. Montoya, who was seeking a third term. Now a consulting geologist, the 41-yr-old Schmitt had a 3-to-2 lead in the returns. CBS commentator Walter Cronkite noted in a news broadcast at 10:25 p.m. EST on 2 Nov. that astronauts constituted only 2 ten-millionths of the U.S. population, yet there would now be two U.S. senators who were astronauts—Schmitt and John Glenn (D-Ohio)—which was "representation out of all proportion." (*W Post*, 3 Nov 76, A-17; broadcast, 2 Nov 76)

4 November: JSC announced selection of the Boeing Aerospace Co. for negotiations leading to award of a contract, valued at about $970 000, funded by NASA and ERDA for the study of space-based solar-power systems. The 12-mo two-phase study would provide specific data on the most effective way to convert solar to electrical energy using satellites, and would determine where in space the various phases of a solar-power satellite should be constructed and assembled; the second phase would define weights and costs of a satellite system. The project would aim at providing uninterrupted energy beamed to earth from large satellites in fixed orbits at altitudes above 35 000 km. (JSC Release 76-69)

- ERDA conducted a demonstration at Ft. Belvoir, Va., for military decisionmakers on the use of solar cells by the armed forces. Arranged in cooperation with the U.S. Army's Mobility Equipment Research and Development Command, the exhibit would show the advantages of solar cells over conventional sources of electricity in applications such as communications and surveillance. Increased purchase by the military could improve availability of solar cells to the general population, ERDA said, by stimulating mass production and thereby lowering costs. (ERDA release 1 Nov 76)

5 November: NASA Hq announced that Robert S. Kraemer, Director of Lunar and Planetary Programs, would become special assistant to Dr. Robert S. Cooper, Director of Goddard Space Flight Center, working on the applications missions of the center. Kraemer had been at NASA Hq for 9 yr, 6 of them in his current position. (NASA Hq. special anno., 5 Nov 76)

- LaRC announced that James S. Martin, Jr., manager of the Viking project, would resign in December to become vice president for advanced programs at Martin Marietta Aerospace. Martin, who came to LaRC in 1964, had managed the Viking project since 1968, directing work at JPL, at Martin Marietta's facility in Denver, and at universities and subcontractor plants throughout the U.S. Martin Marietta had been prime contractor for the Viking project. LaRC said that G. Calvin Broome, Viking mission director, would become Viking project manager 15 Nov., directing the 18-mo extended mission from the control center at JPL in Pasadena, Calif. (NASA special anno., 5 Nov 76; LaRC Release 76-37; *NYT,* 6 Nov 76, 9)

- KSC announced award to Norflor Construction Co., Orlando, Fla., of a $284 000 construction contract for an airlock to provide a clean-room environment in the spin-test facility of Cape Canaveral Air Force Station, to be used in preparing Delta-launched spacecraft and later for Space Shuttle payload processing. Construction would be completed within 160 days after contractor's notification to proceed. The airlock installation would include a 5-ton crane, air conditioning, personnel airlock, and lines for electricity, vacuum, compressed air, and gaseous nitrogen. Present clean-room facilities were occupied by elements of the Mariner Jupiter-Saturn spacecraft scheduled for launch in Aug. and Sept. 1977. Delta payloads scheduled for launch in the spring and summer of 1977 included Geos (ESA), April; Goes, May; OTS (ESA), June; Japan's GMS metesat, July; Sirio, an Italian comsat, August; and Meteosat (ESA), Sept. (KSC Release 490-76)

- NASA Hq and LaRC announced that A. Thomas Young, former mission director for the Viking project, had been appointed Director for Lunar and Planetary Programs at NASA Hq and would begin his new duties 6 Dec. in the Office of Space Science. Young, who had worked with the Viking project since 1968 and helped develop Mars mission objectives, was most recently the mission operations manager and director of the 750-person Viking flight team at JPL in Pasadena, Calif. In his new job he would manage NASA's unmanned planetary programs, including Viking, Pioneer, Mariner, and Helios, as well as studies in planetary astronomy, atmospheres, and geology; advanced scientific planning, programming, and technology; extraterrestrial materials research; and flight-program support. (LaRC Release 76-40; NASA Hq special anno., 5 Nov 76)

- After a series of technical checks, the *Ekran* satellite launched in the USSR 26 Oct. had been activated, Tass annouced. *Ekran,* in synchronous orbit, had transmitted experimental television programs over "a vast territory in Siberia between the Surgut and Yakutsk meridians," Tass said, describing the new service as a "gigantic telebridge." USSR first deputy communications minister V. Shamshin said the new generation of comsats would provide a reliable and economical system of television communications throughout the Soviet Union. *Ekran,* whose transmitters were apparently more powerful than those of the Molniya series, would dispense with the need for 12-meter ground antennas used by the previous Orbita system, Tass said. (FBIS, Moscow Tass in English, 5 Nov 76; Moscow Domestic Service in Russian, 6 Nov 76)

6 November: Scientists at Johnson Space Center had used helium-filled balloons as the preferred means of carrying payloads to study the atmosphere above the ozone layer. Although sounding rockets and airplanes could operate in the medium between 30 and 50 km over the earth's surface, neither could remain stationary for very long, the rocket's time in the stratosphere being measurable in minutes; also, payload weight for an airplane must decrease with the altitude it can maintain, whereas a balloon would be capable of taking an 1800-kg payload as high as 50 km. JSC's Environmental Effects Office had used balloons to study effects on the ozone layer of fluorocarbons and other chemical interactions in the upper atmosphere; recent Natl. Research Council recommendations on reduction of use of freons had been partially based on JSC balloon-flight data. The JSC Space Physics Branch had used balloons to carry cosmic-ray instrumentation for the study of high-energy particles, and had worked with the Utrecht Space Sciences lab in carrying an ultraviolet telescope by balloon for use in a range of light that was opaque to observation from earth. Balloons and payloads prepared by JSC were launched from Palestine, Tex., by the Natl. Center for Atmospheric Research, with parachutes for recoverable experiments. (JSC Release 76-70; *JSC Roundup,* 19 Nov 76, 1)

7 November: Instruments left on the moon's surface during the Apollo program were continuing to transmit information from 13 experiments, JSC announced. The *Apollo 12* mission (Nov. 1969) accounted for 2 of the experiments; *Apollo 15* (July-Aug. 1971), for 3; *Apollo 16* (April 1972), for 3; and *Apollo 17* (Dec. 1972), for 5. Findings included the discovery that the moon did in fact possess an atmosphere, only about a millimeter thick and composed of ions of light minerals, and that the first two layers below the moon's surface consisted largely of basaltic rock and silicon. Data transmitted by 5 of the nuclear transmitters still operating were sent to stations at JPL, Pasadena, Calif., and GSFC, Greenbelt, Md., where they were analyzed by 10 scientists. (*NYT*, 7 Nov 76, 45)

- Atomic reactor waste buried in the Ural Mountains had exploded in 1958, killing hundreds and affecting thousands with radiation sickness, according to an exiled Soviet scientist whose article in the British *New Scientist* was reported by Associated Press. Zhores Medvedev, a biochemist and geneticist who had been a wellknown dissident in the Soviet Union, had been allowed to go to Britain in Jan. 1973 but became an exile when Soviet authorities refused to allow him to return. His article said that reactor waste buried for many years in a deserted area had overheated and erupted "like a violent volcano," and that strong winds had blown the resulting radioactive clouds for a long distance; however, no one had been evacuated from the area until after symptoms of radiation sickness appeared. Many Ural towns with medium to high levels of radiation had never been evacuated, he said, adding that the area was among several that were off limits to Western correspondents. Medvedev said several biology research stations had been built in what was "the largest area of gamma radiation in the world" to study the damage to plant and animal life. In 1974, a Soviet official denied reports of explosions at the the Shavchenko nuclear power station on the Caspian Sea, at the same time that *Pravda*, the Communist Party newspaper, was assuring the public that radioactive wastes had been stored safely.

  Another Soviet disaster had occurred in Oct. 1960, Medvedev said, when then-Premier Nikita Krushchev ordered the launch of a moon rocket to be timed with his arrival in New York City for a session of the UN General Assembly. When the ignition button was pushed, nothing happened; under normal procedures, Medvedev said, workers would drain the fuel from the rocket before inspecting to find the cause of the failure. However, Marshal Mitrofan Nedelin, then head of Soviet rocket forces, felt himself "under an obligation to fulfill the ambitious order" issued by Krushchev and decided to investigate immediately. While dozens of engineers and other experts were examining the rocket and its support systems, the ignition suddenly started to work, but the rocket could not take off because of "the forest of inspection ladders." The rocket toppled over and turned the Soviet cosmodrome into a holocaust, killing many persons in the area, including "some of the best representatives of Soviet space technology," Medvedev said. AP noted that the Soviet media routinely ignore all disasters, whether natural or manmade. (*W Post*, 7 Nov 76, A–1; *W Star*, 7 Nov 76, B–8; *Av Wk*, 15 Nov 76, 20)

8 November: The U.S. Air Force announced it would begin flight tests of a new system called latar (laser-augmented target acquisition/recognition), developed by Northrop Corp. to give pilots of single-seat aircraft enhanced air-to-surface and air-to-air attack capability. The latar pod, mounted in the gun bay of the USAF F–4E, would contain long-range high-resolution electro-optical equipment for target imaging, laser designation, and acquisition and tracking; its field-of-view would be limited only by the fuselage and the externally carried munitions on the aircraft. The system would allow a pilot to acquire a target visually, using either a helmet-mounted sight system or a telescopic radar, with the optical turret following his line of sight; the view seen by the latar would appear on a helmet-mounted display built by Minneapolis Honeywell. The system would be suitable for use in all types of aircraft, including helicopters. (OIP Release 203.76)

- FAA Administrator John L. McLucas announced that the agency would sponsor a conference in Washington, D.C., 16 Nov. to foster the use of metrics in aviation, in accordance with national policy set forth in the Metric Conversion Act of 1975. The conference would hear presentations by FAA, other federal agencies, aviation users, and the public, and would run into a second day if necessary. Subjects to be discussed would include air-traffic control, operations, and aeronautical charts and navigation aids, as well as design of aviation products and the impact on the aviation community of a transition to the metric system. McLucas noted that FAA had already begun to use metric measurements in areas such as standards for airport design and construction. (FAA Release 76-106)

- Federal agencies allocated $4.5 billion to institutions of higher learning in FY 1975, the Natl. Science Foundation reported, a level of support about the same as the previous year's but representing an 8% decline when converted to constant (1972) dollars. HEW allocated the largest amount, $3.2 billion, or 70% of the total; the NSF supplied the second largest, $491 million or 11%. Other sponsoring agencies were the Dept. of Agriculture, DOD, ERDA, and NASA. Of the $2.2 billion earmarked for research and development (about half of all allocations to academic institutions), researchers in life sciences received $1.2 billion, more than half of all federal R&D funds; research in physical sciences received $307 million, about 14%, and engineering and environmental sciences each received about 9%. All other fields together received only about 13%. (NSF Release PR76-91)

9 November: The $100 million spent by the U.S. on a search for life on Mars had produced only disagreement over the findings, Viking project scientists said at a JPL press conference. Four of six Viking scientists said they did not know whether life existed on Mars; one denied that Viking had found life; the sixth said he felt Viking had found "primitive microbes" in the Mars soil samples. All six agreed that Viking had found nothing at either the Chryse or the Utopia landing site representing fossils that would have confirmed a previous existence of life on Mars, *Washington Post* writer Thomas O'Toole reported.

Dr. Klaus Biemann of MIT, who designed the instrument to look for dead organic matter, noted that nothing had been detected, and explained readouts from other instruments as the result of an exotic chemistry catalyzed by some superoxidant in the surface of Mars and activated by solar ultraviolet rays that penetrated the Mars atmosphere all the way to the surface. Dr. Carl Sagan of Cornell Univ. disagreed, saying that the failure to find dead organisms did not outweigh the activity detected by the biology instruments, which produced readings at a lower temperature and lost them when the sample was heated. This result "smells more like biology than any chemistry I can think of," Sagan added. Vance Oyama of ARC favored a chemical explanation; Dr. Gilbert V. Levin of Biospherics said he leaned toward a biological explanation. Dr. Norman H. Horowitz of Calif. Inst. of Technology, who designed the photosynthesis experiment, favored the biology explanation but regretted the absence of fossils. Dr. Harold P. Klein of ARC said the Viking results "do not rigorously prove there is life on Mars, nor do they rigorously disprove it."

New York Times reporter John Noble Wilford described the failure to detect on Mars the organic compounds essential to life processes on earth as "a major surprise," causing most of the confusion over interpretations of the data; a *NYT* editorial noted that the Mars findings had forced a new look at theories about the origins of life on earth, which "seems neither so certain nor so inevitable as it did before the Viking landings . . ." (*W Post*, 10 Nov 76, A-9; *NYT*, 10 Nov 76, A-16; 18 Nov 76, 42)

- The National Aeronautic Assn., at its annual meeting in Washington, D.C., announced its selections for Elder Statesmen of Aviation for 1976. The award, established in 1954, honored outstanding Americans over the age of 60 who had made significant contributions to aeronautics. The three persons chosen were: J. Leland Atwood, 72, leader in aviation for more than 50 yr, who headed North American Rockwell (now Rockwell Intl.) until his retirement in 1970 and originated the design concept for the P-51 Mustang; Clifford W. Henderson, 81, originator of the Bendix and Thompson trophy races and the Power Puff Derby, and manager of the National Air Races 1928-1939; and Blanche Wilcox Noyes, 76, who learned to fly in 1928 and won the national closed-course race in 1929, the Miami All-American air race in 1931, and the Bendix trophy for a cross-country flight in 1936. (NAA release 9 Nov 76)

- INTELSAT announced plans to conduct a 1977 program of research and development costing more than $5.4 million, to advance technology for its global comsat system. The R&D yearly budget, approved by the recent meeting of the board of governors, would consist of $1.3 million for research to identify technology that appeared promising for the future INTELSAT projects; $2.2 million for contract authority for projects with near-term applications; and $1.9 million for in-house support of such projects. Areas of exploratory research would include spacecraft and microwave technology, communications processing, analysis of propagation and transmission, and studies of materials and devices. Near-term projects would include development of antennas, transponders, and other components; investigation of cross polarization; development of NiH batteries, voice-channel decoders, high-power transistors, and improvement of techniques. (INTELSAT Release 76-33-M)

- ESA announced acceptance of the first solid-propellant apogee boost motor produced in Europe—designed and made by SNIA-Viscosa of Italy in collaboration with SEP of France—for use in its Geos (geostationary scientific satellite). The motor, measuring 1.1 m long with maximum diameter of 0.7 m, weighed 305 kg, more than the remainder of the spacecraft, and would serve to inject the satellite into geostationary orbit from the elliptical transfer orbit provided by a Thor-Delta launch vehicle. The satellite was scheduled for launch from Cape Canaveral in April 1977. (ESA release 9 Nov 76)

10 November: The *Marshall Star* reported that a former student research assistant at MSFC, now a member of the faculty at Alabama A&M Univ., would be principal investigator in a study to develop remote-sensing applications for resources study and management, under a $35 000 grant from NASA and MSFC. The research effort would catalog the transportation network of Ala., including highways, railroads, water systems, and airports, and would devise a system for storing information in quickly retrievable form that could readily be updated. Dr. Oscar Montgomery, who had worked on the study at the outset, had graduated from A&M and had received a Ph. D. from Purdue Univ., returning to A&M in the department of natural resources and environmental studies. He and other researchers at A&M would identify and assist users of remotely sensed data in Ala., develop university facilities and skills to handle, process, and interpret such data, and support "ground truth" and related activities of the Earth Resources Office at MSFC. The grant, part of the Minority Institutions Research Program, would provide remotely sensed data obtained by aircraft or spacecraft to the Ala. Development Office and other state agencies. (*Marshall Star,* 10 Nov 76, 4)

- The Air Force Systems Command announced completion of 3 yr of test and evaluation on a production-engineered version of a laser-guided bomb kit designed for easy mounting on standard unguided bombs. AFSC noted that, during the Southeast Asia conflict, laser-guided bombs were 200 times as likely to reach the target as manually released unguided bombs, and had outscored computer-aimed bombs by 50 to 1. The USAF described the LGB as "one of the most effective technological advances" in weaponry, one LGB costing less than 12 standard unguided bombs and offering additional savings in fewer missions flown, fewer crew losses, increased storage life, and greater reliability. (OIP Release 241.76)

10-11 November: The mysterious high-power radio signal that had interrupted international communications for months disappeared suddenly on 2 Nov., officials at the Federal Communications Commission reported, but resumed as unexpectedly 10 Nov., the foreign minister of Norway told his parliament. Personnel at the Rogaland radio station on Norway's west coast began receiving the signals again from a shortwave transmitter they thought was located near Kiev in the Soviet Ukraine. The FCC said in Oct. that hundreds of complaints had been received about the interference, heard through a wide range of high frequencies from about 6 to 28 mhz, like a rapid ticking 10 times per sec. The signal had disrupted ship-to-shore, aeronautical, telephone, and amateur and international broadcast services around the world, particularly in Europe and across the North Atlantic. The USSR had never acknowledged that the transmissions originated there; a spokesman at the Soviet embassy in Washington, D.C., said he did not know what was causing the interference. (*W Star,* 9 Nov 76, 1; *W Post,* 11 Nov 76, A32; *NYT,* 11 Nov 76, 5)

11 November: Ames Research Center announced that a NOAA scientist, Dr. Peter M. Kuhn, had discovered a way to give airplane pilots up to 12 min. warning of air turbulence ahead, during his investigation of infrared radiation emitted from atmospheric water vapor using ARC's Convair aircraft. While measuring background infrared radiation in the atmosphere to define sources of astronomical infrared, Dr. Kuhn had seen sudden drastic changes in water-vapor content followed closely by turbulence. A radiometer in the aircraft wheel well of the Galileo II flying laboratory could detect atmospheric infrared from water vapor, and predict accompanying turbulence with an 81% reliability. Experiments with varying infrared wavelengths had permitted detection of turbulence more than 100 km ahead of the plane. The water-vapor anomalies resulted from wave motions in the atmosphere caused by lee waves over high terrain or by sudden shifts in wind speed or direction with resulting friction between adjacent streams of air. Although water-vapor content would usually be fairly constant, the wave motions would thin the vapor in one place and concentrate it in another, making the turbulence detectable. (ARC Release 76-80; *Av Wk*, 15 Nov 76, 25; NOAA Release 76-251)

- NASA announced appointment of Glynn S. Lunney, manager of the Shuttle payload integration and development program office at JSC, as Deputy Associate Administrator for Space Flight at NASA Hq, replacing Dr. William C. Schneider. Dr. Schneider had been assigned as director of project management at GSFC. Lunney's NASA career began at LeRC in 1958; he transferred to the Space Task Group at LaRC in 1959, and moved with that group to JSC. He was manager of the Apollo spacecraft program in 1972, and served as technical director of the Apollo-Soyuz Test Project. (NASA Release 76-183; JSC Release 76-71; *JSC Roundup*, 19 Nov 76, 1)

- Despite lower profits in its work on the Space Shuttle and increased losses in business aircraft, Rockwell Intl. reported higher sales and earnings for both the last quarter and the full fiscal year ending 30 Sept. 1976. Earnings were $121.1 million, up 29% in fiscal 1976, and sales increased 8% to a record $5.2 billion. The figures did not include discontinued operations in Rockwell's industrial components area, sold in Dec. 1975. Four of the company's five operating areas reflected better management and lower interest costs: automotive operations, electronics operations, consumer operations, and utility and industrial operations. The aerospace operations figures reflected fewer award-fee opportunities, Rockwell said. (Rockwell Release R-40)

12 November: The first Soviet jumbo jet, the IL-86, would begin test flights in Dec., according to a Moscow broadcast. The plane would be able to carry 350 passengers at about 373 kph over distances up to 1460 km, according to claims, which would mean it could make nonstop flights from Moscow to Lisbon. The noise level inside the plane would be lower than that in Boeing jumbo jets, the broadcast added. (FBIS, Moscow to North America in English, 12 Nov 76)

14 November: The USSR would be able by 1980 to produce a thermonuclear reaction by using a powerful laser, according to scientists at the Lebedev Physics Institute quoted in the USSR's Academy of Sciences journal *Priroda*. Institute Director Nikolai Basov (sharer of the 1964 Nobel prize in physics, for work in laser amplification) emphasized that Soviet science had been first to begin work on this problem, which would be the basis of power engineering in the future. Scientists were using two approaches: building powerful installations called Tokamaks, in which a reaction would proceed continuously; or using a laser beam to compress a small quantity of material such as deuterium to "a hundred trillion atmospheres, thousands of times as great as inside the sun," with accompanying rise in temperatures until a microexplosion occurred, to produce current for future power and heat stations. Unlike the Tokamak, this line of research could lead to production of transportable nuclear-power stations for delivery to remote areas, the article said. (FBIS, Moscow Domestic Service in Russian, 14 Nov 76; *A&A 64*, 367)

16 November: MSFC announced it had completed a critical design review for part of a heavy-nuclei experiment scheduled to go into space aboard the third High Energy Astronomy Observatory (HEAO-C) in 1979. The experiment, called C-3, would characterize cosmic-ray flux from detailed measurements of charge and energy spectra, study propagation of cosmic rays, and search for superheavy elements. Principal investigators were Dr. Martin Israel, Washington Univ.; Dr. Edward Stone, Calif. Inst. of Technology; and Dr. Cecil Waddington, Univ. of Minn. The HEAO-C, whose objective would be to map the sky for gamma-ray and cosmic-ray flux, would carry two other experiments besides the C-3. A design review for the experiment hardware would be held in Feb. 1977. (MSFC Release 76-199)

• Three NASA employees were among 16 career federal employees selected as winners of the Natl. Civil Service League's 1976 career service awards. In ceremonies at the Smithsonian Institution, Washington, D.C., those honored included former astronaut Michael Collins, third U.S. man to walk in space, selected for his work with the Smithsonian's Air and Space Museum, of which he had been director since April 1971; Christopher C. Kraft, Jr., director of JSC; and Charles F. Hall, Pioneer project manager at ARC, where he had worked since 1942 beginning with NACA. (*W Star*, 23 Oct 76, C-8; *W Post*, 19 Oct 76, C2)

17 November: MSFC announced it had issued requests for bids on a contract to design, construct, and test on earth a space-based manufacturing facility to make structural members for assembling large structures in space. Proposals would be received until 30 Nov. The machine to be produced under this contract would demonstrate automatic beam fabrication, a necessary step in developing a facility to support requirements for large structures in space, such as space platforms, electrical power-generating plants, or large antennas for communications and astronomy. (MSFC Release 76-200)

18 November: MSFC announced near-completion of an experimental chamber being built for the Atmospheric Cloud Physics Laboratory (ACPL) scheduled to fly on Spacelab for use in developing techniques to predict, alter, and control weather. The chamber would be a ground-test engineering model simulating the expansion of a parcel of air as it rises in the atmosphere, using carefully controlled temperature and pressure to monitor moist air as it would undergo changes like those causing the formation of cumulus clouds. Droplets formed in the simulated clouds would be photographed by cameras outside the chamber; the lack of gravity in space would allow the droplets to remain suspended for longer periods of time, permitting scientists to study them at length. The completed chamber, liquid thermal-control system, and temperature sensors would be tested to obtain uniform cooling of the chamber walls. Data from the experiments would serve to improve computer models of clouds, and to aid in developing methods to increase snowfall or rain, dissipate fogs, or suppress lightning. (MSFC Release 76-201; *Marshall Star*, 24 Nov 76, 1)

21 November: NASA-funded studies conducted in Fla. and Tex. using infrared images from NOAA's Natl. Environmental Satellite Service had provided fruit and vegetable growers with timely information on ground temperatures during periods of crop-killing freezes, NOAA announced. The images, provided every half hour by *Goes 1* from its equatorial station over South America at 35 800-km altitude, were transmitted from heat-radiation sensors to ground stations for conversion to visual imagery. The Fla. study used a computer to display the temperature variations in shades of gray, verified by ground readings taken in the area by Univ. of Fla. personnel. The Texas study converted the satellite data to typewriter characters that produced a remarkably similar printout, transmissible over teletypewriter machines, more widely available than photofax equipment. Both studies aimed at helping growers decide whether to undertake expensive steps to keep crops from being damaged by frost; heating orchards in Florida might cost up to $5 million per night. (NOAA Release 76-257)

22 November: MSFC announced it had shipped to GFSC the experiment payload for SPAR III, third in a series of about 15 space-processing applications rockets being flown over a 15-yr period to obtain space-processing data in near-zero gravity until Space Shuttle flights become available in the 1980s. The scientific payload built by MSFC would be integrated with GFSC components—igniter, separator, control and recovery systems—and shipped to the U.S. Army White Sands Missile Range, N.M., for a scheduled mid-December launch. SPAR payloads recovered by parachute would be forwarded to investigators for analysis. The SPAR flight would provide about 5 min of near-zero gravity in which to perform materials processing. (MSFC Release 76-204)

- MSFC announced that officials from NASA Hq and three of its field centers were participating in a Spacelab preliminary design review being held by ESA at Noordwijk, The Netherlands, 19–26 Nov. Besides representatives of MSFC, lead center for Spacelab design in the U.S., employees of KSC and JSC were attending the review to see that basic design requirements had been met before construction of the engineering model of Spacelab began. A NASA in-house review had been held at MSFC 15–17 Nov. (MSFC Release 76-203)

23 November: NASA was negligent in the 1972 death of Kirby Dupree, supervisor of a support unit for astronauts, a federal judge ruled in awarding $575 000 to the widow and a 4-yr-old daughter. When a battery box exploded in an astronaut training facility at JSC, Dupree and James E. Scott were assisting with water experiments used to simulate space flights. Scott won $100 000 for injuries suffered in the accident. Both Dupree and Scott were employees of a contractor at the facility. (*NYT*, 25 Nov 76, 20)

- A program conducted jointly by NASA's Dryden Flight Research Center and Langley Research Center had been trying literally to work the bugs out of aircraft, by studying materials and methods to keep insects from sticking to the leading edge of aircraft wings. The study was part of NASA's aircraft energy-efficiency program, aimed at developing an advanced long-range aircraft with possible savings of 20 to 40% in fuel consumption. The laminar flow-control technology on which the advanced craft would depend for savings required smooth airflow over the wings; impacted insects sticking to the leading edge of the wing could interrupt the airflow and cause turbulence. The test planes used at LaRC and DFRC had variously coated panels installed on the leading edges, with instrumented probes above them to register airflow changes resulting from adherence of insects and monitor differences attributable to the panel coatings. After a low flight over areas where insects would impact the wing surfaces, the aircraft would land for preliminary measurements and then fly at high altitudes and speeds to measure the effects on airflow. (DFRC Release 20-76; LaRC Release 76-41; NASA Release 76-190)

- ERDA announced selection of Clayton, N.M., as the first municipal utility to field-test its 200-kw wind-turbine generator. The Clayton utility would operate the modern windmill for 2 yr beginning late in 1977, and would assemble test data on the performance and economics of wind-energy systems connected with conventional power plants and providing power through existing utility lines. Clayton, a 3000-inhabitant town located in an area of the Great Plains with high average wind speeds, would be the first U.S. locality in more than 30 yrs to generate electric power for public consumption through the use of wind. Its conventional utility system used both gas and oil generators. Integration of the wind-power system would be supervised by technicians from NASA's LeRC at Cleveland, O., who would train the municipal utility employees in operating the wind turbine. (ERDA Release 76-348)

23–26 November: Reports in two national magazines—"both denied," the *NY Times* noted—charged that the Soviet Union had used laser beams to put U.S. early-warning satellites out of commission, and each predicted that the U.S. had been working on its own hunter-killer satellites. The current issue of *Newsweek* had reviewed the incident of 1975 when a U.S. early warning satellite and a companion relay satellite had been disabled over the Indian Ocean, citing "strong evidence, despite official U.S. denials, that last year's incident over Siberia was the work of a laser beam 10 000 times as strong as any natural blaze." U.S. Secy. of

Defense Donald Rumsfeld had stated that the U.S. satellites had been damaged by the glare of natural gas fires along a pipeline in Western Russia, but did not deny that lasers might have caused the damage. *Newsweek* said the U.S. had already developed chemical lasers requiring no electrical power, and described a possible fleet of U.S. "dark satellites" with radar-absorbing exteriors that would be invisible to Soviet detection. It also predicted that a full-scale war in space—a science-fiction type of clash of hunter-killer satellites, manned orbital shuttles, and laser death rays—could "leap off the drawing boards" in the 1980s.

Reporter Tad Szulc said in the current *Penthouse* magazine that the U.S. government had not publicized the attack on its satellites used to police the 1972 arms pact with the USSR for fear of endangering negotiations with Moscow on a new strategic arms pact. He said Washington was perplexed by the Soviet effort to interfere with satellites that were verifying compliance with the 1972 treaty.

The *NY Times* noted that spokesmen for both the State Dept. and the Dept. of Defense had denied the magazine reports of interference with U.S. satellites, although DOD had verified that the USSR had been attempting to intercept its own satellites already in orbit, not always successfully. Pentagon sources told the *NYT* that the Soviet hunter satellites used conventional explosives, although shockwaves from an explosion would not proliferate in the vacuum of space and a hunter would have to be accurate enough to close in and destroy a target with debris. A *NYT* editorial 26 Nov. said the "vast sums" needed for space weapons would be more wisely spent on peace and good relations on earth. (*C Trib*, 23 Nov 76, 4-15; *NYT*, 23 Nov 76, 15; 24 Nov 76, 12; 26 Nov 76, A-22)

24 November: A lunar scientist in Johnson Space Center's Division of Lunar and Planetary Sciences had suggested that low-viscosity lava rather than water could have caused the large channels seen on Mars, JSC announced. Ernest Schonfeld, in a paper to be presented at the annual meeting of the Am. Geophysical Union, said that the abundant large channels—one of the most puzzling landforms on Mars—were more easily explained by lava than by water; scientists had not been able to identify thick deposits of sediment that should have resulted from channel formation by water. *Viking 1* had landed near the mouth of a large channel where thick sediments should have been deposited, but rocks at that site appeared to be volcanic. Schonfeld proposed that thin basaltic liquid melted below the Mars surface had flowed freely to erode the surface, and said that the coincidence of the age of volcanic activity on Mars and the erosion of the channels supported his idea. (JSC Release 76-72)

30 November: NASA announced establishment of a joint working group with the Agency for Intl. Development (AID) to explore further cooperative programs upon successful completion of the AIDSAT project. Final demonstration of AIDSAT came 30 Oct. in Haiti. NASA's *Ats 6* satellite showed people in 27 developing nations how space could improve their way of life, using small portable transmitter-receiver units to broadcast high-quality color TV directly to low-cost receivers in remote areas. Programs consisted of both filmed and live portions: on film, viewers saw a message from President Ford, followed by one of three films produced by NASA on communications technology, remote-sensing satellites (like Landsat), and prediction and relief in natural disasters. The live segment contained two half-hour programs; the first, originating in the host country, showed officials describing technological challenges and current efforts to meet them, and the second was a two-way question-and-answer session between senior officials in the host country and experts in Washington, D.C., discussing education, health, agriculture, and similar national concerns. Millions of viewers had seen what in many instances were the first color TV broadcasts in the host country. (NASA Release 76-189)

- Wallops Flight Center announced appointment of Dorinda S. Bailey, an equal opportunity specialist in the office of the Center Director, as full-time Equal Opportunity Officer for the center. Before coming to Wallops in 1974, Mrs. Bailey had been an examiner in the Civil Service Commission's Bureau of Personnel Investigations. As EO officer, she would be responsible for developing and managing programs to assure equal opportunity in center employment. (WFC Release 76-16)

- The Federal Aviation Administration would stick to its deadline for modifications to McDonnell Douglas DC-10 jumbo jets to withstand sudden depressurization, the *Washington Post* reported, as FAA Administrator John L. McLucas withdrew from subordinates the authority to extend a 1 Dec. 1977 deadline set at the recommendation of the Natl. Transportation Safety Board. The modification, which applied also to Lockheed L-1011s and Boeing 747s, was ordered after the 1974 crash near Paris of a DC-10 whose cargo door blew off in flight, resulting in the death of 346 persons on board. Depressurization of the plane crushed control cables running beneath the cabin floor. An official at FAA had granted a 1-yr extension of the deadline in Oct. 1976—without McLucas's knowledge—but all U.S. airlines flying DC-10s had agreed to comply with the original deadline. Compliance dates for the L-1011 and 747 were 31 March 1978 and 30 June 1978, respectively. (*W Post*, 30 Nov 76, A-9)

During November: The Natl. Aeronautic Assn. newsletter reported that, during the 69th annual general conference of the Fédération Aeronautique Internationale in Oct. in Iran, the 1975 FAI gold space medal was awarded jointly to Thomas P. Stafford, commander of the Apollo crew on the Apollo-Soyuz Test Project, and to cosmonaut Alexey Leonov, who commanded the *Soyuz 19* crew. Stafford and his crew, Donald K. Slayton and Vance Brand, also received the De La Vaulx medal for establishing during the joint mission the world absolute records for duration, distance, and altitude in group flight, as well as the V.M. Komarov Diploma for outstanding performance in space exploration during the previous year. Their Russian counterparts also received the De La Vaulx medal and the Komarov award. Astronauts Slayton and Brand also each received the Yuri Gagarin gold medal, created by the FAI in memory of the first man in space, for their role in the conquest of space for peaceful purposes. Stafford also accepted on behalf of the U.S. ASTP team the Honorary Group Diploma for Astronautics, for the planning and execution of the first international rendezvous and docking mission in space. (NAA newsletter, Nov 76)

- Maj. Gen. Thomas P. Stafford, currently commander of the USAF Flight Test Center at Edwards AFB, Calif., received the National Geographic Society's Gen. Thomas D. White space trophy for 1975, in recognition of "outstanding contributions" in the Apollo-Soyuz Test Project. (*AF Magazine*, Nov 76, 21)
- Former astronaut Frank Borman, who commanded both the *Gemini 7* and *Apollo 8* missions and who currently was president and chief executive officer of Eastern Air Lines, was elected chairman of the board for EAL replacing Floyd D. Hall, who had resigned to become permanent chairman of the executive committee of the International Air Transport Association (IATA). (*Av Wk*, 29 Nov 76, 20)
- The Air Force Systems Command announced completion of vibration and performance tests at Arnold Engineering Development Center on a small solid-propellant rocket motor scheduled to boost the Intl. Ultraviolet Explorer spacecraft into orbit in 1977. The motor, Thiokol Corp.'s TE-M-604-4, would be fired about 24 hr after launch to move the IUE into a near-synchronous orbit at altitudes of from 24 000 to 48 000 km. IUE, a cooperative research program shared by NASA, ESA, and the U.K. Science Research Council, would carry a 45-mm-diameter ultraviolet telescope to investigate characteristics of stars and planets. (*AFSC Newsreview*, Nov 76, 4)
- The Natl. Science Foundation's Nov. bulletin contained information on new NSF policy on grant renewals, submission of final reports, and other activities. As of 1 Oct., standard NSF awards could be renewed only once, by amendment of the original grant or contract, provided that the cumulative duration of the grant would not exceed 5 yr. Further support for the same or a different project would be awarded through new grants, based on new proposals. Also effective 1 Oct., the final fiscal and technical report on a grant and a summary of the completed project must be submitted within 90 days after expiration of the grant. NSF also announced a program called Research Initiation in Minority Institutions, to encourage research among enrollments predominantly of black, native American, Spanish-speaking, or other ethnic minorities; a program called National Needs Science Faculty Professional Development, to support training for science teachers at both 2- and 4-yr colleges and universities; and a Law and Social Sciences program to fund empirical research by law-review students in collaboration with legal scholars and social scientists. (*NSF Bulletin*, Nov 76, 1)
- "Washington sources" not further identified had revealed that the first underwater test launch of a Soviet intercontinental ballistic missile, the SS-NX-18, had occurred "early in November" between a submarine in the White Sea and an impact area near Plesetsk. The "new missile" had been test-launched previously from a land-based facility at Plesetsk. *Spaceflight*, published by the British Interplanetary Society, quoted its sources as saying that the SS-NX-18—which used storable liquid propellants, and carried multiple independently targetable reentry vehicles (MIRVs)—had traveled about 5150 km, about 70% of its design range, and would replace SS-N-8 in the Delta 1 and Delta 2 class of USSR nuclear submarines. (*Spaceflight*, vol 19 no 2, 41)

December 1976

1 December: Officials of U.S. airlines told Secretary of Transportation William T. Coleman, Jr., in an all-day hearing that lack of government help in financing the costs of government-imposed rules for quieting planes could lead to severe setbacks in service, such as wholesale groundings, diminished competition, fuel waste, less relief from noise, and possible loss of the national lead in aircraft technology. The airlines had suggested a plan last spring to set aside a fourth of the 8% ticket tax for replacement or refitting of planes that exceeded the new noise limits. Opposed to this plan was James C. Miller 3d, assistant director of the administration's council on wage and price stability, who said that he would favor a "pollution tax," imposed on planes inversely to the amount of noise suppression achieved, rather than "a subsidy from the public purse." Secy. Coleman said he would make a recommendation to the outgoing Ford administration by the end of Dec. New noise rules promulgated by the Federal Aviation Administration, DOT, were to take effect on 1 Jan., setting up a 4- to 8-yr timetable for replacing or refitting aircraft that did not comply with stringent limits on noise. (*NYT*, 2 Dec 76, 21)

- NASA announced appointment of Dr. James J. Kramer as acting Associate Administrator in the Office of Aeronautics and Space Technology (OAST), replacing Robert E. Smylie, who would become Deputy Director at GSFC. Dr. Kramer, who came to NASA Hq from LeRC in 1971, had been manager of the refan program office. In a related action, Paul F. Holloway, director for space at LaRC, would begin a temporary appointment as acting deputy Associate Administrator of OAST on 3 Jan. 1977. (NASA anno. 1 Dec 76)

- Launch complex 14 at KSC—used for John Glenn's orbital flight in Feb. 1962 and other manned launches until 1966—was blown up with plastic explosives by an Army demolition team after the Air Force decided the rusty and obsolete towers constituted a hazardous area and should be demolished. The scrap metal would be sold to the highest bidder. News reports noted that a stainless steel memorial would remain to mark the place where the seven original astronauts took their first steps into space. (*W Star*, 2 Dec 76, A-3; CBS News transcript, Cronkite-Dean, 1 Dec 76; NBC Nightly News, Brinkley, 1 Dec 76)

- A Pentagon group charged with making decisions on weapons programs—the Defense Systems Acquisition Committee—held its first meeting to consider production of the controversial B-1 bomber and scheduled another meeting for 6 Jan. 1977 to decide on consolidation of cruise missile programs backed separately by the Air Force and the Navy. Defense Secretary Donald H. Rumsfeld had stated—apparently with approval from the transition team of the incoming Carter administration—that he was responsible until 20 Jan. for making weapons decisions that came up for scheduled consideration;

however, he had also told subordinates that DOD should exercise restraint in passing on programs that the incoming administration might review or change. President-elect Carter had questioned the necessity for a B-1 bomber program. The controversy over cruise missiles arose from Air Force advocacy of an air-launched missile that could be used only with a B-1 or B-52 bomber; this restraint offered the verification possibility, acceptable to the USSR, of controlling the number of cruise missiles by counting the bombers carrying them. However, the Navy had advocated its Tomahawk missile which could be launched from a submarine torpedo tube as well as from a surface ship or a bombing plane. The Navy's preference would thwart verification of future arms-limitation agreements, since all attack submarines as well as surface ships and bombers could be potential platforms for launching the Navy missile. The Carter transition team was concerned, said the *New York Times*, that the Navy might be trying to lock in the new administration with a pre-inauguration decision that would complicate future negotiations with the USSR. Both the Navy and the Air Force programs were started after a 1972 agreement with the USSR that placed a 5-yr limitation on strategic ballistic missiles; the $15 million design studies had grown into $300 million development programs, and had reached the point where a decision was needed on proceeding into costly engineering development. Observers pointed to the Army's plan to award contracts in the next week or so on its $7 million programs for troop-carrier and attack helicopters as a decision that would make it difficult for the incoming administration to re-examine its predecessor's policy. (*NYT*, 2 Dec 76, 23)

- A report in *Izvestia* on Soviet passenger planes to be in service by 1980 failed to mention the Tupolev-144 supersonic transport, the *Washington Post* reported. An official of the USSR Aviation Ministry said the trouble-plagued Tu-144 would not go into passenger service in 1976 as promised last Dec., when the highly publicized "first scheduled supersonic service" began between Moscow and Alma Ata in Soviet Kazakhstan. That service, initially twice a week, carried only mail and cargo, and was viewed by Western observers as an extended test-flight program. The mail service had been cut back to once a week in June 1976, with reports of problems in the passenger cabin such as noise, vibrations, and pressure. The Tu-144, which on 31 Dec. 1968 had been the first supersonic civilian plane to fly, was designed to carry up to 140 passengers up to 5600 km, but had never been known to meet either specification; observers said its fuel consumption was greater than expected. Its service on Aeroflot lines, set to begin in 1974 or 1975, had been set back by a crash in Paris in 1973. Although ignoring the Tu-144 in the *Izvestia* story, Aviation Minister Boris Bugayev said the state-operated Aeroflot fleet would include the 350-passenger Ilyushin-86 airbus, not yet flown, and the medium-range 120-passenger Yak-42 airplane, which he said would be "put into service in the current Five-Year Plan period." (*W Post*, 1 Dec 76, A-12)

- ESA announced award to European industry of two study contracts, one to define a large (about 900 kg) technological and experimental satellite in geostationary orbit, the other to define the communications payload for such a satellite. The 4-mo contracts, each valued at about $450 000 U.S., went to groups headed by Engins Matra and SNIAS of France for the spacecraft, and to a group headed by Germany's AEG-Telefunken for the payload. The new satellite would be designed for launch by ESA's Ariane, possibly on the fourth development flight scheduled for Oct. 1980; it would flight-test equipment and techniques as forerunner of a large platform carrying various communications payloads for direct and semidirect television and sound broadcasting and for propagation experiments in the 20- to 30-ghz range. The new system would differ from ESA's OTS in offering new services required by countries without elaborate ground communications and broadcasting installations; it would carry four or more high-power TV channels for direct home reception, besides providing low-cost facilities for national telecommunications through small "thin route" terminals. (ESA release 1 Dec 76)

2 December: Secretary of Defense Donald H. Rumsfeld announced that the Ford administration had decided to authorize construction of the controversial B-1 bomber, under a contract arrangement that would allow the incoming Carter administration 5 mo to review the $22.8-billion program. Rumsfeld's announcement said that B-1 testing was complete and that it was "in the national interest" to proceed with production. During the campaign, candidate Carter had said he was opposed to immediate production but held out the possibility of further research on the bomber, which the Pentagon now declared completed. The unusual contract arrangement of month-by-month funding until the end of June was an extension of the 1 Feb. deadline set by Congress for commitment of funds for the program; the Air Force awarded contracts totaling $704.9 million to Rockwell Intl., General Electric, and the Boeing Company for the first three production models of the bomber, plus procurement to make eight more. The contracts, however, limited government liability to a monthly total of $87 million through June. Air Force Secretary Thomas C. Reed said at a news conference that keeping the B-1 in the research state was unnecessary and wasteful, and that "it would be irresponsible not to initiate B-1 production at this time" in view of the expansion of Soviet strategic forces. A manned bomber, he said, was the third part of a triad—including land-based intercontinental ballistic missiles and missile-bearing submarines—that would present "an insoluble problem" to the USSR and make it a losing proposition to attack the U.S. because of retaliation that would follow. The swing-wing B-1, although only two-thirds the size of the B-52, could carry twice the payload, was faster in takeoff and flight, and carried electronic systems more difficult to jam than those of the B-52. Opponents claimed that $100 million apiece was too high a price to pay for a bomber in the age of antiaircraft missiles, and suggested alternatives such as loading a commercial Boeing 747 with cruise missiles that could be fired while the plane was a safe distance from enemy defenses. (*NYT,* 3 Dec 76, A18; *W Post,* 3 Dec 76, A-1)

- The Natl. Science Foundation announced that employment of academic scientists and engineers had increased 3% in 1976; the increase had cut across all major fields of study, from 1% in life sciences and engineering to nearly 7% in social and environmental sciences. The number of full-time scientists and engineers at academic institutions went from 224 800 to 230 500, or 3%, and the number of those employed part time went from 55 900 to 58 700 (5%). Over the period 1965 to 1976, there was a significant shift to teaching. Scientists and engineers employed primarily as teachers increased 223 200 from 1975 to 1976 (3%); those working primarily in research and development increased 51 000 (2%). The number of women employed full time as scientists and engineers increased 5% from 1975 to 1976, a rate more than double that for men. (NSF *Highlights,* 2 Dec. 76, 1)

• A solar-powered microwave station, first of its type in commercial use, had been installed on the Navajo settlement of Medicine Hat, Utah, the *Chicago Tribune* announced. The microwave relay tower, working in the hot climate without the need for airconditioning controls, would bring dial telephone service to the Indian community. (*C Trib,* 2 Dec 76, 4–9)

3 December: NASA announced that it would simulate the drop of the main probe of its planned 1978 Pioneer Venus mission into the Venusian atmosphere, by dropping the probe from an Air Force balloon at 32 km altitude over the White Sands Missile Range, N.M. The drop, scheduled for a date between 13 and 16 Dec., would be the first of two tests, a second being scheduled for early 1977. Flight events to be simulated just before the probe's descent into the dense hot lower atmosphere of Venus—deployment of the probe parachute, separation of the heatshield, and (after a 17-min parachute descent) separation of the parachute—would occur at an altitude of 16 km, where the temperature and density of earth's atmosphere and the velocity of the probe would be much the same as those on Venus. The instrumented vessel would contain special equipment to monitor system performance, and movie cameras on the balloon platform as well as telescopic still and movie cameras on the ground would record the test. (NASA Release 76-192; ARC Release 76-84)

• NASA announced two appointments effective 19 Dec. in the Office of General Counsel: Allie B. Latimer as assistant general counsel for general law, and Richard J. Wieland as assistant general counsel for litigation. Ms. Latimer, who since 1972 had been assistant general counsel for automated data and telecommunications for the General Services Administration, was the first national president of Federally Employed Women (FEW). Wieland, who had been at JSC since 1963, was currently assistant Chief Counsel for general legal matters at that center. (NASA anno. 3 Dec 76)

4 December: According to a story in the *Chicago Tribune,* the USSR on an undisclosed date had fired a submarine-launched missile farther than ever before. An SSN-8 missile had landed in mid-Pacific slightly more than 9000 km from its launching point, presumably a submarine in the Barents Sea. (The longest-range U.S. submarine-launched missile could reach just under 4700 km.) U.S. intelligence sources were reported unsure whether the additional range of the SSN-8 resulted from extra rocket power or a lighter warhead; SSN-8 had previously been identified as a single-warhead weapon used in the USSR's 17 new Delta-class missile submarines. Commenting on the report, the *New York Times* said the Soviets' test shot was "good news paradoxically" for the U.S. and the whole world: in contrast to large land-based intercontinental ballistic missiles, the submarine-based missiles were neither large enough nor accurate enough for effective first-strike counter-silo use, and their concealment and mobility would make them ideal second-strike weapons useful only as deterrents. As long as the USSR had no submarine-launched missile, there was little possibility of lowering the ceilings set in a previous agreement to a level that would head off a Soviet first-strike capability. The ceilings had allowed each side to have 2400 strategic bombers and missiles, of which 1320 could carry multiple warheads; if the USSR should agree to deploy half its multiple-warhead missiles at sea as the U.S. did, such an agreement could eliminate the possibility of either side's constructing a high-confidence first-strike force and would help ensure stability of the nuclear balance for a long time. The *NYT* favored such an agreement, which would call a halt to ongoing missile programs. (*C Trib*, 4 Dec 76, 2-5; *NYT*, 6 Dec 76, 32)

5 December: The fundamental cause of the succession of Ice Ages undergone by the earth had been the slight but regular changes in the shape of earth's orbit around the sun, according to a team of U.S. and British scientists headed by Dr. James D. Hays of Columbia Univ. Analysis of layers of fossil microorganisms from cores of sediment taken from the floor of the Indian Ocean produced measurements of the earth's climate over the past 450 000 yr; changes in climate appeared in alternating layers of warm- or cold-preferring microfossils, known as radiolaria. The discovery should permit prediction of the onset of the next Ice Age, although the cooling trend now in process should continue for about 20 000 yr. The last great glaciation 18 000 yr ago brought ice down over most of Canada, northern Europe and Asia, and the northern U.S.; it was one of eight major Ice Ages over the past 700 000 yr. The ice retreated about 11 000 yr ago to its present boundaries, and the earth is now in a warmer period. The investigation, to be reported formally in the 10 Dec. issue of *Science* magazine, was supported by a consortium of universities and the Natl. Science Foundation. (*NYT*, 5 Dec 76, 4-14; *C Trib*, 5 Dec 76, 1-9)

6 December: NASA announced "an unusual agreement" with McDonnell Douglas Corp., under which the firm would build a solid-fuel upper-stage rocket as a commercial venture, to be available for use in future Space Shuttle missions. The firm would build and market the product using its own funding and initiative, at no direct cost to NASA or the U.S. government, except when the government might be a purchaser; NASA agreed that it would not knowingly fund or formally solicit development of competitive or alternate upper-stage systems. McDonnell Douglas would market the stage hardware and services either directly to users, or through NASA; NASA would not be committed to buy any hardware or services from McDonnell Douglas. The Delta rocket, workhorse of NASA's expendable launch-vehicle program, which had carried most of the comsats and metsats now in geosynchronous orbit, would be phased out along with other expendable vehicles after the Space Shuttle became operational. After the Shuttle had carried payloads into low earth orbit, the upper stage would put the payloads (weighing up to 1110 kg) into a transfer orbit, where a kick motor would insert them into geosynchronous orbit. (NASA Release 76-198)

- MSFC announced that two of its employees had manufactured a precision quartz ball that would be used by the National Bureau of Standards as a density measure. John Rasquin and Jack Reed of the Materials and Processes Laboratory in MSFC's Science and Engineering Directorate made the precision sphere, which is 90 mm in diameter and weighs 290 g. NBS had selected quartz as the homogeneous material for its standard because it could be made with a high order of purity, had great temperature stability, and was ideal for use in an interferometer because of its precisely known optical properties. MSFC originally made two clear quartz spheres for NBS that were unsatisfactory because of their high transparency, as the interferometer required a highly reflective surface; NBS exposed the spheres to gamma radiation for 4 mo to make them opaque by changing their molecular structure, but the radiation induced minute internal stresses that altered the contour. The spheres, returned to MSFC, were reworked by Rasquin and Reed to NBS specifications of 1×10^{-6} deviation in radius. One of the spheres had been completed and the other was about 50% complete. (MSFC Release 76-209)

7 December: The mission of the launch of *Comstar B* on 22 July 1976 from the Eastern Test Range on an Atlas-Centaur AC-40 was declared successful by John F. Yardley, Associate Administrator for Space Flight, NASA announced in a postlaunch mission operations report. The spacecraft's apogee motor was successfully fired 24 July 1976, injecting the Comstar into the desired synchronous orbit. (NASA MOR M-491-201-76-02, 13 Dec 76)

- The Peoples Republic of China launched an unidentified satellite from the Shuang-Cheng-Tzu site into an orbit with 483-km apogee, 170-km perigee, 59.5° inclination, and 91.1-min period. *Aviation Wk and Space Technology* magazine reported that mission characteristics indicated the vehicle was in the 2700- to 4500-kg class and that the launch vehicle was a CSS-X-3 booster. Hsinhua News Agency, whose launch announcement the Tass agency picked up, reported 10 Dec. that the satellite had returned to earth "with precision according to plan," with no further details. This was PRC's seventh satellite launch. (FBIS, Tass in English, 7 Dec 76; GSFC SSR, 9 Dec 76; FBIS, Peking NCNA, 10 Dec 76; *Av Wk*, 13 Dec 76, 29; *SBD*, 14 Dec 76, 234)

- The flight of NASA's *Pioneer 11* into unexplored space above the plane of earth's orbit had confirmed the structure of the sun's magnetic field for the first time, NASA announced. Dr. Edward J. Smith of the Jet Propulsion Laboratory—magnetometer experimenter for *Pioneer 11*—reported to the meeting of the Am. Geophysical Union that observations conducted from Feb. to Nov. 1976, while the spacecraft was four times farther from the sun than the earth, showed that the solar magnetic envelope had a simple north pole-south pole structure split into northern and southern portions at the magnetic equator by a "warped" sheet of electric current. This warped sheet had appeared to move as the sun rotated, up and down relative to the earth's orbital plane. As the solar magnetic field extended several billion miles over the north and south solar poles, well beyond the orbit of Saturn, earth spacecraft traveling in earth's orbital plane had passed through the warped electric current and detected contradictory motions, the field in the northern solar hemisphere being carried outward by the solar wind and the field in the southern hemisphere reversing back toward the sun. Earlier spacecraft therefore reported reversals in the direction of the solar magnetic field each time the current sheet was encountered, reports that led to a variety of interpretations. When *Pioneer 11* had passed close to Jupiter in Dec. 1974, the gravitational effect had thrown the spacecraft 62 million km above earth's orbital plane, allowing it to measure solar phenomena at a point 16° above the solar equator—9° higher than previously possible—where it discovered a uniform solar field, pointing away from the sun. The present solar model would exhibit the warped current sheet about 15° each side of the solar equator, accompanied by small-scale random magnetic fields of varying intensity and direction; the north polar region would generate a well-ordered magnetic field in a single direction, and the south pole a similar field in the opposite direction. The solar wind would carry the magnetic field out until it met the interstellar gas, perhaps near the orbit of Pluto, where the outgoing north-polar field would link with incoming south-polar field to "close the magnetic loop," Dr. Smith suggested. (NASA Release 76-199; ARC Release 76-85; *NYT*, 7 Dec 76, 1)

8 December: NASA announced that the proceedings of its Nov. 1976 workshop on space battery technology were available from GSFC, where the sessions had been held annually since 1968. Batteries had proved to be the major limiting factor in long-life space systems used for research in communications, weather, and earth-resources detection and monitoring; current life expectancy for batteries was 3 to 7 yr, depending on orbital altitude. (Spacecraft in synchronous orbit at 35 000 km altitude derived power from electrochemical batteries charged by solar cells on the spacecraft surface; batteries on these craft, continuously in the sun except for two eclipse periods per year, would escape the charge-discharge cycling that wore out similar batteries carried on spacecraft circling the earth every 90 min at lower altitudes, discharging their power in the resultant periods of darkness.) NASA had begun the workshops to exchange information between manufacturers and users for increasing efficiency and lifetime of the batteries, resulting in improvements by industry in batteries used for heart pacemakers, aircraft, and consumer goods such as portable tape recorders, radios, and flashlights.

Gerald Halpert, GSFC chairman of the battery workshop, said the Nov. session had reviewed a new technique of fabricating battery plates electrochemically at temperatures of 100°C, which would reduce the number of steps in the process and provide a more uniform product. Also discussed was a rechargeable nickel-hydrogen battery, a new product that could provide more energy per kg for larger systems, with storage capacities ranging from 25 to 100 ampere-hr. The nickel-hydrogen unit would be used for the first time on the Navigational Technology Satellite-2, part of the joint services research program for a global satellite system. Another topic was the special procedures developed at

GSFC for battery improvement that had been implemented in industry for quality control and evaluation of the product. (NASA Release 76-200)

- ESA concluded its preliminary design review of the manned orbital Spacelab at the premises of the prime Spacelab contractor, VFW-Fokker/ERNO in Bremen, W. Germany. Senior representatives of ESA, NASA, and ERNO said the review was successful and the result exceeded expectations. The review cleared the way for a detailed design phase to begin development of the engineering model of Spacelab, due for delivery by the end of 1978. The flight model would be delivered by the end of 1979 for a first ESA-NASA mission in 1980. (ESA release 9 Dec 76)

9 December: America's Bicentennial year might be called the "Year of the Communications Satellites," NASA Administrator Dr. James C. Fletcher told participants at the Conference on Satellite Communication and Public Service meeting at GSFC. Recalling that Alexander Graham Bell had unveiled the telephone at the Philadelphia Centennial exactly a century ago, Dr. Fletcher quoted an observer's statement that, of all the gifts young America had received on its 100th birthday, the telephone proved to be the most valuable; in 1976, of NASA's 19 launches, 13 had been communications satellites, which might prove the most valuable gift received by the U.S. on its 200th birthday. Comsats were in use by 107 countries or territories on 6 continents, Dr. Fletcher said, and more than a billion people—one of every 4 on earth—could witness an international event as it happened "live by satellite." Not only had comsats been able to do current jobs better and cheaper, they had also opened new possibilities for public service. Development of small inexpensive simple earth stations would permit sending needed information to millions of people without present access to it, Fletcher said; very large antennas and high transmitter power in space would make it possible to use receivers on earth "not 30 feet in diameter, or even 10 feet, but the size of your watch crystal," he added. To get the larger, heavier spacecraft into orbit, Fletcher predicted use of the Space Shuttle, making "runs to space on a regular schedule, carrying people and cargo for communications, scientific research, earth resources inventory, materials processing, and other tasks." Use of the Shuttle would reduce the cost of putting a satellite in orbit from more than half to less than a quarter of the total cost of design, construction, and launch. Fletcher called on members of the conference to help NASA define the public purposes to be served by "an imaginative space program" that would involve users in its development from the outset. (Text; NASA Release 76-202)

- The U.S. Navy, in the last days of the Ford Administration, had awarded a $159.9 million contract to Rohr Marine, Inc., for design of a new type of warship that would ride on an air bubble at three times the speed of conventional ships. The 3000-ton prototype of the "surface effect" ships would ride on an air bubble trapped by sidewalls and bow and stern seals; the high speed would result from not having to push through water as conventional vessels do. The Navy had experimented with a 100-ton test vehicle that reached speeds of more than 165 km per hr, a new Navy surface-speed record, in tests last June near Panama City, Fla. Although critics had questioned the need for a ship that speedy, the Navy had said its primary roles would be antisubmarine warfare and sea-to-shore hauling of people and cargo. The Navy had also successfully fired an antiship missile from an experimental air-cushion vessel traveling at about 111 kph. (*W Post*, 10 Dec 76, A-2; *W Star*, 10 Dec 76, B-12; *B Sun*, 10 Dec 76, 6)

10 December: GSFC Director Dr. Robert S. Cooper presented a Group Achievement Award to the AIDSAT team for "the highly successful international demonstrations in which leaders of over twenty-five underdeveloped countries were shown the potential benefits to be derived from the application of satellite communications and remote earth sensing." The demonstrations had taken place over a 90-day relocation period during which the *Ats 6* was moving from its position over India to a new geosynchronous position over the western Pacific. The team—led by former GSFC engineer Paul McCeney, NASA Hq Office of Applications, as AIDSAT program manager; Al Whalen of GSFC, as project manager; and John Wilhelm of the Agency for Intl. Development, as program coordinator—undertook to produce 27 programs for broadcast to as many nations over the 3-mo period. Three technical teams carried portable terminal equipment to three clusters of nations: A, consisting of Thailand, Oman, Jordan, Sudan, Mali, Cameroon, Uruguay, Ecuador, and Jamaica; B, consisting of Bangladesh, United Arab Emirates, Yemen, Libya, Upper Volta, Central African Republic, Bolivia, Surinam, and Costa Rica; and C, consisting of Pakistan, Kenya, Morocco, Ivory Coast, Liberia, Sierra Leone, Argentina, Peru, and Haiti. Goddard engineers Dave Nace, John Chitwood, and Al Whalen were managers of the A, B, and C clusters respectively. The high-level government officials and businessmen in the participating countries who took part in the programs included eight presidents, three prime ministers, a king, and a sultan. Each country received virtually the same 2.5-hr program format, which began with a personal taped message from U.S. President Ford and presented films on remote sensing and on earth-orbiting communications and meteorological satellites, ending with an hour of two-way direct video communications between panelists in the U.S. and the host country. Millions of viewers in the host countries watched the programs by direct or delayed broadcast over national TV networks, some seeing the first colorcasts ever received in their country. The historic experiment in telecommunications technology transfer fulfilled a promise to make U.S. technology available to developing nations, given by Secretary of State Henry Kissinger in May 1976 at the U.S. Conference on Trade and Development in Nairobi, Kenya. (*Goddard News*, Dec 76, 3-4)

11 December: Eight nations located on the equator had claimed sovereignty over the orbital locations of space satellites stationed above their territory, the *Washington Star* reported. The foreign ministry of Colombia listed the nations as Brazil, Ecuador, Colombia, Congo, Kenya, Uganda, Zaire, and Indonesia. (*W Star*, 11 Dec 76, A-4)

12-20 December: "One of the most important ore-bearing structures in the world" might prove to be located in an Antarctica site known as the Enchanted Valley, the *New York Times* reported, when a U.S. Geological Survey party of six—including two women—had completed their study of the formation after an airlift to the site from the McMurdo Sound station. The body of rock in the Pensacola Mts., first reached by explorers in 1957 and first studied in the 1965–66 season, was a layered structure produced by eruptions from earth's interior and "strikingly similar" to productive formations elsewhere in the world, such as the Bushveld complex of S. Africa, the Stillwater formation of Montana, and the Sudbury region of Ontario. A Soviet party in the Shackleton Range to the northeast, a similar area, reported findings which the leader of the USGS group, Dr. Arthur B. Ford, hoped to visit by air. Dr. Edward S. Grew of UCLA, who had spent the Antarctic winter at the Soviet Molodezhnaya base, had been working with the Russians from a temporary camp in the Lambert Glacier, near which "a mountain of iron" was reported recently.

Meanwhile, a Natl. Science Foundation project that aimed at drilling a hole through the Ross Ice Shelf was proceeding with the participation of 10 nations, seeking to explore the depths below the ice for data on the Antarctic bottom water, indirectly responsible for much of the world's oceanic food, and to learn whether the ice sheet might eventually slip into the sea, raising global sea levels by as much as 10 meters. Countries participating with the U.S. in the Ross Ice Shelf Project (RISP) were Australia, Britain, Denmark, Japan, New Zealand, Norway, the Soviet Union, Switzerland, and West Germany. The *NYT* commented editorially that the Antarctic continent posed a number of scientific problems still unanswered in the age of space that had seen visits to and returns from the moon; close photographs of the moon and planets; and regular reports by man-made instruments on the surfaces of Mars and Venus. The RISP—an example of "continued, steady, quiet international cooperation" in Antarctic research—was "one of the great success stories of the post-1945 world," encountering temporary setbacks that would be overcome in time as were those of the space program, the *NYT* added. (*NYT*, 12 Dec 76, 1-1; 19 Dec 76, 26; 20 Dec 76, A-22)

13 December: Forecasts of NASA activity in 1977 included plans to launch 19 missions, with a possible four backup spacecraft in case of trouble with current missions. The 19 included eight comsats (NATO–3B; Palapa-B; OTS, an operational test satellite developed by ESA; Intelsat IVA-C; Sirio, an Italian experimental spacecraft; Intelsat IVA-D; FLTSATCOM, part of a Navy-Air Force operational near-global system for command and control of all DOD forces; and Japan's CS, a comsat to be placed in synchronous orbit); six scientific satellites (HEAO-A, a high-energy astronomical observatory; GEOS, an ESA mission to study geomagnetic phenomena from synchronous orbit; MJS-A and -B, two Mariner spacecraft scheduled to fly by Jupiter and Saturn, and possibly Uranus; ISEE, a two-spacecraft international sun-earth explorer mission developed by NASA and ESA; and IUE-A, the international ultraviolet explorer spacecraft also developed jointly by NASA and ESA); three metesats (GOES 2, to be launched for NOAA into synchronous orbit; GMS, a Japanese geosynchronous meteorological satellite; and Meteosat, ESA's synchronous-orbit weather spacecraft, scheduled for 0° longitude); one navigational satellite, the Navy's Transit 19, with R&D modifications not common to other Transits; and one applications satellite, Landsat-C, third in its series of earth resources technology spacecraft. The backup launches to be scheduled only if replacements were needed were comsats RCA-C and a spare OTS; navsat Transit 20; and metesat ITOS E-2, on standby in case of trouble with *Noaa 4* or *Noaa 5*, already in orbit. Of the 19 scheduled launches, all but two would be from Cape Canaveral, using 12 Delta rockets, 3 Atlas-Centaurs, and 2 Titan III-E-Centaurs. The other two launches, both from Vandenberg AFB, would use one Delta and one Scout rocket. MSFC forecast for 1977 that facilities for assembling and testing the Space Shuttle orbiter would be completed, as well as test firings of the Shuttle's solid-fuel motors and main engines. (*Av Wk*, 13 Dec 76, 26; KSC Release 492–76; MSFC Release 76–218)

- NASA's permanent-personnel status as of the end of November 1976 showed the agency as having 153 employees over its end-of-year ceiling, 23 816. Center with most employees over allowed number was MSFC, with 143; also over strength were Headquarters with 34, LaRC with 14, ARC with 8, and JSC with 5. However, GSFC had 20 fewer employees than allowable, KSC had 15 fewer, LeRC and Wallops were each 6 below ceiling, and NSTL and DFRC were short three and one employees respectively. The General Management Review Report containing these figures noted that efforts would be made to achieve the FY 1977 ceiling through attrition. (GMRR 13 Dec 76, 13)

- Radio links with the Viking landers on Mars were reestablished, ending a month of silent hibernation during which communications had been blocked by the passage of the sun between Mars and earth. Scientists at the Jet Propulsion Laboratory at Pasadena sent a "wake-up call" to the robot laboratories on Mars that had been collecting a month's supply of data during the blackout period; these data, to be played back beginning 20 Dec., would include images of the area around the two landers, inorganic chemical analysis, additional biology data, and any records to support the (thus far unsuccessful) search for marsquakes, using the seismometer on the *Viking 2* lander.

    Resumption of activity would begin Viking's 18-mo "extended mission," which would provide more images of the Mars surface and polar regions, as well as measurements of water vapor and temperature, some at twice the resolution of previous observations; process more samples of the surface for life-detection tests and inorganic chemical analysis; photograph Deimos and Phobos, the moons of Mars; and monitor daily and seasonal weather changes, origins of planetwide dust storms, and any seismic activity that might occur. As Mars entered its closest approach to the sun in the spring of 1977, scientists would be watching for signs of the dust storms thought to be triggered by perihelion; when *Mariner 9* arrived at Mars in 1971, the entire planet was engulfed in the greatest dust storms in the history of Mars observations and surveys from orbit were limited for several months. Such storms could endanger the Vikings on the surface of Mars, besides impeding the view from orbit. (*W Post*, 15 Dec 76, 2; NASA Release 76–208; *W Star*, 16 Dec 76, A-7; *NYT*, 21 Dec 76, 21)

- NASA announced that a series of aircraft flights from Alaska to Argentina between January and May of 1974 had provided information on the transport and distribution of water vapor in the atmosphere from the tropic seas to the higher latitudes, as well as vertically to the stratosphere, important to an understanding of earth's ozone layer and possibly of earth climate. Previous studies of water in the stratosphere, made from aircraft and balloon flights, had not provided data on latitudinal transport. Ernest Hilsenrath of GSFC, who reported this first mapping of water-vapor distribution in both northern and southern latitudes in a paper at the American Geophysical Union meeting 8 Dec., noted that the 1974 flights had confirmed an increase in the amount of atmospheric water vapor since 1964. The experiments, flown on an Air Force RB-57, were managed by GSFC under contract to DOT as part of its Climatic Impact Assessment Program. (NASA Release 76-201)

- Simultaneous measurements by two polar-orbiting satellites of powerful electric currents causing magnetic disturbances on earth had produced composite diagrams of horizontal current-flow in the ionosphere, vertical currents along magnetic-field lines, and the resulting magnetic perturbations, NOAA announced. A team of scientists representing NOAA and five universities had presented their findings at the annual meeting of the American Geophysical Union in San Francisco. Dr. Yohsuke Kamide of the Cooperative Institute for Research in Environmental Sciences (sponsored by NOAA and the Univ. of Colorado) said that, before the advent of satellites, auroras had been studied from the ground with cameras and magnetometers, which could detect the presence of powerful currents but not their location or direction. The *Triad* satellite launched in 1972 by the U.S. Navy had carried a Johns Hopkins Univ. three-component magnetometer to measure the strength of vertical currents supplying the "auroral electrojets"; using it, together with a satellite launched by the USAF Defense Meteorological Satellite Program and ground-based measurements from stations in Alaska, Canada, and the USSR, the team had produced its composite for 30 instances of perturbation. Other members of the team were Herbert W. Kroehl of NOAA's Environmental Data Service; Dr. Gordon Rostoker, Univ. of Alberta; Dr. Syun-ichi Akasofu, Univ. of Alaska; Dr. Thomas A. Potemra, Johns Hopkins Univ.; and Dr. Ching Meng, Univ. of Calif. at Berkeley. (NOAA Release 76-274)

14 December: NASA announced selection of three contractors to negotiate for provision of various subsystems for the Solar Maximum Mission spacecraft. The SMM, scheduled for launch in the third quarter of 1979, would carry instruments to investigate solar flares; the spacecraft would be built on the modular concept, so that its instrumentation could be readily changed or replaced.

Fairchild Industries, Germantown, Md., would negotiate to provide the communications and data-handling subsystem at a proposed value of $2.5 million, for one protoflight model with options for up to six additional. The C&DH subsystem would provide ground and onboard control of all spacecraft and sensor functions, and of retrieval of housekeeping and experiment data.

McDonnell Douglas Astronautics Co., East St. Louis, Mo., would negotiate to provide the power module subsystem for the SMM at a proposed value of $2.3 million, with options for up to six additional. The power module, an unregulated system operating from a deployable solar array, would provide 1280 w at the beginning of its operating life.

General Electric Space Div., Valley Forge, Pa., would negotiate to provide the attitude-control subsystem at a proposed value of $3.5 million, with options for up to six additional subsystems. The ACS would include reaction-control devices as well as a high-performance gyro inertial-reference unit, a pair of fixed electronic-scanning star trackers, a three-axis magnetometer set, and sun sensors.

The SMM spacecraft, a 381-cm-long cylinder weighing about 1740 kg, would be powered with solar panels and fit atop a modular spacecraft base; it would be capable of launch on a Delta and retrieval by the Space Shuttle. (NASA Releases 76-203, 76-204, 76-205)

- MSFC announced the successful launch on a Black Brant VC research rocket at White Sands Missile Range, N.M., of the first containerless experiments in materials processing ever attempted on a spacecraft. NASA's third space-processing applications rocket, SPAR III, carried a five-experiment payload, two of these using acoustic and electromagnetic suspension devices, or levitators. Materials suspended in a levitator could be melted and resolidified without touching a container, resulting in a degree of purity never before achieved in high-temperature processing, as the melt in a conventional container would be contaminated to some extent by the container itself. The rocket flight provided about 5 min of near-weightlessness for performance of the experiments, which were recovered by parachute for experimenter analysis. Experimenters were Charles Schafer, MSFC; Jerry Wouch, GE; Dr. John Papazian, Grumman Aerospace Corp.; Dr. David Lind, Rockwell Intl. Science Center; and Dr. Donald R. Uhlmann, MIT. Conclusions as to the success of containerless processing would await analysis of telemetry, inflight film, and experiment results. (MSFC Releases 76-207, 76-212)

- The USSR Ministry of Higher and Specialized Secondary Education had announced plans to build an Institute of Nuclear Energetics at Obninsk, in central Russia, site of the "world's first industrial atomic power station," the Tass news agency reported. First Soviet educational institution of this type, the new center would train engineers in design and use of nuclear power stations. In accordance with a new program of constructing nuclear power stations, Tass said the present 6-million-kw capacity would be supplemented by an additional 13 to 15 million kw; new stations were being built at Kursk, Smolensk, Kalinin in central Russia, and in the southern Ukraine. The Obninsk center would be the 860th higher education institution in the USSR, Tass noted. (FBIS, Tass in English, 14 Dec 76)

- Dr. Donald H. Menzel, one of the world's leading authorities on the sun and its corona, died at Mass. General Hospital after a long illness. During his lifetime, he had viewed 15 total solar eclipses, leading expeditions to distant places like the Sahara (1973) and Siberia (1936), where the Soviets provided a parlor car of the Trans Siberian Railroad for his equipment and camping gear. Dr. Menzel had taught for nearly 40 yr at

Harvard; Dr. Fred L. Whipple of Harvard, a colleague and former student of Dr. Menzel, noted that many of the best-known astronomers in the U.S. had also been taught by Dr. Menzel. The Minor Planet Center of the Intl. Astronomical Union recently named an asteroid Menzel in recognition of his contribution to astrophysics. In 1938, Dr. Menzel developed the first coronagraph in the U.S. to permit study of the sun's corona without an eclipse; in 1933, he had collaborated with J.C. Boyce in establishing the presence of oxygen in the solar corona, and in 1941 he had worked with Winfield W. Salisbury on the initial calculations that enabled the first radio contact with the moon in 1946. He had received a Ph.D. in astrophysics at Princeton in 1924, writing science-fiction stories to help pay his expenses, and had taught at the Univ. of Iowa and Ohio State Univ. before becoming professor of astrophysics at Harvard in 1939. He was director of the Harvard College Observatory from 1954 to 1966, when he was named research scientist at the Smithsonian Astrophysical Laboratory in Cambridge. He had retired from the university in 1971, but had continued to do research there. (*NYT*, 16 Dec 76, 50)

15 December: The "hotline" between Washington, D.C. and Moscow had changed into a Direct Communications Link using two independent sets of orbiting satellites, and eliminating any actual wire between the two capitals, the *Christian Science Monitor* reported. A spokesman for the Army Communications Command at Ft. Huachuca, Ariz., where the U.S. portion of system was designed and built, said it could be put into use whenever diplomats from both countries made the final arrangements. The $15-million system would be located at Ft. Detrick, Md., near Washington, the spokesman said. (*CSM*, 15 Dec 76)

• Ithaco Inc. of Ithaca, N.Y., a space-industry subcontractor specializing in low-orbital unmanned-satellite electronics, had discovered a lucrative and largely untapped agricultural market by turning its expertise on animal electronics: specifically, a new device first marketed 19 mo ago to help farmers determine whether their pigs were pregnant and to preview the quality of porkchops the hogs would produce. The device would let farmers determine pig-pregnancy status 50 to 60 days earlier than previously possible, meaning substantial savings for breeders, company officials said, adding that the market would not be attractive to a large company. Overseas farmers accounted for up to 40% of "Scanoprobe" sales, which totaled $3.8 million for the 1976 fiscal year, best in the firm's 14-yr history. Ithaco would not abandon its space contracts, the company said, but employees who worked on intricate satellite parts could be switched over to the "animal electronics" contracts. (*NYT*, 15 Dec 76, A 16)

16 December: NASA announced that it would modify equipment and facilities used in the Apollo, Skylab, and Apollo-Soyuz programs for use in future space programs, and that it would dispose of remaining Saturn-Apollo hardware that had no future application. Disposal would eliminate storage costs, free storage space, and make facilities available for ongoing programs, the agency said. The action marked the transition of U.S.

manned space activity from use of expendable vehicles to use of the reusable Space Shuttle, scheduled to undertake its first mission within 3 yr. Although NASA, at the request of Congress, had kept its remaining Saturn-Apollo flight hardware since April 1975 in minimum-cost storage, to be able to restore it to flight condition if needed, the backup Skylab workshop and Apollo-Soyuz backup docking module had been transferred to the National Air and Space Museum; twenty-two H-1 rocket engines from Saturn stages had been transferred for use in Thor-Delta vehicles. Much of the remaining Saturn-Apollo hardware would probably be turned over to the Smithsonian Institution, and other equipment would be screened for use in the Shuttle or other programs. (NASA Release 76-206)

- NASA announced award to Owen Enterprises, Inc., of Wilmington, Calif., of an exclusive patent license for production of a solar-energy concentrator invented by JPL's Dr. Katsunori Shimada that focuses the sun's rays from almost any angle without the need for a tracking mechanism. The concentrator—a special arrangement of multifaceted Fresnel lenses—magnifies solar energy tenfold, heating a fluid in channels beneath the lenses, which is released through a thermostat when the proper temperature is attained. A series of the devices could be set up, depending on the specific energy requirements of a structure. Suitable for residential, commercial, or industrial applications, the concentrator had proved much more efficient than other solar-collector units on the market. Owen Enterprises, Inc., the licensee, is an American Indian-owned firm that planned to assemble the device at facilities on the Rincon Indian Reservation near Escondido, Calif. The firm would invest $200 000 to develop the concentrator for the market, and pay 1% royalty to the U.S. Treasury on its gross sales. (NASA Release 76-209)
- At the request of the Soviet Union, U.S. satellites took pictures of the predicted bumper crop of grain in the USSR and turned the pictures over to that country to confirm the predictions, said presidential science adviser Dr. H. Guyford Stever at a White House briefing. The Russians had announced 5 Nov. that more than 220 million metric tons of grain had been harvested; U.S. Dept. of Agriculture officials said the final figures would probably exceed 1973's record Soviet harvest of 222.5 million tons. (*NYT*, 17 Dec 76, A-4)

16–22 December: The USSR announced its intention of launching carrier rockets into two areas of the northern Pacific between 20 and 30 Dec., and requested the governments of other countries using sea and air routes in the vicinity not to permit their ships and planes to enter those areas during the launching period. At the same time the Soviets declared other areas free for navigation because the launchings in those areas had been completed. The Tass broadcast gave the locations of the danger spots as having center coordinates of 46°N, 164°E, and 32°27′N, 170°10′E. The nature of the launches was not specified. On 22 Dec. Tass announced it had been authorized to state the locations were "free for sea and air navigation from December 22, 1976." (FBIS, Tass in English, 16 Dec 76, 22 Dec 76)

17 December: Dr. Michael Duke, curator of lunar samples at NASA's Johnson Space Center, returned from Moscow with two half-gram samples of moon soil brought to earth in Aug. 1976 by the USSR's *Luna 24* spacecraft. The samples were part of a 2-meter-long core of lunar material obtained by a hollow drill from the surface of Mare Crisium, a region of the moon from which the U.S. had not had any samples. This was the third time U.S. scientists had obtained a Soviet sample in exchange for material collected by Apollo astronauts from other places on the moon. Other members of the U.S. scientific delegation to Moscow this week were Dr. Charles Simonds, Lunar Science Institute, Houston, Tex., and Prof. Gerald J. Wasserburg of the Calif. Institute of Technology. The small size of the sample would not detract from its usefulness, according to Dr. Bevan French, chief of NASA's program of research on extraterrestrial materials; methods routinely used for extracting information from samples of meteorite or deep-sea basalt could produce hundreds of chemical analyses from a single tiny crystal or determine the age of a fragment smaller than an aspirin tablet. (JSC Release 76-75)

- NASA announced selection of United Space Boosters, Inc., Sunnyvale, Calif., as assembly contractor for the Space Shuttle solid-fuel booster rocket. USBI, a wholly owned subsidiary of United Technologies Corp., received a cost-plus-award fee contract for $122 million, to cover six design, development, and test and evaluation flights up to March 1980, with options for 21 operational flights extending into 1982. The contract would cover all booster assembly activities at MSFC and KSC; MSFC would supervise assembly, checkout, and refurbishment of the boosters, and KSC would supervise final assembly, checkout, launch operations, and postlaunch disassembly of the boosters. First of six orbital test flights had been scheduled for the second quarter of 1979, with operational flights to begin in 1980. At launch, the three main engines and two boosters of the Space Shuttle would operate together; upon burnout at about 43 km altitude, the boosters would separate and descend by parachute to the ocean for retrieval and refurbishment. The boosters were designed for use 20 times. (NASA Release 76-212; MSFC Release 76-215; *WSJ*, 21 Dec 76, 11)

- The Concorde supersonic airliner, making regularly scheduled flights between London, Paris, and the Dulles airport outside Washington, D.C. since 24 May, had been lending its distinctive sonic waves to atmospheric research by Columbia Univ. monitors, the *New York Times* reported. Dr. William L. Donn, senior research associate at Columbia's Lamont-Doherty Geological Observatory, and his colleagues Nambath Balachandran and David Rind had derived surprising data on daily variations in the supposed prevailing winds in the upper atmosphere, and were anticipating new data on dispersal of atmospheric pollutants and the content of the rarefied reaches of earth's atmosphere. The sound waves, inaudible to the human ear but detectable by sensitive microphones registering changes in air pressure, contained atmospheric information up to as far as 120-km altitude, through reflection from the thermosphere. Precise knowledge of the Concorde's elevation and location along its flight path had aided in interpreting data on atmospheric temperatures and wind conditions depending on the exact strength and path of the sound waves, Dr. Donn said. (*NYT,* 17 Dec 76, A-18)

- A British Aircraft Corp. team had successfully launched the second of a series of Skylark 12 high-altitude research rockets from the Andoya Range in arctic Norway to investigate auroral activity for the U.K. Science Research Council and other groups investigating the intense phenomena above the Arctic Circle, the London Press Assn. reported in a broadcast. The rocket reached an altitude of more than 680 km, sending scientific data to ground receivers for 14 min of the flight. First launch of the new three-stage Skylark, capable of carrying payloads to three times the altitude of earlier versions, took place 21 Nov. (FBIS, London Press Assn., 17 Dec 76)

18 December: Astronaut-pilots in training for the Space Shuttle program had used a simulated shuttle—a Grumman Gulfstream 2 twin-engine jet—in approach-and-landing maneuvers over the southern New Mexico desert at White Sands Missile Range, the *New York Times* reported. Dr. Christopher C. Kraft, JSC director, who recently watched the training at White Sands, said that the latter installation might eventually become a space port for launch and recovery of the shuttle. The pilots were to test an actual space shuttle—the Enterprise, unveiled this fall by Rockwell Intl.—some time next year; the first vertical launch of the shuttle had been scheduled for 1977 from KSC at Cape Canaveral, Fla. (*NYT,* 19 Dec 76, 41)

19 December: Three satellites making simultaneous investigations of earth's magnetic field had confirmed that long low-frequency waves spreading the disruptive effects of magnetic storms on earth had been generated far out in space by energetic particles from the sun, NOAA announced. Speaking at the annual meeting of the American Geophysical Union in San Francisco, Calif., earlier in Dec., Dr. Joseph N. Barfield of NOAA, R.L. McPherron of UCLA, and W.J. Hughes of London's Imperial College reported the waves might have practical uses in exploring for oil and minerals, and in aiding undersea communication with submarines, which now had to surface to make contact. Environmental satellites *Sms 1* and *Sms 2*, with research satellite *Ats 6*, in geosynchronous orbit on a line along the earth's equator, had carried magnetometers used to detect the low-frequency waves passing each satellite in a predicted order, the first use of multiple spacecraft to make such measurements simultaneously. Confirmation of the theory also identified the point of origin as the magnetosphere and the existence of a "resonance region" that amplified the wave frequency. The region was described as about 1200 km thick. (NOAA Release 76-281)

20 December: MSFC announced opening of a new solar heating and cooling test facility, part of a nationwide demonstration program conducted by ERDA, to evaluate solar-powered systems and components and determine their efficiency and suitability for residential and commercial applications. Systems tested would mainly be those built by commercial manufacturers, but would include experimental systems provided by ERDA and

NASA. The facility occupied an area formerly used for testing in the Saturn program, and data-acquisition equipment used in that program would be used in the new facility to collect and store information on the effectiveness of the units under test. The site could accommodate 511 m^2 of solar collectors at one time. The facility could test air and liquid units for airconditioning (up to 10 tons) or heating (systems up to 500 000 Btu); it included a test bed for evaluating efficiency of tanks to store heat collected by the systems for use when there was no sunlight. Systems meeting the criteria for operation, maintenance, and performance would be installed in buildings throughout the country for trial use. (NASA Release 76-211)

21 December: The Dept. of Defense had reported McDonnell Douglas as the largest U.S. defense contractor for the year ended 30 June, replacing Lockheed Aircraft Co., the *Wall Street Journal* announced. The leader, which had been fourth on the list in FY 1975, had obtained contracts valued at $2.46 billion (5.9% of all defense contracts of more than $10 000 awarded during FY 1976). Lockheed, which had been the leader for 6 of the previous 7 yr, received contracts in FY 1976 valued at $1.51 billion (3.6% of the total). Third was Northrop Corp. with $1.48 billion (3.5%); fourth was General Electric Co., $1.35 billion (3.2%). Others of the ten largest were United Technologies Corp., $1.23 billion (2.9%); Boeing Co., $1.18 billion (2.8%); General Dynamics Corp., $1.07 billion (2.6%); Grumman Corp., $982 million (2.3%); Litton Industries, Inc., $978 million (2.3%); and Rockwell Intl. Corp., $966 million (2.3%). (*WSJ*, 21 Dec 76, 11; DOD Release 585-76)

- The Natl. Science Foundation announced that U.S. and Canadian astronomers had succeeded in using telescopes more than 800 km apart, linked by satellite, to simulate a telescope nearly as large as the earth. A radio telescope at the Natl. Radio Astronomy Observatory at Green Bank, W. Va., had been linked through a joint U.S.-Canadian satellite to a radio telescope at the Algonquin Radio Observatory in Ontario, allowing the observers to review the results while the observations were in process. The technique would permit astronomers to measure the size and shape of a distant galaxy or quasar with a precision better than 1/1000 arcsec, equivalent to measuring the size and shape of a penny more than 4000 km distant. The satellite would transmit information at a rate equivalent to 10 000 simultaneous telephone calls; the large data rate would increase the sensitivity of measurements, allowing the study of very faint objects in space. Besides the principal investigator, Dr. George W. Swenson, Jr. of the Univ. of Ill., the team included S.H. Knowles and W.B. Walton of NRL; N.W. Broten and D.N. Fort of Canada's Natl. Research Council; K.I. Kellermann and Benno Rayhrer of NRAO; and J.L. Yen of the Univ. of Toronto. (NSF Release PR76-106)

- ESA announced that its council meeting in Paris 16 and 17 Dec. had approved a new program, Earthnet, in which ESA would centralize and coordinate European activities in reception, processing, distribution, and archiving of earth-resources satellite data. Earthnet would give European users access to data from NASA programs such as Landsat, Seasat,

Nimbus-G, and the Heat Capacity Mapping Mission, to provide a basis for defining European requirements for future remote-sensing satellite programs. Earthnet would be integrated into existing data-reception and processing facilities at Fucino, Italy, and would use ESA's computerized data center at Frascati, Italy, with an information-retrieval network RECON to disseminate information on the data available. (ESA release 21 Dec 76)

22 December: Rockwell International—prime contractor for NASA's Space Shuttle—announced that it had purchased more than $18 million of goods and services from minority businesses throughout the U.S. in 1976, a 58% increase over 1975. Rockwell had been cited by the Black Businessmen's Assn. of Los Angeles for its participation in "Black Economic Development," and by the Natl. Assn. of Black Manufacturers for its contributions to industry development. Kenneth B. Gay, Rockwell vice president for purchasing, said that "qualified minority businesses in the more technical areas" had been hard to find in the past, but there were "many highly skilled, competent small companies competing aggressively for our business" at present. The more than 500 minority firms supplying Rockwell received about $5 million for machining and specialty fabrication; more than $2 million for metal structures; another $2 million each for raw materials and supplies and for technical services; and more than $1 million each for electrical equipment and for maintenance and repair services. (Rockwell Release R-46)

- An unmanned Soviet spacecraft that returned to the earth 17 Dec. in Soviet Asia after 18 days in orbit might have been the test vehicle of a new Soyuz designed to carry three cosmonauts into space, instead of the two that have flown on the 13 Soyuz missions during the past 5 yr. Thomas O'Toole, reporting in the *Washington Post*, pointed out that the *Cosmos 869* had flown the same pattern followed by all Soviet manned space flights, and noted that the USSR had long followed the practice of disguising unmanned test flights of Soyuz spacecraft by using the generalized Cosmos designation. At least four unmanned Soyuz vehicles had flown under the name of Cosmos during the past 4 yr, the article said. The USSR had not attempted a three-man flight since 1971, when the three crewmen of *Soyuz 11* suffocated on their return to earth; the crew could have survived had they worn space suits, but the vehicle being used had no room for three men in space suits. Officials said then that the spacecraft would be modified to accommodate three-member crews in space suits, and Western observers predicted that the Inter-Soyuz flights scheduled to begin in 1978 with Russians and non-Russians together would use three cosmonauts again. (*W Post*, 22 Dec 76, A-2)

23 December: Aircraft overflights by Wallops Flight Center, coordinated by LaRC at the request of NOAA, were helping the U.S. Coast Guard track an oil slick from a tanker grounded off Nantucket Island, NASA announced. Flights on 19 and 22 Dec. had obtained data from the ocean around the tanker to use in calculating the slick's trajectory; another flight had been scheduled for 24 Dec. Imagery from *Landsat 2* was providing synoptic views of the spill area. NOAA would use the information to set up analytical models for determining potential environmental damage associated with offshore oil and gas production on the continental shelf. Brookhaven Natl. Laboratory, ERDA, had also requested the data for use in environmental studies. (WFC Release 76-17)

- The U.S. Army awarded a contract to Sikorsky Aircraft, Stratford, Conn., for a new generation of combat assault helicopters—single-rotor craft driven by jet turbine engines, capable of carrying a three-man crew plus 11 heavily armed soldiers—in a decision that would bring a total of $3.4 billion including development and other costs to the depressed area economy over the next 10 yr, the *New York Times* reported. State officials said the contract could produce up to 17 000 new jobs and called it "the perfect Christmas present." Sikorsky officials said the contract would bring it $2 billion directly and $1.4 billion more to subcontractors in 35 states. Also competing for the contract was the Vertol division of Boeing Corp., in Morton, Pa. (*NYT*, 24 Dec 76, A-1)

24 December: The first group of future cosmonauts from socialist countries had been welcomed at the USSR training center named for Yuri Gagarin, Maj. Gen. Georgy Beregovoy, director of the center, told a correspondent for the Tass news agency. The joint flights with Soviet cosmonauts would include representatives from Czechoslovakia, Poland, and the German Democratic Republic (the first three groups to arrive), Bulgaria, Hungary, Cuba, Mongolia, and Romania, and the joint flights were scheduled for 1978 to 1983. Beregovoy said the socialist countries had been cooperating for the past decade on the Intercosmos program, launching satellites and meteorological rockets and more than 20 experiments in various branches of science. He also mentioned the Apollo-Soyuz flight as a successful example of international cooperation in space. (FBIS, Tass in English, 24 Dec 76)

30 December: The Dept. of Transportation announced that the Federal Aviation Administration had awarded a $125 363 contract to York Univ., Ontario, Canada, for a program to measure ozone and the oxides of nitrogen present in all aircraft-engine emissions in the stratosphere. The program, to begin in the spring of 1977 in cooperation with NOAA's Aeronomy Laboratory in Boulder, Colo., would use detection instruments lifted by balloons to an altitude of more than 35 km for the first simultaneous measurement of three types of nitrogen oxides and ozone; determining the high-altitude level of nitric oxide, nitrogen dioxide, and dinitrogen pentoxide would be an important step in finding the role of these oxides in maintaining the ozone balance. The contract would support the High Altitude Pollution Program, started in 1975 as part of DOT's Climatic Impact Assessment Program; York Univ., which participated in CIAP, had in 1972 provided the first data on the stratospheric density of nitric oxide under the direction of Prof. Harold Schiff, a pioneer in the field. (DOT Release 76-122)

31 December: Clocks around the world would be set back one second this New Year's Eve, because the earth's spin was that much longer in 1976 than it was last year. The backward move would be the sixth in the last 5 yr because of a slowing in the earth's spin, the *Washington Post* explained;

the changes were made twice in 1972 and once in each of the last 4 yr, each time on New Year's Eve. The standard U.S. timepiece (at the Natl. Bureau of Standards Laboratory at Boulder, Colo.) would be reset at midnight GMT (7 pm in Washington, D.C.), and similar atomic clocks would be reset at the same moment in more than 80 countries including the Soviet Union and the Peoples Republic of China. The earth's spin had been slowing for at least 250 yr, when it was first noticed, except for a 10-yr period before 1900 when the rotation speeded up by 25 sec. In 1878, Simon Newcomb, then chief astronomer at the Naval Observatory, first systematically observed a fluctuation in the rate of rotation, which still is verified by 80 observatories around the world who send their findings to the Intl. Time Bureau in Paris, established in 1972 (the year the clocks lost 2 sec). Scientists attributed the slowdown to the moon's pull on the tides, braking the motion of the earth's mantle. (*W Post*, 31 Dec 76, A2)

During December: The Air Force Systems Command announced that former astronaut Maj. Gen. Michael Collins had been appointed mobilization assistant to the AFSC commander. The position was the top Air Force Reserve post in AFSC. Collins, who was confirmed by the U.S. Senate as a reserve major general 10 Mar. 1976, would continue to serve as director of the National Air and Space Museum. (*AFSC Newsreview*, Dec 76, 10)

- Images from Landsat had reinforced the theory that mankind had created his own deserts through overgrazing of livestock, according to Joseph Otterman, a Landsat data user at Tel Aviv University. Satellite and other data had shown that overgrazed land on one side of the fence along the Israel-Egypt armistice line consistently measured several degrees cooler than the other side; the denuded land, reflecting more sunshine, retained less heat than vegetated land. The warmer land heated the overlying air, causing it to rise, form clouds, and rain; lack of this "heat mountain" effect over the denuded land encouraged formation of deserts. (*CSM*, 15 Dec 76)
- Further press comment on Soviet "hunter-killer" satellites appeared in the *Chicago Tribune, Washington Post*, and *New York Times*, prompted by reports of the piggyback launch of *Cosmos 881* and *882* on 15 Dec. Columnist Nicholas von Hoffman noted Pentagon denials that the USSR had used laser beams to destroy an American satellite and claims that the Soviets had been practicing knockouts only on their own targets; but the physical safety of the U.S. was "more in jeopardy now than ever," he concluded, adding that "of course, we'll put our money into aggressive basic research efforts to stop the Russian hunter-killer satellites." Thomas O'Toole in the *Washington Post* described the launch and recovery of two Soviet satellites said by intelligence experts to be larger, newer versions of the hunter-killer craft, flown into orbit and back to base after circling the earth once, a move the experts said was to prevent China's tracking the craft by radar. O'Toole said the Soviets were believed to have carried out 16 experiments in the past 9 yr in which unmanned spacecraft followed target satellites into space and blew them-

selves up to remove the targets from orbit; these tests stopped in 1971, but resumed in Feb. 1976 after the launches of *China 5* and *China 6*. The two Soviet spacecraft followed an unusual flight path, similar to those taken by manned Soyuz satellites, and were also unusual in returning to earth after so brief a flight. The *New York Times* reported the launch on 27 Dec. of *Cosmos 886*, which "apparently attempted to intercept" *Cosmos 880*, launched 9 Dec. "The hunter satellite never came closer than one mile to its target . . . and finally disintegrated on its third orbit," the *NYT* noted, adding that *Cosmos 880* was still in orbit. The *NYT* story went on to describe the *Cosmos 881* and *882* launch, "at first thought to have been such a test," as actually connected with the manned space program. No effort had been made to interfere with U.S. space vehicles, and the 27 Dec. launch—the USSR's fourth test of a satellite interceptor this year—had been classified as a failure by intelligence sources, the *NYT* said. (*C Trib*, 4 Dec 76, 1-7; *W Post*, 18 Dec 76, A-1; *NYT*, 31 Dec 76, A6)

- "One of the environmental disasters of the century" was the appearance of water hyacinth in the Sudan's Nile river basin, reported an international conference of scientists representing the U.S., the Democratic Republic of the Sudan, Egypt, Ethiopia, the Federal Republic of Germany, Indonesia, and Mozambique, after a meeting sponsored by the Natl. Academy of Science's Board on Science and Technology for International Development, and by the Sudan Agricultural Research Council. Besides choking watercourses and interfering with fishing and navigation, water hyacinth flourishing in warm river waters around the world had proved to harbor carriers of bilharzia and malaria. Measures suggested for control included herbicides, biological intervention, exchange of information and trade restrictions, and stimulation of economic incentive to harvest the plants. The report cited a NASA discovery that water hyacinth was a possible source of biogas production, and recommended harvesting to exploit this resource as well as using the plant for water-buffalo fodder. (NAS *News Report*, Dec 76, 2)

Appendix A

SATELLITES, SPACE PROBES, AND MANNED SPACE FLIGHTS, 1976

World space activity increased only slightly in 1976; total launches, 128, increased from 125 in 1975. Of this total, the U.S. had 26 plus one space probe, described in the introduction to Appendix B. The USSR had 99 launches with 121 payloads, 79 of these in the Cosmos series; the 101 Cosmos payloads included such specialized spacecraft as 8 navigation satellites (*Cosmos 789, 800, 823, 842, 846, 864, 883,* and *887*), and the ocean-reconnaissance satellites *Cosmos 860* and *861* working as a team. The Soviets also launched 7 spacecraft in the Molniya comsat series, 3 in the Meteor series, 2 Intercosmos spacecraft carrying experiments from other countries, a moon probe (*Luna 24*) that returned a surface sample to earth, the magnetospheric investigator *Prognoz 5*, 2 additional geostationary domestic color–TV satellites (*Statsionar 1B* and *1C*), and 3 manned Soyuz spacecraft, as well as another orbiting space station (*Salyut 5*).

The number and variety of launches by other countries decreased sharply to a total of 3. In Feb., Japan launched *Ume*, an ionosphere-sounding satellite, using its N rocket for the second time. The Peoples Republic of China in Aug. launched its sixth satellite, still in orbit; in Dec., it launched *China 7*, which reentered in Jan. 77.

Sources of these data include the United Nations Public Registry of Space Flights; the *Satellite Situation Report* compiled by Goddard Space Flight Center's Operations Control Center; and press releases of NASA, Department of Defense, National Oceanic and Atmospheric Administration, and other government agencies, as well as the Communications Satellite Corporation. Soviet data derive from statements in the Soviet press, translations from the Tass news agency, international news-service reports, and announcements and briefings by Soviet officials. Data on satellites of other nations also come from the announcements of their respective governments and international news services.

SATELLITES, SPACE PROBES, AND MANNED SPACE FLIGHTS, 1976

Launch Date	Spacecraft, Country, Int'l Designation, Vehicle, Launch Site	Payload Data	Apogee (km)	Perigee (km)	Period (min)	Inclination (degrees)	Remarks
6 Jan.	Cosmos 787 USSR 1976–1A C–1 Plesetsk	Total weight: 875 kg. ? Objective: "Continuation of outer space investigations." Description: Unavailable.	545	517	95.3	74.0	Still in orbit; probable electronic ferret.
7 Jan.	Cosmos 788 USSR 1976–2A A–2 Plesetsk	Total weight: 6300 kg. ? Objective: "Continuation of outer space investigations." Description: Unavailable.	383	120	89.5	62.9	Reentered 20 Jan.
15 Jan.	Helios 2 FRG 1976–3A Titan IIIE–Centaur ETR	Total weight: 370 kg. Objective: Investigation of solar processes and solar-terrestrial relationships. Description: Spool-shaped body with 2 conical solar arrays attached to ends of 16-sided cylinder 2.1 m long; deployable booms on central body carrying radio and magnetometer experiments; 13 experiments on board.	Heliocentric orbit.				Launched by NASA for West Germany; aimed at 43 million km from sun, closer than any manmade object; all 10 active experiments functioning.
17 Jan.	Cts 1 Canada 1976–4A Thor-Delta ETR	Total weight: 500 kg. Objective: Advancement in technology of spacecraft-mounted and related ground-based components and systems applicable to satellites with high radiated RF power. Description: Rectangular box 188 cm long and 183 cm in outside diameter; 2 curved sides carrying extendable solar arrays; forward platform carrying high-frequency antennas and earth sensors, after platform carrying reaction-control system; earth-oriented satellite receiving signals at 14 Ghz and converting them to 12 Ghz for retransmission.	36 022	33 814	1392.0	0.7	Communications technology satellite, most powerful to date, launched by NASA for Canadian Department of Communications; still in orbit.

Date	Satellite	Description				Remarks
20 Jan.	Cosmos 789 USSR 1976-5A C-1 Plesetsk	Total weight: 680 kg. ? Objective: "Continuation of outer space investigations." Description: Unavailable.	1007	981	105.0	83.0	Probable navigational satellite; still in orbit.
22 Jan.	Molniya 1-AH USSR 1976-6A A-2-e Baikonur-Tyuratam	Total weight: 1750 kg. ? Objective: Operation of long-range telephone and telegraph communications system; transmission of TV programs to stations in Orbita network. Description: Unavailable.	38 927	466	698.4	62.9	Thirty-second in series; still in orbit.
	Cosmos 790 USSR 1976-7A C-1 Plesetsk	Total weight: 875 kg. ? Objective: "Continuation of outer space investigations." Description: Unavailable.	548	509	95.2	74.0	Still in orbit: probable electronic ferret.
28 Jan.	Cosmos 791 USSR 1976-8A C-1 Plesetsk	Total weight: 50 kg. ? Objective: "Continuation of outer space investigations." Description: Unavailable.	1488	1401	114.8	74.0	Probable 8 comsats on single carrier; all still in orbit.
and	Cosmos 792 USSR 1976-8B C-1 Plesetsk	Total weight: 50 kg. ? Objective: "Continuation of outer space investigations." Description: Unavailable.	1492	1435	118.2	74.0	
and	Cosmos 793 USSR 1976-8C C-1 Plesetsk	Total weight: 50 kg. ? Objective: "Continuation of outer space investigations." Description: Unavailable.	1492	1417	115.0	74.0	

Appendix A

Launch Date	Spacecraft, Country, Int'l Designation, Vehicle, Launch Site	Payload Data	Apogee (km)	Perigee (km)	Period (min)	Inclination (degrees)	Remarks
	Cosmos 794 USSR 1976-8D C-1 Plesetsk	Total weight: 50 kg. ? Objective: "Continuation of outer space investigations." Description: Unavailable.	1496	1450	115.4	74.0	
and							
	Cosmos 795 USSR 1976-8E C-1 Plesetsk	Total weight: 50 kg. ? Objective: "Continuation of outer space investigations." Description: Unavailable.	1501	1465	115.6	74.0	
and							
	Cosmos 796 USSR 1976-8F C-1 Plesetsk	Total weight: 50 kg. ? Objective: "Continuation of outer space investigations." Description: Unavailable.	1516	1472	115.9	74.0	
and							
	Cosmos 797 USSR 1976-8G C-1 Plesetsk	Total weight: 50 kg. ? Objective: "Continuation of outer space investigations." Description: Unavailable.	1531	1478	116.1	74.0	
	Cosmos 798 USSR 1976-8H C-1 Plesetsk	Total weight: 50 kg. ? Objective: "Continuation of outer space investigations." Description: Unavailable.	1555	1479	116.4	74.0	

ASTRONAUTICS AND AERONAUTICS, 1976 — Appendix A

Date	Satellite / Designation	Details	Col4	Col5	Col6	Remarks
29 Jan.	*Cosmos 799* USSR 1976-9A A-2 Baikonur–Tyuratam	Total weight: 5700 kg. ? Objective: "Continuation of outer space investigations." Description: Unavailable.	303	202	89.6	71.4	Reentered 10 Feb.; probable military photoreconnaissance.
30 Jan.	*Intelsat IVA F2* U.S. 1976-10A Atlas-Centaur ETR	Total weight: 825.5 kg. (in orbit). Objective: To provide 6250 2-way voice and 2 TV channels simultaneously, or a combination of telephone, TV, and other forms of communications traffic. Description: Solar-cell-covered rotating cylinder 2.4m in dia. containing power, orientation, and despin subsystems; earth-oriented platform on cylinder carrying 20 communications repeaters (transponders) plus reflectors and telemetry; 7m long overall.	35 901	35 216	1424.5	0.1	Second in series; launched by NASA for ComSatCorp.; stationed over Atlantic at 29.5°W; still in orbit.
3 Feb.	*Cosmos 800* USSR 1976-11A C-1 Plesetsk	Total weight: 680 kg. ? Objective: "Continuation of outer space investigations." Description: Unavailable.	1013	982	105.1	83.0	Probable navigational satellite; still in orbit.
5 Feb.	*Cosmos 801* USSR 1976-12A B-1 Plesetsk	Total weight: 300 kg. ? Objective: "Continuation of outer space investigations." Description: Unavailable.	723	290	94.8	70.6	Reentered 5 Jan. 78; probable radar-calibration satellite.
11 Feb.	*Cosmos 802* USSR 1976-13A A-2 Baikonur–Tyuratam	Total weight: 6300 kg. ? Objective: "Continuation of outer space investigations." Description: Unavailable.	333	172	89.6	65.0	Reentered 25 Feb.; probable military photoreconnaissance
12 Feb.	*Cosmos 803* USSR 1976-14A C-1 Plesetsk	Total weight: 750 kg. ? Objective: "Continuation of outer space investigations." Description: Unavailable.	617	553	96.4	65.9	Still in orbit.

Launch Date	Spacecraft, Country, Int'l Designation, Vehicle, Launch Site	Payload Data	Apogee (km)	Perigee (km)	Period (min)	Inclination (degrees)	Remarks
16 Feb.	Cosmos 804 USSR 1976–15A F-1-m Baikonur-Tyuratam	Total weight: Unavailable. Objective: "Continuation of outer space investigations." Description: Unavailable.	614	555	96.4	65.9	Reentered 16 Feb.
19 Feb.	DOD spacecraft U.S. 1976–16A Thor-Burner 2 WTR	Total weight: Unavailable. Objective: "Development of spaceflight techniques and technology." Description: Unavailable.	360	81	88.9	98.9	Reentered 19 Feb., the day of launch; considered unsuccessful by DOD (although it fulfilled Cospar rules for successful orbit).
	Marisat 1 U.S. 1976–17A Delta ETR	Total weight: 327 kg. (in orbit). Objective: To provide 24-hour real-time communications between ships and shore. Description: Spin-stabilized cylinder 380 cm high, 215 cm in dia., covered with solar cells; pole from body carrying 3 UHF helical antennas, quad L-band array.	35 867	35 703	1436.1	02.4	First commercial maritime satellite; launched by NASA for ComSatCorp.; stationed over Atlantic at 15°W; U.S. Navy to be prime UHF user for first 2 years.
20 Feb.	Cosmos 805 USSR 1976–18A A-2 Plesetsk	Total weight: Unavailable. Objective: "Continuation of outer space investigations." Description: Unavailable.	339	171	89.6	67.1	Reentered 11 Mar.
29 Feb.	Ume (ISS-1) Japan 1976–19A N Tanegashima	Total weight: 85 kg. (in orbit). Objective: Ionosphere observations. Description: Cylinder, 82 cm long and 94 cm high.	1010	994	105.2	69.7	Ionosphere sounding satellite; 2nd launch by new N rocket; still in orbit.

Date	Spacecraft	Details				Notes
10 Mar.	Cosmos 806 USSR 1976-20A A-2 Baikonur-Tyuratam	Total weight: Unavailable. Objective: "Continuation of outer space investigations." Description: Unavailable.	334	177	89.6	71.4	Reentered 23 Mar.
11 Mar.	Molniya 1-AJ USSR 1976-21A A-2-e Plesetsk	Total weight: Unavailable. Objective: Operation of long-range telephone and telegraph communications system; transmission of TV programs to stations in Orbita network. Description: Unavailable.	40 676	495	734.4	62.8	Thirty-third in series; still in orbit.
12 Mar.	Cosmos 807 USSR 1976-22A C-1 Plesetsk	Total weight: Unavailable. Objective: Continuation of outer space investigations." Description: Unavailable.	1970	395	109.1	83.0	Still in orbit.
15 Mar.	DOD spacecraft (LES 8) U.S. 1976-23A Titan IIIC ETR	Total weight: 454 kg. Objective: Evaluation of advanced satellite communications techniques. Description: Unavailable.	35 822	35 748	1436.0	25.3	Four payloads on one carrier; all still in orbit.
and	DOD spacecraft (LES 9) U.S. 1976-23B Titan IIIC ETR	Total weight: 454 kg. Objective: Evaluation of advanced satellite communications techniques. Description: Unavailable.	35 806	35 768	1436.1	25.3	
and	DOD spacecraft (Solrad 11a) U.S. 1976-23C Titan IIIC ETR	Total weight: 181 kg. Objective: Measurement of solar x-rays, ultraviolet light, and particle emissions. Description: Unavailable	118 985	18 109	7323.6	25.3	
and							

Launch Date	Spacecraft, Country, Int'l Designation, Vehicle, Launch Site	Payload Data	Apogee (km)	Perigee (km)	Period (min)	Inclination (degrees)	Remarks
	DOD spacecraft (Solrad IIb) U.S. 1976–23D Titan IIIC ETR	Total weight: 181 kg. Objective: Measurement of solar x-rays, ultraviolet light, and particle emissions. Description: Unavailable.	116 940	115 562	7122.6	25.4	
16 Mar.	Cosmos 808 USSR 1976–24A A-1 Plesetsk	Total weight: Unavailable. Objective: "Continuation of outer space investigations." Description: Unavailable.	632	601	97.0	81.2	Still in orbit.
18 Mar.	Cosmos 809 USSR 1976–25A A-2 Baikonur-Tyuratam	Total weight: Unavailable. Objective: "Continuation of outer space investigations." Description: Unavailable.	297	202	89.5	65.0	Reentered 30 Mar.
19 Mar.	Molniya 1-AK USSR 1976–26A A-2 Baikonur-Tyuratam	Total weight: Unavailable. Objective: Operation of long-range telephone and telegraph communications system; transmission of TV programs to stations in Orbita network. Description: Unavailable.	38 883	418	696.5	62.9	Thirty-fourth of series; still in orbit.
22 Mar.	DOD spacecraft U.S. 1976–27A Titan IIIB-Agena D WTR	Total weight: Unavailable. Objective: "Development of spaceflight techniques and technology." Description: Unavailable.	352	123	89.3	96.4	Reentered 18 May.
26 Mar.	Cosmos 810 USSR 1976–28A A-2 Plesetsk	Total weight: Unavailable. Objective: "Continuation of outer space investigations." Description: Unavailable.	310	168	89.3	62.8	Reentered 8 Apr.

Date	Spacecraft	Details	Col1	Col2	Col3	Col4	Remarks
26 Mar.	Satcom 2 U.S. 1976–29A Delta ETR	Total weight: 463 kg. (in orbit). Objective: Transmission of television, voice channels, and high-speed data to Hawaii, Alaska, and continental U.S. Description: Three-axis-stabilized box 1.2m by 1.2m, 1.6m high; 2 solar panels on short booms; body panels carrying hydrazine tanks, communications and housekeeping instrumentation.	35 882	36 642	1434.9	0.3	Second of three commercial comsats launched by NASA for RCA into stationary orbit at 135°W over equator; still in orbit.
31 Mar.	Cosmos 811 USSR 1976–30A A–2 Plesetsk	Total weight: Unavailable. Objective: "Continuation of outer space investigations." Description: Unavailable.	335	204	89.9	72.9	Reentered 12 Apr.
6 Apr.	Cosmos 812 USSR 1976–31A C–1 Plesetsk	Total weight: Unavailable. Objective: "Continuation of outer space investigations." Description: Unavailable.	546	507	95.2	74.0	Still in orbit.
7 Apr.	Meteor 24 USSR 1976–32A A–1 Plesetsk	Total weight: Unavailable. Objective: Acquisition of meteorological information for use in operational weather system. Description: Unavailable.	891	841	102.3	81.3	Still in orbit.
9 Apr.	Cosmos 813 USSR 1976–33A A–2 Plesetsk	Total weight: Unavailable. Objective: "Continuation of outer space investigations." Description: Unavailable.	230	205	88.9	81.3	Reentered 21 Apr.
13 Apr.	Cosmos 814 USSR 1976–34A F–1–m Baikonur-Tyuratam	Total weight: Unavailable. Objective: "Continuation of outer space investigations." Description: Unavailable.	480	118	90.4	65.1	Reentered 13 Apr.; in orbit only 0.28 day.
22 Apr.	Nato IIIA NATO 1976–35A Delta ETR	Total weight: 310 kg. (in orbit). Objective: Position comsat in synchronous orbit over equator for use by NATO. Description: Drum-shaped craft 2.2m dia., 3.1m long including antenna	35 863	35 209	1423.4	02.9	First of 3 comsats to be launched by NASA for NATO; stationed at 15.5°W in synchronous orbit over equator near Africa; still in orbit.

Appendix A

Launch Date	Spacecraft, Country, Int'l Designation, Vehicle, Launch Site	Payload Data	Apogee (km)	Perigee (km)	Period (min)	Inclination (degrees)	Remarks
28 Apr.	Cosmos 815 USSR 1976-36A A-2 Plesetsk	Total weight: Unavailable. Objective: "Continuation of outer space investigations." Description: Unavailable.	229	215	89.0	81.3	Reentered 11 May.
	Cosmos 816 USSR 1976-37A C-1 Plesetsk	Total weight: Unavailable. Objective: "Continuation of outer space investigations." Description: Unavailable.	513	480	94.6	65.8	Still in orbit.
30 Apr.	DOD spacecraft (NOSS 1) U.S. 1976-38A Atlas WTR and	Total weight: Unavailable. Objective: Ocean survey for Navy. Description: Unavailable.	1122	1100	107.5	63.5	Four spacecraft on a single carrier; still in orbit. Naval ocean-survey satellite part of "Whitecloud" system.
	DOD spacecraft (SSU-1) U.S. 1976-38C Atlas WTR and	Total weight: Unavailable. Objective: "Development of spaceflight techniques and technology." Description: Unavailable.	1129	1093	107.5	63.4	Spacecraft support unit; still in orbit.
	DOD spacecraft (SSU-2) U.S. 1976-38D Atlas WTR and	Total weight: Unavailable. Objective: "Development of spaceflight techniques and technology." Description: Unavailable.	1129	1093	107.5	63.4	Spacecraft support unit; still in orbit.

Date	Spacecraft	Description	(col4)	(col5)	(col6)	Remarks
	DOD spacecraft (SSU-3) U.S. 1976-38J Atlas WTR	Total weight: Unavailable. Objective: "Development of spaceflight techniques and technology." Description: Unavailable.	1139	1083	107.4	63.4	Spacecraft support unit; still in orbit.
4 May	Lageos U.S. 1976-39A Thor-Delta ETR	Total weight: 411 kg. Objective: To provide permanent reference point in orbit for precision earth-dynamics measurement, demonstrate laser tracking techniques, and measure earth rotation and movement of crust. Description: Aluminum hemispheres bolted around solid brass core to form 60cm sphere with 426 laser reflectors mounted on surface.	5944	5835	225.4	109.9	Laser geodynamics satellite, NASA's first passive craft dedicated to laser ranging, with useful life up to 50 yrs; expected to remain in orbit up to 8 million yrs.
5 May	Cosmos 817 USSR 1976-40A A-2 Baikonur-Tyuratam	Total weight: Unavailable. Objective: "Continuation of outer space investigations." Description: Unavailable.	322	172	89.5	65.0	Reentered 18 May.
12 May	Molniya 3E U.S.S.R 1976-41A A-2-e Plesetsk	Total weight: Unavailable. Objective: To operate long-range radio communications system; transmit Soviet TV programs to stations in Orbita network. Description: Unavailable.	39 731	629	717.9	62.9	Fifth of series; still in orbit.
13 May	Comstar 1A U.S. 1976-42A Atlas-Centaur ETR	Total weight: 816.5 kg (in orbit). Objective: To provide 14 400 high-quality 2-way voice circuits for Hawaii, Alaska, Puerto Rico, and contiguous U.S. Description: Spin-stabilized solar-cell-covered cylinder 244cm in dia., 610cm long, carrying 24 transponders, half polarized horizontally and half vertically, doubling capacity of each frequency.	35 802	35 269	1423.3	0.1	First of planned 3-comsat network to be launched by NASA for Comsat General; in stationary orbit at 128°W over Pacific; still in orbit.
15 May	Meteor 25 USSR 1976-43A A-1 Plesetsk	Total weight: Unavailable. Objective: To acquire meteorological data for use in operational weather system. Description: Unavailable.	893	844	102.3	81.2	Still in orbit.

Appendix A ASTRONAUTICS AND AERONAUTICS, 1976

Launch Date	Spacecraft, Country, Int'l Designation, Vehicle, Launch Site	Payload Data	Apogee (km)	Perigee (km)	Period (min)	Inclination (degrees)	Remarks
18 May	Cosmos 818 USSR 1976–44A B–1 Plesetsk	Total weight: Unavailable. Objective: "Continuation of outer space investigations." Description: Unavailable.	479	270	92.1	71.1	Reentered 7 Mar. 77.
20 May	Cosmos 819 USSR 1976–45A A–2 Baikonur-Tyuratam	Total weight: Unavailable. Objective: "Continuation of outer space investigations." Description: Unavailable.	286	200	89.4	65.0	Reentered 1 June.
21 May	Cosmos 820 U.S.S.R. 1976–46A A–2 Plesetsk	Total weight: Unavailable. Objective: "Continuation of outer space investigations." Description: Unavailable.	231	206	88.9	81.4	Reentered 2 June.
22 May	DOD spacecraft (P–76–5) U.S. 1976–47A Scout WTR	Total weight: 72.6 kg. Objective: Evaluation of effects of disturbed plasma on radar and communications systems. Description: Box with deployable solar arrays; experiment package with multiple-antenna array.	1058	994	105.7	99.7	Launched by NASA for U.S. Air Force; still in orbit.
26 May	Cosmos 821 USSR 1976–48A A–2 Plesetsk	Total weight: Unavailable. Objective: "Continuation of outer space investigations." Description: Unavailable.	312	203	89.7	72.8	Reentered 8 June.
28 May	Cosmos 822 USSR 1976–49A C–1 Plesetsk	Total weight: Unavailable. Objective: "Continuation of outer space investigations." Description: Unavailable.	709	279	94.5	74.0	Reentered 8 Aug. 78.

Date	Spacecraft	Details				Remarks
2 June	DOD spacecraft U.S. 1976-50A Titan IIIB–Agena D WTR	Total weight: Unavailable. Objective: "Development of spaceflight techniques and technology." Description: Unavailable.	39 259	312	700.0	62.5	Still in orbit.
	Cosmos 823 USSR 1976-51A C-1 Plesetsk	Total weight: Unavailable. Objective: "Continuation of outer space investigations." Description: Unavailable.	1009	978	105.0	83.0	Probable navsat; still in orbit.
8 June	Cosmos 824 USSR 1976-52A A-2 Baikonur-Tyuratam	Total weight: Unavailable. Objective: "Continuation of outer space investigations." Description: Unavailable.	316	169	89.4	71.4	Reentered 21 June.
10 June	Marisat 2 U.S. 1976-53A Thor-Delta ETR	Total weight: 327 kg (in orbit). Objective: To provide 24-hr real-time communications between ships and shore. Description: Solar-cell-covered spin-stabilized cylinder 380cm long, 215cm in dia., carrying 3 helical UHF antennas and quad L–band antenna on pole extending from body; 44 duplex teleprinter channels and one 2–way voice circuit.	35 807	35 788	1436.6	02.5	Second commercial maritime comsat, launched by NASA for Comsat General Corp.; stationed at 176.5°E over Pacific; prime UHF user for first 2 yrs to be U.S. Navy; still in orbit.
15 June	Cosmos 825 USSR 1976-54A C-1 Plesetsk and	Total weight: Unavailable. Objective: "Continuation of outer space investigations." Description: Unavailable.	1487	1395	114.7	74.0	Probable 8 comsats on single carrier; all still in orbit.
	Cosmos 826 USSR 1976-54B C-1 Plesetsk and	Total weight: Unavailable. Objective: "Continuation of outer space investigations." Description: Unavailable.	1545	1482	116.3	74.0	

Appendix A

Launch Date	Spacecraft, Country, Int'l Designation, Vehicle, Launch Site	Payload Data	Apogee (km)	Perigee (km)	Period (min)	Inclination (degrees)	Remarks
	Cosmos 827 USSR 1976-54C C-1 Plesetsk and	Total weight: Unavailable. Objective: "Continuation of outer space investigations." Description: Unavailable.	1490	1413	114.9	74.0	
	Cosmos 828 USSR 1976-54D C-1 Plesetsk and	Total weight: Unavailable. Objective: "Continuation of outer space investigations." Description: Unavailable.	1490	1433	115.1	74.0	
	Cosmos 829 USSR 1976-54E C-1 Plesetsk and	Total weight: Unavailable. Objective: "Continuation of outer space investigations." Description: Unavailable.	1491	1451	115.1	74.0	
	Cosmos 830 USSR 1976-54F C-1 Plesetsk and	Total weight: Unavailable. Objective: "Continuation of outer space investigations." Description: Unavailable.	1493	1469	115.6	74.0	
	Cosmos 831 USSR 1976-54G C-1 Plesetsk and	Total weight: Unavailable. Objective: "Continuation of outer space investigations." Description: Unavailable.	1507	1476	115.8	74.0	

Date	Spacecraft					
16 June	Cosmos 832 USSR 1976-54H C-1 Plesetsk	Total weight: Unavailable. Objective: "Continuation of outer space investigations." Description: Unavailable.	1521	1452	116.0	74.0	
	Cosmos 833 USSR 1976-55A A-2 Plesetsk	Total weight: Unavailable. Objective: "Continuation of outer space investigations." Description: Unavailable.	315	177	89.4	62.8	Reentered 29 June.
18 June	Gravity Probe A (Red Shift Experiment) U.S. Scout ETR	Total weight: 102 kg. Objective: To test Einstein's "equivalence principle," foundation of relativity theory. Description: Suborbital probe 114cm long, 96cm in dia, carrying extremely accurate hydrogen-maser atomic clock with S-band telemetry.	Maximum altitude, 10 000 km.				Launched by NASA into elliptical flight of 1 hr 55 min over Atlantic Ocean; data substantially confirmed equivalence aspect of theory.
19 June	Intercosmos 15 USSR 1976-56A C-1 Plesetsk	Total weight: Unavailable. Objective: Flight-testing of spacecraft systems designed by specialists in Eastern Bloc countries. Description: Unavailable.	517	483	94.6	74.0	Science data from comsat to be received by stations in Hungary, GDR, and Czechoslovakia as well as USSR; still in orbit.
22 June	Salyut 5 USSR 1976-57A D-1 Baikonur-Tyuratam	Total weight: Unavailable. Objective: "Continuation of outer space investigations." Description: Unavailable.	244	213	89.1	51.6	Still in orbit.
24 June	Cosmos 834 USSR 1976-58A A-2 Plesetsk	Total weight: Unavailable. Objective: "Continuation of outer space investigations." Description: Unavailable.	231	211	88.9	81.4	Reentered 6 July.
26 June	DOD spacecraft U.S. 1976-59A Titan IIIC ETR	Total weight: Unavailable. Objective: "Development of spaceflight techniques and technology." Description: Unavailable.	35 840	295	633.0	26.3	Still in orbit.

Launch Date	Spacecraft, Country, Int'l Designation, Vehicle, Launch Site	Payload Data	Apogee (km)	Perigee (km)	Period (min)	Inclination (degrees)	Remarks
29 June	Cosmos 835 USSR 1976-60A A-2 Baikonur-Tyuratam	Total weight: Unavailable. Objective: "Continuation of outer space investigations." Description: Unavailable.	316	174	89.4	65.0	Reentered 12 July.
	Cosmos 836 USSR 1976-61A C-1 Plesetsk	Total weight: Unavailable. Objective: "Continuation of outer space investigations." Description: Unavailable.	817	789	100.9	74.1	Probable comsat; still in orbit.
1 July	Cosmos 837 USSR 1976-62A A-2 Plesetsk	Total weight: Unavailable. Objective: "Continuation of outer space investigations." Description: Unavailable.	936	438	98.5	62.8	Probably a failed *Molniya 2* launch; still in orbit.
2 July	Cosmos 838 USSR 1976-63A F-1-m Baikonur-Tyuratam	Total weight: Unavailable. Objective: "Continuation of outer space investigations." Description: Unavailable.	442	429	93.3	65.1	Reentered 23 Aug. 77.
6 July	Soyuz 21 USSR 1976-64A A-2 Baikonur-Tyuratam	Total weight: 6570 kg. Objective: Rendezvous and docking with *Salyut 5* space station; materials processing and environmental experiments. Description: Unavailable.	274	245	89.7	51.6	Boris Volynov, Vitaly Zholobov first crew to visit *Salyut 5*; docked 7 July, undocked 24 Aug.; crew performed materials processing and environmental experiments; reentered 24 Aug.
8 July	DOD spacecraft U.S. 1976-65A Titan IIID WTR and	Total weight: Unavailable. Objective: "Development of spaceflight techniques and technology." Description: Unavailable.	237	157	88.4	97.0	Reentered 13 Dec.

ASTRONAUTICS AND AERONAUTICS, 1976 — Appendix A

Date	Spacecraft	Details					Remarks
	DOD spacecraft (SESP 74-2) U.S. 1975-65B Titan IIID WTR	Total weight: Unavailable. Objective: To measure intensity, distribution, and effects of protons, electrons, and alpha particles in space. Description: Unavailable.	8049	234	179.0	97.5	Carried 5 experiments for USAF Geophysics Laboratory, 3 for ONR; still in orbit.
	and						
	DOD spacecraft U.S. 1976-65C Titan IIID WTR	Total weight: Unavailable Objective: "Development of spaceflight techniques and technology." Description: Unavailable.	630	627	97.3	96.4	Still in orbit.
	Palapa 1 Indonesia 1976-66A Thor-Delta ETR	Total weight: 281 kg in orbit. Objective: To provide television, voice, and other data transmission throughout Indonesia. Description: Solar-cell-covered cylinder 3.3m high, 1.8m in dia., with despun earth-oriented antenna assembly on top; 12 RF channels, 10 for traffic and 2 protective.	35 809	35 764	1436.1	0.0	Indonesian comsat, launched by NASA into synchronous orbit at 83°E over equator; still in orbit.
	Cosmos 839 USSR 1976-67A C-1 Plesetsk	Total weight: Unavailable. Objective: "Continuation of outer space investigations." Description: Unavailable.	2098	983	116.9	65.9	Still in orbit.
14 July	Cosmos 840 USSR 1976-68A A-2 Plesetsk	Total weight: Unavailable. Objective: "Continuation of outer space investigations." Description: Unavailable.	312	201	89.6	72.9	Reentered 26 July.
15 July	Cosmos 841 USSR 1976-69A C-1 Plesetsk	Total weight: Unavailable. Objective: "Continuation of outer space investigations." Description: Unavailable.	806	785	100.8	74.0	Possible comsat; still in orbit.

Launch Date	Spacecraft, Country, Int'l Designation, Vehicle, Launch Site	Payload Data	Apogee (km)	Perigee (km)	Period (min)	Inclination (degrees)	Remarks
21 July	Cosmos 842 USSR 1976-70A C-1 Plesetsk	Total weight: Unavailable. Objective: "Continuation of outer space investigations." Description: Unavailable.	1006	974	104.9	83.0	Possible navsat; still in orbit.
	Cosmos 843 USSR 1976-71A F-1-m Plesetsk	Total weight: Unavailable. Objective: "Continuation of outer space investigations." Description: Unavailable.	345	132	89.3	65.1	Ignition failed, reentered 21 July within 8 hrs of launch.
22 July	Cosmos 844 USSR 1976-72A A-2 Plesetsk	Total weight: Unavailable. Objective: "Continuation of outer space investigations." Description: Unavailable.	344	180	89.8	67.1	Reentered 30 Aug.
	Comstar 2 U.S. 1976-73A Atlas-Centaur ETR	Total weight: 816.5 kg (in orbit). Objective: To provide 14 400 2-way high-quality voice circuits to Alaska, Hawaii, P.R., and contiguous U.S. Description: Spin-stabilized solar-cell-covered cylinder 610cm high, 244cm in dia.; carrying 24 transponders, 12 polarized horizontally and 12 vertically, doubling capacity of each frequency.	32 747	35 388	1425.0	0.2	Second of planned 3-satellite network owned by Comsat, leased to AT&T; launched by NASA into stationary orbit at 95°W over southwestern U.S.; still in orbit.
23 July	Molniya 1-AL USSR 1976-74A A-2-e Baikonur-Tyuratam	Total weight: Unavailable. Objective: Operation of long-range telephone and telegraph communications system; transmission of TV programs to stations in Orbita network. Description: Unavailable.	38 917	475	698.3	63.0	Thirty-fifth of series; still in orbit.

Date	Spacecraft	Details				
27 July	Cosmos 845 USSR 1976-75A C-1 Plesetsk	Total weight: Unavailable. Objective: "Continuation of outer space investigations." Description: Unavailable.	543	95.2	74.1	Still in orbit.
	Intercosmos 16 USSR 1976-76A C-1 Kapustin Yar	Total weight: Unavailable. Objective: Continuation of international cooperative science program. Description: Unavailable.	516	94.4	50.6	Still in orbit; data received from instrumentation built by Czechoslovakia, GDR, Sweden.
29 July	Noaa 5 (ITOS-H) U.S. 1976-77A Thor-Delta ETR	Total weight: 345 kg. Objective: To provide meteorological data from sun-synchronous orbit in both direct and stored readout modes. Description: Earth-oriented 3-axis-stabilized rectangular box 102cm by 102cm, 122 cm long, with 3-panel solar array; carrying 2 VHR radiometers, 2 scanning radiometers, solar proton monitor, and vertical temperature profile radiometers, for day and night coverage.	1518	116.2	102.1	Fifth of a series of 2nd-generation operational metsats launched by NASA; turned over to NOAA 20 Aug.; replaced Noaa 4 in global weatherwatch program; still in orbit.
	Cosmos 846 USSR 1976-78A C-1 Plesetsk	Total weight: Unavailable. Objective: "Continuation of outer space investigations." Description: Unavailable.	1013	104.8	82.9	Possible navsat; still in orbit.
4 Aug.	Cosmos 847 USSR 1976-79A A-2 Plesetsk	Total weight: Unavailable. Objective: "Continuation of outer space investigations." Description: Unavailable.	307	89.2	62.8	Reentered 17 Aug.
6 Aug.	DOD spacecraft (SDS-2) U.S. 1976-80A Titan IIIB-Agena D WTR	Total weight: Unavailable. Objective: "Development of spaceflight techniques and technology." Description: Unavailable.	39 315	703.8	63.3	Satellite data system spacecraft, still in orbit.

Appendix A

Launch Date	Spacecraft, Country, Int'l Designation, Vehicle, Launch Site	Payload Data	Apogee (km)	Perigee (km)	Period (min)	Inclination (degrees)	Remarks
9 Aug.	*Luna 24* USSR 1976-81A D-1-e Baikonur-Tyuratam	Total weight: Unavailable. Objective: Continuation of lunar investigations. Description: Unavailable.	Translunar trajectory				Reported landing on moon, 18 Aug. Return capsule reported to have brought moon-surface sample back to earth 22 Aug.
12 Aug.	*Cosmos 848* USSR 1976-82A A-2 Plesetsk	Total weight: Unavailable. Objective: "Continuation of outer space investigations." Description: Unavailable.	300	203	89.5	62.8	Reentered 25 Aug.
18 Aug.	*Cosmos 849* USSR 1976-83A C-1 Plesetsk	Total weight: Unavailable. Objective: "Continuation of outer space investigations." Description: Unavailable.	863	262	95.9	71.0	Reentered 24 April 78.
26 Aug.	*Cosmos 850* USSR 1976-84A B-1 Plesetsk	Total weight: Unavailable. Objective: "Continuation of outer space investigations." Description: Unavailable.	489	269	92.1	70.9	Reentered 16 May 77.
27 Aug.	*Cosmos 851* USSR 1976-85A A-1 Plesetsk	Total weight: Unavailable. Objective: "Continuation of outer space investigations." Description: Unavailable.	635	565	96.7	81.2	Still in orbit.
28 Aug.	*Cosmos 852* USSR 1976-86A A-2 Baikonur-Tyuratam	Total weight: Unavailable. Objective: "Continuation of outer space investigations." Description: Unavailable.	317	170	89.4	65.0	Reentered 10 Sept.

Date	Name/Designation	Description	Weight (kg)			Notes
30 Aug.	China 6 PRC 1976-87A Vehicle not announced. Shuang-Cheng-Tze	Total weight: Unavailable. Objective: Unavailable. Description: Unavailable.	2147	194	108.8	69.1	Still in orbit.
1 Sept.	Cosmos 853 USSR 1976-88A A-2-e Plesetsk	Total weight: Unavailable. Objective: "Continuation of outer space investigations." Description: Unavailable.	487	238	91.8	62.8	Probably a failed Molniya 2 launching; reentered 31 Dec.
	DOD spacecraft (TIP-3) U.S. 1976-89A Scout WTR	Total weight: 166 kg. Objective: To test improved navsat before integration into operational Transit system. Description: Main body carrying solar panels and accordionlike boom to deploy after insertion into orbit.	787	341	96.0	90.8	2rd in series of improved Transit navsats; launched by NASA for U.S. Navy into elliptical transfer orbit; solar panels did not deploy; salvage attempts continue; still in orbit.
3 Sept.	Cosmos 854 USSR 1976-90A A-2 Plesetsk	Total weight: Unavailable. Objective: "Continuation of outer space investigations." Description: Unavailable.	270	162	88.8	81.3	Reentered 16 Sept.
11 Sept.	DOD spacecraft (AMS 1) U.S. 1976-91A Thor-Burner 2 WTR	Total weight: Unavailable. Objective: "Development of spaceflight techniques and technology." Description: Unavailable.	846	817	101.5	98.7	Still in orbit.
	Raduga (Statsionar 1B) USSR 1976-92A D-1-3 Baikonur-Tyuratam	Total weight: Unavailable. Objective: To provide transmissions of color TV to outlying stations in Orbita network. Description: Unavailable.	35 809	35 766	1436.1	01.3	2nd operational Statsionar satellite, instrumented for color TV; still in orbit.

Launch Date	Spacecraft, Country, Int'l Designation, Vehicle, Launch Site	Payload Data	Apogee (km)	Perigee (km)	Period (min)	Inclination (degrees)	Remarks
15 Sept.	*Soyuz 22* USSR 1976-93A A-2 Baikonur-Tyuratam	Total weight: 6570 kg (estimated). Objective: To continue program of manned space flight in cooperation with socialist countries. Description: Unavailable.	259	248	89.6	64.7	Valery Bykovsky and Vladimir Aksenov, crew, failed to link with *Salyut 5*; used GDR-made multispectral camera to photograph GDR and USSR from orbit; reentered safely 23 Sept.
	DOD spacecraft U.S. 1976-94A Titan IIIB-Agena D WTR	Total weight: Unavailable. Objective: "Development of spaceflight techniques and technology." Description: Unavailable.	336	134	89.2	96.4	Reentered 5 Nov.
21 Sept.	*Cosmos 855* USSR 1976-95A A-2 Plesetsk	Total weight: Unavailable. Objective: "Continuation of outer space investigations." Description: Unavailable.	339	200	89.9	72.9	Reentered 3 Oct.
22 Sept.	*Cosmos 856* USSR 1976-96A A-2 Baikonur-Tyuratam	Total weight: Unavailable. Objective: "Continuation of outer space investigations." Description: Unavailable.	299	203	89.5	65.0	Reentered 5 Oct.
24 Sept.	*Cosmos 857* USSR 1976-97A A-2 Plesetsk	Total weight: Unavailable. Objective: "Continuation of outer space investigations." Description: Unavailable.	336	176	89.6	62.8	Reentered 7 Oct.
29 Sept.	*Cosmos 858* USSR 1976-98A C-1 Plesetsk	Total weight: Unavailable. Objective: "Continuation of outer space investigations." Description: Unavailable.	813	790	100.8	74.0	Still in orbit.

Date	Name/Designation	Details			Notes
10 Oct.	Cosmos 859 USSR 1976-99A A-2 Baikonur-Tyuratam	Total weight: Unavailable. Objective: "Continuation of outer space investigations." Description: Unavailable.	312	171	89.3	65.0	Reentered 21 Oct.
14 Oct.	Soyuz 23 USSR 1976-100A A-2 Baikonur-Tyuratam	Total weight: 6570 kg (estimated) Objective: To continue work begun by crew of Soyuz 21 in July-Aug. 76 on board orbiting station Salyut 5. Description: Unavailable.	224	188	88.6	51.6	Vyacheslav Zudov and Valery Rozhdestvensky, crew, making 2nd attempt to board Salyut 5 space station; automatic rendezvous system malfunction prevented linkup; crew reentered 16 Oct. at night in blizzard, first Soviet crew to land in water; recovered safely.
	Marisat 3 U.S. 1976-101A Thor-Delta ETR	Total weight: 327 kg (in orbit). Objective: To provide 24-hr real-time communications between ships and shore. Description: Spin-stabilized solar-cell-covered cylinder 215cm in dia., 380cm long, carrying 3-antenna UHF array and quad L-band array.	36 525	35 051	1436.2	02.6	Final link in 3-comsat military-commercial system, launched by NASA for ComSatCorp.; stationed over Pacific for checkout, over Indian Ocean at 73°E in Dec. 76; spare for use if Marisat 1 or 2 fail.
15 Oct.	Meteor 26 USSR 1976-102A A-1 Plesetsk	Total weight: Unavailable. Objective: To acquire meteorological data for use in operational weather system. Description: Unavailable.	890	855	102.4	81.3	Still in orbit.
17 Oct.	Cosmos 860 USSR 1976-103A F-1-m Baikonur-Tyuratam	Total weight: Unavailable. Objective: "Continuation of outer space investigations." Description: Unavailable.	263	250	89.6	65.0	Probable ocean-reconnaissance satellite, working with Cosmos 861; still in orbit.
21 Oct.	Cosmos 861 USSR 1976-104A F-1-m Baikonur-Tyuratam	Total weight: Unavailable. Objective: "Continuation of outer space investigations." Description: Unavailable.	263	250	89.6	65.0	Operational with Cosmos 860; still in orbit.

Appendix A

Launch Date	Spacecraft, Country, Int'l Designation, Vehicle, Launch Site	Payload Data	Apogee (km)	Perigee (km)	Period (min)	Inclination (degrees)	Remarks
22 Oct.	Cosmos 862 USSR 1976-105A A-2-e Plesetsk	Total weight: Unavailable. Objective: "Continuation of outer space investigations." Description: Unavailable.	39 514	579	712.5	62.8	Still in orbit.
25 Oct.	Cosmos 863 USSR 1976-106A A-2 Plesetsk	Total weight: Unavailable. Objective: "Continuation of outer space investigations." Description: Unavailable.	319	168	89.4	62.8	Reentered 5 Nov.
26 Oct.	Ekran (Statsionar 1C) USSR 1976-107A D-1 Baikonur-Tyuratam	Total weight: Unavailable. Objective: To transmit TV directly to isolated communities in USSR. Description: Solar-oriented 3-axis spacecraft.	36 086	35 507	1436.6	01.2	Third Soviet geostationary domestic comsat; still in orbit.
29 Oct.	Cosmos 864 USSR 1976-108A C-1 Plesetsk	Total weight: Unavailable. Objective: "Continuation of outer space investigations." Description: Unavailable.	1009	964	104.8	82.9	Probable navsat; still in orbit.
1 Nov.	Cosmos 865 USSR 1976-109A A-2 Plesetsk	Total weight: Unavailable. Objective: "Continuation of outer space investigations." Description: Unavailable.	323	201	89.8	72.9	Reentered 13 Nov.
11 Nov.	Cosmos 866 USSR 1976-110A A-2 Baikonur-Tyuratam	Total weight: Unavailable. Objective: "Continuation of outer space investigations." Description: Unavailable.	309	174	89.4	65.0	Reentered 23 Nov.

Date	Satellite	Description	Perigee/Apogee (km)	Period (min)	Inclination (deg)	Remarks
23 Nov.	Cosmos 867 USSR 1976-111A A-2 Plesetsk	Total weight: Unavailable. Objective: "Continuation of outer space investigations." Description: Unavailable.	401	251	91.1	62.8	Reentered 6 Dec.
25 Nov.	Prognoz 5 USSR 1976-112A A-2-e Baikonur-Tyuratam	Total weight: Unavailable. Objective: Further study of solar radiation and its effects on the magnetosphere. Description: Unavailable.	195 482	3869	5728.6	66.5	Still in orbit; instrumentation partly supplied by France and Czechoslovakia.
26 Nov.	Cosmos 868 USSR 1976-113A A-2-e Baikonur-Tyuratam	Total weight: Unavailable. Objective: "Continuation of outer space investigations." Description: Unavailable.	443	427	93.3	65.0	Reentered 8 July 78.
29 Nov.	Cosmos 869 USSR 1976-114A A-2 Baikonur-Tyuratam	Total weight: Unavailable. Objective: "Continuation of outer space investigations." Description: Unavailable.	286	194	89.3	51.8	Reentered 17 Dec.
2 Dec.	Cosmos 870 USSR 1976-115A C-1 Plesetsk	Total weight: Unavailable. Objective: "Continuation of outer space investigations." Description: Unavailable.	548	511	95.2	74.0	Still in orbit.
	Molniya 2-R USSR 1976-116A A-2-e Plesetsk	Total weight: Unavailable. Objective: Operation of long-range telephone and telegraph radiocommunications system; transmission of TV programs to stations in Orbita network. Description: Unavailable.	40 533	630	734.2	62.8	Still in orbit.

Launch Date	Spacecraft, Country, Int'l Designation, Vehicle, Launch Site	Payload Data	Apogee (km)	Perigee (km)	Period (min)	Inclination (degrees)	Remarks
7 Dec.	China 7 PRC 1976-117A Vehicle not announced Shuang-Cheng-Tze	Total weight: Unavailable. Objective: Unavailable. Description: Unavailable.	483	170	91.1	59.5	Reentered 2 Jan. 77.
	Cosmos 871 USSR 1976-118A C-1 Plesetsk and	Total weight: Unavailable. Objective: "Continuation of outer space investigations." Description: Unavailable.	1464	1418	114.7	74.0	Eight comsats on single carrier; all still in orbit.
	Cosmos 872 USSR 1976-118B C-1 Plesetsk and	Total weight: Unavailable. Objective: "Continuation of outer space investigations." Description: Unavailable.	1464	1399	114.5	74.0	
	Cosmos 873 USSR 1976-118C C-1 Plesetsk and	Total weight: Unavailable. Objective: "Continuation of outer space investigations." Description: Unavailable.	1498	1462	115.6	74.0	
	Cosmos 874 USSR 1976-118D C-1 Plesetsk and	Total weight: Unavailable. Objective: "Continuation of outer space investigations." Description: Unavailable.	1517	1464	115.8	74.0	

Cosmos 875 USSR 1976–118E C-1 Plesetsk		Total weight: Unavailable. Objective: "Continuation of outer space investigations." Description: Unavailable.	1448	1440	114.8	74.0	
and							
Cosmos 876 USSR 1976–118F C-1 Plesetsk		Total weight: Unavailable. Objective: "Continuation of outer space investigations." Description: Unavailable.	1540	1463	116.0	74.0	
and							
Cosmos 877 USSR 1976–118G C-1 Plesetsk		Total weight: Unavailable. Objective: "Continuation of outer space investigations." Description: Unavailable.	1464	1455	115.1	74.0	
Cosmos 878 USSR 1976–118H C-1 Plesetsk		Total weight: Unavailable. Objective: "Continuation of outer space investigations." Description: Unavailable.	1480	1460	115.3	74.0	
9 Dec.	*Cosmos 879* USSR 1976–119A A-2 Plesetsk	Total weight: Unavailable. Objective: "Continuation of outer space investigations." Description: Unavailable.	215	208	88.7	81.4	Reentered 22 Dec.

Appendix A ASTRONAUTICS AND AERONAUTICS, 1976

Launch Date	Spacecraft, Country, Int'l Designation, Vehicle, Launch Site	Payload Data	Apogee (km)	Perigee (km)	Period (min)	Inclination (degrees)	Remarks
15 Dec.	Cosmos 880 USSR 1976-120A C-1 Plesetsk	Total weight: Unavailable. Objective: "Continuation of outer space investigations." Description: Unavailable.	615	560	96.4	65.9	Probable target for satellite intercept tests; still in orbit.
	Cosmos 881 USSR 1976-121A A-2 Plesetsk and	Total weight: Unavailable. Objective: "Continuation of outer space investigations." Description: Unavailable.	No elements reported.				Probable objects to test effects of lasers in space; reentered within 3 hrs 15 Dec.
	Cosmos 882 USSR 1976-121B A-2 Plesetsk	Total weight: Unavailable. Objective: "Continuation of outer space investigations." Description: Unavailable.	No elements reported.				Probable objects to test effects of lasers in space; reentered within 3 hrs 15 Dec.
	Cosmos 883 USSR 1976-122A C-1 Plesetsk	Total weight: Unavailable. Objective: "Continuation of outer space investigations." Description: Unavailable.	1012	957	104.8	83.0	Probable navsat; still in orbit.
17 Dec.	Cosmos 884 USSR 1976-123A A-2 Plesetsk	Total weight: Unavailable. Objective: "Continuation of outer space investigations." Description: Unavailable.	314	164	89.3	65.0	Reentered 29 Dec.
	Cosmos 885 USSR 1976-124A C-1 Plesetsk	Total weight: Unavailable. Objective: "Continuation of outer space investigations." Description: Unavailable.	511	466	94.4	65.8	Still in orbit.

Date	Spacecraft		Details				Notes
19 Dec.	DOD spacecraft U.S. 1976–125A Titan IIIB WTR	Total weight: Unavailable. Objective: "Development of spaceflight techniques and technology." Description: Unavailable.	530	244	92.3	96.9	Operational satellite of the "Big Bird" type; still in orbit.
27 Dec.	Cosmos 886 USSR 1976–126A F-1-m Baikonur-Tyuratam	Total weight: Unavailable. Objective: "Continuation of outer space investigations." Description: Unavailable.	2296	595	114.8	65.8	Probable interceptor spacecraft; passed near Cosmos 880, reported to have exploded into about 20 fragments 27 Dec.
28 Dec.	Molniya 3F USSR 1976–127A A-2-e Plesetsk	Total weight: Unavailable. Objective: To operate long-range telephone and telegraph radiocommunications system; to transmit TV programs to stations in Orbita network. Description: Unavailable.	40 625	613	735.8	62.8	Still in orbit.
	Cosmos 887 USSR 1976–128A C-1 Plesetsk	Total weight: Unavailable. Objective: "Continuation of outer space investigations." Description: Unavailable.	1016	951	104.8	82.9	Probable navsat; still in orbit.

Appendix B

MAJOR NASA LAUNCHES, 1976

The following table of major NASA launches includes payloads carried by all rocket vehicles larger than sounding rockets launched in 1976 by NASA or under NASA direction.

During 1976, the U.S. had 27 launches (with 35 payloads) including a suborbital gravity probe, all of which were successful. Of these, 11 (with 19 payloads) were launches by DOD. The remaining 16 were NASA launches, only two for its own programs: *Lageos*, the laser geodynamics "golfball" satellite launched in May as a permanent reference point for earth-dynamics studies, and the gravity probe launched in June carrying a red-shift experiment that apparently confirmed the "equivalence principle" of Einstein's theory of relativity. (In 1975, NASA had launched 10 all–U.S. scientific payloads.) The 14 launches for others in 1976 included 2 spacecraft for DOD; 7 comsats—*Intelsat IV-A F-2* in Jan.; *Marisats 1, 2,* and *3* in Feb., June, and Oct.; RCA's *Satcom 2* in Mar.; and *Comstars 1* and *2* in May and July—; 1 for NOAA, the Noaa 5 in July; and 4 international satellites—*Helios 2* for FRG in Jan.; *Cts 1* for the Canadian Dept. of Communications in Jan.; *Nato IIIA* for NATO in Mar.; and *Palapa 1* in July, a domestic comsat for Indonesia.

These tables usually categorize vehicle and payload performance as S for successful, P for partially successful, or U for unsuccessful. A fourth category (Unk) would indicate payloads that did not operate because of vehicle failure. These categories, which are unofficial, do not take into account that U missions might produce valuable information, or that payloads with a long-life design might later fail to meet the design requirements and might then become officially unsuccessful. Further information on these launches appears in Appendix A and in the indexed entries in the text.

Appendix B

MAJOR NASA LAUNCHES, 1976

Date	Name (NASA Code)	General Mission	Launch Vehicle (Site)	Performance Vehicle	Performance Payload	Remarks
15 Jan.	*Helios 2*	Investigation of fundamental solar processes and solar-terrestrial relationships.	Titan IIIE–Centaur (ETR)	S	S	Elliptical solar orbit, carrying spacecraft closer to sun than any manmade object (43 million km); second West German-built satellite successfully launched by NASA.
17 Jan.	*Cts 1*	Advancement in technology of spacecraft-mounted and ground-based components and systems applicable to high-power radio-frequency satellites.	Thor-Delta (ETR)	S	S	Launched by NASA for Canadian Dept. of Communications; most powerful communications satellite to date; stationed at 116°W to provide 2-way TV and voice communications across Canada.
29 Jan.	*Intelsat IV-A F-2*	As backup for *Intelsat IV-A* launched in Sept. 76, provision of 2 TV channels and 6250 2-way voice circuits simultaneously between Europe and North America.	Atlas-Centaur (ETR)	S	S	Second in series of improved comsats launched by NASA for ComSatCorp., mgr. of Intelsat; twice the communications capacity of *Intelsat IV* series; stationed at 29.5°W.
19 Feb.	*Marisat 1*	Provision of 24-hour real-time telex and telephone communications between ships and shore.	Thor-Delta (ETR)	S	S	First in series of maritime comsats to be launched by NASA for ComSatCorp.; stationed at 15°W over Atlantic; U.S. Navy to be prime user for first 2 yr.
26 Mar.	*Satcom 2*	Provision of high-speed data, TV, and voice channels between mainland U.S. and Hawaii; operation of all 24 transponder channels throughout 8-yr lifetime.	Thor-Delta (ETR)	S	S	Second in series of commercial comsats to be launched by NASA for RCA; stationed at 135°W over equator.
22 Apr.	*Nato III-A*	Stationing of synchronous satellite over equator for use by NATO.	Thor-Delta (ETR)	S	S	First of three NATO comsats to be launched by NASA; stationed at 15.5°W south of western Africa.

Date	Name	Purpose	Launch vehicle	Orbit	Remarks
4 May	Lageos	Provision of permanent reference point in space for precision measurement of earth dynamics, and demonstration of laser satellite-tracking techniques.	Thor-Delta (WTR)	S	Laser geodynamic "golfball" satellite designed to remain in orbit for 8 million yr.
13 May	Comstar 1	Provision of 14 400 high-quality 2-way voice circuits between Hawaii, Alaska, Puerto Rico, and contiguous U.S. by using satellite in synchronous orbit.	Atlas-Centaur (ETR)	S	First of 3-comsat commercial network to be launched by NASA for AT&T, under lease from Comsat General Corp. (third operational domestic comsat system, after Westar and Satcom); stationed at 128°W over Pacific.
22 May	P76-5 (DOD)	Evaluation of propagation effects of disturbed plasma on radar and communications systems.	Scout (WTR)	S	Transit navsat modified for ionospheric study, launched by NASA for U.S. Air Force; mission accomplished.
10 June	Marisat 2	Provision of 24-hr real-time telex and telephone communications between ships and shore.	Thor-Delta (ETR)	S	Second in series of maritime comsats to be launched by NASA for ComSatCorp.; stationed at 176.5°E over Pacific; U.S. Navy to be prime user of UHF for first 2 yr.
18 June	Gravity Probe A	Test of Einstein's "equivalence principle," foundation of theory of relativity.	Scout (ETR)	S	Launched by NASA into elliptical flight, 10 000 km maximum altitude, duration 1 hr 55 min.; data substantially confirmed equivalence aspect of theory.
8 July	Palapa 1	Provision of TV, voice, and other communications throughout Indonesia.	Thor-Delta (ETR)	S	Indonesian domestic comsat launched by NASA; stationed over equator at 83°E.
22 July	Comstar 2	Provision of 14 400 high-quality 2-way voice circuits between Hawaii, Alaska, Puerto Rico, and contiguous U.S., by using satellite in synchronous orbit.	Atlas-Centaur (ETR)	S	Second of 3-comsat commercial network to be launched by NASA for AT&T, under lease from Comsat General Corp.; stationed at 95°W below southwestern U.S.
29 July	Noaa 5 (ITOS-H)	As part of NOAA global weather-watch program, provision of regular 24-hr meteorological observations from sun-synchronous orbit in direct-readout and stored modes.	Thor-Delta (ETR)	S	Fifth operational craft of a series of second-general metesats launched by NASA (27th U.S.-launched metesat, built by RCA for NOAA); replaced Noaa 4, which became backup spacecraft.

Date	Name (NASA Code)	General Mission	Launch Vehicle (Site)	Performance Vehicle	Performance Payload	Remarks
1 Sept.	TIP 3	Test of developmental satellite with improved operation for the Transit navigation system, to be integrated into that system.	Scout (WTR)	S	P	Third in a series of improved Transit satellites; launched by NASA for U.S. Navy; solar panels did not deploy, Navy salvage attempts continuing.
14 Oct.	Marisat 3	Provision of 24-hr real-time telex and telephone communications between ships and shore.	Thor-Delta (ETR)	S	S	Third in a series of maritime comsats to be launched by NASA for ComSatCorp.; after checkout, stationed over Indian Ocean at 73°E; to be used as spare by Navy in case Marisat 1 or 2 should fail.

Appendix C

MANNED SPACE FLIGHTS, 1976

Manned flight activity worldwide in 1976 consisted of only 3 launches, all in the USSR Soyuz series. The new Soviet space station, *Salyut 5,* was not launched until June; in July, the crew of *Soyuz 21* docked successfully with it and remained 7 weeks. Neither the crew of *Soyuz 22* nor that of *Soyuz 23* was able to dock successfully, although each returned to earth safely.

At the end of 1976, the U.S. had made 31 manned space flights: 2 suborbital, 20 in earth orbit, 3 in lunar orbit, and 6 lunar landings, with a total of 43 different crewmen. The USSR had made 30 manned flights, all in earth orbit, with 38 cosmonauts. The U.S. total of manned-spacecraft hours in flight was 7681 hrs 10 min; the USSR total, 6685 hr 27 min. Total cumulative man-hours in space for the U.S. were 22 503 hr 39 min; for the USSR, 13 575 hr 36 min.

MANNED SPACE FLIGHTS, 1976

Date Launched	Date Recovered	Designation Crew	Weight (km)	Duration, Revolutions	Remarks
6 July	24 Aug.	*Soyuz 21* Boris Volynov Vitaly Zholobov	6570 (est.)	1182 hr 24 min 13.18 rev.	Docked with *Salyut 5* on 7 July; launched from Baikonur-Tyuratam on A–2; crew performed materials-processing and environmental experiments; undocked and reentered 24 Aug.
15 Sept.	23 Sept.	*Soyuz 22* Valery Bykovsky Vladimir Aksenov	6570 (est.)	189 hr 54 min 2.12 rev.	Launched from Baikonur-Tyuratam on A–2; failed to link with *Salyut 5*; used GDR-made camera to photograph GDR and USSR from orbit; reentered safely 23 Sept.
14 Oct.	16 Oct.	*Soyuz 23* Vyacheslav Zudov Valery Rozhdestvensky	6570 (est.)	48 hr 6 min 0.5 rev.	Launched from Baikonur-Tyuratam on A–1; malfunction in automatic rendezvous system prevented linkup with *Salyut 5*; crew reentered 16 Oct. at night, in blizzard, first Soviet crew to land in water; recovered safely.

Appendix D

NASA SOUNDING ROCKET LAUNCHES, 1976

The following table lists the 63 sounding rockets of the Arcas class and above launched by NASA in 1976. The launches took place in Norway, Sweden, Greenland (for Denmark), and Canada, as well as in the United States (including Alaska and Hawaii). The rockets carried payloads for the Naval Research Laboratory, the Energy Research and Development Administration, and the Smithsonian Astrophysical Observatory, as well as for 5 NASA centers, 13 universities, 3 foreign governments, and 2 U.S. corporations. Types of onboard experiments included 28 in plasma physics; 8 in solar physics; 4 in astrophysics; 3 each in aeronomy, cometary physics, and meteorology, and 2 in galactic astronomy, as well as 5 vehicle systems tests, 2 flight systems tests, 1 engineering test, and 3 space processing tests, plus 1 unidentified payload from Norway.

Information for the table came from Goddard Space Flight Center's Quick Look Sounding Rocket data sheets, issued after launches, with additional information from some of the experimenters concerned. Launch dates are in local time, with the date in Greenwich Mean Time added if different.

Appendix D

Launch Date	Rocket NASA Designation, Launch Site	Apogee (km)	Remarks
10 Jan.	Black Brant VC 12.029CT/IT Wallops Flight Center, Va.	290	GSFC vehicle systems test.
15 Jan.	Terrier Malemute 12.03GT/CT Wallops Flight Center, Va.	208	GSFC flight systems test; Malemute failed at T+30 sec.
18 Jan.	Nike Tomahawk 18.1007UE, 18.169UE Wallops Flight Center, Va.	120	Univ. of Ill.–Univ. of Bern (Switzerland) plasma physics experiments; Swiss portion failed.
18 Jan.	Astrobee D 23.003UE Wallops Flight Center, Va.	78	Penn. State plasma physics experiment (magnetosphere).
23 Jan.	Astrobee D 23.004UE Wallops Flight Center, Va.	76	Penn. State plasma physics experiment (magnetosphere).
30 Jan.	Super Arcas 15.049UE Wallops Flight Center, Va.	86	Univ. of Texas plasma physics experiment (magnetosphere).
30 Jan. (31 Jan. Z)	Black Brant VC 21.027UE Poker Flat Range, Al.	225	Univ. of Minn. plasma physics experiment (electrical fields).
1 Feb.	Nike Tomahawk 18.208IE Andoya Park Range, Norway		Norwegian experiments successful; no other information available.
17 Feb.	Nike Tomahawk 18.1001UE Poker Flat Range, Al.	232	Rice Univ. plasma physics experiment (magnetosphere).
18 Feb.	Black Brant VC 21.030US/IS White Sands Missile Range, N.M.	282	Univ. of Colo. solar physics experiment.

Date	Vehicle / ID / Location	Apogee	Experiment
26 Feb. (27 Feb. Z)	Super Arcas 15.147CM Greenland	83	GSFC meteorology experiment (ozone).
28 Feb.	Nike Tomahawk 18.164UE; 18.1003UE Poker Flat Range, Al.	174	Univ. of Texas plasma physics experiment (radiation in E and F regions).
1 Mar.	Nike Tomahawk 18.1002UE Poker Flat Range, Al.	242.4	Rice Univ. plasma physics experiment (field-aligned currents and auroral particles).
1 Mar. (2 Mar. Z)	Nike Apache 14.461UE Ft. Churchill, Canada	1	Univ. of Pitt. plasma physics experiment; failed.
4 Mar.	Super Arcas 15.148CM Greenland	83	GSFC meteorology experiment (ozone).
5 Mar.	Aerobee 170 13.128UC White Sands Missile Range, N.M.	215	Univ. of Colo. cometary physics experiment (UV spectrometer).
	Aerobee 200A 26.050UC White Sands Missile Range, N.M.	129.6	Johns Hopkins Univ. cometary physics experiment (UV spectrometer).
10 Mar.	Aerobee 170 13.129CC White Sands Missile Range. N.M.	189	GSFC cometary physics experiment.
19 Mar.	Nike Cajun 10.425UM Wallops Flight Center, Va.	190	Univ. of Mich. meteorology (nitric oxide measurement) experiment.
22 Mar.	Aerobee 170A 13.116UE Ft. Churchill, Canada	182.4	Univ. of Mich. plasma physics (auroral) experiment.
22 Mar. (23 Mar. Z)	Aerobee 200 26.055GG White Sands Missile Range, N.M.	216.4	GSFC galactic astronomy (spectrograms of stars) experiment.

Appendix D

Launch Date	Rocket NASA Designation, Launch Site	Apogee (km)	Remarks
25 Mar. (26 Mar. Z)	Aerobee 170 13.077UA Ft. Churchill, Canada	175.2	Johns Hopkins Univ. aeronomy (auroral) experiment.
26 Mar.	Aerobee 200A 26.017AS White Sands Missile Range, N.M.	203	ERDA (Los Alamos) solar physics (radiation) experiment.
26 Mar. (27 Mar. Z)	Nike Tomahawk 18.176GE Poker Flat Range, Al.		GSFC plasma physics experiment; released 4 barium, 1 trimethylaluminum clouds; radar not tracking.
27 Mar.	Nike Tomahawk 18.194GE Poker Flat Range, Al.		GSFC plasma physics experiment; released 4 barium/lithium/sodium, 1 lithium clouds; radar not tracking.
28 Mar.	Nike Tomahawk 18.195GE Poker Flat Range, Al.		GSFC plasma physics experiment; released 4 barium/lithium/sodium, 1 lithium clouds; radar not tracking.
30 Mar.	Nike Tomahawk 18.177GE Poker Flat Range, Al.	306 (est.)	GSFC plasma physics experiment; released 4 barium, 1 trimethylaluminum clouds; results observed by camera.
	Astrobee F 25.013NT/CS White Sands Missile Range, N.M.	231	ARC/Lockheed engineering test.
	Aries 24.001IE Esrange, Sweden		Swedish (MPE) plasma physics (intl. magnetospherics studies) experiment; rocket failed.

NASA-LAUNCHED SOUNDING ROCKETS, 1976

Launch Date	Rocket NASA Designation, Launch Site	Apogee (km)	Remarks
31 Mar.	Black Brant VC 21.003UI Ft. Churchill, Canada	231	Univ. of Texas plasma physics (auroral) experiment.
21 Apr.	Super Arcas 15.128UE Ft. Churchill, Canada	61.2	Univ. of Houston plasma physics (magnetospheric) experiment.
	Aerobee 200 26.049DH White Sands Missile Range, N.M.	202	NRL high-energy astrophysics (x-ray astronomy) experiment.
11 May	Aerobee 200A 26.038GH White Sands Missile Range, N.M.	175.4	GSFC high-energy astrophysics (x-ray astronomy) experiment.
	Aerobee 200 26.045CS White Sands Missile Range, N.M.	247	Aerospace Corp. solar physics experiment.
13 May	Astrobee D 23.006UE White Sands Missile Range, N.M.	121	Univ. of Alaska plasma physics (magnetospheric) experiment.
17 May	Black Brant VC 21.033NP White Sands Missile Range, N.M.	187.6	MSFC space processing (SPARZ) experiment.
22 May (23 May Z)	Black Brant VC 21.048NP Kauai Test Range, Hawaii	195.6	JPL space processing (superfluid helium in zero gravity) experiment.

Launch Date	Rocket NASA Designation, Launch Site	Apogee (km)	Remarks
11 June	Nike Javelin 32.004WT Wallops Flight Center, Va.	115	WFC engineering test of Javelin motor, separable nosecone; did not reach predicted altitude.
17 June	Aerobee 350 17.017UH White Sands Missile Range, N.M.	231.5	Smithsonian and Harvard Observatories high-energy astrophysics (cosmic x-ray) experiment.
29 June	Nike Tomahawk 18.181GA White Sands Missile Range, N.M.	222	GSFC aeronomy (solar UV) experiment.
	Aerobee 170 13.130GS White Sands Missile Range, N.M.	225.3	GSFC solar physics (irradiance calibration) experiment.
13 July	Aerobee 170A 12.032GT White Sands Missile Range, N.M.	182.9	GSFC vehicle systems test.
17 July (18 July Z)	Javelin 12.177CT Wallops Flight Center, Va.	24.4	GSFC vehicle systems test; vehicle broke apart at 2d-stage ignition.
26 July	Aerobee 170A 12.033GT White Sands Missile Range, N.M.	170	GSFC vehicle systems test.
4 Aug.	Aerobee 200A 26.043US White Sands Missile Range, N.M.	222.2	Univ. of Hawaii solar physics experiment.

ASTRONAUTICS AND AERONAUTICS, 1976 — Appendix D

Date	Vehicle	Altitude (km)	Experiment/Result
11 Aug.	Astrobee D 23.007UE White Sands Missile Range, N.M.	84.6	Penn State Univ. plasma physics experiment.
12 Aug.	Terrier Malemute 12.036GT Wallops Flight Center, Va.	611.6	GSFC vehicle systems test.
12 Aug.	Nike Tomahawk 18.1006UE Wallops Flight Center, Va.	148.3	Univ. of Ill.–Univ. of Bern plasma physics experiments.
22 Aug.	Nike Tomahawk 18.209IE Greenland		Danish Space Research Inst. plasma physics (magnetospheric) experiment; no information.
27 Aug.	Nike Tomahawk 18.210IE Greenland	— —	Danish Space Research Inst. plasma physics experiment; no information.
16 Sept.	Aerobee 170A 13.134CS White Sands Missile Range, N.M.	177.3	Am. Science and Engrg. Corp. solar physics experiment.
21 Sept.	Nike Tomahawk 18.178GE Poker Flat Range, Al.	235 (predicted)	GSFC plasma physics (magnetospheric) experiment; payload did not separate; experiment mostly successful.
23 Sept.	Nike Tomahawk 18.179GE Poker Flat Range, Al.	235 (predicted)	GSFC plasma physics (magnetospheric) experiment.
30 Sept.	Nike Tomahawk 18.180GE Poker Flat Range, Al.	65	GSFC plasma physics experiment; structural failure; no data recovered.
6 Oct.	Nike Cajun 10.317UI Wallops Flight Center, Va.	22.8	Penn. State plasma physics experiment; 2nd-stage ignition failure; no payload data.
22 Oct.	Black Brant VC 21.036DS White Sands Missile Range, N.M.	238.5	NRL solar physics (UV at solar minimum) experiment.

Appendix D ASTRONAUTICS AND AERONAUTICS, 1976

Launch Date	Rocket NASA Designation, Launch Site	Apogee (km)	Remarks
29 Oct.	Nike Tomahawk 18.155UA Wallops Flight Center, Va.	1228.	Univ. of Colo. aeronomy (nitric oxide density) experiment.
29 Oct. (30 Oct. Z)	Aerobee 200A 26.056DG White Sands Missile Range, N.M.	182.1	NRL galactic astronomy (far–UV images of Andromeda) experiment; parachute failure damaged payload.
17 Nov.	Nike Black Brant VC 27.001CS White Sands Missile Range, N.M.	260.6	American Sci. and Engrg. solar physics experiment.
19 Nov.	Nike Black Brant VC 27.002UH White Sands Missile Range, N.M.	294.5	Univ. of Cal. at Berkeley high-energy astrophysics experiment (nonsolar extreme UV emissions).
27 Nov.	Terrier Malemute 29.005 IE Norway	550 (predicted)	Norwegian plasma physics (auroral) experiment.
	Nike Tomahawk 18.206IE Norway	215 (predicted)	Norwegian plasma physics (auroral) experiment.
14 Dec.	Black Brant VC 21.034NP White Sands Missile Range	174.4	MSFC space processing experiment.

Appendix E

ABBREVIATIONS OF REFERENCES

Listed here are the abbreviations used for citing sources in the text. Not all the sources are listed, only those that are abbreviated.

AAAS Bull	American Association for the Advancement of Science's *AAAS Bulletin*
A&A	American Institute of Aeronautics and Astronautics' magazine, *Astronautics & Aeronautics*
A&A 1976	NASA's *Astronautics and Aeronautics, 1976: A Chronology* (this publication)
ABC	American Broadcasting Company
AEC Release	Atomic Energy Commission News Release
Aero Daily	*Aerospace Daily* newsletter
Aero Med	*Aerospace Medicine* magazine
AF Mag	Air Force Association's *Air Force Magazine*
AFHF Newsletter	*Air Force Historical Foundation Newsletter*
AFJ	*Armed Forces Journal* magazine
AFSC *Newsreview*	Air Force Systems Command's *Newsreview*
AFSC Release	Air Force Systems Command News Release
AIA Release	Aerospace Industries Association of America News Release
AIAA *Facts*	American Institute of Aeronautics and Astronautics' *Facts*
AIAA Release	American Institute of Aeronautics and Astronautics News Release
AIP *Newsletter*	American Institute of Physics *Newsletter*
AP	Associated Press news service
ARC *Astrogram*	NASA Ames Research Center's *Astrogram*
Astro Journ	American Astronomical Society's *Astrophysical Journal*
Atlanta JC	*Atlanta Journal Constitution* newspaper
Av Wk	*Aviation Week & Space Technology* magazine
B News	*Birmingham News* newspaper
B Sun	*Baltimore Sun* newspaper
Bull Atom Sci	Education Foundation for Nuclear Science's *Bulletin of the Atomic Scientists*
Bus Wk	*Business Week* magazine
C Daily News	*Chicago Daily News* newspaper
C Trib	*Chicago Tribune* newspaper
Can Press	Canadian Press news service
CBS	Columbia Broadcasting System
C&E News	*Chemical & Engineering News* magazine
Cl *PD*	Cleveland *Plain Dealer* newspaper
Cl Press	*Cleveland Press* newspaper
Columbia J Rev	*Columbia Journalism Review* magazine

Appendix E

ComSatCorp Release	Communications Satellite Corporation News Release
CQ	*Congressional Quarterly*
CR	*Congressional Record*
CSM	*Christian Science Monitor* newspaper
CTNS	Chicago Tribune News Service
D News	*Detroit News* newspaper
D Post	*Denver Post* newspaper
DASA Release	Defense Atomic Support Agency News Release
DFRC	See FRC.
DJ	Dow Jones news service
DOC PIO	Department of Commerce Public Information Office
DOD Release	Department of Defense News Release
DOT Release	Department of Transportation News Release
EOP Release	Executive Office of the President News Release
FAA release	Federal Aviation Administration News Release
FBIS—Sov	Foreign Broadcast Information Service, Soviet number
FonF	*Facts on File*
FRC Release	Flight Research Center News Release (after 8 Jan. 1976, became Dryden Flight Research Center News Release)
FRC *X–Press*	NASA Flight Research Center's *X–Press*
GE Forum	*General Electric Forum* magazine
Goddard News	NASA Goddard Space Flight Center's *Goddard News*
GSFC Release	NASA Goddard Space Flight Center News Release
GSFC *SSR*	NASA Goddard Space Flight Center's *Satellite Situation Report*
GT&E Release	General Telephone & Electronics News Release
H Chron	*Houston Chronicle* newspaper
H Post	*Houston Post* newspaper
JA	*Journal of Aircraft* magazine
JPL *Lab-Oratory*	Jet Propulsion Laboratory's *Lab-Oratory*
JPL Release	Jet Propulsion Laboratory News Release
JPRS	Department of Commerce Joint Publications Research Service
JSC Release	NASA Lyndon B. Johnson Space Center (Manned Spacecraft Center until 17 Feb. 1973) News Release
JSC *Roundup*	NASA Lyndon B. Johnson Space Center's *Space News Roundup*
JSR	American Institute of Aeronautics and Astronautics' *Journal of Spacecraft and Rockets* magazine
KC Star	*Kansas City Star* newspaper
KC Times	*Kansas City Times* newspaper
KSC Release	NASA John F. Kennedy Space Center News Release
LA *Her-Exam*	Los Angeles *Herald-Examiner* newspaper
LA Times	*Los Angeles Times* newspaper
Langley Researcher	NASA Langley Research Center's *Langley Researcher*
LARC Release	NASA Langley Research Center News Release
LATNS	Los Angeles Times News Service
LERC Release	NASA Lewis Research Center News Release
Lewis News	NASA Lewis Research Center's *Lewis News*
M HER	*Miami Herald* newspaper
M News	*Miami News* newspaper
M Trib	*Minneapolis Tribune* newspaper
Marshall Star	NASA George C. Marshall Space Flight Center's *Marshall Star*
MJ	*Milwaukee Journal* newspaper
MSFC Release	NASA George C. Marshall Space Flight Center News Release
N Hav Reg	*New Haven Register* newspaper
N News	*Newark News* newspaper

N Va Sun	*Northern Virginia Sun* newspaper
NAA *News*	National Aeronautic Association *News*
NAA Record Book	National Aeronautic Association's *World and U.S.A. National World Aviation-Space Records*
NAC Release	National Aviation Club News Release
NAE Release	National Academy of Engineering News Release
NANA	North American Newspaper Alliance
NAS Release	National Academy of Sciences News Release
NAS–NRC Release	National Academy of Sciences–National Research Council News Release
NAS–NRC–NAE *News Rpt*	National Academy of Sciences—National Research Council—National Academy of Engineering *News Report*
NASA anno.	NASA announcement
NASA GMR	NASA Headquarters "General Management Review Report"
NASA HHR—39	NASA Historical Report No. 39
NASA Hist Off	NASA History Office
NASA Hq *WB*	NASA Headquarters *Weekly Bulletin*
NASA Int Aff	NASA Office of International Affairs
NASA *LAR*, XIII/8	NASA *Legislative Activities Report*, Vol. XIII, No. 8
NASA Leg Off	NASA Office of Legislative Affairs
NASA MOR	NASA Headquarters Mission Operations Report, preliminary prelaunch and postlaunch report series; information may be revised and refined before publication
NASA prog off	NASA program office (for the program reported)
NASA proj off	NASA project office (for the project reported)
NASA Release	NASA Headquarters News Release
NASA Rpt SRL	NASA report of sounding rocket launching
NASA SP–4019	NASA Special Publication No. 4019
Natl Obs	*National Observer* magazine
Nature	*Nature Physical Science* magazine
NBC	National Broadcasting Company
NGS Release	National Geographic Society News Release
NMI	NASA Management Instruction
NN	NASA Notice
NOAA Release	National Oceanic and Atmospheric Administration News Release
NRL Release	Naval Research Laboratory Release
NSC Release	National Space Club News Release
NSC *News*	National Space Club *News*
NSC *Letter*	National Space Club *Letter*
NSF *Highlights*	National Science Foundation's *Science Resources Studies Highlights*
NSF Release	National Science Foundation News Release
NSTL Release	NASA National Space Technology Laboratories News Release
NY News	*New York Daily News* newspaper
NYT	*New York Times* newspaper
NYTNS	New York Times News Service
O Sen Star	*Orlando Sentinel Star* newspaper
Oakland Trib	*Oakland Tribune* newspaper
Omaha W–H	*Omaha World-Herald* newspaper
ONR *Rev*	Navy's Office of Naval Research *Reviews*
P *Bull*	Philadelphia *Evening* and *Sunday Bulletin* newspaper
P *Inq*	*Philadelphia Inquirer* newspaper
PAO	Public Affairs Office
PD	National Archives and Records Service's *Weekly Compilation of Presidential Documents*

PIO	Public Information Office
PMR *Missile*	USN Pacific Missile Range's *Missile*
PMR Release	USN Pacific Missile Range News Release
Pres Rpt 74	*Aeronautics and Space Report of the President: 1974 Activities*
SAO Release	Smithsonian Astrophysical Observatory News Release
SBD	*Defense/Space Business Daily* newspaper
Sci Amer	*Scientific American* magazine
Sci & Govt Rpt	*Science & Government Report*, independent bulletin of science policy
SciServ	Science Service News service
SD	*Space Digest* magazine
SD Union	*San Diego Union* newspaper
SET Manpower Comments	Scientific Manpower Commission's *Scientific, Engineering, Technical Manpower Comments*
Sf	British Interplanetary Society's *Spaceflight* magazine
SF Chron	*San Francisco Chronicle* newspaper
SF Exam	*San Franciso Examiner* newspaper
Sov Aero	*Soviet Aerospace* newsletter
Sov Rpt	Center for Foreign Technology's *Soviet Report* (translations)
SP	*Space Propulsion* newsletter
Spaceport News	NASA John F. Kennedy Space Center's *Spaceport News*
Spacewarn	IUWDS World Data Center A for Rockets and Satellites' *Spacewarn Bulletin*
SR list	NASA compendium of sounding rocket launches
SSN	*Soviet Sciences in the News*, publication of Electro-Optical Systems, Inc.
St Louis G–D	*St. Louis Globe-Democrat* newspaper
St Louis P–D	*St. Louis Post-Dispatch* newspaper
T-Picayune	New Orleans *Times-Picayune* newspaper
Tech Rev	Massachusetts Institute of Technology's *Technology Review*
Testimony	Congressional testimony, prepared statement
Text	Prepared report or speech text
Transcript	Official transcript of news conference or congressional hearing
UN Reg	United Nations Public Registry of Space Flight
UPI	United Press International news service
USGS Release	U.S. Geological Survey News Release
USPS Release	U.S. Postal Service News Release
W Post	*Washington Post* newspaper
W Star-News	*Washington Star-News* newspaper
WFC Release	NASA Wallops Flight Center News Release
WH Release	White House News Release
WJT	*World Journal Tribune* newspaper
WSJ	*Wall Street Journal* newspaper

INDEX AND LIST OF ABBREVIATIONS AND ACRONYMS

A300. See Airbus.
AAAS. See American Association for the Advancement of Science.
Abilene, Texas, 116
ablation, 37
aborigine, 251
abort system, 89, 244, 251
abrasion, 60
absorption, 46, 141, 227, 250, 281
AC-40 (Atlas-Centaur launch vehicle), 290
Academy of Sciences, USSR. See Soviet Academy of Sciences.
acceleration, 25
acceptance testing, 97, 259
accident, 15, 16, 44, 53, 59, 94, 135, 153, 209, 214, 244, 251, 253, 263, 273, 280, 282, 303
accretion, 56
accuracy, 126, 142, 151, 209, 281
acoustical holography, 106
acoustics, 27, 33, 168, 170, 240, 254, 267, 297
ACPL. See atmospheric cloud physics laboratory.
acrylic, 227
Advance II (research ship), 119
advanced manned strategic aircraft, 158
AE-C (Atmosphere Explorer C), 27
AE-D (Atmosphere Explorer D), 27
AE-E (Atmosphere Explorer D), 27
AEG-Telefunken, 136, 287
aerial photography, 105, 115, 133, 151, 269
Aeritalia, 165, 187, 209
aerodynamics, 23, 24, 33, 194
aerodyne (wingless aircraft), 29
aeroelasticity, 18
Aeroflot, 286
aeromechanics, 117
aeronautical research
—military, 84, 113
—NASA, centers, 5, 48, 54, 95, 117, 118, 241
—NASA-industry, 81, 97, 230
—overseas, 187, 246
aeronautics, 17, 39, 79, 85, 87, 115, 124, 193
aeronomy, 343
Aeronomy Laboratory, NOAA, 17, 304
Aeronutronic Ford Corp., 186, 232
Aerosat, 14, 130, 211, 214
Aerosat Space Segment Board, 14, 211
aerosols, 17, 77, 216, 261
Aerospace Corp. See Grumman.
Aerospace Industries Association, 3, 17, 81
aerospace industry, 18, 61, 72, 81, 82, 87, 135, 165, 166, 180, 187, 192, 211, 249, 277
Aerospace Medical Division, USAF, 172
aerospace sciences, 18, 63, 256
aerospace technology, 13, 18, 80, 126, 256
Aerospatiale (French contractor), 97, 130, 180, 187, 188, 195
Aetna Insurance, 78
AF. See Air Force, U.S.
AFML. See Air Force Materiel Laboratory.

Africa, 18, 73, 79, 84, 106, 136, 146, 176, 210
AFSATCOM. See Air Force Satellite Communications System.
Agency for International Development (AID), 105–106, 175, 281, 293
aging, 55, 111, 113, 255
agriculture, 40, 47, 73, 81, 137, 141, 178, 198, 221, 228, 237, 245, 262, 279, 282, 298, 299, 305
Agriculture, U.S. Dept. of (USDA), 141, 150, 198, 201, 254, 274, 299
agronomy, 254
Ahlborn, Dr. H., 96
Ahmadabad, India, 228
AIAA. See American Institute of Aeronautics and Astronautics.
AID. See Agency for International Development.
AIDSAT. See AID space-age technology.
AID space-age technology (AIDSAT) program, 175, 281, 293
aileron, 118
air-breathing engine, 47
Air Commerce Act of 1926, 85
air-cushion system, 84, 292
air-drag balance measurement, 20
Air Force, U.S. (USAF)
—contracts, 99, 108, 149, 186, 248, 271, 287
—cooperative programs, 27, 49, 90, 125, 233, 242
—flight testing, 14, 44, 47, 53, 84, 95, 123, 273, 276, 283, 288
—people, 19, 29, 44, 69, 70, 80, 99, 129, 145, 178, 231
—planes, 58, 81, 92, 184, 246, 285, 287
—policy, 158, 193, 199, 257, 285, 287
—satellites, 50, 79, 102, 133, 141, 195, 214, 294
—technological utilization, 17, 24, 61, 130, 134, 162, 172
Air Force Academy Foundation, 145
Air Force Association, 125, 161
Air Force Flight Test Center, 14
Air Force Geophysics Laboratory, 233
Air Force Materials Laboratory, 162
Air Force Satellite Communications System (AFSATCOM), 118
Air Force Space and Missile Systems Organization. See Space and Missile Systems Organization (SAMSO).
Air Force Systems Command (AFSC), 129, 130, 159, 170, 257, 267, 276, 283, 305
Air France, 3, 103, 269
air-launched missile, 286
Air Line Pilots Association, 16
Air Mail Act of 1925, 67
Air Mobility Research and Development Laboratory (U.S. Army), 258
Air National Guard, 109
air pollution, 3, 17, 19, 34, 39, 64, 76, 88, 102, 105, 117, 160, 189, 216, 221, 227, 261

air safety, 17, 66, 84, 89, 94, 97, 114, 134, 135, 149, 167, 209, 214, 229, 257, 263, 277, 282
air stewardess, 229
air-traffic control, 16, 33, 66, 68, 84, 94, 167, 181, 198, 214, 229, 263, 274
Air Transport Association, 124, 134, 166
Airborne Observatory. See Kuiper observatory, Galileo II.
Airbus (European passenger transport), 134, 188
airconditioning, 19, 47, 78, 108, 271, 288, 302
aircraft
—communications, 4, 212, 263, 276
—design, 8, 33, 71, 84, 87, 90, 95, 113, 123, 144, 181, 229, 231, 246, 248, 253, 258, 273, 274, 280, 282
—flight control, 61, 68, 95, 113, 167, 170, 194, 214
—fuel conservation, 89, 230, 280
—improvement, 33, 66, 84, 87, 95, 113, 125, 134, 149, 160, 164, 273, 280, 282, 285
—manufacturing, 17, 81, 89, 125, 135, 149, 161, 180, 190, 211
—technology, 8, 14, 33, 285
—testing, 84, 95, 164, 229
—use for research:
AF, 14, 28
Army, 40, 258
Coast Guard, 2, 236
FAA, 67
NASA, 33, 44, 46, 76, 95, 132, 144, 145–146, 194, 249, 256, 261, 272, 280, 295–296
NOAA, 17, 204, 277
university, 300
USSR, 6, 8, 73
AiResearch Manufacturing Co. of Calif., 142
airfoil, 5, 29, 81
airlines, 15, 16, 26, 85, 124, 125, 134, 149, 166, 180, 181, 182, 198, 212, 256, 269, 277, 285, 286
airlock, 271
airmail, 67
Airplane Research Institute, Vienna, 28
Akasofu, Dr. Syun-ichi, 296
airport, 4, 33, 66, 68, 84, 89, 118, 124, 134, 167, 181, 198, 229, 257, 274, 276
airships, 6, 41, 97
airstream, 33, 130, 132
Aksenov, Vladimir, 220–221, 342
Alabama, 106, 241, 276
Alamogordo, N.M., 27, 238, 259
alarm system, 102
Alaska, 29, 37, 62, 63, 69, 93, 94, 247, 261, 295, 296, 343
Alaska Native Claims Settlement Act of 1971, 69
Alaska, Univ. of, 69, 296
Alberta, Univ. of, 296
Alcator (fusion machine), 36
Aldrin, Edwin E., Jr., 178
Aleksandrov, Anatoly P., 76, 221
Aleuts, 69
Alfred P. Sloan Foundation, 166
Algernon Blair Co., 262

Algonquin Radio Observatory, Ontario, 302
alien, 252, 264
All-American air race, 275
Allegheny mountains, 109
Allen, H. Julian, award, 48
Allende meteorite, 56
alloys, 95–96
alluvial deposits, see geology.
Alma Ata, USSR, 8, 286
Alpha Building Corp., 265
ALSEP. See Apollo lunar scientific experiment package.
ALT: approach and landing tests
Altai territory, USSR, 177
altimeter, 46
altitude
—aircraft, 47, 76, 123, 164, 165, 170, 280, 300
—balloon, 216, 229, 230, 272, 304
—missile, 53
—over (earth, moon, planet) surface, 46, 64, 82, 119, 122, 131, 138, 291
—rocket, 126, 250
—spacecraft, 66, 79, 166, 218, 227, 271, 279, 282
aluminum, 17, 61, 67, 96, 134, 146, 157, 160, 170, 197, 222, 225
amateur radio, 195, 248, 263, 276
Amazon river, 240
ambassador, 79
American Association for the Advancement of Science (AAAS), 1, 32, 33, 35
American Astronautical Society (AAS), 241
American Broadcasting Co. (ABC), 152
American Chemical Society, 216
American Federation of Government Employees, 191
American Geophysical Union, 77, 281, 290, 296, 301
American Heart Association, 259
American Indians, 69, 288, 299
American Institute for Chemical Engineering, 186
American Institute of Aeronautics and Astronautics (AIAA), 3, 65, 106
American Institute of Industrial Engineers, 99
American National Standards Institute, 78
American Physical Society, 25, 238, 242
American Radio Relay League, 196
American Revolution Bicentennial Assn., 96
American Rocket Society, 241
American Society for Mechanical Engineering, 186
American Society of Engineering Education, 241
American Telephone and Telegraph (AT&T), 18, 94, 157, 186
Ames Research Center (ARC)
—contracts, 94, 97
—meetings, 103
—missions, 49, 58
—people, 48, 144, 156, 175, 177, 218, 231, 275, 277, 278, 295
—programs, 33, 38, 45, 61, 77, 117, 127, 168, 194, 231, 258, 261, 269–270
amphibious craft, 90, 251, 265

amplifier, 6
Ams 1 (DOD advanced meteorological satellite), 214, 327
AMSA: advanced manned strategic aircraft
Amsat (amateur radio comsat, W. Germany), 248
An Analysis of Federal R&D Funding by Function, Fiscal Years 1967–1977, 192
anaerobic, 40
Anders, William A., 48, 79
Anderson, Arthur T., 151
Anderson, Dr. Don L., 237
Anderson, Jack, 78
Anderson, Dr. James G., 216
Andes, 45
Andover, Me., 200
Andoya range, Norway, 301
Andrews Air Force Base, 129
animals, 38, 64, 113, 172, 298, 305
anomaly, 15, 17, 71, 78, 183, 250–251, 277
anorthosite, 55
Antarctic, 17, 79, 175, 200, 209, 261, 293
Antares (star), 115
antenna, 31, 62, 79, 84, 123, 141, 150, 165, 181, 197, 205, 207, 213, 228, 229, 233, 268, 269, 272, 275, 278, 292
antiaircraft missile, 287
antiballistic missile, 151, 230
antigen, 251
Antilles, 83
antisatellite spacecraft. See interceptor satellite.
antisubmarine warfare, 292
Apollo (program)
—history, 10, 29, 51, 66, 76, 80, 90, 112, 119, 126, 129, 147, 148, 231, 285, 298–299
—people, 18, 19, 27, 37, 39, 48, 50, 53, 54, 57, 70, 79, 91, 96, 137, 178, 231, 238, 259, 270, 277, 278, 282–283, 305
—results, 11, 20, 36, 37, 57, 60, 63, 66, 74, 78, 97, 111, 116, 158, 179, 246, 256, 264, 272, 300
Apollo 8, 48, 79, 283
Apollo 9, 19
Apollo 10, 51
Apollo 11, 18, 35, 54, 112, 137, 148, 150, 178, 246, 305
Apollo 12, 11, 37, 272
Apollo 13, 37
Apollo 14, 11, 19, 36, 74, 150
Apollo 15, 11, 66, 272
Apollo 16, 11, 272
Apollo 17, 11, 50, 131, 246, 270, 272
Apollo 204 fire, 54
Apollo lunar scientific experiment package (ALSEP), 11, 36, 74, 272
Apollo-Soyuz Test Project (ASTP), 20, 27, 29, 51, 53, 54, 73, 76, 80, 119, 126, 129, 163, 190, 193, 221, 231, 244, 277, 282–283, 298, 304
Apollo telescope mount (ATM), 179
Apple (Ariane passenger payload experiment), India comsat, 248
Applications Technology Satellite (ATS) program (see also Ats 6), 47, 228
approach and landing (Shuttle tests), 53, 62, 66, 98, 204, 210, 220, 222, 229, 250–251, 301
APT: automatic picture transmission
Aquila (constellation), 150
Arab states, 69, 210
ARC. See Ames Research Center.
ARCO. See Atlantic Richfield Co.
Arctic research, 6, 73, 209, 247, 301
Arcturus (star), 155
Ardashat, Armenia, 231
Arecibo, P.R., 33, 77, 84, 148, 213
Argentina, 149, 293, 295
argon, 83, 139, 146, 155, 181
Argyre (region of Mars), 162
Ariane (ESA launch vehicle), 80, 88, 97, 192, 239, 248, 249, 287
Aries (rocket), 49
Arizona, 7, 71, 121, 139, 209, 225, 240
Arizona, Univ. of, 82, 132
Arkalyk, USSR, 251
Arlington, Tex., 258
Armament Development and Test Center, USAF, 170
Armenia, 231
arms limitation agreement, 230, 286
arms race, 1, 30, 32, 47, 238, 281, 286, 289
Armstrong, Neil A., 18, 63, 147, 238, 255
Army, U.S., 25, 28, 40, 72, 113, 193, 237, 261, 262, 271, 279, 285, 286, 304
Army Communications Command, 298
Army Missile Command, 25
Army Signal Corps, 40
Arnold Engineering Development Center, USAF, 24, 130, 283
Arnold, Dr. James R., 131
ARO, Inc., 131
arrays. See solar conversion devices.
arsenic, 157
art, 137
artificial gravity, 38
artificial limbs, organs, 37, 63
Artsimovich, Dr. Lev, 36
AS–W17 (sailplane), 109
asepsis, 23
Asia, 74, 106, 177, 215, 238, 276, 289, 303
Asimov, Isaac, 176
Askins, Barbara S., 158
Aspin, Rep. Les (D-Wis), 7, 58, 160
assembly, 97, 112, 169, 198, 222, 245, 258, 270, 278, 295, 300
ASSESS (airborne science Spacelab-experiment systems simulation), ESA program, 194, 227
ASSESS II, 194, 227
Associated Universities, Inc., 32
Association of Universities for Research in Astronomy (AURA), 7
asteroid, 25, 35, 105, 168
ASTP. See Apollo-Soyuz Test Project.
Astrain, Santiago, 21, 238
astrogeology, 77
astronaut (see also names of individuals)
—administration, 19, 46, 57, 95, 224, 270
—history, 4–5, 10, 25, 27, 51, 66, 119, 137, 147, 222, 231, 238, 255, 282–283, 285

—research, 11, 14–15, 20–21, 26, 36, 52, 59–60, 63, 66, 74, 80, 90, 111, 127, 145, 159, 232–233, 255, 256, 264, 272, 300
—training, 118, 142, 162, 194, 220, 227, 229, 257, 301
Astronautics Engineer Award, 49
astronomers, 7, 23, 27, 46, 74
astronomy, 7, 32, 33, 35, 77, 93, 108, 158, 170, 179, 198, 220, 227, 244, 272, 277, 283, 294, 298, 302, 305
astrophysics, 148, 193, 198, 298, 343
asylum, political, 210
ATA. See Air Transport Association.
Atkinson, Dr. Richard T., 185
Atlanta, Ga., 35, 247
Atlanta International Airport, 257
Atlantic City, N.J., 257
Atlantic ocean, 14, 18, 31, 47, 60, 66, 83, 89, 103, 115, 118, 129, 143, 145, 169, 181, 186, 195, 200, 212, 223, 239, 242, 249, 256, 267, 276
Atlantic Richfield Co., 37
Atlantic Science Corporation, 132
Atlas (launch vehicle), 99
Atlas-Centaur, 18, 49, 94, 157, 211, 218, 290, 295
Atlas MA–5, 149
ATM. See Apollo telescope mount project.
atmosphere (see also individual planets)
—ionosphere, 20, 32, 142, 233, 239, 250
—magnetosphere, 11, 49, 68, 142, 301
—planetary (see individual planets)
—protonosphere, 250
—research, 15, 27, 32, 53, 76, 77, 119, 142, 148, 187, 189, 249, 272, 277, 279, 288, 295, 300, 304
—stratosphere, 17, 32, 53, 64, 76, 216, 272, 295, 304
—thermosphere, 19, 27, 300
—upper atmosphere, 27, 46, 53, 64, 73, 148, 163, 187, 221, 230, 233, 250, 261, 272, 300
Atmosphere Explorer (AE) project, 19, 27
Atmospheric Cloud Physics Laboratory, 249, 278
atmospheric pollution, 3, 17, 64, 105, 189, 232, 235, 261, 300
atmospheric pressure, 8, 19, 104, 146
Atomic Energy Commission. See ERDA; NRC.
atomic physics, 254, 273
atomic power, 36, 74, 127, 146, 149, 252, 254, 297
atomic research, 7
ATS. See Applications Technology Satellite.
Ats 3, 10
Ats 6, 49, 93, 105, 111, 175, 228, 281, 293, 301
AT&T. See American Telephone and Telegraph Co.
attitude control, 20, 59, 218, 297
Atwood, J. Leland, 275
Auburn, Mass., 51
Auburn Univ., Ala., 241
AURA. See Association of Universities for Research in Astronomy.

Auriga (constellation), 108
aurora, 37, 49, 63, 68, 233, 296, 301
Australia, 35, 93, 150, 151, 172, 251, 261, 263, 294
Austria, 143, 237, 238
automatic beam fabrication, 278
automatic picture transmission (APT), 133
automated systems, 60, 76, 79, 144, 148, 160, 168, 177, 182, 197, 227, 288, 295, 342
automobile, 29, 39, 135, 146
autoradiographic film intensification, 158
AV8 "Harrier" (STOL aircraft), 164
Avco Corporation, 248
Avco Everett Research Laboratory (AVCO Electronics), 35, 39, 80
Aviation Historical Foundation, 67, 97
Aviation Ministry, USSR, 286
aviation policy, 125, 212, 274
aviation safety. See air safety.
avionics, 28, 81, 117, 247
Aviation Space Writers Association, 96
awards, 18, 40, 48, 49, 65, 79, 90, 99, 133, 143, 165, 186, 196, 204, 229, 231, 235, 237, 241, 247, 254, 256, 259, 275, 280, 282, 283, 293
Azores Is., 83, 242
Azur (joint U.S.–FRG satellite), 7

B–1 (bomber), 51, 81, 92, 108, 123, 158, 161, 185, 248, 257, 285, 287
B–52 (bomber), 47, 58, 97, 286, 287
B–747. See Boeing 747.
background radiation, 15, 217, 277
Backus, John W., 254
bacteria, 44, 45, 60, 149, 216, 267
Bahamas, 19, 57
Bahrein, 3, 269
Bailey, Dorinda S., 282
Baker, Dr. William O., 9
Balachandran, Nambath, 300
Baldwin, Jack, 124
ballistic missiles, 286–287
balloon, 6, 11, 17, 29, 68, 88, 115, 117, 136, 138, 145, 160, 171, 190, 212, 216, 229, 230, 242, 272, 288, 304
balloon-borne telecommunications and broadcast system, 136
balloon-borne ultraviolet stellar spectrometer (BUSS), 115
Baltic Sea, 263
Baltimore, Md., 36, 37, 73, 80, 99
Banana River, 169
Bandanaraike International Airport (Colombo, Sri Lanka), 96
Bangladesh, 176, 293
banks, 142, 191
Barents Sea, 288
Barfield, Dr. Joseph N., 301
barium, 49, 63
Barron, James, 127
Barstow, Calif., 197
basalt, 55, 122, 148, 215, 272, 281, 300
basic research, 238, 251
Basov, Nikolai, 278
Battelle Memorial Institute, 94

Battelle-Northwest Research Institute, 95
battery, 15, 144, 181, 213, 219, 275, 280, 291
Baykonur cosmodrome, USSR, 140, 204, 220, 221, 250, 273
beacons, 66, 233
beam separation, 18
Beckman, Dr. John, 194
Bedford, Mass., 190
Beech Aircraft Corp., 145
Beer, Dr. Reinhold, 46
behavior, 148, 159, 199
Belenko, Lt. Victor I., 210
Belgium, 4, 130
Bell, Alexander Graham, 139, 292
Bell Laboratories, 9
Bell X–1 (supersonic aircraft), 137
Bell Helicopter Textron, 258
Bendix Corporation, 36, 262
Bendix trophy, 275
Benedict, Manson, 254
Benefield, Doug, 124
Benjamin Franklin (research sub), 159
Bennett (comet), 45
Beregovoy, Maj. Gen. Georgy, 304
Bering Strait, 143
Bering, Vitus, 143
Berkeley, Calif., 56
Berlin, Germany, 224
Bermuda, 46, 83, 181, 212
Bern, Univ. of, 187
beryllium, 25
Bethe, Hans W., 254
Bethpage, N.Y., 118
Bicentennial, U.S., 66, 135, 243, 292
—NAA, 85, 97, 98
—NASA and centers, 25, 29, 70, 107, 123, 132, 139, 144, 145, 150, 210, 270
—NASM, 137
—USAF, 80
Biemann, Dr. Klaus, 157, 236, 274
"big bang" theory, 32
"Big Bird" (reconnaissance satellite), 141
Bignier, Michel, 241, 269
Bijvoet, Jan A., 48
bilharzia, 306
Biliu, David S., 227
Bimini, 57
Binder, Alan S., 155
biochemistry, 1, 40, 273
bioengineering, 15, 20, 23, 37, 63, 65, 172
biogas, 40, 306
biohazard, 11
biology
—international research, 148, 212, 306
—Mars, 17, 100, 139, 154–157, 177, 199–200, 203–204, 208, 215, 217, 236–237, 260, 274, 295
—NASA research, 35, 126, 169
—USSR research, 38, 119, 159, 193, 231
biomedicine, 71, 148, 188, 190
biosat, 38
biosensor, 60
Biospherics, Inc., 156, 175, 275
biotechnology, 172, 199
Bir, James E., 98

birds, 134, 172
Black Brant (sounding rocket), 95, 297
Black Businessmen's Assn. of Los Angeles, 303
black holes, 26, 32
Black Sea, 195
Bledsoe, Maj. Adolphus (Pat), 164
Blinov, Nikolay, 6
Block 5D (USAF metesat program), 214
block grants, 13
blood, 20, 23, 47, 63, 171, 251, 255
Blumberg, Dr. Baruch S., 251, 255
Board on Science and Technology for International Development, NAS, 306
boats, 50, 117
Boeing 7X7 (proposed jet aircraft), 187
Boeing 707, 124, 180, 191, 209
Boeing 727, 124, 188
Boeing 131, 180, 188
Boeing 747, 37, 62, 66, 134, 137, 170, 180, 191, 210, 282, 287
Boeing Aerospace Corp. (see also Boeing Commercial Airplane Co., Boeing Co.), 47, 53, 60, 75, 186, 187, 193, 270
Boeing Commercial Airplane Co., 33
Boeing Company, 60, 75, 180, 186, 187, 188, 198, 229, 230, 240, 277, 287, 302, 304
Boldt, Dr. Elihu, 101
Bolivia, 191, 293
bomb. See weapons systems.
Bombay, India, 48, 128
bomber aircraft, 47, 58, 158, 161, 191, 257
Bonnet, Dr. Roger, 178
booster. See launch vehicle.
Bootes (constellation), 115
boranes, 254
Borman, Frank, 283
boron-epoxy composite, 222
Boston, Mass., 33, 36, 51, 96, 98, 102, 143
Boston Univ. School of Medicine, 176
Boulder, Colo., 179, 304
boundary layer, 5, 231
Boyce, J.C., 298
Bradsher, Henry S., 229
Bradt, Dr. Hale, 101
Braginsky, Prof. Vladimir B., 45
brain, 252
braking, 121
Brand, Vance, 27, 282
Brazil, 5, 117, 238, 240, 293
breakthrough, 36, 80
Bredt, James H., 65
breeder reactor, 11
Brekhovskikh, Leonid, 15
Bremen, W. Germany, 113, 292
Brigham City, Utah, 188
Britain. See United Kingdom.
British Aircraft Corp., 165, 239, 301
British Airways, 3, 103, 180, 190, 212, 214, 269
British Interplanetary Society, 108, 283
bromine compounds, 64
Brookhaven National Laboratory, 66, 242, 304
Brooks, Rep. Jack (D-Tex.), 218
Broome, G. Calvin, 271

359

Broten, N.W., 302
Brown University, 166
Broyles, Howard, 172–173
Buchwald, Art, 154
Buck Rogers, 154
budget (federal), 1, 12, 14, 16, 43, 58, 134, 138, 151, 160–161, 193, 201, 274
—NASA. See NASA, funding.
Budget Reform Act of 1974, 12
Buffalo (transport aircraft), 33
Buffalo, N.Y., 36
Bugayev, Boris, 286
Bulgaria, 128, 163, 219, 224, 250, 304
Bumper 8 (captured V–2 rocket), 261
buoys, 29, 119, 145, 267
Burch, Dr. James L., 142
Burger, Jan J., 91
Burroughs, Edgar Rice, 100
Bushveld area, S. Africa, 294
BUSS: balloon-borne ultraviolet stellar spectrometer
Bykovsky, Col. Valery, 220–221, 342

C–3 experiment (HEAO), 278
C–5A galaxy, 134
C–8 "Buffalo" (transport plane), 33
C–47 "Gooney Bird" (transport plane), 66, 115
C–54 (research aircraft), 119, 133
C-band telemetry, 115
CAB. See Civil Aeronautics Board.
cadmium, 41
Cal-Doran Metallurgical Services, 82
calcium, 157, 187, 224
calendar, 12
calibration, 62
caliche, 224
California, 24, 38, 115, 121, 124, 150, 175, 198, 218, 224, 225, 240, 243
California Institute of Technology (Caltech), 23, 25, 45, 64, 69, 72, 101, 119, 132, 133, 151, 225, 237, 275, 278, 300
California, Univ. of
—Berkeley, 254, 296
—Livermore Laboratory, 216
—Los Angeles, 132, 226, 301
—San Diego, 1, 101, 131
—San Francisco, 45
Calio, Dr. Anthony, 100
Calspan Corp., 67
Caltech. See California Institute of Technology.
Cambridge Light Co., 125
Cambridge Research Laboratory, USAF, 69
camera, 43, 101, 119, 121, 130, 133, 152, 154, 220–221, 224, 225, 288, 296, 342
Cameroon, 181, 293
Canada, 1, 9, 14, 48, 76, 84, 114, 117, 141, 149, 179, 200, 212, 214, 247, 249, 289, 296, 302, 304, 337, 338, 343
Canadian Department of Communications (CDC), 10, 337, 338
Canadian Department of Mines and Resources, 244

Canadian Department of National Defence, 114
canals of Mars (see also Mars), 100, 154
Canaveral. See Cape Canaveral.
cancer, 11, 35, 99, 137, 216
Cannon AFB, N.M., 184
Cannon, Sen. Howard (D-Nev.), 67
Canoga Park, Calif., 63, 219
capacity, 118, 120, 162, 291
Cape Canaveral, Fla., 1, 31, 107, 131, 133, 157, 193, 223, 224, 261, 271, 276, 295, 301
Cape Fear Technical Institute, 119
Cape Kennedy. See Cape Canaveral.
Capella (star), 108
Capri Chasm (Mars), 139
capsule, 251
"capture cell," 167
carbon, 37, 56, 60, 61, 64, 224
carbon–14, 199
carbon-based life forms, 215, 224, 236
carbon dioxide, 83, 139, 146, 155, 156, 199–200, 215, 226
carbon monoxide, 46, 64, 146, 189, 199–200, 215, 264
carcinogenic materials, 9, 60, 80
cardiovascular research, 255
career, 72, 164
Carey, Governor Hugh (N.Y.), 3
Cargill Investor Services, 141
cargo, 59, 97, 166, 257, 292
Caribbean Sea, 66, 76, 176, 182
Carr. Gerald P., 129
Carr, Dr. Michael, 218
carrier rocket, 299
cartel, 182
Carter, President Jimmy, 91, 161, 185, 238, 285, 287
cartography, 73, 262
Casani, John R., 43, 71, 245
Case, Sen. Clifford P. (D-N.J.), 128
Case Institute of Technology, 62
Caspian Sea, 41, 177, 273
Castor (rocket booster), 65
catalyst, 200, 216, 274
catastrophe, 154
Cavaillé, Marcel, 187, 269
CDC. See Canadian Dept. of Communications.
celestial mechanics, 8
cells, 64
CEMA (international socialist standing committee for post and telecommunications), 225
censor, 29
census, 262
Centaur (rocket), 49, 128
Centennial exposition, U.S., 138, 292
Central African Republic, 293
Central America, 76, 106, 176
Central Asian power grid, 29
Central Intelligence Agency (CIA), 141
Centre National de la Récherche Scientifique (CRNS), Paris, 178
Centre National d'Études Spatiales (CNES), Paris, 88, 212, 241, 269
centrifuge, 38

ceramic, 53, 126, 130, 166, 236
CERN. See European Center for Nuclear Research.
Cernan, Eugene A., 50, 51, 57
Cerro Tololo Inter-American Observatory, 7
certification, 114, 134, 233, 235
cervit, 7, 53
cesium ion thruster, 181
Cessna 402 (aircraft), 231
CF6 (GE engine), 134
CFM 56 (GE–Snecma engine), 188
challenge, 22
channels, on Mars. See Mars.
Chappell, Dr. Charles R., 142
charged particle, 36, 58, 278
Charleston, S.C., 143
charter flights, 182, 213
Charyk, Joseph V., 92
Chattanooga, Tenn., 77
chemical analysis, 300
chemistry
—awards, 254
—Mars, 146, 147, 154, 178, 199–200, 203, 205, 215–216, 224, 236–237, 274, 295
—moon, 54, 183–184, 300
—pollution, 35, 60–61, 80, 88, 117, 160, 235, 272
—technology utilization, 37, 81, 131, 143, 158, 167, 193, 280
chemotherapy, 254
Chern, Shiing-Shen, 254
Chetveryakov, S.D., 246
Chicago, Ill., 35, 67, 68, 85, 141, 167, 257
Chicago, Univ. of, 105, 255
"chicken soup," 156
Chile, 7, 45, 191, 238
China. See Peoples Republic of China; Republic of China (Taiwan).
China 4, 5
China 5, 306
China 6, 196, 306, 307, 327
China 7, 290, 307, 332
Chitwood, John S., 293
chlorine compounds, 17, 61, 64, 216
chloroform, 60
chlorophyll, 45
chrome, 141
chromosphere, 88
Chryse (region of Mars), 77, 121, 138–140, 146, 148, 153, 155, 203, 205, 208, 215–216, 217, 226, 274
CIA. See Central Intelligence Agency.
Cicerone, R.J., 216
Cincinnati, O., 36
Cincinnati, Univ. of, 63
circuits, satellite/airplane, 18, 30, 153, 247
citizen groups, 124
Civil Aeronautics Administration (CAA), 66
Civil Aeronautics Board (CAB), 26, 182
civil aviation, 15, 26, 33, 61, 67, 81, 84, 85, 113, 134–135, 167, 180–181, 181–182, 187, 195, 198, 209, 212, 214, 229, 230–231, 242, 253, 257, 263, 269, 274, 275, 277, 282, 283, 285, 286, 300, 304

civil service. See federal employees; personnel, U.S. govt.
Civil Service Commission (CSC), 167, 191, 192, 198, 267, 282
Civil Transport Development Corporation (Japan), 187
civilian space program, 21, 37, 66, 71, 185
Clark, Dr. Benton, 157
Clark, Dr. John F., 51, 77, 133
Clark, Leonard, 261
Clarke, Arthur C., 21, 159
Clayton, N.Mex., 280
clean room, 271
cleaning, 112, 227
Clemson Univ., 93
Cleveland, O., 36, 67, 143, 145, 280
Cleveland Pneumatic Co., 81
climate. See meteorology.
Climatic Impact Assessment Program, 296, 304
clocks, 45, 126, 304
cloud, 132, 143, 148, 150, 177, 249, 278, 279, 305
cloudcover, 10, 14, 39, 46, 165, 192, 213, 247
Clovis, N.Mex., 16
CNES. See Centre National d'Études Spatiales.
coal, 11, 35, 58, 125, 151, 176
Coast Guard, U.S., 4, 130, 236, 247, 303
Coleman, William T., Jr., 3, 102, 124, 212, 285
Collet, Jacques, 91
Collier trophy, 40
collimator, 101
Collins, Michael J., 137, 231, 278, 305
Collins Radio, 29
collision, 25, 26, 214
Colom, Audrey Rowe, 270
Colombia, 10, 238, 293
colonization of space, 80, 107, 127, 139, 159, 176, 244
color, 10, 106, 133, 141, 143, 153, 155, 186, 204, 209, 215, 221, 279, 281, 293
Colorado, 66, 150, 179, 296
Colorado, Univ. of, 31, 296
Colson, Jack, 196
Columbia Broadcasting System (CBS), 152, 270
Columbia Univ., 25, 254, 289, 300
combat aircraft. See military aircraft.
combustion, 58, 134, 216
comet (see also individual comets), 8, 31, 45, 168, 252, 343
command and data acquisition, 8
commemoratives, 67
Commerce, U.S. Dept. of, 26, 85, 107, 177, 201
commercial aviation, 15, 26, 67, 81, 85, 89, 102, 124, 134, 230, 261, 269
commercial broadcasting, 18, 37, 152
commercial spacecraft, 28, 31, 75, 117, 196
commercialization, 29, 188, 203, 211, 223, 227, 239, 248, 268, 269, 288, 289, 299, 301
Committee on Peaceful Uses of Outer Space, 128

361

Committee on Space Research (COSPAR), 171, 222
Commonwealth Scientific and Industrial Research Organization, 261
communications
—commercial, 9, 30, 35–36, 77, 82, 92, 125, 252, 288
—galactic, 268, 269–270
—Intelsat, 18, 20, 21, 73, 119, 186
—international, 9, 10, 69, 105–106, 114, 135, 136, 141, 175–176, 181, 200–201, 209–210, 211–212, 213, 228, 239, 240, 248, 275, 281–282, 286–287, 293
—relay, 4, 146, 205, 215, 223, 225, 226, 233, 249, 264, 280, 288 (see also data relay; satellite relay)
—satellites, 1, 2, 6, 18, 29, 31, 37, 57, 60, 61, 79, 89, 93, 94, 115, 132, 157, 195, 197, 223, 226–227, 233, 249, 263–264, 292, 294–295
(See also individual comsats: Aerosat, AIDSAT, Ats, Comstar, *Copernicus*, *Cts*, *Ekran*, *Goes*, *Intelsat*, *Marisat*, *Nato IIIA*, *Oscar*, *Palapa*, *Raduga*, *Satcom*, Seasat, *Symphonie*, TDRSS, *Westar*)
—USSR, 59, 117–118, 214–215, 224–225, 263, 272, 276, 298
Communications Research Center, Canada, 114
Communications Satellite Act, of 1952, 21
Communications Satellite Corporation (ComSatCorp), 6, 9, 21, 30, 76, 77, 92, 115, 149, 181, 196, 201, 245, 252, 307
Communications Technology Satellite (*Cts 1*), Canadian comsat, 1, 9, 247, 249, 308, 337, 338
compatibility, 113
competition, 182, 188, 204, 265, 275, 285, 303
components, 259
composite materials, 61, 62, 67, 89, 96, 135, 222
computer, 8, 43, 61, 78, 80, 81, 89, 112, 114, 131, 141, 154, 155, 159, 161, 172, 177, 189, 193, 204, 224, 229, 232, 254, 263, 268, 279, 303
Computer Sciences Corp., 91, 114
COMSAT; ComSatCorp. See Communications Satellite Corporation.
Comsat General Corp., 2, 14, 31, 37, 60, 82, 89, 94, 157, 212, 223, 249, 252
comsats. See communications satellites.
Comstar D–1 (1 A–1), AT&T satellite, 2, 94, 252, 317, 337, 339
Comstar D–2, 157, 290, 324, 337, 339
Concorde (SST aircraft), 3, 102, 125, 149, 188, 189, 195, 269, 300
condensation, 56, 121, 139
Conference on Applied Science and Technology in the Arab World, 176
Conference on Trade and Development, 293
confirmation, 129
Congo, 293
Congress, U.S., 12, 16, 39, 50, 58, 124, 138, 165, 185, 192, 193, 228, 235, 246, 247, 253, 287, 299
—House of Representatives, 58, 75, 111, 149, 160, 161, 185, 222
—House Committee on:
Government Operations, 218–219
Science and Technology, 19, 222, 224, 235
—Joint Committee on Defense Production, 57
—Senate, 58, 67, 91, 128, 129, 151, 161, 184, 185, 195, 222, 270, 305
—Senate Committee on:
Aeronautical and Space Sciences, 13, 62, 112, 126
Appropriations, 1, 151, 161
Armed Services, 158
Commerce, 3
Foreign Relations, 128
Congressional Research Division. See Library of Congress.
Connecticut, 113, 115, 150, 304
Conn. Dept. of Commerce, 188
Conn. Product Development Corp., 188
Conn., Univ. of, 188
conservation, 11, 30, 41, 137, 230, 256
consortium, 21, 101, 187, 209, 214, 249, 289
Constellation (transport aircraft), 72
Constitution (suggested name for Space Shuttle orbiter), 211
construction, 58, 59, 62, 77, 248, 270
construction of facilities, 72, 225, 265, 271, 278
consumables, 142
consumers, 94, 182, 216
contained reaction, 36
containerless processing, 297
contamination, 23, 57, 105, 216, 232, 236, 249, 267, 297, 304
Contemporary Arts Museum, 127
contractors
—airmail, 67
—ComSatCorp, 91
—DOD, 81, 92, 99, 108, 141, 158, 161, 162, 164, 184, 186, 248, 266, 287, 292, 302, 304
—ERDA, 52, 94, 116, 124, 142, 167, 181, 225, 270
—European, 6, 113, 130, 190, 195, 209, 212, 214, 249, 286, 292
—FAA, 257, 304
—FCC, 125
—Intelsat, 15, 20, 179, 181, 186, 239
—NASA, 1, 47, 57, 65, 78, 81, 94, 97, 101, 107, 111, 112, 113, 114, 130, 149, 168, 170, 179, 190, 191, 198, 211, 226, 230, 232, 259–260, 265, 267, 278, 280, 296, 298, 299
—NOAA, 242
—NSF, 7
—Shuttle, 19, 25, 73, 74, 77, 78, 82, 98, 145, 150, 161, 162, 168, 188, 192, 198, 203, 218, 262, 271, 277, 289, 300, 303
Control Data Corp., 112
control systems, 168, 188, 193–194, 199–200, 225, 240, 251, 257, 262, 279, 282

Convair (division of General Dynamics), 211, 253
Convair 990 ("Galileo II," NASA research aircraft), 194, 277
convection, 186
conventional weapons, 113, 257, 281
converter, 264, 270
cooling systems. See airconditioning; heat transfer; solar heating and cooling.
Cooper, Dr. Robert S., 77, 133, 271, 293
Cooperative Institute for Research in Environmental Sciences, 296
coordination, 44, 53, 59, 91, 142, 293
Copernicus. See Orbiting Astronomical Observatory.
copper, 41
core sampling, 32, 289, 300
Cornell Univ., 32, 33, 82, 166, 208, 254, 274
corona. See solar corona.
coronagraph, 298
Corporation for Public Broadcasting, 18, 125, 270
Corps of Engineers. See Army, U.S.
correlation, 67
corrosion, 17, 81, 169, 236
cortisone, 254
Cortright, Edgar M., 18
cosmic dust, 8, 15, 105
cosmic event, 45, 253
"cosmic golfball," 23, 87, 337
cosmic radiation, 8, 11, 33, 88, 190, 212, 221, 229, 230, 249, 272, 278
Cosmic Research Institute, USSR, 212
"cosmic soccerball" (rescue system), 59
cosmonaut, 5, 73, 76, 129, 140, 176, 190, 193, 197, 204, 219, 221, 245, 250–251, 260, 282, 303, 304, 341
COSMOS (consortium of European aerospace companies), 130, 214, 249
Cosmos (series of USSR satellites), 73, 307
Cosmos 782, 38, 119, 212
Cosmos 787, 308
Cosmos 788, 308
Cosmos 789, 307, 309
Cosmos 790, 309
Cosmos 791, 309
Cosmos 792, 309
Cosmos 793, 309
Cosmos 794, 310
Cosmos 795, 310
Cosmos 796, 310
Cosmos 797, 310
Cosmos 798, 310
Cosmos 799, 311
Cosmos 800, 307, 311
Cosmos 801, 311
Cosmos 802, 311
Cosmos 803, 34, 311
Cosmos 804, 34, 312
Cosmos 805, 312
Cosmos 806, 313
Cosmos 807, 313
Cosmos 808, 314
Cosmos 809, 314
Cosmos 810, 314
Cosmos 811, 315
Cosmos 812, 315
Cosmos 813, 315
Cosmos 814, 315
Cosmos 815, 316
Cosmos 816, 316
Cosmos 817, 317
Cosmos 818, 318
Cosmos 819, 318
Cosmos 820, 318
Cosmos 821, 318
Cosmos 822, 318
Cosmos 823, 307, 319
Cosmos 824, 319
Cosmos 825, 319
Cosmos 826, 319
Cosmos 827, 320
Cosmos 828, 320
Cosmos 829, 320
Cosmos 830, 320
Cosmos 831, 320
Cosmos 832, 321
Cosmos 833, 321
Cosmos 834, 321
Cosmos 835, 322
Cosmos 836, 322
Cosmos 837, 322
Cosmos 838, 322
Cosmos 839 (target spacecraft), 163, 323
Cosmos 840, 323
Cosmos 841, 323
Cosmos 842, 307, 324
Cosmos 843, 324
Cosmos 844, 324
Cosmos 845, 325
Cosmos 846, 325
Cosmos 847, 325
Cosmos 848, 326
Cosmos 849, 326
Cosmos 850, 326
Cosmos 851, 326
Cosmos 852, 326
Cosmos 853, 327
Cosmos 854, 214, 327
Cosmos 855, 328
Cosmos 856, 328
Cosmos 857, 328
Cosmos 858, 328
Cosmos 859, 329
Cosmos 860, 307, 329
Cosmos 861, 307, 329
Cosmos 862, 330
Cosmos 863, 330
Cosmos 864, 307, 330
Cosmos 865, 330
Cosmos 866, 330
Cosmos 867, 331
Cosmos 868, 331
Cosmos 869, 303, 331
Cosmos 870, 331
Cosmos 871, 332
Cosmos 872, 332
Cosmos 873, 332
Cosmos 874, 332
Cosmos 875, 333

Cosmos 876, 333
Cosmos 877, 338
Cosmos 878, 333
Cosmos 879, 333
Cosmos 880, 306, 334
Cosmos 881, 305, 334
Cosmos 882, 305, 334
Cosmos 883, 307, 334
Cosmos 884, 334
Cosmos 885, 334
Cosmos 886, 306, 335
Cosmos 887, 307, 335
COSPAR. See Committee on Space Research.
cost control, 1, 53, 58, 161, 193, 228, 260, 265, 270, 279
cost reduction, 6, 7, 66, 67, 70, 102, 111, 114, 124, 125, 131, 135, 141, 160, 166, 167, 169, 183, 187, 194, 213, 219, 229, 239, 247, 268–269, 271, 276, 292
Costa Rica, 293
Coty, Ugo, 102
Council on Wage and Price Stability, 285
counterfeiting, 259
court cases, 191, 209, 253, 267, 280
coverage, 152, 165, 269
CPB. See Corporation for Public Broadcasting.
Crabbe, Buster, 154
Craig, Jimmie, 88, 160
crash, 120, 253, 282, 286
crater, 25, 35, 122, 162, 218, 270
crew safety, 5, 59, 89, 92, 172
crew training. See training; Space Shuttle training.
Crimean Astrophysical Observatory, 151
Croatia, 143
Cronkite, Walter, 270
crop survey, 35, 141, 171, 198, 279, 299
cross polarization, 94, 275
cruise missile, 47, 53, 158, 285, 287
crustal activity, 23, 146, 155, 305
Cruzen, Paul J., 32
cryogenics, 50, 63, 77, 145
crystals, 126, 140, 225, 300
CS (Japanese comsat), 290
CSS–X–3 (PRC booster), 290
CSC. See Civil Service Commission.
CTS. See Communications Technology Satellite.
Cuba, 219, 224, 304
culturing, 20
current-machine chamber (Tokamak), 36
Currie, Dr. Malcolm R., 57, 129, 164, 193
Cutler Hammer Corp., 98
cyclone, 74, 83
Cyclops (joint ARC–Stanford project), 270
cyclotron, 235
Cydonia (region of Mars), 77, 153, 155
Cygnus (constellation), 115
cytology, 193
Czechoslovakia, 128, 163, 219, 224, 250, 304
Da Vinci I (balloon), 88, 160
Da Vinci II, 88, 117, 160
Da Vinci III, 160
Dallas, Tex., 35, 73
DAM. See detection and mapping system.

damage, 16, 33, 127, 135, 151, 168, 171, 183, 208, 215, 279, 281, 304
Dana, William H., 241
danger, 128, 216, 230, 251, 268, 277, 299, 305
Dantzig, George B., 254
Dartmouth College, 28
Dassault-Breguet (French contractor), 187–188
data
—acquisition, 5, 6, 72, 80, 168, 229, 248, 296, 302, 303
—analysis, 96, 131, 133, 142, 144, 160, 221, 230, 232, 268, 272, 274, 275, 279, 295, 297
—collection, 20, 82, 101, 113, 115, 160, 171, 247, 267, 270, 280, 300, 302
—distribution, 6, 128, 179, 198, 225, 235, 242, 276, 302
—management, 44, 82, 296, 303
—processing, 43, 60, 79, 81, 112, 114, 127, 133, 153, 158, 162, 171, 232, 239–240, 242, 248, 254, 263, 268, 275, 276, 302, 303
—relay, 4, 5, 10, 37, 93, 120, 146, 151, 154, 205, 207, 229
—transmission, 36, 62, 82, 101, 147, 148, 155, 165, 205, 223, 225, 235
—utilization, 4, 22, 24, 119, 171, 279, 304, 305
Davies, Merton E., 119, 132
Davis, Hallowell, 254
daylight, 36, 45, 74, 165, 215, 245
DC–3 (G–47; Gooney Bird), 66
DC–8, 124
DC–9, 124, 214, 222
DC–10, 134, 180, 253, 282
death, 5, 15, 16, 28, 29, 39, 44, 51, 71, 135, 237, 251, 253, 273, 280, 281, 282, 303
Death Valley, Calif., 121
debris, 25, 187, 214, 281
decision-making, 39
Declaration of Independence, 98, 132
decontamination, 147
Deep Space Network (DSN), 24, 71, 151, 205
defector, 210
Defense, Department of (DOD)
—budget, 13, 185, 192–193, 201, 257, 274
—contracts, 19, 53, 57, 81, 92, 99, 107, 108, 149, 184, 186, 248, 259, 260, 292, 302, 304
—history, 25, 199, 231, 233, 242, 246, 261, 295
—people, 19, 37, 44, 50, 70, 83, 96, 99, 129, 145, 163, 164, 178, 193, 255–256, 283, 305
—policy, 30, 32, 57, 129, 135, 158, 161, 165, 172, 185, 192, 228, 257, 259, 271, 285, 287
—programs, 46, 47, 58, 60, 79, 80, 89, 90, 95, 99, 102, 115, 117, 123–124, 133, 158, 160, 161, 164, 184, 203, 214, 219, 223, 229–230, 276, 280–281, 283, 294–295, 298, 305
—research, 14, 24, 27, 40, 50, 61, 84, 87, 106, 113, 130, 135, 150, 162, 170, 172, 229, 247, 258, 265, 267, 273, 279, 296
Defense Electronics Supply Center, DOD, 259
Defense Meteorological Satellite Program, USAF, 296

Defense Supply Agency, DOD, 163
Defense Systems Acquisition Review Committee, DOD, 172, 285
deficit, 12
dehydration, 19
Deimos (moon of Mars), 220, 264, 295
Deissler, Robert G., 186
deJager, Cees, 116
De La Vaulx medal, 282
De Loach, Dr. Anthony C., 194
Deloffre, Bernard, 91, 133, 134
Delpech, J.-L., 7
Delta (launch vehicle), 9, 28, 61, 63, 65, 73, 79, 87, 99, 115, 131, 133, 141, 165, 223, 249, 271, 289, 295, 297
Delta 1, Delta 2 (Soviet nuclear-submarine classes), 283, 288
Deming, Wash., 38
demolition, 285
dendrite, 95, 225
Deneb (star), 115
Denmark, 4, 45, 130, 230, 294, 343
density, 36, 288, 290
Denver, Colo., 96, 98, 116, 257, 265
Denver, Univ. of, 68
deployment, 136, 257
depreciation, 78
Des Moines, Iowa, 263
desalination, 45
DESC. See Defense Electronics Supply Center.
desert, 120, 153, 154, 155, 157, 305
design, 133, 134, 162, 168, 169, 172, 218, 220, 241, 242, 247, 248, 249, 251, 253, 262, 265, 268, 269, 275, 278, 279, 283, 286, 292
destruction, 34, 118, 163, 176, 215, 229–230, 236, 277, 280, 285, 289, 305
detection, 41, 47, 57, 61, 88, 93, 118, 130, 132, 139, 154, 156, 171, 215, 225, 230, 255, 267, 268, 269–270, 274, 277, 281, 300, 304
detection and mapping system (DAMS), 4
detente, 119
deterrent, 7
Detroit, Mich., 23, 36, 67, 85, 257
Detroit, Univ. of., 193
deuterium, 149, 278
developing nations, 106, 112, 119, 281, 293
de Waard, Johannes, 194
DFRC. See Dryden Flight Research Center.
DGA International, 195
diagnostic aid, 236
diet, 11, 40
dinitrogen pentoxide, 304
direct-broadcast satellites, 18, 243–244
Direct Communications Link (hotline), 298
dirigible, 6, 41, 97
dirt, 208
disarmament, 16, 30, 47, 149
disaster, 10, 22, 38, 142, 171, 175, 214, 252, 253, 273, 282, 306
Discoverer 36 (USAF project), 195
discrimination, 111
disease, 251, 267, 306
Disher, John H., 106

display, 133, 245, 274, 279
distance, 151, 160, 162, 196, 220, 252, 264, 268, 270, 277, 282, 288, 291, 302
Distinguished Service Medal (NASA), 39, 99, 241
distress signal, 114, 171, 205
docking, 38, 53, 129, 140, 162, 163, 220–221, 251, 260, 282, 299
documents, 127, 150, 163, 253
DOD. See Defense, (U.S.) Department of.
—satellite (launched 19 Feb.), 312
—satellite (launched 22 Mar.), 314
—satellite (launched 2 June), 319
—satellite (launched 26 June), 321
—satellite (launched 8 July), 322
—satellite (launched 15 Sept.), 328
—satellite (launched 19 Dec.), 335
DOI. See Interior, Department of.
domestic communications, 119, 157, 287
Domestic Council, White House, 92
domestic satellites, 1, 10, 35, 61, 65, 73, 239, 240
Donn, Dr. William L., 300
Doolittle, Gen. James M., 255
Doppler shift, 114
Dora (solar-power generator), 136
d'Ornano, Michel, 88
DOT. See Transportation, Department of.
Douglas Division, McDonnell Douglas Corp., 253
Doyon, Inc., 69
draft, 13
drag reduction, 81, 231
Drake, Dr. Frank, 32, 33
Draper Laboratories, 28
drogue parachute, 24
drone, 47
drought, 141
drugs, 39, 250
Dryden Flight Research Center (DFRC)
—missions, 5, 37
—people, 66, 70, 99, 241, 295
—programs, 95, 96, 97, 98, 144, 164, 193, 204, 210, 280
Dryden, Dr. Hugh L., 5, 62
drying, 127
DSCS: defense satellite communications system
DSCS-2 (military comsat), 117
DSCS-3 (military comsat), 118, 228
DSN. See Deep Space Network.
dual-satellite mission, 143
Duewer, William H., 217
Duke, Dr. Michael G., 246, 300
Dulles International Airport, Va., 3, 16, 102, 257, 269, 300
DuPont, 24
Dupree, Kirby, 280
Durant, Dr. Frederick, 238
duration, 96, 129, 133, 140, 171, 190, 221, 229, 272, 282, 306, 341
Dusey, Michael, 102
dust, 76, 105, 148, 171, 295
dust storms, 101, 295
Dutch. See The Netherlands.

365

Duxbury, Dr. Thomas C., 162
dynamics, 15, 24, 166, 169–170, 262
dynamite, 255
Dyson, Dr. Freeman, 35

Early Bird (Intelsat I), 21, 92
early-warning satellites, 133, 280
earth, 71, 114, 139, 151, 156, 218, 226–227, 275, 289, 304, 337
earth resources, 69, 77, 106, 170, 203, 227, 262, 276, 291, 294, 302
Earth Resources Laboratory, La., 179, 204
earth sensing, 10, 76, 229, 235, 293
earth station (see also individual listings)
—international, 21, 31, 69, 76, 119–120, 128, 163, 179, 181, 200, 205–207, 213, 224, 235, 239, 240, 252, 272, 281, 286, 302–303
—monitoring, 38, 46–47, 79, 114, 165–166, 171, 213–214, 229, 242, 247, 264, 267, 272, 279, 296
—remote, 5, 10, 29, 105–106, 111–112, 117–118, 214–215, 223, 228, 244, 276, 293
—unmanned, 6, 82, 98–99
—use in research, 23, 35–36, 38, 45, 87, 93, 115, 120, 197, 211–212, 226–227, 233, 249, 261, 288, 292, 302
Earthnet (ESA program), 302
earthquake, 1, 23, 38, 140
East Coast, 38, 111, 143
East Germany. See German Democratic Republic.
Eastern Airlines, 15, 198, 283
Eastern Test Range (ETR), 9, 60, 61, 79, 94, 115, 141, 249, 290
Ecastar (project), 241
Echo 1, 28, 149
echo canceler, 181
eclipse, 52, 80, 291, 297
ecology, 33, 67, 176, 204, 256
Economic Development Administration, 93
economics, 12, 18, 75, 84, 107, 139, 141, 149, 166, 167, 176, 181, 182, 221, 244, 250, 255, 269, 272, 277, 280, 304, 306
ECS (European communications satellite), 209
Ecuador, 293
Edison Company, 143
Edison Light Co., 125
education, 10, 13, 22, 39, 72, 106, 111, 175, 195, 228, 241, 274, 282, 283, 288, 297
Edwards AFB, Calif., 5, 14, 44, 98, 106, 123, 164, 248, 283
EEC. See European Economic Community.
EEO. See equal employment opportunity.
efficiency, 50, 83, 159, 169, 181, 197, 212, 213, 240, 248, 262, 280, 291, 299, 301, 302
Eglin AFB, Fla., 170
Egypt, 20, 69, 122, 149, 305, 306
Ehricke, Dr. Krafft, 107, 245
Einstein theory, 52, 126, 337
Eisenhower, President Dwight D., 21, 92, 185
ejection, 44, 80, 214, 246
Ekran (Statsionar 1C, Soviet comsat), 272, 330
elastic-loop mobility system, 264
Elder Statesman of Aviation, 275

elderly, 11, 111
ELDO: European Launcher Development Organization.
elections, U.S., 12, 91, 161, 185, 211, 238, 270, 285, 287
electric car, 213
electricity, 34, 36, 45, 63, 75, 102, 112, 125, 131, 143, 144, 167, 173, 197, 219, 225, 240, 249, 259, 268, 271, 278, 280, 281, 290, 296, 303
electrochemistry, 291
Electrodynamics Explorer satellite, 142
electromagnetism, 75, 172, 297
electron, 25, 63, 105, 242
electron-beam fusion, 9
electronics, 1, 11, 46, 58, 60, 61, 132, 153, 171, 184, 208, 215, 228, 230, 240, 244, 254, 259, 267, 287, 297, 298
Electronics Industries Assn., 260
electrophoresis, 20, 126
elementary particle, 25, 56
Elisabeth Bolton (W. Ger. tanker), 242
Ellington AFB, Tex., 118, 170, 220
Elliott, Maj. Larry A., 164
Elmira Heights, N.Y., 109
ELT: emergency locator-transmitter.
emergency, 20, 59, 66, 93, 102, 114, 171, 194, 195, 232, 255
emissions, automobile, 189, 216
employment (see also personnel, U.S. govt.; NASA management), 72, 79, 81, 82, 161, 172, 262, 287, 304
Enchanted Valley, Antarctica, 293
Encke (comet), 8
endangered species, 32
endurance, 129
energetic particles, 49, 105, 301
energy
—conservation, 11, 58, 68, 71, 97, 115, 144, 241, 265, 280
—research, 11, 13, 22, 27, 30, 49, 58, 71, 99, 101, 102, 116, 133–134, 213, 247–248, 278
—Energy Research and Development Administration (ERDA):
contracts, 94, 102, 116, 124, 142, 270
history, 11, 219, 247, 255
participation with NASA, 13, 27, 29, 67, 72, 112, 116, 124, 141, 144, 167, 181, 239, 280, 301–302
policy, 11, 13–14, 48, 133–134, 201, 271, 274
pollution, 88, 117, 160, 304
solar power, 13, 27, 29, 52, 72, 75, 94, 111–112, 116–117, 124, 141, 142, 144, 219, 225, 239, 240, 247–248, 270, 301
wind power, 67, 102, 124–125, 167, 181, 280
—source, 36, 101, 102, 137, 176, 181, 225, 268, 271, 301
Engelmann, Dr. Rudolf J., 88, 160
engineering, 30, 35, 64, 65, 72, 81, 97, 113, 118, 133, 156, 172, 204, 241, 256, 274, 278, 279, 287
Engineers, Corps of. See Army, U.S.

engines (see also Space Shuttle main engine), 58, 160, 167
Engins Matra (France), 286
England. See United Kingdom.
Engle, Joe H., 222
Enterprise (TV space ship), 159, 211
Enterprise (Space Shuttle orbiter prototype OV–101), 210, 211, 222, 301
environment, 16, 22, 24, 40, 45, 48, 53, 60, 70, 71, 80, 82, 84, 87, 104, 113, 127, 140, 165, 168, 169, 192, 195, 208, 222, 241, 255, 267, 271, 274, 276, 287, 304
Environmental Data Services, NOAA, 296
Environmental Effects Office, JSC, 272
environmental impact, 3, 18, 80, 84, 237, 269, 306
environmental monitoring, 5, 82, 267, 272, 301
Environmental Protection Agency (EPA), 3, 61, 105, 117, 124, 133, 160, 189, 201, 267
enzyme, 20
EPA. See Environmental Protection Agency.
epoxy, 62
Epsilon (star in Gemini), 83
Epstein, William, 30
equal employment opportunity (EEO), 250, 282
equator, 5, 10, 19, 38, 59, 62, 79, 94, 121, 128, 139, 141, 153, 157, 279, 293, 301
equivalence, 52, 126
ERDA. See Energy Research and Development Administration.
ERNO (Raumfahrttechnik GmbH), VFW-Fokker subsidiary, 113, 209, 292
error, 153–154, 156, 218
ESA. See European Space Agency.
Escondido, Calif., 299
Eskimo, 69
ET: external tank (see Space Shuttle).
ETR. See Eastern Test Range.
Ethiopia, 306
Europa (ESA launch vehicle), 80
Europe, 18, 238, 270, 289
European aerospace industry, 87, 186
European Center for Nuclear Research (CERN), 242
European Common Market, 182
European Economic Community, 268
European Launcher Development Corporation (ELDO), 6
European Southern Observatory, 31
European Space Agency (ESA)
—cooperative programs, 14, 68, 95, 130, 194, 211–212, 227, 239, 248, 271, 275, 283, 286–287, 294
—facilities, 88, 97, 213, 302
—people, 48, 133, 134, 241, 248, 269
—policy, 4, 80, 209, 214, 244, 249, 279, 292
—Spacelab, 44, 53, 78, 91, 113, 123, 192, 197–198
European Space Conference 1975, 88
European Space Research and Technology Center (ESTEC), 194
eutectic (alloys), 95
EVA. See extravehicular activity.

Evans, Ronald E., 57
Evans, R. Adm. Stuart J., 83
Evans, Gen. William J., 158, 257
evaporation, 45, 50, 224
Everest, Mt., 218
evolution, 54, 56, 82
Exceptional Service Award, 186, 241
exercise, 92, 140
exhaust gases, 216, 261, 267
exhibit, 25, 137, 163, 215, 246, 270
exobiology, 100, 146, 203, 208, 215–216, 217–218, 236–238
Exosat (ESA cosmic-ray satellite project), 249
expedition, 143
expendable equipment, 90, 218, 229, 289, 299
experiment
—data, 26, 177, 199, 236, 260, 272, 279, 296, 297
—design, 36, 44, 52, 66, 101, 114, 126, 132, 139, 140, 148, 154, 155, 166, 178, 197, 224, 225, 226, 239, 275, 278, 280, 297, 305
experimental aircraft, 164, 258
exploration, 22, 139, 147, 149, 205, 225, 254, 256, 264, 301
Explorer 1, 25, 65
Explorer 51, 19
Explorers Club, 256
explosion, 56, 134, 149, 163, 252, 255, 273, 280, 281, 305
explosive, 268, 285
exports, 81, 268
exposure, 172, 216
extinction, 33
extrapolation, 56
extraterrestrial life
—indications, 22, 24, 56, 104, 199, 252
—Mars, 15, 17, 49, 77, 100, 121, 139, 146, 149, 152–154, 177, 200, 203–204, 205, 207–208, 215, 217–218, 224, 225–226, 236–237, 274–275, 295
—radio detection, 33, 268, 269–270
extraterrestrial materials, 250, 272, 300
extravehicular activity, 73, 76, 232, 278

F3 (Orion CP140, Canadian aircraft), 179
F–4 (fighter aircraft). See Phantom.
F–4c (fighter aircraft), 44
F–4e (fighter aircraft), 273
F–4EJ (Japanese interceptor plane), 247
F–5e (fighter jet aircraft), 14, 165
F–14 (Tomcat, fighter aircraft), 160, 184
F–14a (fighter jet aircraft), 184
F–18 (fighter aircraft), 184
F–104 (Starfighter, fighter aircraft), 165
F–111 (fighter jet aircraft), 184
FAA. See Federal Aviation Administration.
fabrication, 73, 135, 188, 248, 278, 291, 303
fabrics, 23, 24, 26, 59–60, 93
facilities modification, 78, 98, 298
facsimile, 62, 79, 115, 206
Faget, Dr. Maxime A., 241
FAI: Fédération Aeronautique Internationale (Intl. Aeronautical Federation, Paris)

failure, 15, 39, 49, 59, 156, 163, 183, 200, 203, 207, 214, 217, 239, 251, 255, 259, 260, 275, 306, 337
Fairchild Industries, Inc., 99, 211, 296
fallout, 172
family planning, 105, 228
fan, 134, 184
Far East, 262
fare systems, 124, 166, 182, 213
Farley, Clare F., 163
Farmer, Dr. Barney, 122
Farnborough Intl. Aerospace Exhibition, 192
fatality, 15, 16, 44, 209, 214, 216
fatigue, 15, 17, 247
fault, geological, 84
fauna, 38
FCC. See Federal Communications Commission.
Federal Aviation Act of 1958, 229
Federal Aviation Administration (FAA), 14, 16, 48, 66, 89, 94, 103, 124, 125, 134, 149, 167, 212, 231, 253, 257, 262, 274, 282, 285, 304
Federal Communications Commission (FCC), 18, 30, 35, 78, 92, 125, 157, 252, 263, 276
Federal Coordinating Council for Science, Engineering, and Technology, 185
federal employees, 13, 57, 191, 193, 267, 278, 282
Federal Republic of Germany (West Germany), 1, 4, 7, 29, 43, 49, 50, 69, 113, 117, 123, 130, 135, 165, 186, 187, 190, 194, 209, 210, 213, 238, 242, 249, 263, 287, 292, 294, 306, 308, 337, 338
federal spending, 1, 7, 12, 58, 75, 134, 185, 191, 192, 201, 222, 224, 225, 228, 229, 238, 265, 274, 281, 285
Federal Women's Program, 164, 270
Federally Employed Women (FEW), 164, 288
federally sponsored research. See government-sponsored research.
Fédération Aeronautique Internationale (FAI), 109, 164, 282
female physiology, 255
Feoktistov, Konstantin, 5, 92
Fermi National Accelerator Laboratory, 25, 242
fertilizer, 40, 64, 216
FEW. See Federally Employed Women.
Fiat, 165
fiber research, 93
fighter aircraft (see also individual aircraft), 14, 95, 160, 165, 172, 179, 184, 191, 210
film, 45, 158, 281
filter, 146, 153–154, 156
Finland, 59, 179
fire, 38, 57, 59, 93, 127, 177, 184, 199, 273, 280
firefighting equipment, 26, 102, 130
firefly, 1
fiscal year, 12, 75
fish, 24, 34, 45, 67, 74, 119, 140, 204, 267, 306
Fisher, Adrian, 149
Fisk Univ., 247

fission, 36
Fitzmorris, Gov. James (La.), 150
flag, American, 70
Flash Gordon, 154
Fleet Satellite Communications System (FLTSATCOM), 31, 118, 211, 249
Fletcher, Dr. James C., 21, 28, 62, 117, 129, 138, 143, 147, 148, 163, 171, 211, 222, 243, 292
flexibility, 72
flight control. See aircraft; missile control.
Flight Inspection Natl. Field Office, FAA, 66
flight operations, 15, 17, 53, 121, 140, 146, 160, 232, 263, 272
flight readiness, 123, 218
Flight Research Center. See Dryden Flight Research Center.
flight safety. See air safety.
flight simulator (at JSC), 91
flight test, 37, 44, 47, 96, 97, 118, 123, 144, 248, 258, 273, 277, 287, 343
Flight Test Center, USAF, 283
floating platforms, 6
flooding, 38, 121, 139, 177
flora, 38
Florida, 40, 108, 132, 151, 169, 171, 224, 225, 240, 242, 264, 271, 279
Florida State Univ., 159, 209
Florida, Univ. of, 132, 279
FLTSATCOM. See fleet satellite communications system, DOD.
Fluid Dynamics Laboratory, Princeton Univ., 189
fluid-transfer system, 23, 64
fluids, 126, 186, 225, 227, 299
fluorocarbons, 17, 64, 216, 261, 272
fluorosensing, 61, 105, 155, 216
fly-by-wire (aircraft control system), 193
flying lunar-excursion experimental platform (training device), 66
Flying Yacht (aircraft), 40
foam insulation, 170
fog, 139, 205, 249, 251, 279
foil sampler, 77
Fokker-VFW, 209
food, 9, 11, 16, 22, 34, 105, 111, 123, 141, 156, 176, 204, 279, 294, 306
footprint, 150
Ford (automobile), 29
Ford, Dr. Arthur B., 294
Ford, President Gerald M., 12, 16, 48, 50, 57, 70, 91, 107, 126, 129, 137, 143, 147, 148, 151, 176, 185, 211, 212, 238, 243, 254, 262, 282, 285, 287, 292, 293
Ford, Henry, 85
Ford, Sen. Wendell H., 13
Ford Motor Co., 92
Fore (tribe of New Guinea), 252
"A Forecast of Space Technology 1980–2000" (NASA report), 21
forecasting. See meteorology.
Foreign Agricultural Service, 198
forest monitoring (see also timber), 38, 73, 177, 221, 237, 268–269
Fort, D.N., 302

Fort Belvoir, Va. 271
Fort Detrick, Md., 298
Fort Huachuca, Ariz., 298
Fort Worth, Tex., 116
fossils, 11, 44, 58, 121, 138, 167, 237, 274, 289
"Foundations of Space Biology and Medicine" (report), 159
Fourth of July, 120, 138
Fowler, Dr. William A., 238
Foxbat (aircraft). See MiG-25.
France, 3, 4, 6, 7, 50, 69, 80, 87, 88, 117, 128, 130, 135, 178, 180, 186, 187, 188, 195, 209, 212, 230, 238, 241, 249, 259, 269, 275
Frank, Dr. Neil, 177
Frascati, Italy, 211, 303
free flight, 37, 66
freeze, 19, 70, 279
freight, 135
French, Dr. Bevan, 300
freon, 77, 216, 272
Fresnel lens, 227, 299
Friedman, Dr. Herbert, 101
Friedman, Prof. Milton, 255
Friendship 7, 223
"frog suit," 23
frontier, 80, 137
frost, 139, 205, 279
Frutkin, Arnold W., 91, 95
FSCS. See Fleet Satellite Communications System.
Fucino, Italy, 303
fuel, 26, 30, 36, 61, 68, 89, 99, 103, 120, 123, 134, 144, 145, 149, 170, 172, 187, 216, 230, 248, 267, 268, 280, 285, 286
Fuller, Maj. John, 164
Fullerton, Lt. Col. Charles G., 37, 222
Fulton, Fitzhugh L. Jr., 66
funding, federal, 1, 58, 59, 129, 134, 135, 151, 158, 160, 185, 192–193, 201, 222, 246, 265
Fuqua, Rep. Don (D-Fla.), 224
fusion (nuclear), 9, 11, 26, 36, 58, 99
future, 12, 18, 21, 22, 35, 75, 80, 99, 107, 137, 144, 148, 160, 168, 171, 176, 203, 211, 225, 229, 235, 245, 265, 268, 269, 275, 292, 294–295, 298

Gagarin, Yuri, 73, 76, 238, 282, 304
Gaja Mada, 141
Gajdusek, Dr. D. Carlton, 252, 255
Galaxy (aircraft). See C-5A.
galaxies, 23, 24, 32, 33, 74, 82, 150, 302, 343
Galilean satellites (of Jupiter), 104
Galileo Galilei, 52, 100
Galileo II (NASA research aircraft; see Convair 990), 194, 261, 277
Gamma Boötes (star), 132
gamma rays, 8, 101, 131, 190, 212, 229, 230, 273, 278, 290
GAO. See General Accounting Office.
Gardner, Mass., 190
Garmire, Dr. Gordon, 101

Garrick, I.E., 18
Garriott, Owen K., 176
Gartrell, Harold E., 91
gas chromatograph, 49, 61, 146, 155, 156, 157, 200, 217
gas-exchange experiment, 15, 17, 156, 175, 200, 215, 217, 261
gases, 82, 104, 108, 123, 126, 135, 155, 167, 169, 170, 177, 181, 193, 199, 215, 236, 250, 267, 280, 304
Gaviria, Fernando, 238
Gay, Kenneth B., 303
geiger counter, 177
Gelles, Dr. S.H., 95
Gemini (constellation), 83
Gemini program, 39, 50, 70, 90
Gemini 6, 90
Gemini 7, 283
Gemini 9, 50
General Accounting Office (GAO), 7, 57, 184, 228
General Dynamics Corp., 53, 70, 184, 186, 211, 253, 302
General Electric Co., 19, 91, 92, 93, 134, 167, 181, 188, 212, 214, 287, 297, 302
General Electric Space Division, 142, 297
General Services Administration (GSA), 28, 218, 288
General Telephone & Electronics Corporation (GTE), 94, 186
generation (of power; of signals), 67, 81, 136, 172–173, 263, 268, 270–271, 278, 280
genetics, 11, 193, 251
Geneva, Switzerland, 31, 45, 243, 263
Geochemistry and Analytical Chemistry Institute (Sov. Acad. of Sci.), 183
geodynamic experimental ocean satellite (GEOS), 5–6, 87, 271, 294
geography, 44
Geological Survey, U.S., 77, 82, 121, 208, 218, 293–294
geology, 20, 35, 55, 69, 73, 79, 84, 121, 155, 177, 208, 218, 221, 224, 237, 262, 270, 272, 300
geomagnetism. See magnetic phenomena.
geometry, 254
geophysical research, 23, 27, 68, 69, 170, 171, 193, 208, 250
George A. Towns elementary school, 247
George Washington University (D.C.), 65, 75, 190
Georgia, 161, 247
Georgia Institute of Technology, 248
Georgia Tech. See Georgia Inst. of Technology.
Geos (proposed ESA satellite), 68, 213, 271, 275
GEOS: geodynamic experimental ocean satellite.
Geos 3, 5, 46, 119
Geosari (proposed ESA program), 239, 248
geostationary operational environmental satellite (GOES), 1, 4, 5, 10, 29, 38, 93, 242, 247, 271, 275, 286, 289
geostationary orbit, 14, 35, 50, 68, 135, 192, 212, 213, 275, 286

369

geosynchronous orbit, 62, 79, 265, 289, 293, 301
geothermal energy, 58, 137, 176
German Democratic Republic (East Germany), 79, 128, 163, 219, 221, 224, 250, 304
German Institute for Space Research, 136
German Space Operations Center, 8
Germany, East. See German Democratic Republic.
Germany, West. See Federal Republic of Germany.
germination, 150
getaway special (Space Shuttle payload), 245
Gibraltar, 242
Gibson, Edward G., 129
Gibson, Roy, 88, 91, 134, 213, 241, 244
Gilinsky, Victor, 268
Gillam, Isaac, IV, 99
Gilruth, Dr. Robert R., 259
glacier, 79, 289
Glacier (U.S. Coast Guard icebreaker), 247
glass, 55, 123, 167, 171, 250
Glenn, Sen. John (D-Ohio), 223, 270, 285
Glennan, Dr. T. Keith, 62
global climate, 34
Global Communications, Inc. (RCA), 62, 226
global comsat system, 21, 252, 275
Global Information Services Satellite System (future project), 235
global satellite system, 171, 200, 291, 294
Globcom. See Global Communications, Inc.
GMS (Japanese meteorological satellite), 271, 294
Goddard, Dr. Robert, 49, 51, 138
Goddard Space Flight Center (GSFC)
—contracts, 73, 226, 296
—meetings, 291, 292
—missions, 27
—people, 77, 133, 171, 178, 190, 196, 237, 248, 269, 271, 277, 285, 291, 293, 295, 296
—programs, 23, 28, 29, 46, 51, 60, 63, 101, 132, 133, 143, 145, 151, 187, 204, 226, 272, 279, 307, 343
Godfrey, Arthur, 98
GOES. See geostationary operational environmental satellite.
Goes 1, 4, 5, 10, 38, 279
Goes 2 (Goes-B), 1, 294
Goes 3, 93
gold, 41
Goldstone tracking station, 84, 120, 197
Goldwater, Sen. Barry M. (R-Ariz.), 222
Goodell, Charles E., 195
Goodlette, John D., 120
Goodyear Aerospace Corp., 97
"Gooney Bird" (C-47; DC-3), 67
Gordon, Robert V., 142
Gorove, Dr. Stephen, 244
government-furnished materials, 77
government-industry cooperation. See industry-government.
Government Printing Office (GPO), 159
Government Radio Administration (Republic of China), 20
government regulation, 9

government-sponsored research, U.S., 7, 93, 111, 113, 117, 162
GPA. See gravity probe.
Grace, Donald, 102
grain, 141, 299
Grand Canyon, 121, 122, 139
Grand Tour (exploration of outer planets), 265
Grant, Carl E., 193
grants, 184
gratuities, 57
gravitational waves, 45
gravity, 20, 38, 43, 52, 82, 104, 105, 126, 131, 132, 162, 166, 225, 249, 256, 270, 291
Gravity Probe-A (GPA) red-shift experiment, 126, 321, 337, 339
Gray, Laurence F., 6
Great Britain. See United Kingdom.
Great Galactic Ghoul, 15
Great Lakes, 4, 236, 247
Great Plains, 151, 280
Great Race of 1908, 145
Great Red Spot (on Jupiter), 104, 189
Greece, 49
Green Bank, W. Va., 302
Greenland, 55, 343
Grew, Dr. Edward S., 294
Griffin, Gerald D., 70
Griggs, David, 229
Griner, Carolyn, 95, 225
Groo, E.S. (Todd), 63
gross national product, 50
ground control, 17, 43, 98, 144, 190, 224, 229, 230, 264, 296
ground station. See earth station.
ground support, 15, 78, 189, 223, 226–227, 233, 242, 279, 300
ground tracking, 29, 143, 240
grounding, 184
"group flight," 282
Grumman Aerospace Corp., 46, 47, 96, 118, 125, 184, 211, 218, 220, 229, 297, 301, 302
GSFC. See Goddard Space Flight Center.
GSR-A. See *Azur*.
GTE. See General Telephone & Electronics.
Guest, Dr. John, 218
Guiana Space Centre, 88
guidance system, 47, 53, 99, 120, 230, 232, 276
Guidry, Louis E., 66
Gulf Coast, 10, 60
Gulf of Suez, 123
Gulf Stream, 118
Gulfstream (Grumman twin-engine jet), 118, 220, 229, 301
Gursky, Dr. Herbert, 101
Gyorgy, Paul, 254
gyroscope, 34, 297

H−1 (Hughes-designed aircraft), 71
H−1 (rocket engine), 299
Haack, Robert W., 179, 191
Hady, Ahmed Abdel, 122
Hagood, Carlos C., 194
Hague (The), Netherlands, 119

Haise, Fred W., Jr., 37, 222
Haiti, 281, 293
Haley, Andrew G., award, 241
Hall, Charles F., 278
Hall, Floyd D., 283
Halley's comet, 168
halocarbons, 77, 261
Halpert, Gerald, 291
ham radio. See amateur radio.
Hamburg, Univ. of, 96
Hamilton, Mass., 233
Hamilton-Standard Division, United Techn. Corp., 108, 145, 167, 181, 232
handicapped, 66
Hanscom AFB, Mass, 190
hardware, 25, 58, 72, 78, 81, 83, 96, 101, 112, 131, 172
Hargrave, Dr. Robert B., 156
Harrier (STOL aircraft). See AV8.
Harrington, Rep. Michael J. (D-Mass.), 111
Harris, Ruth Bates, 250
Harvard College Observatory, 298
Harvard Univ., 39, 83, 90, 254, 298
harvest, 141, 299
Haskins, Dr. Larry, 54
Hauzeur, Max, 44
Hawaii, 5, 62, 94, 96, 132, 196, 261, 343
Hawaii, Univ. of, 71, 102, 105, 125, 179
Hawaiian Electric Co., 125
Hawker Siddeley Dynamics Ltd., 164, 209
Hayman, Dr. Deter, 54
Hays, Dr. James D., 289
hazard, 26, 59, 64, 121, 128, 132, 134, 153, 158, 172, 193, 208, 285
haze, 121
health, 13, 22, 34, 92, 105, 175, 221, 228, 267, 282
Health, Education, and Welfare, U.S. Dept. of (HEW), 201, 274
HEAO. See High-Energy Astronomy Observatory.
HEAO-A, 294
heart, 137
Hearth, Donald P., 22
heat, 23, 26, 37, 48, 54, 56, 61, 70, 71, 74, 104, 107, 130, 131, 146, 147, 156, 170, 193, 203, 208, 215, 220, 226, 227, 235, 236, 274, 278, 288, 297, 299, 305
Heat Capacity Mapping Mission, 303
heat mapping, 24, 67, 115, 122, 303
heat sensing, 67, 105, 115, 279
heat transfer, 70, 72, 94, 104, 124, 130, 141, 186, 189, 198, 227, 278
Heathrow airport, London, 89, 103
heavy lifter, 97
heavy metals, 41
heavy particles, 278
helicopter, 51, 97, 113, 117, 130, 251, 258, 274, 286, 304
Helios program, 272
Helios 1, 8, 49
Helios 2, 1, 7, 49, 272, 308, 337, 338
heliostat, 240
helium, 9, 26, 78, 88, 97, 171, 230, 242, 272
Hellas (region of Mars), 76

helmet, 172, 274
Helt, Capt. Robert C., 164
hemoglobin, 63, 171
Henderson, Clifford W., 275
Hendricks, Sterling Brown, 254
Henebry, John P., 229
Henry, Joseph, 138
hepatitis, 251
herbicide, 306
Heronemus, William, 102
Herrington, Preston B., 160
Hess, Dr. Seymour L., 209
HEW. See Health, Education, and Welfare, U.S. Dept. of.
Heyman, Joseph S., 47
High, Richard W., 167
High Altitude Observatory, Colo., 179
High-Altitude Pollution Program, 304
high-altitude research, 142, 170, 229, 301
High-Energy Astronomy Observatory (HEAO), 62, 101, 149, 211, 230, 278, 294
High Flight Foundation, 66
high frequency, 37, 106, 143, 276
higher education, 274
highly maneuverable aircraft technology (HIMAT) program, 95
Hilsenrath, Ernest, 296
HIMAT. See highly maneuverable aircraft technology (program).
Hindenburg (airship), 41
Hinners, Dr. Noel W., 53, 100, 142
Hirsch, Robert, 52
Hirschfelder, Joseph O., 254
Hispanics, 18, 72
historical exhibit. See exhibit.
history, 28, 35, 54, 79, 222
HL-10 (lifting body), 241
Hlass, Jerry, 190
HMS *Bounty*, 173
Hokkaido, Japan, 210
hold, 170
Holland. See The Netherlands.
Holloway, Paul F., 285
holography, 106
Holzer, Dr. Thomas E., 32
home terminals, 235, 287
Honduras, 177
Honeywell, Inc., 116, 142, 240, 274
Hong Kong, 5, 182
Honolulu, Hawaii, 96
Honorary Group Diploma for Astronautics, 282
honors, 18, 238, 251, 259, 282
Hoover, Richard, 93
Horowitz, Dr. Norman, 156, 226, 237, 275
Horton, Victor W., 66
hotline, 59, 298
Hotz, Robert B., 87
House Committee on. See Congress; general subject.
housekeeping, 15
Housing and Urban Development (HUD), U.S. Dept. of, 30
Houston, Texas, 35, 71, 127–128
Houston Baptist Univ., 116

Houston Intl. Airport, 257
Houston, Univ. of, 128
Howard, Paul W., 33
Hq: Natl. Aeronautics and Space Administration, Headquarters
HSD. See Hawker Siddeley Dynamics.
HUD. See Housing and Urban Development.
Hughes, Howard R., 71
Hughes, W.J., 301
Hughes Aircraft Co., 15, 71, 92, 94, 141, 183, 186, 249
human factors, 22, 172
human relations, 250
humanitarian, 48
Humboldt State Univ., 24
humidity, 38, 179, 248
Humphrey, Marshall, 61
Hungary, 128, 163, 219, 224, 238, 304
hunter satellite. See interceptor.
hunting, 19
hurricane, 10, 74, 177
hydraulic systems, 169, 170, 259
hydrazine, 80, 99, 172
hydrocarbons, 189, 216
hydrochloric acid, 172
hydroelectric, 123
hydrofluoric acid, 56
hydrogen, 26, 36, 50, 97, 104, 145, 170, 171, 240, 250, 259, 262, 268
hydrogen bomb, 12, 36
hydrogeology, 20
hydrology, 5, 6, 67, 73, 82

IAF: International Astronautical Federation
IBM. See International Business Machines.
ICAO. See International Civil Aviation Organization.
ICAS: Interdepartmental Committee for Atmospheric Sciences
ICBM. See intercontinental ballistic missiles.
ice, 4, 6, 35, 46, 70, 73, 74, 138, 145, 225, 236, 247, 251, 294
Ice Age, 289
Iceland, 122
Icewarn (project), 4
Idaho, 214, 240
ignition, 267, 273
IL-86 (Soviet jumbo jet), 277, 286
Illinois, 23, 145
Illinois, Univ. of, 187, 302
illiteracy, 228
imaging, 131, 133, 151, 153–155, 158, 165, 171, 177, 207, 242, 247, 262, 263, 273, 279, 295, 299, 303, 305
Imboden, Otis, 88
IMCO. See Intergovernmental Maritime Consultative Organization.
immigrant, 143
IMOS. See inadvertent modification of the stratosphere.
impact, 55, 84, 134, 140, 148, 162, 167, 214, 239, 251, 252
Imperial College, London, 301
implants, 37

improvements. See new products; new techniques.
inadvertent modification of the stratosphere (IMOS), 64
incubation, 156, 199–200, 215, 217, 261
India, 47, 78, 106, 128, 149, 175, 228, 248, 293
Indian Ocean, 79, 181, 186, 200, 223, 249, 280, 289
Indian Space Research Organization (ISRO), 111, 228
Indiana, 117
Indians. See American Indians.
Indonesia, 1, 2, 96, 141, 293, 306, 337, 339
Industrial Contractors, Inc., 240
Industrial Research (magazine), 47, 235
industrial waste, 41
industry-government cooperation, 29, 61, 65, 75, 81, 93, 117, 125, 135, 141, 148, 149, 159, 162, 188, 204, 212, 225, 227, 247, 253, 269, 285, 289, 291
inertia, 52, 297
inert gases, 181
Inex Adria Airways, 214
infection, 252
inflation, 18, 50, 58, 124, 171
inflation, economic, 141, 201
influence, 161
information retrieval, 63, 68, 303
information transfer laboratory, GSFC, 151
infrared, 5, 7, 24, 46, 69, 119, 166, 189, 192, 227, 247, 277, 279
—cameras, 5, 7, 10, 130, 132–133, 220
—mapping, 67, 69
—radiometry, 67, 277
—spectrometer, 43
—telescope facility, Hawaii, 53, 54
injury, 26, 280
inland waterways, 4
INMARSAT: International Maritime Satellite Organization
innovations. See new products; new techniques.
Innsbruck, Austria, 23
inorganic matter, 155, 157, 213, 216, 236, 295
insects, 113, 280, 306
inspection, 106, 118, 188
Institute for Solar Research, 52
Institute of Nuclear Energetics, USSR, 297
instrument landing, 66, 98, 257
Instrument Society of America, 241
instrumentation, 35, 44, 58, 62, 74, 77, 88, 93, 115, 117, 123, 131, 132, 145, 146, 150, 153, 156, 165, 175, 178, 203, 208, 213, 217, 221, 222, 223, 224, 229, 230, 239, 241, 247, 250, 252, 257, 262, 267, 272, 273, 274, 275, 279, 288, 294, 296, 302
insulation, 50, 61, 115, 170, 220, 236
integrated subsystem test bed, 25, 64
integration, 44, 53, 97, 117, 148, 159, 166, 169
intelligence, 24, 34, 163, 263, 268, 269, 288, 305–306
INTELSAT. See International Telecommunications Satellite Corporation.

Intelsat (comsat) program, 49, 59, 118, 186, 200, 252
Intelsat 1, 92
Intelsat IV program, 186
Intelsat IV-A, 21, 92, 181, 186, 252
Intelsat IVA-C, 294
Intelsat IV-A F-1, 18, 78
Intelsat IV-A F-2, 18, 78, 311, 337, 338
Intelsat V program, 21, 73, 186, 211, 238
interceptor (aircraft; satellites), 34, 163, 164, 229, 246, 280–281, 305
intercontinental ballistic missiles (ICBM), 1, 32, 49, 106, 149, 257, 283, 287, 289
Intercosmos program (USSR), 212, 219, 220, 225, 244, 245, 304, 307
Intercosmos 15, 128, 321
Intercosmos 16, 163, 325
Interdepartmental Committee for Atmospheric Sciences (ICAS), 64
interference, 29, 31, 146, 213, 263, 270, 276, 281, 306
interferometry, 78, 213, 290
intergalactic space, 75
Intergovernmental Maritime Consultative Organization (IMCO), 205
Intergovernmental Science, Engineering, and Technology Panel, 185
interim test stand, 25
interim upper stage (IUS) for Space Shuttle, 90, 186, 193
Interior, Dept. of, 116, 144, 201
International Academy of Astronautics, Paris, 238
International Aeronautical Federation. See Fédération Aeronautique Internationale.
International Air Transport Association (IATA), 283
International Amateur Radio Union, 263
International Astronautical Federation (IAF), 65, 241, 243
International Astronomical Union, 298
International Business Machines (IBM), 78, 112, 232, 248, 254
International Civil Aviation Organization (ICAO), 14, 212
international communications, 21, 47, 92, 176, 276
international cooperation
—general, 1, 10, 21, 27, 28, 44, 48, 49, 68, 69, 73, 77, 79, 119, 128, 168, 175, 176, 205–207, 237, 242, 243, 246, 281, 306
—European, 4, 6, 44, 54, 78, 87, 128, 130, 165, 249, 292
—USSR, 78, 80, 95, 119, 191, 212, 219, 299, 303, 304
—U.S.-Africa, 20, 41, 122
—U.S.-Asia, 1, 47, 111, 187
—U.S.-Canada, 1, 9, 10, 14, 302
—U.S.-South America, 7
—U.S.-USSR, 15, 38, 39, 45, 59, 73, 83, 159, 163, 195, 231, 294, 298, 304
—U.S.-W. Europe, 1, 7, 19, 23, 31, 45, 48, 49, 53, 93, 95, 113, 116, 125, 135, 149, 164, 165, 198, 239, 302

International Council of Scientific Unions, 68, 171
International Geophysical Year, 68
International Institute of Space Law, 244
International Magnetospheric Study, 49, 68
International Maritime Satellite Organization (INMARSAT), 205
International Organization for Radio and Television. See OIRT.
International Radio Amateur Satellite Organization, 248
International Solar System Decade, 65
International Sun-Earth Explorer (ISEE), 68, 294
International Telecommunications Satellite Organization (INTELSAT), 15, 18, 20, 21, 73, 119, 179, 181, 186, 200, 206, 237, 239, 275
International Telecommunications Union, 21, 73, 263
International Time Bureau, Paris, 305
international trade, 141, 268
International Ultraviolet Explorer (IUE) project, 283, 294
International Union of Pure and Applied Physics, 242
interplanetary medium, 24
interplanetary spacecraft, 24, 80, 197
Inter-Soyuz (USSR program), 303
Intersputnik (International Organization for Cosmic Telecommunications), 224
interstellar medium, 32, 82, 291
invention, 39, 50, 90, 102
Inventions and Contributions Board, NASA, 87
inversion, atmospheric, 105
Io (moon of Jupiter), 105
ion, 36, 45, 142, 272
ion-beam fusion, 9
ionosphere research (see also atmosphere), 20, 44, 233, 239, 250, 296
Iowa, 264
Iowa, Univ. of, 105, 298
IR–100 (award), 235
Iran, 44, 76, 96, 117, 149, 269, 282
Ireland, Republic of, 4
iridium, 162
iron, 157, 175, 183, 187, 209, 236, 294
Irwin, Col. James B., 66
Isabelle (proton collider), 243
Isaksen, Dr. I.S.A., 32
ISEE. See International Sun-Earth Explorer.
Isle Royale National Park, 116
isotopes, 158, 230, 235, 254
Israel, 122, 210, 305
Israel, Dr. Martin, 278
ISRO. See Indian Space Research Organization.
Istanbul, Turkey, 145
Italy, 4, 20, 46, 130, 165, 187, 190, 209, 210, 211, 238, 249, 271, 275, 294
Ithaco, Inc., 298
Itos program, 133, 165
ITOS E-2, 295

ITOS-H. See *Noaa 5*.
IUE-A. See International Ultraviolet Explorer.
IUS (interim upper stage for Space Shuttle). See Space Shuttle, upper stage.
Ivey's Plumbing and Electrical Co., 72
Ivory Coast, 293

J particle, 255
Jaenicke, Joachim, 238
Jaffe, Dr. Leonard, 244
Jakarta, Indonesia, 96
Jakob, Max, memorial award, 186
Jakobowski, Walter, 100
Jamaica, 293
James, Jack N., 22
jamming, 50, 118, 153, 287
Japan, 44, 50, 68, 179, 186, 187, 210, 239, 244, 246–247, 271, 294, 307, 312
Javelin (sounding rocket), 143
JBF Scientific Corp., 125
Jeppesen Sanderson, Inc., 209
jet, 47, 53, 67, 81, 85, 118, 124, 165, 170, 180, 187, 189, 190, 191, 194, 220, 227, 261, 269, 277, 282, 304
Jet Propulsion Laboratory (JPL)
—contracts, 63, 219
—meetings, 226, 274
—missions, 16, 43, 49, 100, 120, 131, 138–140, 146, 148, 152, 155, 160, 168, 175, 177, 203, 205, 207–209, 223, 224
—patents, 23, 299
—people, 64, 69, 70, 71, 133, 139, 162, 227, 254, 271, 290, 299
—programs, 22, 46, 47, 61, 63, 71, 75, 84, 132, 144, 166, 172, 217, 237, 245, 264, 268, 272, 295
JFK airport. See Kennedy International Airport.
Jodrell Bank Observatory, U.K., 233
Joersz, Capt. Elden, 164
John F. Kennedy Airport. See Kennedy International Airport.
Johns Hopkins Applied Physics Laboratory, 196
Johns Hopkins Hospital, 99
Johns Hopkins Univ., 31, 95, 296
Johnson, Dr. Bruce B., 161
Johnson, Clarence "Kelly," 18
Johnson, David S., 38
Johnson Gage, 78
Johnson, President Lyndon B., 39
Johnson, Dr. T.V., 132
Johnson Space Center (JSC)
—contracts, 27, 47, 91, 112, 145, 161, 162, 170, 179, 199, 220, 232–233, 265, 270, 280
—meetings, 35, 54, 65, 72
—missions, 11, 36, 53, 74, 112, 142
—patents, 167
—people, 18, 37, 39, 54, 66, 70, 116, 241, 246, 259, 277, 278, 281, 288, 295, 300, 301
—programs, 4, 11, 14, 54, 59, 60, 90, 93, 115, 118, 127, 132, 161, 188, 220, 224, 249, 255, 256, 257, 272, 279, 281
Johnston, Bradford, 90
Johnston, Dr. Mary Helen, 95, 225, 270
Johnston, Moira, 253

Jones, Gen. David C., 231
Jones, Dr. Robert T., 144
Jordan, 176, 293
JPL. See Jet Propulsion Laboratory.
JSC. See Johnson Space Center.
judicial ruling, 191
jumbo jet. See jet aircraft.
Jupiter (planet), 13, 24, 35, 43, 46, 58, 69, 71, 103, 168, 189, 245, 271, 291, 294
Jupiter (rocket), 25
Justice, U.S. Dept. of, 195

Kalinin, USSR, 297
Kamchatka, 143
Kamide, Dr. Yohsuke, 296
Kamunay, Mt., Philippine Is., 209
Kansas, 55
Kansas City, Mo., 242
Kansas City International Airport, 257
Kantrowitz, Dr. Arthur, 35, 39
kapton, 24
Karl Marx collective farm, USSR, 193
Karl Zeiss Jena, 220
Kaufman, Gerald, 269
Kazakhstan, USSR, 8, 38, 193, 221, 250, 286
Keefe, Dr. J. Richard, 152
Kellermann, K.I., 302
Kemayoran airport (Indonesia), 96
Kennedy International Airport, N.Y., 3, 15, 89, 102, 134, 257, 269
Kennedy Space Center (KSC)
—contracts, 78, 113, 145, 271, 300
—meetings, 107, 279
—missions, 98
—people, 159, 164, 295
—programs, 50, 70, 78, 90, 96, 113, 123, 132, 148, 164, 168–169, 199, 210, 211, 261, 285, 300, 301
Kentucky, 151
Kentucky State Univ., 113
Kenya, 106, 176, 237, 293
kerosene, 51, 80, 111
Kerr, Joseph H., 144
Kerwin, Joseph P., 111
Ketchum, Rep. William M. (R-Calif.), 222
kevlar, 24, 25, 59
Khabarovsk, USSR, 262
Khrushchev, Nikita, 273
kidney, 20
Kieffer, Dr. Hugh H., 122, 226
Kiev, USSR, 276
killer (satellite) system. See interceptor.
Killian, Dr. James R., 185
Kingsland, Dr. Louis, 153, 156, 236
Kissinger, Henry, 34, 106, 128, 293
Kistiakowsky, Dr. George B., 185
kites, 138
Kitt Peak National Observatory, Ariz., 46, 53, 71, 82
Kitty Hawk Flyer, 137, 229
Klate Holt Co., 170
Klein, Dr. Harold P., 139, 175, 177, 217, 275
Klimuk, Pyotr, 92, 129, 171, 222
Knowles, S.H., 302
Kohoutek (comet), 31

Komarov, V.M., 282
Kondo, Yoji, 116
Korean War, 145, 158, 178
Kourou, New Guinea, 6, 88
Kraemer, Robert S., 100, 271
Kraft, Dr. Christopher C., Jr., 259, 278, 301
Kramer, Dr. James J., 285
Kratzer, Myron, 149
Krier, Gary, 95, 194
Kroehl, Herbert W., 296
Krstic, Dr. D., 245
krypton, 226
KSC. See Kennedy Space Center.
Kuhn, Dr. Peter M., 277
Kuiper airborne observatory, 82
Kurchatov Institute, 36
Kursk, USSR, 297
kuru (laughing disease), 252

L-1011 (TriStar, Lockheed jet transport), 179, 180, 190, 191, 282
L-band telemetry, 115, 229
labeled-release experiment (Viking), 156, 175, 177, 199, 216, 260
labor unions, 120, 198, 206
LACIE. See large-area crop inventory experiment.
Ladish Co., 82
Lageos (laser geodynamic satellite), 1, 23, 87, 317, 337, 339
LaGuardia airport, N.Y., 198
Laird, Melvin, 158
Lake Cowichan, B.C., 201
Lake Superior, 117
Lambda-Airtron Div., 162
Lambert glacier, Antarctica, 294
laminar-flow control, 231, 280
Lamont-Doherty Geological Observatory, 300
Land, Elwood W., 89
land lines, 18
land use, 105, 237, 262, 305
lander, 146, 147, 153, 160, 207, 217, 223, 224, 260, 263, 264, 295
landing, 84, 98, 120-122, 138-140, 142, 146, 147, 148, 151, 153, 155, 160, 162, 168, 203, 205, 208, 221, 232, 251, 257, 260
landing rights, 3, 103, 195
Landsat program, 4, 20, 69, 80, 133, 144, 151, 171, 204, 237, 262, 268-269, 282, 302, 305
Landsat 1, 69, 204
Landsat 2, 303
Landsat-C, 294
Langley medal, 40
Langley Research Center (LaRC)
—contracts, 101, 231
—missions, 100
—people, 18, 38, 47, 87, 100, 148, 171, 271, 277, 285, 295
—programs, 30, 47, 60, 66, 101, 117, 267, 280, 303
language, 119, 207, 214
LaRC. See Langley Research Center.
large-area crop inventory experiment (LACIE), 198

Large Space Telescope (future project), 230, 265
LaRock, Ralph I., 75
laser (light amplification by stimulated emission of radiation), 9, 14, 17, 23, 35, 52, 83, 87, 105, 144, 162, 172, 197, 227, 273, 276, 278, 280, 305
laser-augmented target acquisition and recognition (latar), 14, 273
latar. See laser-augmented target acquisition and recognition.
Latimer, Allie D., 288
Latin America, 106
latitude, 17, 261, 296
Latter, Robert, 157
launch. See individual mission; appendixes.
launch-abort system (for Space Shuttle), 89, 169
launch complex 14, KSC, 285
launch complex 41, KSC, 148
launch facility, 1, 6, 7, 9, 18, 19, 27, 29, 32, 88, 97, 168, 230, 267, 283
launch registration, 128
launch vehicle (see also Ariane; Delta; Saturn; Titan), 7, 9, 18, 25, 28, 31, 44, 49, 50, 51, 64, 65, 73, 79, 80, 82, 87, 88, 90, 94, 95, 97, 99, 102, 149, 150, 198, 211, 218, 235, 239, 240, 248, 249, 250, 261, 262, 268, 275, 276, 279, 283, 289, 290, 295, 297, 299, 300, 301
launches, DOD:
—*AMSI*, 214, 327
—*Les-8, -9*, 50, 313
—*Noss 1*, 316
—*P-76-5*, 102, 318, 339
—*SDS-2*, 325
—*SESP 74-2*, 323
—*Solrad 11a, 11b*, 50, 313
—*SSU-1, -2, -3*, 316-317
—*TIP-3*, 203, 327, 340
—unspecified, 133, 141, 219, 312, 314, 319, 321, 322, 323, 328, 335
launches, NASA:
—*Comstar 1*, 2, 94, 252, 317, 339
—*Comstar 2*, 157, 290, 324, 339
—*Cts 1*, 1, 9, 249, 308, 338
—*Gravity Probe A*, 126, 321, 339
—*Helios 2*, 1, 7, 49, 272, 308, 338
—*Intelsat IV-A F-2*, 18, 78, 311, 338
—*Lageos*, 1, 23, 87, 317, 339
—*Marisat 1*, 2, 31, 60, 89, 115, 249, 312, 338
—*Marisat 2*, 115, 249, 319, 339
—*Marisat 3*, 223, 249, 329, 340
—*Nato III-A*, 2, 79, 315, 338
—*Noaa 5* (ITOS-H), 165, 192, 295, 325, 339
—*P-76-5*(DOD), 102, 318, 339
—*Palapa 1*, 1-2, 133, 141, 323, 339
—*Satcom 2*, 61, 65, 315, 338
—*TIP 3* (DOD), 203, 327, 340
lava, 122, 146, 148, 213-214, 281
law, 65, 161, 165, 172, 195, 229, 244
Lawson, Dr. James R., 247
lawsuit. See court cases.
LDEF. See long-duration exposure facility.
lead (element), 41

375

leakage, 120, 170
Lear 4 (1974 Spacelab mission), 227
Learjet, 96, 98, 231
Lebedev Physics Institute, USSR, 278
Le Bourget International Airport, Paris, 96
Lederberg, Dr. Joshua K., 204
Lee, Chester M., 91, 245
Lee, Gentry, 236
Legostayev, V.P., 246
Lemonnier crater (moon), 183, 197
Leningrad, USSR, 212
Leningrad Arctic and Antarctic Institute, 6
Leninsk (USSR launch site), 29, 221
Leonov, Aleksey, 73, 282
Les 8, *Les 9* (Lincoln experimental satellites, DOD comsats), 50, 313
leukemia, 20
Levin, Dr. Gilbert, 156, 175, 178, 275
levitators, 297
Lewin, Dr. Walter, 101
Lewis Research Center (LeRC)
—contracts, 149, 181, 211
—people, 37, 49, 54, 186, 277, 285, 295
—programs, 4, 10, 81, 112, 115, 116, 117, 144, 167, 213, 235, 247, 261, 280, 285
lexan (material), 59, 201
Lexington, Ky., 160
Leyden, John F., 167, 198
LGB (laser-guided bomb), 276
liability, 244, 253, 280, 287
liaison, 48, 65
Libbey-Owens-Ford Co., 116
Liberia, 31, 293
Library of Congress, 111, 112
Libya, 20, 176, 293
licensing, 188, 231
life forms, 32
life in space. See extraterrestrial life.
life-detection instruments, 17, 146, 203, 205, 208, 236–237, 274, 295
life sciences, 28, 104, 197, 274, 287
life support, 59, 60, 145, 160, 232
lifetime (satellite; instrument), 44, 62, 80, 118, 123, 159, 212, 213, 236, 291
lifting-body program, 44
light, 1, 14, 26, 45, 60, 71, 83, 101, 115, 132, 200, 215, 221, 227, 235, 245, 272
lighter-than-air vehicles. See airships; balloons.
light years, 23
lightning, 132, 249, 279
Lin, Robert P., 132
Lincoln, President Abraham, 138
Lincoln Laboratory, MIT, 233
Lind, Dr. David, 297
Lindbergh, Charles A., 137, 255
Lindley, Robert N., 248
Lippisch, Dr. Alexander M., 28
Lipscomb, Prof. William N., 254
liquid, 156, 193, 302
liquid fuel, 51, 138, 170, 172, 240, 259, 262, 283
liquid metallic hydrogen, 104
Lisbon, Portugal, 277
Littlefield, Robert, 150
Litton Industries, Inc., 162, 302

Livermore Laboratory (Univ. of Calif.), 216
loading, 24, 77
lobbying, 161, 195
location-reporting system, 184, 251
Lock Haven, Pa., 109
Lockheed Aircraft Corp., 47, 70, 72, 102, 164, 179, 180, 186, 190, 191, 219, 231, 282, 302
Lockheed Electronics Co. of Houston, 179
Lockheed Missiles and Space Co., 141, 186, 220
Loening, Grover, 39, 90
Logan International Airport, Boston, 51, 96
logistics, 163, 232
London, England, 3, 102, 181, 205–206, 269, 300, 301
London, Univ. of, 218
Long, Rep. Clarence D. (D-Md.), 149
long-duration exposure facility, 101
Long Island, N.Y., 133, 242
long-range planning, 12
Lop Nor (PRC test and launch facility), 255
Los Alamos, N.M., 36, 83
Los Angeles, Calif., 18, 35, 37, 46, 60, 82, 120, 257
Los Angeles International Airport, 198, 257
Louisiana, 40, 204
Louisville, Univ. of, 152
Love, Lt. Col. Michael A., 44
Lovelace, Dr. Alan M., 129
Low, Dr. George M., 13, 28, 54, 80, 129, 163
low-temperature research, 170
Lowell, Percival, 100
Lowell Observatory, 76, 100
LTV Corp., 53
Lucas, Dr. William R., 126, 150
Lucey Boiler Co., 77
luminescence, 1, 14
Luna 16, 183, 196
Luna 17, 183
Luna 20, 183, 196
Luna 21, 183
Luna 23, 183
Luna 24, 182, 196, 300, 307, 326
lunar landing, 5, 11, 19, 35, 36, 39, 50, 51, 131, 183, 194, 205
Lunar Orbiter 5, 183
lunar research, 11, 35, 36, 52, 54, 69, 74, 76, 131, 168, 171, 183–184, 192, 205, 212, 246, 264, 272, 281, 300
lunar sample curatorial facility (at JSC), 28, 57, 246, 300
lunar samples, 54, 76, 119, 196, 215, 246, 300
Lunar Science Conference, 35, 54
Lunar Science Institute, Houston, Tex., 35, 54, 300
Lundin, Dr. Bruce T., 49
Lunetta (proposed space reflector), 245
Lunney, Glynn S., 277
Lunokhod (Soviet mobile moon lander), 183, 197, 209
Lynchburg, Va., 239
Lyra (constellation), 115

M-2 (lifting body), 44, 241
M-1008 (India rockets), 128

McCeney, Paul, 293
McCord, Dr. Thomas B., 132
McCoy, Gene, 159
MacCracken, William P., Jr., 85
McDonald, Rob Roy, 22
McDonnell Douglas Astronautics Corp., 47, 50, 53, 91, 94, 124, 127, 163, 164, 180, 184, 187, 191, 198, 214, 230, 240, 253, 258, 282, 289, 296, 302
McDowell, Ed, 124
McElroy, Dr. William D., 1
McIntyre, Sen. Thomas J. (D-N.H.), 158
McLucas, Dr. John L., 16, 125, 134, 149, 257, 274, 282
McKinley Climatic Laboratory, USAF, 170
McMaster, L. Roy, 109
McMurdo Sound, Antarctica, 79, 294
McMurtry, Thomas C., 66
McPherron, R.L., 301
McNamara, Robert, 158
macrosegregation, 95
Madison Square Garden, N.Y., 120
Madrid, Spain, 46
Magellanic Cloud, 7
magnesium, 187
magnetic phenomena, 14, 15, 33, 36, 43, 45, 49, 54, 58, 68, 75, 131, 143, 160, 197, 230, 290, 294, 296, 301
magnetometer, 132, 290, 296, 297
magnetosphere. See atmosphere; see also Jupiter.
mail carriers, 135, 286
Maine, 242
maintenance, 15, 74, 112, 113, 190, 220, 231, 302, 303
malaria, 306
malfunction, 60, 89, 93, 135, 140, 143, 169, 170, 205, 208, 214, 224, 239, 260, 263
Mali, 293
management, 44, 90, 153, 159, 161, 164, 166, 204, 224, 232, 241, 244, 257, 268, 276, 277, 293, 298
Management and Technical Services Co., 242
Manchester, Univ. of (U.K.), 132
maneuver, 95, 120, 264
Manicouagan, Quebec, 55
Manila, P.I., 96, 209, 239
manmade materials, 25, 64, 135
manned space flight
—future, 34, 35, 65, 74, 92, 107, 113, 127, 160, 166, 176, 188, 198, 203, 222, 232, 289
—history, 25, 37, 38–39, 54, 75, 91, 96, 130, 147, 162, 178, 231, 238, 241, 243–246, 264, 270, 277, 281, 282, 283, 285, 298–299, 305
—policy, 8, 89–90, 142, 145, 159, 192, 270
—training, 118, 163, 194, 220, 224, 227, 229, 255, 256, 257, 259, 301
—USSR, 5, 12, 15, 41, 73, 76, 92, 112, 129, 140, 171, 188, 190, 193, 204, 219, 220–222, 250–251, 255, 260, 303, 304
—other countries, 4, 95, 123, 148
Manned Spacecraft Center, 54
manpower. See employment.
man-powered flight, 79
Manson, Al, 229

maons (particles), 243
mapping, 4, 20, 32, 67, 69, 84, 120, 131, 151, 168, 209, 213, 247
marble, 137
Mare Crisium. See Sea of Crises.
maria ("seas" on moon), 214
Marine Corps, U.S., 98, 164, 184, 265
marine life, 67, 169, 237
Mariner project, 15, 43, 49, 65, 71, 100, 245, 271, 272, 294
Mariner 7, 16
Mariner 9, 52, 76, 121, 162, 295
Marisat (maritime satellite system), 1, 31, 37, 60, 78, 89, 223, 249, 252
Marisat 1, 2, 31, 60, 89, 115, 223, 249, 312, 337, 338
Marisat 2, 115, 223, 249, 319, 337, 339
Marisat 3 (Marisat-C), 223, 249, 329, 337, 340
maritime communications, 89, 115, 205–206, 223, 249, 263
Maritime Safety Committee, IMCO, 206
Marots (ESA maritime comsat), 130, 192, 209
Mars (planet). See also Viking.
—data, 35, 46, 55, 76, 83, 84, 119, 162, 194, 220, 246, 263–264, 281, 294
—future, 13, 160, 166, 168, 196, 211, 225–226, 235, 264–265
—manned mission, 12, 34
—Viking mission, 15, 17, 43, 49, 77, 100–101, 120–122, 137, 138–140, 146–147, 148–149, 151–157, 175, 177, 178, 192, 199–200, 203, 205, 207–209, 215–216, 217–218, 223–224, 236–237, 245, 260–261, 271, 274–275, 295
Mars 1 (USSR spacecraft), 15
Marshall Space Flight Center (MSFC)
—contracts, 47, 72, 77, 82, 107, 116, 130, 142, 150, 168, 169, 188, 190, 191, 198, 203, 211, 218, 232, 240, 248, 258, 262, 278, 300
—meetings, 75, 106, 148, 179, 218, 279
—missions, 23, 123, 148, 194
—patents, 33, 70, 144, 158
—people, 81, 142, 158, 170, 194, 225, 227, 267, 270, 276, 278, 290, 295, 297
—programs, 25, 26, 27, 29, 37, 47, 48, 50, 62, 63, 72, 80, 82, 90, 93, 95, 97, 101, 115, 124, 126, 141, 144, 150, 166, 168, 169, 179, 198–199, 204, 210, 225, 227, 232, 235, 239, 240, 241, 249, 262, 265, 267, 276
marsquake, 153, 154, 155, 208, 224, 237, 295
Martin, James, Jr., 43, 100, 121, 138, 148, 153, 156, 207–209, 217, 224, 271
Martin Marietta Aerospace Co., 73, 108, 116, 150, 170, 186, 209, 240, 271
Marvin, Dr. Henry H., 219
Maryland, 19, 57, 111, 151, 262
Maryland Point, Md., 151
Maryland, Univ. of, 104
mascons (mass concentrations); 131, 183
maser (microwave amplification by stimulated emission of radiation), 127, 150
Mason-Reguard, 199

377

Masri, Sid, 191
mass production, 227, 271
mass spectrometer, 17, 142, 146, 155, 156
Massachusetts, 124, 233
Mass. Institute of Technology, 23, 36, 69, 101, 132, 208, 233, 236, 254, 268, 274, 297
Mass., Univ. of, 74, 102
Masursky, Dr. Harold, 77, 121
materials engineering, 24, 61, 88, 93, 117, 166
materials processing in space, 22, 65, 95, 96, 107, 126, 203, 225, 250, 279, 292, 297, 342, 343
materials (see also ceramics; composites; crystals; metals; synthetics), 7, 60, 131, 140, 162, 166, 169, 184, .225, 236, 272, 275, 290
mathematics, 254
Mathews, Charles W., 6, 38, 90
matter, 25, 82
Matthofer, Hans, 213
Mauna Kea, Hawaii, 53
Maxwell (area of Venus), 214
Maxwell, James Clark, 214
ME 163B (aircraft), 28
Meal Systems for the Elderly, 11, 111
mechanical problems, 143, 154, 156, 200, 203, 217, 223, 224, 251, 253, 260, 261, 290
media, 39, 100, 105, 151, 154, 187, 195, 220, 230, 250, 262, 269, 270, 273, 280
medical-data communication, 93, 106
Medicare/Medicaid, 13
medicine (see also technology utilization), 10, 13, 20, 37, 64, 66, 93, 106, 126, 130, 148, 159, 167, 171–172, 188, 203, 236, 251, 254, 255
Medicine Hat, Utah, 288
Mediterranean, 20, 45, 123
Medvedev, Zhores, 273
melting, 55, 297
memorial, 5, 25, 41, 51, 255, 282, 285
Mendeleyev elements, 184
Mendenhall, Ted, 229
Mendota, Va., 109
Meng, Dr. Ching, 296
Menges, Constantine, 182
Menzel, Dr. Donald H., 297
Menzies, Dr. Robert T., 227
Merck & Co., 254
Mercure (French jet aircraft), 180, 187
mercury (element), 41
Mercury (planet), 35, 52, 55, 105
Mercury (program), 39, 90
Mercury 6, 91
Mercury capsule, 90
Mercury-Redstone (rocket), 25
Merrill, Dick, 98
MESH: European consortium comprising Matra (France), ERNO (W. Germany), Saab-Scania (Sweden), Hawker Siddeley-British Aerospace, INTA (Spain), Fokker-VFW (Holland), and Aeritalia (Italy), 209
Messerschmitt-Bölkow-Blohm, 28, 165, 249
metabolism, 156, 177, 203, 208, 216–217, 237
metallurgy, 184, 270

metals, 60, 95, 104, 126, 140, 141, 147, 183, 187, 222, 236, 250, 268, 285, 294, 303
meteor, 37, 140, 187, 214, 236
Meteor (Soviet weather-satellite program), 73, 307
Meteor 24 (USSR metesat), 315
Meteor 25, 317
Meteor 26, 329
meteorite, 35, 55, 56, 162, 171, 300
meteoroid, 105, 107, 252
meteorology, 5, 6, 10, 19, 34, 35, 44, 46, 51, 65, 74, 76, 77, 79, 97, 106, 119, 132, 143, 145, 165, 177, 189, 190, 192, 195, 203, 209, 214, 217, 224, 225, 235, 236, 249, 250, 257, 261, 262, 279, 289, 291, 295, 296, 343
Meteosat (proposed European satellite), 130, 192, 213, 248, 271, 294
metesats, 192, 214, 271, 289, 293, 294
methane, 40, 70, 77
Methodist Hospital, Houston, Tex., 127
methodology, 57, 228
Metric Conversion Act of 1975, 274
metric system, 135, 274
Metropolitan Wayne County Airport, 257
Mexico, 5, 10, 56, 205
Miami, Univ. of, 132
Michigan, 116, 171
Michigan, Univ. of, 123, 216, 247
Michoud Assembly Facility, La., 72, 150, 169, 170, 258
microbe, 156, 274
microcomputer, 229
microdetection, 47
microexplosion, 278
micrometeorites, 8
microorganisms, 156, 217, 226, 274–275, 289
microphones, 300
microprocessors, 229, 240
microshock, 140
microwave scanning-beam landing system, 98
microwaves, 10, 46, 80, 132, 150, 197, 219, 247, 275, 288
Middendorf, J. William, 230
Middle East, 27
MiG–25 (Foxbat, Soviet aircraft), 210, 246
military aircraft, 18, 81, 117, 138, 210, 257, 273, 304
military communications, 50, 117
military procurement, 268, 271
military space activity, 7, 13, 50, 79, 112, 117, 133, 255, 257, 260
Milky Way (galaxy), 33, 75
Mill Village, N.S., 201
Miller, Charles O., 253
Miller, James C., III, 285
Milwaukee, Wis., 36
minerals, 69, 122, 131, 155, 171, 175, 209, 221, 223, 224, 237, 245, 272, 293, 301
miniaturization, 147, 292
mining, 41, 127, 151, 191
Ministry of Higher and Specialized Secondary Education, USSR, 297
Ministry of Information and Broadcasting, India, 228

Ministry of Posts and Telecommunications, USSR, 263
Minneapolis, Minn., 116
Minneapolis Honeywell. See Honeywell Inc.
Minnesota, Univ. of, 278
Minor Planet Center (IAU), 298
minority, 142, 283, 303
Minority Institutions Research Program, 276
Minuteman (ICBM), 49
mirror, 7, 53, 62, 94, 173, 240
MIRV. See multiple independently targeted reentry vehicle.
missile (see also individual missiles; types of missile; missile systems), 32, 47, 53, 106, 151, 158, 165, 184, 229, 261, 286, 287, 288
mission control, 61, 112, 142, 146, 148, 153, 155, 200, 224, 232, 261, 264
mission specialists, 46, 59, 142, 222
Mississippi, 40, 225, 232, 240, 244
Mississippi River, 117
Missouri, Univ. of, 145
MIT. See Mass. Institute of Technology.
Mitchell, Cdr. Edgar D., 11, 19
MJS-A and B (Mariner Jupiter-Saturn missions), 294
mobile landers, 208, 264
mobility, 31, 60, 118, 264, 289
Mobility Equipment Research and Development Command, U.S. Army, 271
mockup, 15, 113, 270
model, 25, 27, 41, 43, 58, 78, 113, 131, 146, 159, 168, 172, 189, 195, 218, 267, 279, 287, 291, 292
Modisette, Jerry L., 115
modular construction, 47, 60, 198, 278, 296
Mogavero, Louis, 75
Mohr, Col. George C., 172
moisture, 57, 179, 226
Moldavia, 177
molecular structure, 104
molecule, 139, 150, 199, 215, 225, 236, 254, 290
Molniya (Soviet comsat) program, 59, 262, 272, 307
Molniya 1–AH, 309
Molniya 1–AJ, 313
Molniya 1–AK, 314
Molniya 1–AL, 324
Molniya 2–R, 331
Molniya III, 59
Molniya 3–E, 317
Molniya 3–F, 335
Molodezhnaya (USSR) base, Antarctica, 294
molybdenum, 157
Mondale, Sen. Walter F. (D-Minn.), 161
Mongolia, 219, 224, 304
monitoring, 7, 24, 38, 41, 50, 60, 67, 71, 98, 105, 119, 145, 151, 165, 171, 175, 176, 229, 230, 237, 240, 244, 250, 255, 267, 279, 280, 288, 291, 300, 303, 305
Montana, 214, 294
Montgomery, Dr. Oscar, 276

Montgomery, Ala., 262
Montgomery Ward, 70
Montoya, Sen. Joseph M. (D-N.M.), 270
Montreal, Canada, 200, 205, 218
monument, 51
moon, earth's (see also lunar research)
—history, 5, 18, 19, 37, 50–51, 54, 76, 90, 97, 137, 146, 147, 148, 150, 211, 215, 233, 243, 245, 256, 294, 298–299, 305
—proposed missions, 131, 168, 203, 212, 264
—USSR missions, 112, 119, 182–184, 196–197, 204–205, 209, 300
—use in research, 25, 35, 46, 49, 54–57, 63, 69, 128, 131, 171, 213–214, 246
ALSEPs (Apollo lunar-surface experiment packages), 11, 36, 52, 74, 272
rocks, 55, 57, 171, 183, 215, 246, 272, 300
tree, 150
moonquakes, 11, 37, 74
Moore, R. Gilbert, 245
Moran, John, 105
Morgan, Maj. George T. Jr., 164
Morgan, Thomas H., 116
Morley, Lawrence, 244
Morocco, 176, 293
Morris, Dr. Elliott C., 208
Morrison, Dr. David, 71, 105
Morton, Louis, 28
Mory, Robert, 48
mosaic, 138
Moscow, USSR, 8, 36, 76, 78, 80, 129, 140, 183, 212, 219, 220, 262, 277, 286, 298, 300
Moscow Institute of State and Law, 244
Moscow State Univ., 45
Mosely, Terrell E., Inc., 239
Moss, Sen. Frank E. (D-Utah), 62, 151
Mossinghoff, Gerald A., 65
motion pictures, 41, 71, 154
motion sickness, 256
Motoren-und-Turbinen Union, 165
mountains, 11, 31, 35, 84, 218
Mt. Kamunay, Philippine Is., 209
Mt. Palomar Observatory, 25
Mozambique, 306
MRCA: multirole combat aircraft
MSFC. See Marshall Space Flight Center.
MSBLS. See microwave scanning-beam landing system.
Mueller, Dr. George E., 243
Muhleman, Dr. Duane O., 132
multichannel spectral analyzer, 268
multinational convention (on registration of objects launched into space), 128
multiple independently targeted reentry vehicle (MIRV), 283
multiple sclerosis, 252
multirole combat aircraft. See Tornado.
Munich, Federal Republic of Germany, 200
municipality, 280
Murray, Dr. Bruce C., 69, 71, 119, 139, 245
museums, 66, 127, 137, 159, 163, 215
Musgrave, Dr. Story, 14, 132
music, 190
Mustang (P–51 aircraft), 275
Mutch, Dr. Thomas, 153, 166, 207

MX (intercontinental ballistic missile), 1, 32
mylar, 171

N (Japanese liquid-fuel rocket), 44, 307
NAA. See National Aeronautic Association.
NACA. See National Advisory Committee for Aeronautics.
Nace, David A., 293
Naeher, Willis K., 59
Nairobi, Kenya, 106, 237
Nantucket Island, 303
NaPO. See NASA Pasadena Office.
NASA. See National Aeronautics and Space Administration.
NASA Pasadena Office, 63
NASA Planetary Geology Office, 166
NASA Tech Briefs, 63
NASA Technology Utilization Office, 63, 75
NASA-unique requirements, 90
Nashville, Tenn., 73
Nathanael Greene (nuclear submarine), 159
National Academy of Sciences, 52, 68, 101, 216, 306
National Advisory Committee for Aeronautics (NACA), 5, 38, 54, 62, 186
National Aeronautic Association, 79, 85, 96, 98, 109, 164, 229, 275, 282
National Aeronautics and Space Administration (NASA). See also aeronautical research; aircraft; Apollo program; astronaut; biology; communications; contractors; international cooperation; launches, NASA: (separate) listings for NASA centers (ARC, DFRC, GSFC, JPL, JSC, KSC, LaRC, LeRC, MSFC, WFC); manned space flight; Mars; moon; power sources; safety; Space Shuttle; space station; Viking project.
—commemorations, 5, 25, 28, 62, 70, 107, 132, 137, 143, 147, 187, 210, 222, 238, 243, 285, 292
—development, 11, 23, 26, 33, 38, 40–41, 50, 59–60, 61, 69, 81, 83, 87, 89, 90, 93, 97, 102, 111, 113, 117, 145, 166, 188, 225, 258, 262, 291
—funding, 1, 12, 13, 43, 58, 75, 91, 129, 192, 201, 224, 230, 267, 274
—management, 1, 23, 53, 57, 63, 76, 89, 93, 96, 99, 111, 117, 129, 133, 145, 224, 228, 267, 271, 277, 280, 285, 289, 295, 298, 303
—people, 6, 12, 13, 17, 19, 28, 38, 48, 49, 50, 51, 53, 54, 64, 65, 69, 70, 71, 75, 77, 79, 83, 90, 96, 99, 133, 145, 163, 164, 178, 190, 193, 231, 237, 241, 247, 248, 250, 259, 271, 277, 278, 285, 288
—plans, 1, 21–22, 34, 46, 53, 65, 68, 75, 80, 84, 126, 131, 142, 168, 171, 197, 230, 235, 288, 294, 296
—programs, 13, 17, 35, 47, 66, 72, 75, 82, 91, 95, 100, 103, 111, 126, 142, 166, 194, 210, 213, 235, 263, 290, 296
—research (joint projects):
AID, 106
Air Force, 49, 50, 61, 90, 99, 102, 170
Army, 113
Coast Guard, 4, 130
Commerce, 26
DOD, 129
EDA, 93
ERDA, 13, 27, 29, 67, 75, 83, 94, 102, 112, 116, 124, 141, 142, 144, 167
FAA, 94, 103
HUD, 30
Interior, 116
NAS, 101
Navy, 46, 95, 97, 115, 132, 151
NIH, 23
NOAA, 119, 132, 165
NSF, 63
states, 38, 44, 60, 69, 93, 111, 113
user communities, 57, 101
National Aeronautics and Space Council, 48
National Air and Space Museum, 41, 137, 163, 215, 231, 278, 299, 305
National Air Races, 275
National Archives and Records Service (NARS), 28, 199
National Association of Black Manufacturers, 303
National Association of Home Builders, 30
National Astronomy and Ionosphere Center, 33
National Broadcasting Company (NBC), 152
National Bureau of Standards, 30, 290, 305
National Center for Atmospheric Research (NCAR), 32, 77, 190, 272
National Civil Service League, 259, 278
National Committee for Space Research (India), 111
National Environmental Policy Act, 172
National Environmental Satellite Service (NESS), 24, 38, 192, 279
National Fire Prevention and Control Administration, 26
National Gallery of Art, 137
National Geographic Society, 88, 231, 283
National Heat Transfer Conference, 186
National Hurricane Center, 177
National Institute for Neurological Diseases, 255
National Institutes of Health, 23
National Medal of Science, 133, 254
National Needs Science Faculty Professional Development (NSF program), 283
National Oceanic and Atmospheric Administration (NOAA), 5, 10, 17, 24, 32, 38, 47, 57, 88, 117, 119, 132, 160, 165, 177, 189, 192, 204, 241, 261, 277, 279, 294, 296, 301, 304, 307, 337, 339
National Operational Meteorological Satellite System (NOMSS), 165
National Parachute Test Range, 98
National Park Service, 116
National Patterns of R&D Resources, 192
National Plan for Energy Research, A, 133
national policy, 18, 22, 102, 107, 191, 228, 229, 250, 263, 268, 274, 286, 287
national priorities, 9, 16, 34, 119, 265, 282
National Public Radio (NPR), 125
National Radio Astronomy Observatory, 32, 33, 302

national research centers, 7
National Research Council (NRC), 272
National Research Council, Canada, 302
National Safety Council, 79
National Science and Technology Policy, Organization, and Priorities Act, 185
National Science Foundation (NSF), 1, 6, 7, 13, 17, 32, 50, 63, 74, 82, 151, 184, 191, 192, 201, 274, 283, 287, 289, 294, 302
National Security Council, 92, 185
National Security Industrial Association, 158
National Space Club, 49
National Space Institute, 12
National Space Technology Laboratories (NSTL), 25, 40, 64, 93, 114, 169, 190, 232, 235, 240, 295
National Transportation Safety Board, 15, 16, 134, 253, 263, 282
National Weather Service, 190, 247
National Women's Political Caucus, 270
native Americans, 72
NATO. See North Atlantic Treaty Organization.
NATO IIIA, 2, 79, 315, 337, 338
NATO IIIB, 79, 294
natural gas, 41, 50, 58, 281
Naugle, Dr. John E., 235
Navajo, 288
Naval Observatory, U.S., 305
Naval Research Laboratory, U.S., 31, 47, 99, 101, 132, 150, 229, 302, 343
Naval Weapons Center, 88, 160
navigation, 24, 50, 51, 61, 66, 74, 106, 124, 146, 148, 181, 205-206, 229, 235, 247, 274, 294, 299, 306
Navigation Technology Satellite (SN), 181, 291
nebula, 56
Navy, U.S. (see also Naval Observatory; Naval Research Lab; Naval Weapons Center), 46, 50, 53, 60, 83, 89, 95, 96, 97, 115, 160, 181, 184, 203, 223, 230, 247, 285, 292, 294
NCAR. See National Center for Atmospheric Research.
Nedelin, Marshal Mitrofan, 273
Nellis AFB, Nev., 184
neoprene, 60
Neptune (planet), 43
NESS. See National Environmental Satellite Service.
Netherlands, 4, 23, 48, 119, 194, 207, 209, 279
neurology, 37, 254
neutral-buoyancy simulator, 270
neutrinos, 26
neutron stars, 93
Nevada, 121
New Bedford Gas Co., 125
New England, 51
New England Research Applications Center, 188
New Guinea, 252
new metals, 104
New Mexico, 9, 16, 27, 31, 51, 88, 238, 241, 270, 280

New Mexico Institute of Mining & Technology, 132
New Mexico State Univ., 72
New Orleans, 41, 72, 132, 144, 150
new products, 6, 23, 24, 26, 37, 47, 50, 59, 61, 70, 75, 102, 135, 147, 162, 166, 167, 169, 188, 203, 213, 235, 246, 273, 276, 291, 298, 299
new techniques, 6, 11, 17, 19, 20, 24, 30, 33, 38, 40, 45, 46, 59, 63, 126, 135, 140, 158, 159, 167, 188, 203, 205, 225, 249, 254, 257, 267, 270, 275, 279, 291, 302
New York, 19, 29, 35, 102, 121, 145, 152, 187, 195, 218, 257, 267, 269, 273
New York, State Univ. of, 25, 155
New York, Univ. of, 132
New Zealand, 64, 261, 294
Newcomb, Simon, 305
Newfoundland, 46
Newman, Robert A., 145
news (see also media), 36, 151, 154, 220, 262
Niagara Falls, N.Y., 143
nickel, 41, 141, 181, 213, 291
Nigeria, 73, 136
nighttime operations, 10, 66, 74, 115, 165, 215, 245, 250
NIH. See National Institutes of Health.
Nike-Cajun (rocket), 239
Nike-Tomahawk, 187
Nile River, 20, 41, 123, 306
Nimbus 6, 29, 146
Nimbus-G, 303
1976-AA (asteroid), 25
Nippon Electric Company, 239
nitrogen, 78, 153, 155, 170, 271
nitrogen compounds, 17, 33, 57, 76, 80, 189, 216, 236, 304
Nitze, Paul, 32
Nixon, President Richard M., 48, 91, 151, 158, 185
NOAA. See National Oceanic and Atmospheric Administration.
Noaa 4, 119, 192, 295
Noaa 5 (ITOS-H), 165, 192, 295, 325, 337, 339
Nobel prizes, 204, 251, 254, 278
noise, 3, 27, 33, 103, 113, 117, 124, 125, 144, 149, 168, 172, 240, 258, 267, 277, 285, 286
Nolan, Bernard T., 195
nominations, 129, 151
NOMSS. See National Operational Meteorological Satellite System.
nonprofit groups, 52, 196
Noordwijk, The Netherlands, 194, 207, 279
Nordberg, Dr. William, 237
Norflor Construction Co., 271
North America, 18, 144
North American Air Defense Command (NORAD), 214
North American Div., Rockwell Intl. Corp., 107, 137, 275
North Atlantic Treaty Organization (NATO), 1, 2, 48, 117, 315, 337, 338
North Carolina State Univ., 119, 190

North Pole, 6, 27, 74
Northern Research & Engineering Co., 130
Northrop Corp., 19, 27, 57, 95, 165, 273, 302
Northrup Inc., 116
"Northwest Territory" (on Mars), 138
Northwestern Univ., 37
Norway, 48, 79, 117, 263, 276, 294, 301, 343
NOSS 1 (Navy ocean-survey satellite), 316
nova, 56
Noyes, Blanche Wilcox, 275
NPR. See National Public Radio.
NRAO. See National Radio Astronomy Observatory.
NRC. See National Research Council.
NRL. See Naval Research Laboratory.
NSTL. See National Space Technology Laboratories.
nuclear arms agreement, 1, 34
nuclear energy, 9, 11, 12, 24, 48, 50, 57, 83, 99, 101, 118, 236, 238, 253, 254, 268, 273, 278, 297
nuclear fuel, 9, 11, 24, 48, 50, 83, 99, 125, 149, 167, 238, 252, 268, 278
Nuclear Regulatory Commission, U.S., 48, 79, 149, 268
nuclear waste, 11, 36, 41, 273
nuclear weapons, 32, 47, 83, 149, 154, 238, 255, 257, 268, 283, 288–289
nutrition, 9, 11, 13, 105, 111, 175, 199, 208, 216, 254
nylon, 24, 25, 84

OA. See Office of Applications, NASA Hq.
OAO. See orbiting astronomical observatory.
OASI. See Office of Aeronautics and Space Technology, NASA Hq.
objectives, 22
Obninsk, USSR, 297
observatory, 82, 101, 108, 151, 220, 305
occultation (stellar eclipse), 80, 82
Occupational Health and Safety Act, 172
ocean, 14, 61, 74, 97, 99, 119, 137, 171, 239, 244, 267, 294, 303
oceanography, 5, 15, 24, 35, 46, 57, 83, 119, 195, 204, 237
Odenwald (W. Ger. earth station), 213
odor, 255
Office of Aeronautics and Space Technology, NASA Hq, 96, 166, 285
Office of Applications, NASA Hq, 96, 119, 194, 293
Office of Energy Programs, NASA Hq, 75
Office of Lunar and Planetary Programs, NASA Hq, 271
Office of Management and Budget (OMB), 16, 124, 129, 185
Office of Manned Space Flight, NASA Hq, 75, 106
Office of Science, Engineering, and Technology (White House), 9, 16, 39, 58, 92, 151, 184–185
Office of Space Flight, NASA Hq, 70, 245, 277
Office of Space Science, NASA Hq, 77, 96, 101, 271

offshore installations, 6, 31, 89, 93, 118, 304
O'Hare International Airport, Chicago, 68, 89, 167, 257
Ohio, 54, 91, 151, 270
Ohio State Univ., 298
oil, 6, 11, 37, 41, 58, 102, 111, 122, 123, 171, 176, 177, 181, 213, 237, 245, 280, 301, 303
oil spills, 105, 303
OIRT (International Organization for Radio and Television), 225
O'Keefe, Dr. John A., 171
Okero, Isaac Omolo, 238
Okhotsk, USSR, 143
Oklahoma, 66, 213, 257
Oklahoma State Univ., 123
Olympic games, 10, 23, 154, 200
Olympus Mons (on Mars), 218
Oman, 176, 293
OMB. See Office of Management and Budget.
omnidirectional equipment, 66
ONA: Overseas National Airways
onboard computer (processor), 98, 224, 240
O'Neill, Gerard K., 127
O'Neill, William J., 120
Ontario, 294, 302, 304
OOE. See out-of-ecliptic mission.
Operation Paperclip, 29
operations research, 61, 166, 254
optical navigation, 100, 273–274
optics, 53, 144, 150, 213, 256, 273, 290
orbit, 6, 10, 16, 23, 25, 34, 35, 44, 48, 58, 59, 114, 120, 121, 128, 135, 140, 141, 142, 146, 148, 163, 183, 186, 196, 218, 244, 251, 270–271, 289, 293
Orbita (USSR communications network), 76, 272
orbital plane (earth's), 290
orbiter, Viking (see also Space Shuttle orbiter), 162, 215
orbiting astronomical observatory (OAO), 49, 132
orchards, 270, 279
organic matter, 17, 37, 41, 49, 61, 104, 121, 146, 155, 156, 158, 199–200, 207, 215, 217, 236–237, 274
organization, 159, 205–207
orientation, satellite, 34, 62, 165, 221, 229
origins, 35, 275
Orion (Canadian aircraft). See F3.
Orion 2 observatory (USSR instrument), 108
OSCAR (orbiting satellite carrying amateur radio), 195
Oscar 1, 195
Oscar 4, 196
Oscar 6, 114, 196
Oscar 7, 196
osmosis, 45
Oso 8 (orbiting solar observatory), 88, 178
OSS. See Office of Space Science, NASA Hq.
Otay Mt., Calif., 264
otolith, 256
OTS (ESA orbiting test satellite), 130, 192, 209, 271, 287, 294
Otterman, Joseph, 305

Oulu, Univ. of (Finland), 179
out-of-eclipitc (OOE) mission (proposed ESA-NASA joint project), 239
outer planets, 70, 71
Outer Space Affairs Division, UN, 238, 244
"Outlook for Aeronautics" (report), 84
"Outlook for Space" (report), 21
OV–101 (orbiter vehicle). See Shuttle orbiter; Enterprise.
Overseas National Airways, 134
Owen Enterprises Inc., 299
Owen, Dr. Tobias, 155
Owens-Illinois Inc., 18, 53
Owens Valley, Calif. 151
oxygen, 51, 80, 113, 145, 155, 175, 200, 215, 217, 232, 236, 240, 250, 259, 262, 268, 274, 298
Oyama, Vance I., 136, 218, 275
ozone, 3, 17, 27, 32, 38, 76, 160, 216, 261, 272, 296, 304

P–5 "Mustang" aircraft, 275
P–38 (fighter aircraft), 72
P–76–5 (USAF satellite), 102, 318, 339
pacemakers, 291
Pacific coast, 38
Pacific Ocean, 5, 28, 31, 60, 76, 115, 118, 172, 181, 186, 191, 200, 223, 249, 261, 288, 293, 299
paint, 70, 131, 169
Pakistan, 176, 293
Pal, Dr. Yash, 228
Palapa 1 (Indonesian comsat), 133, 141, 323, 337, 339
Palapa-B, 294
Paleozoic era, 44
Palestine, Tex., 190
Palmer, Arnold, 96, 98
Pan American Technical Services, Inc., 78
Pan American World Airways, 40, 74, 166, 209, 213, 261
Panama City, Fla., 292
Panavia, 165
Papago Indian reservation, N.Mex., 93
Papazian, Dr. John, 297
parachute, 24, 82, 93, 97, 116, 146, 148, 150, 193, 199, 230, 262, 272, 279, 288, 297, 300
paralysis, 37
Paris, France, 3, 4, 29, 41, 44, 48, 91, 95, 96, 97, 98, 102, 130, 164, 238, 239, 248, 253, 256, 269, 282, 286, 300, 302
particle accelerator, 25, 242–243
particle physics, 25, 64, 77, 104, 242–243, 255, 272
Pasadena, Calif., 43, 63
passengers, 8, 26, 59, 67, 84, 85, 103, 134, 166, 180, 198, 212, 232, 253, 269, 277, 286
passive systems, 72, 87, 171
PATCO. See Professional Air Traffic Controllers Organization.
Patent Office, U.S., 50
patents, 6, 23, 33, 40, 50, 65, 70, 87, 102, 158, 167, 188, 299
Patten, Dr. James W., 95

Patterson, William Allen, 229
payload, 51, 95, 96, 148, 169, 194, 195, 244, 267, 271, 272, 277, 279, 286, 287, 301
—recovery, 150, 230, 257, 272, 279
—specialist, 59
PBS. See Public Broadcasting Service.
peace, 147, 238, 281, 282
Pecora, William T., award, 237
Peking, PRC, 4
Pennsylvania, 151, 304
Pennsylvania, Univ. of, 254, 255
Pensacola Mountains, Antarctica, 294
Pentagon (see also Defense, U.S. Dept. of), 19, 117, 158, 164, 165, 281, 285, 287, 305
Penthouse magazine, 281
Peoples Republic of China, 4, 128, 196, 198, 238, 255, 290, 305, 307, 332
PEP (positive-electron project), 243
Pepin, Dr. Robert, 54
Perek, Dr. Lubas, 238, 244
perihelion (Mars), 295
permafrost, 122, 208, 253
permanent space station, 130
Perseids (meteor shower), 187
Pershing (U.S. Army missile), 261
personnel, U.S. government (civil service), 193, 263, 267, 282, 295
Peru, 191, 293
pesticide, 9, 39
Peterson, Dr. Laurence, 101
petroleum. See oil.
Petrov, Dr. Boris N., 244, 246, 250
Petrov, Georgy, 204
Phantom (bomber aircraft). See F–4.
pharmaceuticals. See drugs.
Phi Beta Kappa, 247
Philadelphia, Pa., 19, 36, 129, 138, 150
Philadelphia Centennial Exposition. See centennial exposition, U.S.
Philippine Islands, 209, 239
Phillips, Dr. Roger J., 132
phlebitis, 20
Phobos (moon of Mars), 162, 220, 264, 295
photofax, 279
photography, 5, 39, 43, 80, 122, 131, 138–140, 143, 146, 148, 151, 153, 156, 158, 162, 168, 171, 177, 194, 205, 207, 213, 215, 217, 219, 220–221, 224, 226, 247, 269
photosynthesis, 44, 156, 203, 208, 215, 226, 275
photovoltaic technology, 14, 116, 144
physical sciences, 274
physiology, 254, 255
physics, 25, 35, 51, 68, 77, 142, 166, 173, 204, 242, 254, 278
Pickard, Robert H., 29
Pickering, Dr. William H., 64, 69, 133, 173, 254
pigap (USSR experiment), 15
piggyback launch, 37, 50, 210
pilot, 16, 44, 66, 71, 94, 95, 96, 98, 142, 194, 225, 231, 240, 251, 256, 274
—training (see also training, Space Shuttle), 229, 273–274
pilotless craft, 47

pinkeye, 45
Pioneer project, 39, 49, 103, 272, 278
Pioneer 10, 24, 58
Pioneer 11, 104, 189, 290
Pioneer Parachute Co., 150
Pioneer Venus, 211, 288
pipeline, 281
Pitcairn Is., 172
Pittsburgh, Pa., 35
planetary atmosphere. See atmosphere, planetary.
planetary nebula, 108
planetary research (see also individual planets), 14, 15, 34, 35, 39, 59, 70, 76, 90, 104, 120, 127, 137, 138, 146, 147, 148, 152, 154, 160, 162, 166, 168, 192, 193, 196, 204, 212, 217–218, 225, 235–236, 245, 264, 271–272, 283
planets (see also individual planets), 25, 45, 54, 69, 70, 76, 225, 265
planning, 44, 142, 272, 282
plants, 38
plasmas, 49, 92, 102, 127, 142, 148, 250, 343
plastic, 61, 63, 88, 147, 160, 227, 245
Plateau of the Moon (region of Mars), 140
platinum, 56
Plesetsk, USSR, 283
Plum Brook Station, LeRC, 67, 181
Pluto (planet), 35, 43, 70, 291
plutonium, 48, 149, 268
pneumatic testing, 170
Pogue, Col. William R., 66, 129
Poker Flat Range, Alaska, 63
Poland, 128, 219, 224, 304
Polar Experiment North–76 (USSR), 73
polar research, 6, 73, 79, 139, 225, 261
polar-orbiting spacecraft, 24, 54, 131, 142, 165, 192, 203, 215, 219, 225, 239, 296
polarity, 33
polarization, 88, 162
police, 199, 210
politics, 141, 151, 161, 163, 181, 195, 210, 286
Pollack, Dr. James B., 48
Pollack, Louis, 6
pollution (see also air pollution; heat sensing; noise reduction; water pollution), 3, 34, 39, 40, 76, 133, 137, 160, 171, 195, 227, 235, 261, 267, 285, 300, 303
"pollution tax," 285
polyethylene, 167, 230
Polymode (USSR experiment), 23, 83
polyurethane foam, 50
polyvinyl chloride, 24
Pond, Robert B., 95
Ponnamperuma, Dr. Cyril, 104
Pope Paul, 10
population, 34, 47, 111, 154, 159, 171, 176, 228, 230, 270
Porcupine (project), 49
Port Matilda, Pa., 109
portable equipment, 10, 24, 60, 93, 232, 278, 281, 293
Portland (Ore.) International Airport, 257
Porz-Wahn, W. Germany, 44, 194

position determination, 114, 205
positron, 25
Postal Service, U.S., 67, 213
potassium, 187
Potemra, Dr. Thomas A., 296
Poulos, Leah, 23
Powder Puff Derby, 275
power sources, 11, 22, 58, 127, 133–134, 176, 203, 241
—chemical, 80, 145, 213, 291
—electric, 143
—galactic, 101, 105, 132, 301
—nuclear, 36, 48, 83, 99, 118, 149, 181, 242–243, 252, 268, 297
—solar, 13, 26, 29, 52, 70, 72, 75, 80, 94, 111, 116–117, 124, 136, 141, 142, 144, 160, 172–173, 197, 219, 225, 227, 239, 240, 243, 247, 265, 270, 271, 288, 291, 299, 301
—steam, 35
—thermal, 115, 189, 278
—wind, 67, 102, 124–125, 167, 181, 280
"power tower," 240
Pratt & Whitney Aircraft Group (United Technologies), 134
precipitation, 132
precision, 290, 302
prediction (see also future; meteorology), 142, 176, 277, 289, 299
preservation, 111
Presidential Medal of Merit, 40
President's Committee on Science and Technology, 185
presolar particles, 56
press conference. See media.
pressure, 39, 60, 78, 92, 95, 123, 168, 170, 198, 229, 255, 262, 278, 279, 282, 286, 300
prestige, 265
prices, 141, 219
Princeton Institute for Advanced Study, 35
Princeton Univ., 189, 298
printing, 262, 279
Priroda (USSR science journal), 278
private enterprise, 67, 134, 191, 212, 267, 289
probe (see also rockets), 126, 128, 280, 288
process. See data processing; materials processing; new techniques; sewage.
procurement, 63, 78, 83, 111, 158, 163, 164, 206, 226, 260, 287, 289, 303
productivity, 9, 50, 81
Professional Air Traffic Controllers Organization (PATCO), 16, 167, 198
Prognoz (Soviet science satellites), 73, 212, 307
Prognoz 5, 307, 331
program planning, 22, 268
programming, 254
progress, 137
Project Blue Book (USAF), 199
Project Icewarn, 4
propagation, 44, 102, 187, 287
propane, 259
—propellant, 82, 83, 106, 181, 188, 216, 259, 262, 283
propellers, 5, 67, 81
propulsion system, 123, 149, 169, 181, 259

propulsive-lift technology, 33
protection, 130, 236, 280
proton, 25, 26, 165, 242, 250
protostar, 104
prototype, 47, 53, 108, 123, 142, 164, 185, 248, 249, 287, 292
provinces (planetary regions), 218
Proxmire, Sen. William A. (D-Wis.), 1, 228
Pryor, Harold E., 163
psi particle, 255
Public Broadcasting Service, 18, 125
public communications, 44, 63, 195, 228, 235, 251, 253, 273, 293
public health, 60, 216
public lands, 69
public opinion, 12, 51, 54, 56, 75, 119, 152, 158, 195, 208, 211, 246
public policy, 17, 22, 39, 50, 124, 125, 152, 172, 176, 228, 244, 250, 268, 272, 292
public radio, 125
Public Registry of Space Flights (UN), 307
public television, 18, 125, 228
public transportation, 18
public utilities, 125, 225, 240, 268, 280
Puerto Rico, 33, 77, 94, 213, 225, 240
Pulitzer prize, 28
pulsar, 23, 74
pulse, 26, 63, 263
pump, 130, 146
Puppis A (supernova remnant), 93
Purdue Univ., 276
Purkey, L.L., 98
"purple pigeons," 168
Puthoff, Richard L., 181
pyrolytic reaction, 61
pyrolytic-release experiment (Viking), 156, 200, 215, 217, 260

Qantas Airways Ltd., 261
Qattara Depression, 123
QSRA. See quiet short-haul research airplane.
quality control, 78, 259, 267, 292
quantum mechanics, 254
quarks (particles), 243
quartz, 290
quasars, 31, 45, 52, 82, 302
quiet short-haul research airplane, 33
Quonset Point, R.I., 79

R&D. See research and development.
Rabat, Morocco, 176
racing, 145
radar, 4, 14, 16, 47, 52, 73, 77, 84, 102, 108, 132, 140, 153, 165, 168, 171, 177, 184, 213, 214, 225, 227, 233, 236, 247, 251, 274, 281, 305
radiation, 26, 33, 35, 39, 105, 122, 150, 172, 229, 236, 253, 273, 277, 290
radio, 16, 34, 38, 45, 66, 75, 105, 125, 127, 136, 143, 144, 146, 148, 154, 225, 250, 263, 268, 276, 291, 295, 298
Radio Amateur Satellite Corp., 114, 196
radioastronomy, 32, 41, 77, 150–151, 213–214, 233, 270, 302

Radio Corporation of America (RCA), 34, 62, 65, 186, 212, 214, 226, 337
radioactivity, 36, 156, 158, 175, 177, 178, 199, 208, 215, 226, 236, 253, 254, 273
radioisotope thermal generator (RTG), 118
radiolaria, 289
radiometer, 44, 67, 165, 277
radiotelescope, 31, 33, 46, 77, 84, 150, 213, 268, 302
Radojlovic, Zika, 238
Raduga (Soviet comsat, Statsionar 1B), 214, 327
RAF: Royal Air Force
Ragusa, Dr. James, 159
railroads, 142, 219, 276
rain, 162, 177, 226, 249, 279, 305
Rainey, Ed, 229
Raleigh, N.C., 119
Ramo, Dr. Simon, 9
RAMS: random-access measuring system
Ramstein, W. Germany, 242
Rancho Los Amigos, 37
RAND Corp., 119, 132
random-access measuring system, 29
Ranger (project), 65
ranging, 133
Rasquin, John, 290
ratification, 128
Rayhrer, Benno, 302
Raymond, Dr. Louis, 96
RB–57 (USAF research aircraft), 296
RCA. See Radio Corporation of America.
RCA-B. See *Satcom 2*.
RCA-C, 295
reaction control, 20, 297
reactor, 149, 268, 273, 278
real property, 218
receiver (communications), 6, 106, 242, 263–264, 292
recession, 12, 26, 166, 181
reclamation, 151
RECON (information-retrieval system), 303
reconnaissance (see also surveillance), 7, 141, 164, 219, 230
record systems, 103, 127, 131, 133, 147, 150, 221, 230
recovery (equipment; payload; vehicle), 5, 82, 97, 169, 189, 229, 239, 251, 272, 279, 297, 300, 301, 305
rectenna, 197
recycling, 61, 83, 291
red shift, 126, 337
Redstone test site, 25, 170
redundancy, 90, 181
redwoods, 38
Reed, Jack, 290
Reed, John H., 15
Reed, Thomas C., 161, 287
reentry, 33, 87, 168, 221, 303
refan program, LeRC, 285
refitting, 285
reflector, 23, 52, 62, 87, 157, 171, 245, 290
refraction, 46
registration (of objects launched into space), 128, 244, 307

rehabilitation, 25, 39, 66
Reid, Dr. George C., 32
reimbursable launch, 1, 28, 62, 149, 223
relativity (see also Einstein), 52, 126, 337
relay. See communications relay; data relay; satellite relay.
reliability, 6, 20, 28, 61, 67, 81, 144, 149, 161, 232, 240, 259, 272, 276, 277
remote control, 71, 198, 219
remote manipulator, 59, 198, 257
remote sensing, 20, 43, 46, 54, 82, 105, 118, 122, 131, 204, 235, 237, 244, 261, 267, 276, 282, 293, 303
remote transmission, 10, 93, 281
remotely piloted research vehicle (RPRV), 144
rendezvous, 162, 251, 260, 268, 282
Rensselaer Polytechnic Institute, 54, 89
reorganizations, NASA, 63
repair, 118, 160, 247, 253, 303
repeaters, communications, 94, 264
replication, 104, 113
reprocessing, 11, 83, 268
Republic of China (Taiwan), 20, 128
rescue, 59, 114, 171, 196, 232, 242, 243–244
research and development (R&D), 50, 58, 59, 113, 149, 191, 201, 219, 221, 229–230, 252, 265, 274, 275, 276
Research Initiation in Minority Institutions (NSF program), 283
residence, 142, 301
resolution (optical; spectral), 32, 84, 88, 166, 262, 273
"resonance region" (of magnetosphere), 301
resource survey, 123, 175, 192, 204, 244, 276, 306
retirement, 12, 13, 18, 19, 38, 48, 50, 54, 64, 67, 77, 83, 96, 133
retrieval, 101, 160, 169, 199, 257, 276, 297, 300
reusable equipment, 37, 61, 82, 118, 130, 145, 169, 188, 199, 220, 229, 235, 239, 299, 300
reverse polarity, 33
Rho Sigma Company, 116
Rhode Island, 225, 240
rhodopsin, 45
Rice Institute, 72
Rice Univ., 54
Richter, Prof. Burton, 254
Richter scale, 11
Rincon Indian reservation, 299
Rind, David, 300
Rio de Janeiro, Brazil, 3, 269
Ripley, S. Dillon, 137
rivers, 121, 138, 177, 306
RMI Co., 81
Robinson, Prof. Alan, 83
Robinson, Charles, 212
robot, 47, 139, 160, 183
Rockefeller, Vice President Nelson, 92, 137
rocket (see also launch vehicles; probe), 6, 27, 35, 46, 51, 61, 68, 73, 79, 93, 95, 137, 138, 140, 160, 163, 172, 267, 288
Rocketdyne Division, Rockwell Intl. Corp., 25, 73, 94, 149, 169, 211, 232, 235

rocket launches, 6, 17, 28, 49, 51, 63, 128, 187, 239, 250, 273, 297, 301
rocket-powered aircraft, 28, 241
Rocket Propulsion Laboratory (USAF), 106
rocks, 57, 139, 148, 155, 207, 223, 236, 260, 294
Rockwell International Corp., 25, 47, 50, 57, 66, 73, 78, 81, 94, 99, 107, 108, 123, 149, 158, 161, 168, 187, 203, 204, 210, 218, 222, 232, 235, 240, 245, 248, 257, 259, 275, 277, 287, 301, 302, 303
Rockwell Intl. Science Center, 297
Rockwell, Willard F., Jr., 99, 222
Roddenberry, Gene, 159, 222
Rodina (Soviet ship), 143
Roederer, Juan G., 68
Rogaland, Norway, 276
Rohr Industries, Inc., 82, 292
Rohr Marine, Inc. See Rohr Industries.
Rolls Royce (Aero Engines Inc.), 134, 165, 180
Romania, 219, 238, 304
Roosa, Col. Stuart A., 19
Roosevelt, President Franklin D., 92
Ross Ice Shelf Project (RISP), 294
Rossiya (Soviet ship), 143
Rostoker, Dr. Gordon, 296
Rotary Club, 51
rotation, 189, 290, 304
rotor, 67, 102, 113, 167
rotor aircraft, 84, 113
round-the-world flight, 72
roving vehicle (planet exploration), 168, 235, 264
Royal Academy of Sciences (Sweden), 254
Royal Air Force (U.K.), 165
Royal Navy (U.K.), 164
Rozhdestvensky, Lt. Col. Valery, 250, 255, 342
RPI. See Rensselaer Polytechnic Institute.
RPRV. See remotely piloted research vehicle.
RS–27 (engine), 73
RSRA: rotor systems research aircraft
Rumania. See Romania.
Rumsfeld, Donald H., 129, 161, 281, 285, 287
Russell, Dr. Christopher T., 132
Russia (see also USSR), 59, 69, 76, 250, 281
Rutland, Mass., 190
Rutland, Vt., 167

Saab-Scania AB, 209
Sabreliner, 66
Sacramento, Calif., 38
Sacramento Peak Observatory, N.M., 27
Sadat, Anwar, 20
safety
—aircraft/spacecraft, 15, 16, 66, 77, 78, 89, 94, 100, 120, 138, 148, 153, 243, 257, 277
—airport. See air safety.
—environmental, 158, 168, 172
—equipment, 59, 89–90, 102, 169, 172, 205, 232–233, 277
—manned flight, 59, 231, 243
—maritime, 205

—NASA contract, 199
—nuclear, 9, 268, 273
Sagamore Hill Observatory, 233
Sagan, Dr. Carl, 33, 153, 166, 208, 274
Sagdeyev, R. Z., 246
Sagittarius (constellation), 23
Sahara, 122, 154, 297
sailplanes, 109
St. Augustine, N.M., 31
St. Louis, Mo., 36, 65, 88, 99, 117, 160, 186
St. Joseph's Hospital, Houston, 128
St. Peter (Bering ship), 144
Saito, Shigebumi, 244
salinity, 20, 45
Salisbury, Winfield W., 298
SALT: strategic arms limitation talks
Salt Lake City, Utah, 80, 218
Salyut program (USSR), 49, 73, 80, 245, 260, 307
Salyut 3, 41, 260
Salyut 4, 5, 38, 92, 129, 140, 171, 222
Salyut 5, 129, 140, 177, 190, 193, 204, 221, 250, 255, 260, 307, 321, 341, 342
Samoa, 261
sampling, 60, 61, 154, 155, 156, 160, 183, 199–200, 204, 208, 223, 300
SAMSO. See Space and Missile Systems Organization, USAF.
San Antonio Logistics Center (USAF), 99
San Diego, Calif., 60, 73, 264
San Francisco, Calif., 35, 94, 145, 187, 242, 301
San Francisco International Airport, 257
San Gabriel mountains, 43
San Marco C–2, 19
San Mateo, Calif., College of, 231
Sanchez, Maria-Elena, 231
sand, 220
Sandia Laboratories, ERDA, 9, 160, 241
Sandusky, Ohio, 67
Santiago, Chile, 7
SAO. See Smithsonian Astrophysical Observatory.
Sarett, Lewis H., 254
Sarychev, Vasily A., 246
Sas 3, 23
Satcom 1, 34, 65
Satcom 2, 61, 65, 315, 337, 338
"Satellite Access Cities," 35
Satellite Business Systems, 78
satellite communications, 6, 14, 21, 37, 59, 65, 82, 117, 135, 175, 179, 181, 212, 213, 214, 225, 226, 227, 239, 240, 262
—control, 240, 250–251
—imagery, 24, 120, 143, 144, 198, 225, 262, 268–269, 279
Satellite Instructional Television Experiment (SITE), 111, 175, 228
satellite interception. See interceptor.
satellite power systems (see also power sources), 75, 118, 120, 127, 203, 270–271
satellite relay (see also communications relay; data relay), 47, 93
Satellite Telecommunications Experiments Project (STEP), 248

Saturn (planet), 13, 24, 43, 58, 71, 104, 168, 189, 245, 271, 291, 294
Saturn (program), 64, 72, 302
Saturn (rocket), 50, 51, 64, 137, 150, 240, 262, 299
Saturn-Apollo, 298
Saudi Arabia, 27, 165, 179
SCAD: strategic armed cruise-missile decoy
scandal, 191
scanning, 93, 146, 165, 268
Scanoprobe, 298
Schafer, Charles F., 297
schedule, 68, 73, 85, 162, 167, 207, 300
Schempp-Hirth Standard Cirrus (sailplane), 109
Schiaparelli, Giovanni, 100
Schiff, Prof. Harold, 304
Schmitt, Sen. Harrison H. (R-N.M.), 51, 270
Schneider, Dr. William C., 248, 277
Schnittker Associates, 141
Schonfeld, Ernest, 281
Schuler, Ed, 145
Schurmeier, Harris M. "Bud," 71
Schutzenhofer, Luke A., 33
Science Academy, USSR, 129
science adviser, White House, 151, 184–185, 239, 299
Science Applications, Inc., 203
science court, 9, 39
science fiction, 20, 100, 152, 154, 176, 281, 298
Science Research Council, U.K., 283, 301
Scientific and Technical Information Office, NASA Hq, 163, 194
scientific community, 1, 7, 9, 16, 27, 30, 57, 107, 126, 230, 238
scientific equipment, 117, 264
scientists, 30, 35, 36, 48, 133, 142, 148, 151, 153, 160, 166, 168, 193, 265, 287
scintillation, atmospheric, 20, 179
Scorpio (constellation), 115
Scott, James E., 280
Scott, Dr. Ronald, 156
Scott Aviation, Inc., 102
Scout (launch vehicle), 99, 102, 126, 203, 295
Scoville, Herbert, 149
SDAS: site data acquisition system
SDS–2 (DOD satellite data system spacecraft), 325
Sea of Crises (Mare Crisium) on moon, 183, 300
Sea of Felicity, on moon, 183
Sea of Serenity, on moon, 183
sea state, 46
sea-surface temperature, 24
seagulls, 134
Seamans, Dr. Robert C., Jr., 11, 134, 231
search and rescue, 114, 171, 196, 205, 242, 251
search for extraterrestrial intelligence (SETI), 268, 269
Seasat program, 57, 144, 302
seasonal variations, 17
Seattle, Wash., 166, 187
Seattle-Tacoma International Airport, 257

security, 22, 199, 210, 215, 229, 245, 246, 257, 273, 305
sediment, 138, 281, 289
Sedov, Leonid, 41
seed, 150
Seep project, 173
seismic activity (seismology, seismometer), 11, 23, 27, 37, 84, 140, 153, 155, 177, 178, 208, 224, 237, 281, 294, 295
Semeiz, USSR, 151
Semenov, Y.P., 246
semiconductors, 259
senior citizens, 12, 111
Senate Committee on. See Congress; general subject.
sensors, 8, 14, 43, 60, 88, 131, 155, 229, 235, 256, 267, 279, 296, 297, 302
sensory deprivation, 190
SEP. See Société Européenne de Propulsion.
Serling, Robert J., 96, 98
SESP 74–2 (DOD spacecraft), 323
SETI. See search for extraterrestrial intelligence.
Seuss, Hans, 56
Sevastyanov, Vitaly, 92, 129, 171
747 aircraft. See Boeing 747.
sewage treatment, 40, 60, 61, 267
Shackleton Range, Antarctica, 294
shale, 11
Shamsin, V., 272
SHAPE. See Supreme Headquarters, Allied Powers in Europe.
Shatalov, Lt. Gen. Vladimir, 219, 221, 251
Shavchenko, USSR, 273
Sheldon, Dr. Charles S., II, 112
Shepard, Capt. Alan B., Jr., 11, 19, 25
Shimada, Dr. Katsunori, 299
ship, 129, 130, 143, 145, 154, 159, 173, 204, 205, 230, 303
shipping, 4, 31, 38, 50, 60, 73, 89, 115, 142, 236, 247, 249
shipset (aircraft; spacecraft), 220, 248
shock waves, 33, 267, 281
short takeoff and landing (STOL) aircraft, 33, 84, 164
shortwave, 196, 250, 263, 264, 276
Shuang-Cheng-Tze (PRC launch site), 196, 290
Shuttle. See Space Shuttle for all entries.
Siberia, 90, 215, 252, 272, 280, 297
Sicily, 190
side-looking airborne radar system (SLAR), 4, 171, 247
Sidewinder (missile), 165
Sierra Club, 150
Sierra Leone, 293
Signal Corps, U.S. Army, 40, 237
signal-relay satellites, 10
Signe 3 (proposed French satellite), 212
Sikorsky Aircraft Division, 113, 304
silica (insulation), 220
silo (launch complex), 1
silicon, 56, 109, 123, 157, 272
silver, 122, 158
Silverman, Jack, 94

Simonds, Dr. Charles, 300
Simons, Vera, 88, 160
Simpson, Dr. John, 105
simulation, 15, 25, 43, 44, 64, 72, 91, 113, 114, 117, 118, 127, 130, 162, 169, 194, 199, 204, 220, 226, 227, 240, 257, 262, 267, 270, 280, 288, 301, 302
simultaneous measurement, 46, 67, 121, 131, 163, 261, 268–269, 296, 301, 304
Sinai, 122
Singapore, 73
Singer Co. Simulations Products Div., 91, 162
Sioux Falls, S.D., 230, 242
Sirio (Satellite Italiano Ricerca Industriale Orientata), Italian comsat, 249, 271, 294
SITE. See Satellite Instructional Television Experiment.
site data-acquisition system (SDAS), 247
Sivertson, Gene, 171
skating, 23
skin, 172, 231
Skylab project, 11, 26, 39, 49, 60, 96, 109, 126, 137, 140, 172, 256, 267, 298
Skylab 2, 96
Skylab 3, 66
Skylab 4, 129, 190
Skylab solar workshop, 179
Skylark (U.K. rocket), 93, 301
SLAR. See side-looking airborne radar.
Slayton, Donald K. "Deke," 27, 53, 259, 282
SLBM (surface-launched ballistic missile), 257
Slidell Computer Complex. See Earth Resources Laboratory, La.
slingshot effect, 43, 291
Sloan Foundation. See Alfred P. Sloan Foundation.
slowdown, 167, 198
small business, 116
Small Business Administration, 193
Smith, Cliff, 264
Smith, Dr. Edward J., 290
Smithsonian Astrophysical Observatory, 101, 127, 166, 298, 343
Smithsonian Institution, 40, 90, 103, 137, 193, 238, 278, 299
smog/smoke, 46, 105
Smolensk, USSR, 297
SMS. See synchronous meteorological satellite.
Smylie, Robert E., 285
SNECMA: Société National d'Étude et de Construction de Moteurs d'Aviation
SNIA-Viscosa (Italy), 275
SNIAS: Société Nationale Industrielle Aerospatiale (France)
snow, 51, 74, 138, 139, 249, 251, 279
soaring, 109
social goals, 119, 149, 159, 199, 262, 287
social security, 13
social services, 13
Société des Télécommunications Internationales du Cameroun, 181
Société Européenne de Propulsion (SEP), 275
Société Nationale d'Étude et de Construction de Moteurs d'Aviation (SNECMA), 188

Société Nationale Industrielle Aerospatiale (SNIAS), 286
Society of Automotive Engineers, 81
Soffen, Dr. Gerald S., 100, 156, 157, 215, 226
soft landing, 148, 221
software, 28, 112, 162, 194, 232
soil, 142, 155, 157, 209, 224, 254, 281
—erosion, 76
—sampling and analysis, 49, 146, 154, 155, 156, 160, 175, 177, 178, 183, 199–200, 208, 215, 217, 223, 224, 226, 236, 253, 260, 274
Sokolov, M.V., 246
sol (Mars day), 120
solar activity, 5, 8, 12, 50, 88, 92, 180, 229, 278, 295, 296
solar array. See solar conversion devices.
solar atmosphere, 178
solar cells. See solar conversion devices.
solar conjunction, 261
solar conversion devices
—array, 8, 10, 26, 75, 94, 96, 111, 136, 144, 296
—cell, 45, 111, 116, 117, 136, 144, 145, 146, 165, 186, 219, 271, 291
—collector, 124, 141, 227, 235, 247, 302
—concentrator, 227, 299
—generator, 172, 240
—panel, 19, 70, 96, 118, 165, 186, 203, 227, 297
—receiver, 94, 240
solar corona, 8, 80, 179, 297
solar emission. See solar radiation.
solar energy (see also solar conversion devices; solar heating and cooling; solar radiation; solar wind), 7, 8, 10, 11, 12, 13, 14, 26, 52, 58, 62, 70, 75, 80, 94, 96, 99, 107, 111, 116, 124, 127, 136, 137, 141, 144, 156, 172–173, 176, 197, 225, 227, 235, 240, 243, 248, 254, 265, 270, 288, 301
Solar Energy Products Co., 116
Solar Energy Research, Development, and Demonstration Act of 1974, 52
Solar Energy Systems, Inc., 116
solar flares. See solar activity.
solar heating and cooling, 11, 14, 29, 30, 72, 116, 141, 142, 227, 239, 247, 254, 301
solar magnetic field, 290, 301
Solar Maximum Mission, 296
solar nebula, 56
solar observatories, 163, 290–291
solar orbit, 8, 31
solar panel. See solar conversion devices.
solar physics, 12, 26, 88, 92, 127, 198, 254, 278, 343
solar polar mission (ESA), 239
solar power. See solar energy; see also solar conversion devices; solar heating and cooling; solar radiation; solar wind.
solar-power satellite, 265, 270
solar radiation, 5, 33, 88, 124, 127, 160, 163, 168, 175, 193, 196, 200, 235, 236, 250, 274, 291
solar receiver. See solar conversion devices.

solar research, 8, 12, 26, 27, 52, 80, 88, 92, 127, 178, 179–180, 187, 239, 294
Solar Sailor (future project), 160, 168, 235
solar simulator, 124
solar system, 24, 25, 35, 43, 55, 56, 58, 65, 69, 70, 71, 76, 104, 139, 218, 254
solar wind, 8, 11, 14, 58, 74, 235, 236, 291
Solargenics, Inc., 116
Solaron Corp., 116
Soletta (proposed space reflector system), 245
solid-fuel (rocket) booster (SRB), 33, 65, 82, 89, 97, 168, 188, 198, 210, 232, 262, 275, 289, 300
solid-fuel (rocket) motor (SRM), 82, 89, 188, 198–199, 267, 283, 295, 300
solid-propellant rocket. See solid-fuel rocket.
solid-rocket booster. See solid-fuel rocket.
solid waste disposal. See sewage treatment.
Solrad 11–A, 11–B, 50, 313
sonar, 106
Sones, William K., 6
sonics, 27, 144, 267, 300
sounding rocket (see also Appendix D; individual rockets), 49, 128, 143, 166, 250, 272
South Africa, 149, 294
South America, 18, 76, 106, 176, 279
South Carolina, 151
South Dakota School of Mines, 132
South Pole, 27
Southern California, Univ. of, 37
Southwest United States, 18
sovereignty, 293
Soviet Academy of Sciences, 76, 160, 183, 212, 221, 246, 278
Soviet Intercosmos Council, 212, 244
Soviet space activity (see also USSR), 5, 14, 38, 39, 41, 59, 73, 95, 112, 119, 120, 127, 129, 140, 146, 171, 177, 182–184, 190, 193, 196, 204, 214, 220–222, 230, 250–251, 255, 256, 260, 272, 280–281, 303, 305, 307
Soviet Union. See Union of Soviet Socialist Republics (USSR).
Soyuz program, 80, 245, 303, 306, 307, 341, 342
Soyuz 1, 251
Soyuz 2, 260
Soyuz 3, 251, 260
Soyuz 5, 140
Soyuz 7, 260
Soyuz 8, 251, 260
Soyuz 11, 251, 303
Soyuz 13, 108
Soyuz 15, 251, 260
Soyuz 18, 90, 171, 222
Soyuz 19 (ASTP), 282
Soyuz 20, 38
Soyuz 21, 140, 190, 193, 204, 221, 250, 255, 322, 341, 342
Soyuz 22, 220, 221, 328, 341, 342
Soyuz 23, 250, 255, 260, 329, 341, 342
Space Activities Committee of Japan, 244
Space Age, 171
Space and Missile Systems Organization (SAMSO), USAF, 90, 106, 193, 267
"Space as a Habitat," 160

space biology and medicine, 14, 15, 17, 38, 49, 66, 100, 126, 148, 153, 155–156, 159, 177, 178, 199–200, 212, 215, 218, 231
—animal experiments, 140
—human research, 5, 20, 23, 171, 172, 190, 231, 255, 256
space colonies. See colonization.
space exploration, 7, 12, 14, 17, 22, 23, 24, 25, 26, 27, 31, 32, 33, 35, 39, 43, 44, 45, 46, 48, 49, 50, 52, 54–57, 58, 59, 65, 80, 127, 139, 150, 152, 238, 241, 282
Space Exploration Day, 147
space facilities, 112
Space Hall of Fame, 238, 241, 259
space industrialization, 75, 80, 106–107, 203, 235
"Space Industrialization as a Concept," 107
space law, 243–244
space manufacturing, 80, 107, 184, 193, 203, 250, 278
space medicine, 14, 159, 203
"Space Medicine and Biotechnology," 160
space physics, 14, 272
space platform. See space station.
space power source, 83, 107, 127, 237, 270, 278
space processing applications rockets (SPAR), 95, 126, 225, 279, 297
space processing of materials. See materials processing in space.
Space Sailer, 160
space science, 68, 77, 95, 142, 195, 265
Space Sciences Board, 225
Space Science Steering Committee, 148
Space Shuttle
—administration, 12, 33, 53, 61, 64, 75, 80, 91, 192, 211, 224, 227, 243, 245, 259, 289
—budget, 13, 59, 95, 192, 193, 222, 224, 228, 246
—cargo bay, 218
—contractors, 19, 28, 77, 78, 80, 81, 112, 145, 150, 162, 188, 192, 232, 259, 277, 300, 303
—crew, 14, 46, 59, 66, 89, 91, 95, 96, 118, 123, 142, 145, 162, 222, 232, 255, 301
—data, 107, 112, 232
—design, 130, 220
—development, 33, 35, 72, 90, 118, 130, 161, 192, 194, 198, 222, 228, 232, 257, 262
—experiments, 65, 116, 126, 230
—external tank (ET), 72, 73, 77, 81, 82, 168, 169, 170, 198–199, 210, 222, 240, 258, 262
—launch, 27, 64, 65, 66, 82, 89, 160, 301
—main engine (SSME), 25, 33, 64, 89, 168, 169, 172, 198–199, 204, 210–211, 232, 235, 240, 245, 262, 295, 300
—operations, 17, 53, 89, 91, 99, 118, 142, 162, 227, 232, 245, 249, 257–258, 265, 289, 292, 297, 299, 300, 301
—orbiter, 14, 27, 28, 59, 62, 66, 78, 90, 98, 118, 123, 129, 130, 145, 166, 169, 187, 193, 199, 204, 210, 220, 222, 229, 240, 245, 257, 295
—payload, 96, 101, 113, 126, 142, 159, 160, 169, 203, 218, 222, 245, 249, 257, 270, 271, 277, 279, 289
—propulsion, 33, 81, 82, 90, 150, 168, 186, 188, 193, 232, 240, 295, 300
—safety, 78, 130, 169, 245
—testing, 14, 25, 27, 33, 37, 53, 61, 62, 64, 77, 80, 81, 90, 96, 98, 99, 113, 130, 161, 168–171, 192, 210, 232–233, 235, 240, 262, 267
—training, 91, 118, 162, 220, 222, 224, 229, 257–258, 301
—upper stage, 90, 186, 193, 218, 289
space station
—assembly, 166, 204, 265, 278
—crew, 92, 159, 190, 204, 255
—mission, 22, 35, 75, 129, 140, 148, 204, 250–251, 260
—observations, 41, 177, 204
—planning, 47, 193, 203, 204, 265, 278
—power, 83, 265
—research, 5, 15, 38, 73, 159, 204
space suit (see also rescue), 59, 63, 145, 255, 303
Space Task Force Group, NASA, 241, 277
space technology, 21, 64, 69, 76, 83, 197, 205, 235, 265
Space Telescope (future project), 53, 80, 239
space tourism, 59, 107, 203, 244
space tracking facilities, 128, 227, 240
Space Transportation System (STS, Shuttle), 75, 80, 91, 201, 222, 243, 245
space vehicle debris, 214, 252–253, 281
spacecraft
—communications, 127, 146, 291
—design, 18, 54, 65, 83, 186, 220, 275, 303
—handling, 23, 127, 152, 220, 221
—operations, 121, 221, 245
Spaceflight Tracking and Data Network (STDN), 227
Spacelab (ESA program), 14, 48, 53, 65, 66, 78, 91, 113, 123, 133, 134, 136, 148, 159, 192, 194, 197, 203, 227, 241, 244, 248, 249, 269, 270, 279, 292
Spacelab 1, 148, 197–198, 270
Spacelab 2, 194, 197
Spacelab payload integration and coordination in Europe (SPICE), 44
spaceport, 132, 301
spacewalk. See extravehicular activity (EVA).
Spain, 4
SPAR. See space processing applications rockets.
SPEAR (Stanford positron-electron accelerator), 243
Special Committee on Solar-Terrestrial Physics, 68
spectrum (spectrography, spectrometry, spectroscopy), 17, 63, 88, 108, 115, 131, 142, 146, 155, 217, 221, 227, 247, 268, 278
speed, 23, 44, 51, 52, 71, 74, 87, 96, 108, 144, 164, 165, 220, 231, 267, 275, 277, 287, 292, 302
Sperry, Albert F., award, 241
Sperry Rand, 161, 190
Spica (star), 115

SPICE. See Spacelab payload integration and coordination in Europe.
spin-scan camera, 10
spinning solid upper stage (SSUS), 218
spinoff, 20, 41, 111
Spirit of St. Louis, 137, 256
splashdown, 251
spare, 24
sports, 23, 36
spray cans, 216
Sputnik (Soviet satellite), 76, 92
spy satellite, 7
SR-71 (reconnaissance aircraft), 164
SRB. See solid-fueled rocket booster.
Sri Lanka, 96
SRM. See solid-fueled rocket motor.
S.S. *Arco Prudhoe Bay*, 37
SSME. See Space Shuttle main engine.
SSN-8 (Soviet missile), 283, 288
SS-NX-18 (Soviet ICBM), 283
SST (supersonic transport). See Concorde.
SSU-1, SSU-2, SSU-3 (spacecraft support unit, DOD satellite), 316
STA: Shuttle training aircraft
stability; stabilization, 10, 20, 24, 62, 108, 135, 186, 200, 205, 208, 210, 218, 221, 249, 289, 290
Stafford, Maj. Gen. Thomas P., 27, 51, 231, 282, 283
standardized parts list, 1
Standing Committee for Post and Telecommunications. See CEMA.
Stanford Research Institute, 269-270
Stanford Univ., 127, 203, 243, 254
Stapleton Airport, Denver, 89, 257
star, 32, 33, 52, 62, 69, 82, 93, 104, 108, 115, 132, 150, 297
"Star Trek" (TV series), 159, 211, 222
state governments, 52, 219, 246, 276
State, U.S. Dept. of, 59, 128, 149, 212, 281
static testing, 114, 262
statistics, 141, 263
Statsionar 1 series (Soviet comsat. See Ekran; Raduga), 214, 307
status review, 218
STDN. See Spaceflight Tracking and Data Network.
steady-state operation, 25
steam, 27, 35, 94, 115, 173, 225, 240
steel, 67, 137, 210, 246, 285
stellar eclipse. See occultation.
stellar object, 23
STEP. See Satellite Telecommunications Experiments Project.
Stephens, Col. R.L., 164
Sterba, James P., 246
stereoscopy, 131
sterilization, 23, 24, 147, 157, 200
Stever, Dr. H. Guyford, 27, 151, 184-185, 299
stewardess, 229
Stoeckenius, Dr. Walter, 45
Stoewer, Heinz, 91
STOL. See short takeoff and landing.
Stone Age, 252
Stone, Dr. Edward, 278

storage, 23, 50, 72, 162, 165, 262, 268, 273, 276, 291, 298, 302
straddle carrier transporter, 258
strategic armed cruise missile decoy system, 158
strategic arms limitation talks (SALT), 47, 230
strategic weapons, 47, 81, 108, 158, 229-230, 257, 281, 286, 287, 289
stratosphere. See atmosphere.
stress, 47, 167, 170, 172, 210, 222, 290
Striedieck, Karl, 109
strike, 26, 150, 198, 288-289
strip mining, 105, 151
structure, 61, 71, 80, 81, 117, 123, 130, 137, 142, 159, 225, 262, 278, 303
STS. See Space Transportation System.
student, 166, 172, 231, 245, 276
Sturtevant Aeroplane Co., 40
submarine, 47, 53, 159, 249, 283, 286, 287, 288, 301
subsatellite, 131
subsidy, 125, 212, 285
subsonics, 84, 123, 165, 189, 220, 230, 269
subsystem testing, 123
Sudan, 41, 176, 293, 306
Sudbury, Ont., 294
Suez, 123
Suitland, Md., 38, 242
Sukhanetr, Sribhumi, 238
Sukhoi-22 (Soviet aircraft), 191
sulfur compounds, 117, 158, 160
sulfuric acid, 48
sun-earth physics, 68, 105, 150
Sunbelt, 111
Sunfire project, 172
sunlight, 45, 111, 116, 122, 189, 216, 227, 240, 248, 302, 305
Sunworks, Inc., 116
Superdome, 132
supernova, 56, 93
superoxides, 175
supersonics, 3, 8, 51, 58, 84, 87, 102, 123, 125, 130, 144, 149, 151, 168, 185, 189, 195, 217, 269, 286, 300
support facilities, systems, 6, 8, 145, 162, 165, 179, 199, 249, 265, 267, 269, 272, 273, 275, 278, 280, 289, 303
Supreme Headquarters, Allied Powers in Europe (SHAPE), 48
surface-effect ship, 292
surfaces, planetary; lunar, 49, 71, 119, 121, 146, 148, 154, 155, 157, 160, 168, 175, 183, 189, 196, 199-200, 203, 216, 218, 221, 223, 261, 274, 281, 294, 307
surgery, 28, 47, 63
surplus, 66, 219
Surinam, 293
surveillance (see also reconnaissance), 7, 99, 101, 106, 131, 133, 142, 170, 204, 219, 271, 295
Surveyor (spacecraft), 49, 65, 100
survival, 204
suspension, 297
Sweden, 4, 49, 130, 163, 209, 254, 263, 343
Swenson, Dr. George W., 302

Swider, Nancy, 23
swing-wing aircraft, 123, 165, 287
Switzerland, 4, 23, 46, 130, 238, 243, 263, 294
Symphonie 1, 2 (Fr.-German comsats), 69, 135, 186, 249
Synchronous Meteorological Satellites (*Sms 1, Sms 2*), 4, 5, 10, 38, 301
synchronous orbit, 6, 10, 21, 34, 47, 60, 61, 65, 115, 165, 218, 226–227, 229, 265, 272, 290, 294
synchrotron, 242
synthetics (see also manmade materials; new products), 24, 56, 134, 156, 254
System Development Corp. 243
systems analysis, 47, 116, 166, 265

tactics, 47, 257
Tanegashima, Japan, 44
tape recorder, 147, 205, 214, 221, 230, 237, 264, 291, 293
target, 120, 163, 229, 273, 281, 305
—spacecraft, 34
Tarzan, 154
Taurus-Littrow (on moon), 51
taxes, 12, 106, 160, 269, 285
TCOM Corp., 136
TDRSS. See tracking and data-relay satellite system.
teacher in the sky. See *Ats 6*.
Teague, Rep. Olin E. (D-Tex.), 19, 222
Tech House, 30
technicians, 120
technological innovation (see also new products; new techniques), 9, 50, 70, 147, 235, 256
technology applications, 4, 5, 17, 71, 148, 152, 188, 251, 265, 267
technology exchange, 48, 93, 106, 175, 245, 291, 293
Technology Utilization Office, NASA Hq, 63, 66, 93, 130, 163
tectonics, 84, 208, 218
Teem, Dr. John M., 13
Tehran, Iran, 44, 96, 98
tektites, 171
Tel Aviv Univ., 305
telecommunications, 6, 209, 213, 244, 263, 287, 288, 293
Teledyne Brown Engineering, 77
Teledyne McCormick Selph, 99
Teleglobe Canada, 200
telegraph, 206, 215, 225
telephone, 21, 31, 69, 92, 94, 120, 136, 137, 141, 152, 157, 186, 206, 215, 225, 276, 288, 292, 302
Telesat (Canadian comsat), 141
telescope, 7, 31, 35, 48, 53, 62, 77, 82, 83, 93, 123, 148, 213, 230, 272, 283, 288, 302
teletype, 279
television, 10, 18, 36, 70, 92, 106, 111, 115, 136, 141, 144, 147, 151, 154, 159, 200, 215, 222, 225, 227, 250, 257, 264, 270, 272, 281, 287, 293
telex, 31, 62, 79, 115

temperature
—air, 8, 38, 70, 305
—atmospheric, 27, 170, 301
—building, 11, 29, 30, 71, 115, 116, 227, 239, 247–248, 301–302
—earth, 19, 34–35, 74, 165, 279, 289
—moon, 36, 54, 74
—planetary, 39, 43, 48, 70, 76, 104, 108, 120, 122, 139, 146, 189, 199–200, 205, 208, 209, 226, 288, 295
—simulation, 8, 72, 124, 127, 274, 288
—solar, 26, 56
—space, 165, 245, 291
—utilization, 19, 21, 23–24, 32, 35, 36, 49, 50, 60, 83, 94, 130–131, 142, 145, 166–167, 220, 236, 240, 278, 279, 290, 299
—water, 4, 6, 24, 67–68, 118–119, 192, 247
Tengiz, Lake, 251
Tennessee, 137, 151
Tennessee State Univ., 99
Tereshkova, Valentina, 46
Terman, Frederick E., 254
terracing, on Mars, 225
terrain, 122, 140, 146, 148, 165, 171, 207–209, 218, 230, 264, 277, 281
terrestrial applications, 112, 244
Tesla, Nikola, 143
test facility, 8, 25, 27, 64, 72, 77, 78, 124, 169, 170, 188, 227, 240, 249, 255, 258, 262, 271, 278–279, 295, 301–302
test pilot. See pilot.
testing, 24, 25, 64, 78, 80, 95, 99, 113, 116, 159, 167, 171, 194, 204, 222, 231, 249, 255, 259–260, 261, 276, 277, 280, 283, 288, 303, 306, 343
tethered satellite, 166
Texas, 111, 115, 224, 225, 240, 262, 264, 279
Texas Governor's Committee on Aging, 111
textiles, 23, 93
Thailand, 176, 238, 293
theoretical physics, 26, 51, 52, 56, 291
thermal coating. See thermal protection.
thermal mapping. See heat mapping.
thermal protection, 26, 59, 236, 245
thermal scanning, 115
thermodynamics, 56, 104
thermonuclear reaction, 26, 273, 278
thermosphere (earth). See atmosphere.
Thiokol Corp., 82, 188, 199, 232, 245, 283
Third Century America (exposition), KSC, 70, 107, 147, 210
Thomas, Colin, 263
Thomas, Dr. Roger, 178
Thompson trophy, 275
Thor-Agena (launch vehicle), 195
Thor-Burner 2, 214
Thor-Delta, 218, 276, 299
Thorne, Dr. Kip S., 45
thrust-vector control, 232
Thumba, India, 128
thunderstorm, 10, 132
tidal wave, 38
Tidbinbilla, Australia, 151
tiles (Shuttle insulation), 130, 220

392

tilt-rotor aircraft, 258
timber, 34, 69, 252–253
time, 104, 126, 127, 129, 146, 152, 159, 167, 170, 305
time capsule, 150, 210
Ting, Prof. Samuel C.C., 254
TIP: Transit Improvement Program (U.S. Navy)
TIP-3, 203, 327, 340
Tiros (metesat) program, 165
Titan (moon of Saturn), 168
Titan (missile), 99
Titan IIIB, 219
Titan IIIC, 50, 133, 228
Titan IIID, 141
Titan IIIE-Centaur, 295
Titan-Centaur, 7, 49
titanium, 81, 157, 222
Titanium Metals Corp., 81
Tito (Josip Broz), 143
Todd, Webster B., Jr., 134
Tokamak (fusion machine), 36, 278
Toksoz, Dr. Nafi, 208
Tokyo, Japan, 187, 256
Toledo, Ohio, 116
Tomahawk (Navy solid-fuel rocket), 267, 286
Tomcat. See F–14 aircraft.
tools, 135
topography, topology, 218, 254, 281
tornado, 10
Tornado MRCA (multirole combat aircraft), 165
Toronto, Univ. of, 302
torpedo, 53, 286
tourism, 107, 182, 203, 215
Toth, Robert, 36
Toulmin, Dr. Priestly, III, 157, 236
Toulouse, France, 187
toxic substances, 11, 40, 172
trace elements, 41, 157, 216
tracking and data-relay satellite system (TDRSS), 226–227
tracker station. See earth station.
tracking systems, 14, 29, 31, 67, 100, 119, 144, 145, 165, 181, 240, 273, 297, 305
training, 14, 44, 51, 66, 91, 118, 140, 142, 158, 162, 164, 190, 214, 219, 220, 222, 224, 227, 231, 251, 256, 270, 280, 301
traffic, 61, 120, 144, 166, 198
trajectory, 43, 118
Trans Siberian Railroad, 297
TransWorld Airlines, 16, 26, 166
transatlantic communications, 18, 59, 87, 195, 233
transatlantic flights, 102, 182, 212, 256, 269
transcontinental flight, 71, 229, 275
transfer orbit, 18, 94, 276, 290
transformer, 143
transfusion, 251
transistor, 259, 275
Transit 19 (Navy navsat), 294
Transit 20, 295
Transit Improvement Program (TIP), U.S. Navy navsat, 203, 327, 340
transition period (budget), 12, 75, 285

transmitter, 6, 138, 147, 152, 155, 196, 217, 225, 260, 262, 272, 281, 292
transoceanic aircraft, 14, 212
transonics, 24, 33, 168
transponder, 18, 62, 125, 275
transportation, 16, 18, 71, 72, 81, 84, 89, 97, 118, 138, 144, 149, 166, 276, 293, 294
Transportation, U.S. Dept. of, 3, 102, 124, 201, 212, 285, 296, 304
traveling-wave tube (TWT), 6, 10, 62
Treasury, U.S. Dept. of, 299
treaty, 128, 243, 263, 281
Trella, Prof. Massimo, 134, 241
Triad (U.S. Navy satellite), 296
Tricentennial, U.S., 80, 150, 154, 210
Trident 3, 214
trisonic facility (MSFC), 168
TriStar (jet transport). See L–1011.
Tritonis Lacus (region of Mars), 121
Trombka, Dr. Jacob L., 132
trucking, 38, 62, 72
Truly, Richard H., 222
TRW Systems, Inc., 20, 81, 92, 101, 145, 186, 212, 214
TRW/Aerojet Defense Support Program, 133
tumors, 130
tungsten, 124
Tsarev, A.I., 246
Tselinograd, USSR, 193, 221, 251
Tucson, Ariz., 7
Tucumcari, N.M., 16
Tunguska, Siberia, 252–253
Tunney, Sen. John B. (D-Calif.), 222
Tupolev–144 (TU–144, Soviet supersonic transport), 4, 8, 286
turbines, 236, 240
Turbo-Union, 165
turbulence, 5, 64, 186, 231, 267, 277, 280
Turkey, 253
Turkish Airlines, 253
TWA. See TransWorld Airlines.
200 Yankee (aircraft), 98
TWT: traveling-wave tube.
Tyuratam (Soviet launch site), 29, 34, 250, 273

U–2 (earth-survey aircraft), 76, 158, 170
UFO. See unidentified flying objects.
Uganda, 293
Uhlmann, Dr. Donald R., 297
U.K. See United Kingdom
Ukraine, 177, 276, 297
ultrahigh frequency (UHF), 223
ultrasonic equipment, 47, 102
ultraviolet, 105, 109, 115, 131, 163, 175, 216, 223, 236, 272, 274, 283
Ume (ISS-1, Japanese ionosphere-sounding satellite), 44, 307, 312
U.N. See United Nations.
U.N. Committee on the Peaceful Uses of Outer Space, 128
underground testing, 255
underwater communications, 301
underwater testing, 283
Underwood, Maj. E.B., Jr., 44
unemployment, 12

unidentified flying objects, 16, 199
Union of Soviet Socialist Republics (USSR). See also Russia; Soviet listings.
—cooperative activities, 45, 68, 78, 80, 83, 119, 128, 141, 150, 159–160, 162, 191, 195, 198, 212, 219, 224, 231, 243–246, 250, 293– 294, 296, 299, 300, 304
—history, 1, 29, 46, 47, 59, 76, 90, 92, 128, 143, 149, 158, 163, 207, 214, 238, 252–253, 265, 273
—technology, 4, 6, 8, 9, 15, 30, 32, 34, 36, 50, 73, 79, 108–109, 117–118, 176–177, 210, 214–215, 220, 229, 242–243, 246, 262, 263, 276, 277, 278, 280–281, 283, 286, 287, 288–289, 297, 298, 299, 305–306
—space. See Soviet space activity.
United Air Lines, 229, 261
United Aircraft Corp. See United Technologies.
United Arab Emirates, 27, 176, 293
United Kingdom (U.K.), 4, 87, 93, 115, 117, 130, 164, 165, 181, 187, 192, 209, 212, 233, 238, 263, 269, 273, 283, 294
United Nations (U.N.), 21, 30, 106, 107, 128, 205, 238, 244, 273, 307
United Space Boosters, Inc., 198, 300
United States Air Force (USAF). See Air Force, U.S.
United States Arms Control and Disarmament Agency, 149
United States Army. See Army, U.S.
United States Bicentennial. See Bicentennial, U.S.
United States Coast Guard. See Coast Guard, U.S.
United States Department of Defense. See Defense, U.S. Dept. of.
U.S. Geological Survey. See Geological Survey, U.S.
United States Marine Corps. See Marine Corps, U.S.
United States Navy. See Navy, U.S.
United States Postal Service, 67
United Technologies Corp., 108, 134, 145, 167, 198, 232, 300, 302
Univac 110/46 (computer), 161
universe, 32, 56, 82, 139
Universities Space Research Assn (USRA), 101, 148
university community, 1, 52, 56, 101, 104, 166, 179, 191, 241, 247, 274, 276, 287, 296
Univ. of Calif. See California, Univ. of.
unmanned space flight, 35, 131, 146, 152, 182, 196, 219, 264, 272, 298, 303, 305
unpowered flight, 37
unsymmetrical dimethyl hydrazine, 99
upper atmosphere (earth). See atmosphere.
Upper Volta, 293
"upside down" aircraft wing, 87
upsilon particle, 25
upwelling, 24
Ural mountains, 273
uranium, 48, 122, 123, 254, 268
Uranus (planet), 24, 43, 294

urethane, 59
Urey, Dr. Harold C., 56
urokinase, 20
Uruguay, 293
USAF. See Air Force, U.S.
USDA. See Agriculture, U.S. Dept. of.
user-operated communications, 4, 10
USGS. See Geological Survey, U.S.
USRA. See Universities Space Research Assn.
USSR. See Union of Soviet Socialist Republics.
Utah, 80, 82, 188, 216, 232
utilities, 167, 225, 240, 265, 280
Utopia Planitia (region of Mars), 203, 205, 208, 215–216, 217, 224, 237, 274
Uzbekistan, 41

VAB. See vehicle assembly building, KSC.
Vachnadze, V.C., 246
vacuum, 19, 51, 70, 105, 107, 127, 170, 183, 215, 235, 247, 271, 281
Valles Marineris (region of Mars), 121, 139
Van Allen, Dr. James, 105
vanadium, 157
vandalism, 215
Vandenberg AFB, 1, 23, 87, 99, 102, 141, 165, 168, 172, 193, 195, 203, 214, 219, 267, 295
Vanguard project, 29
Vaughn, O.H., 249
VC–10. See Vickers.
Vega (star), 115
vegetation, 5, 105, 305, 306
vehicle assembly building (VAB) at KSC, 70, 107
vehicle systems, 142, 144, 160, 207, 213, 232, 264, 343
velocity. See speed.
Venera 9, 10, 39, 59, 120, 127, 212
ventilation, 71
Venus (planet), 14, 39, 46, 48, 52, 55, 59, 69, 84, 120, 127, 168, 197, 211, 212, 214, 288, 294
Vereshchetin, Dr. V.S., 244, 246
verification, 154, 164, 262, 281, 286
Vermont, 102
vertical takeoff and landing (VTOL), 97
Vertikal 4 (Soviet geophysical rocket), 250
Vertol Div., Boeing Corp., 304
Very Large Array (telescope), 31
very long-baseline interferometry (VLBI), 150
Vessot, Dr. R.F.C., 127
veterans, 13, 161
VFW-Fokker/ERNO, 113, 292
vibration, 15, 26, 113, 140, 155, 168, 170, 204, 210, 237, 240, 286
Vickers VC–10 (British jet aircraft), 180, 191
video broadcasts, 36
videotape, 150
Vietnam, 13, 138
Viking (mission), 80, 83, 100, 146, 152, 154, 160, 166, 175, 194, 235, 245, 271, 295
Viking (project), 15, 49, 76, 77, 80, 83, 100, 138, 153, 166, 192, 194, 200, 260–261, 263, 264, 271, 272, 274, 295

Viking 1, 17, 43, 49, 76, 120, 137, 138, 139, 146, 147, 148, 151–152, 154, 155, 160, 162, 168, 175, 177, 178, 194, 196, 199, 203, 208, 215, 217, 224, 225–226, 236, 246, 263, 281
Viking 2, 15, 17, 43, 49, 77, 121, 148, 153, 155, 203, 205, 207, 215, 217, 223, 224, 225–226, 236–237, 246, 260
"Viking 1 Early Results," 194
Viking Undergraduate Intern Program, 166
village programs, 111
Villar Furtado, Romulo, 238
Virgin Islands, 246
Virginia, 3
Virginia, Univ. of, 101
Virgo (constellation), 115
virus, 61, 252
visibility, 247, 257, 274, 281
vision, 45
vitamins, 254
VLA. See Very Large Array.
Vladivostok, USSR, 143
VLBI. See very long-baseline interferometer.
Vogel, Orville A., 254
voice, 36, 62, 79, 89, 93, 94, 115, 120, 223, 275
volcano, 35, 55, 139, 148, 171, 218, 273, 281
Volynov, Boris, 140, 177, 190, 193, 204, 342
von Braun, Dr. Wernher, 12, 49, 51, 99, 159
von Eckardt, Wolf, 137
von Kármán lecture, 18
vortexes, ocean, 83
Voskhod 1, 92
Voskhod 2, 73
Vostok (Soviet spacecraft series), 38, 46
Vought Corp., 53
Voyager mission, 100
vulnerability, 163

Wabash College, 90
WAC Corporal (rocket), 261
Waddington, Dr. Cecil, 278
Waddy, Judge Joseph, 191
Wadi el Arish, 123
Wake Island, 96, 98
wake vortexes, 89
Wall Street, 19
Wallops Flight Center
—people, 282, 295
—programs, 47, 79, 118–119, 126, 133, 143, 187, 239, 242, 247, 270, 303
Walsh, John B., 158
Walterhouse, Mae, 164
Walton, Bob, 263
Walton, W.B., 302
war, 154, 230, 276, 281
Wasatch Div., Thiokol Corp., 82, 188, 232
Washington, D.C., 17, 35, 137, 150, 242, 257, 298
Washington State, 38, 214
Washington, State Univ. of, 254
Washington Univ. (St. Louis, Mo.), 254, 278
Washington, Univ. of, 166
Wasserburg, Prof. Gerald J., 300
waste disposal, 11, 36, 60, 61, 107, 142, 144, 273

water, 27, 35, 61, 70, 77, 78, 83, 120, 122, 130, 138, 141, 153, 155, 157, 169, 171, 178, 179, 199, 204, 209, 224, 226, 239, 240, 251, 268, 276, 280, 281
—hyacinths, 10, 306
—pollution, 40, 60, 61, 67, 105, 267, 303
—resources, 4, 20, 40, 60, 61, 67, 83, 123, 171, 267, 276, 306
—vapor, 122, 150, 183, 200, 203, 216, 226, 261, 277, 295, 296, 305
"waterhole," 268
Waters, Dr. Joe W., 46
wave measurements, 119, 277
weapons systems, 13, 14, 135, 149, 165, 172, 257, 273, 276, 280, 283, 285, 288, 304, 305
—testing, 299, 305
weather. See meteorology.
Weatherfax (weather facsimile broadcast system, NOAA), 10
Weaver, Leon B., 227
Webb, James, E., 62, 90
weight, 15, 20, 26, 61, 64, 65, 66, 67, 75, 89, 140, 169, 222, 240, 246, 248, 264, 270, 272
weightlessness (see also zero gravity), 5, 20, 92, 95, 140, 150, 225, 249, 257, 297
Weitz, Paul J., 96
welding, 247
West (comet), 31, 45
West Ford (project), 41
West Germany. See Federal Republic of Germany.
West-Northwest region (of Mars), 140
West, Richard M., 31, 45
West Virginia, 151, 302
Westar (comsat system), 35, 125, 141
Western Hemisphere, 106
Western Test Range. See Vandenberg AFB.
Western Union Telegraph Co., 18, 35, 125, 226
Westford, Mass., 233
Westinghouse Aerospace Electric Div., 81
Westinghouse Electric Corp., 136, 248
Westphal, Prof. James, 69
Weyerhaeuser Corp., 269
Weymann, Dr. Ray J., 82
Whalen, Albert A., 293
Whipple, Dr. Fred L., 298
Whitcomb, Richard T., 87
White, Gen. Thomas D., trophy, 231, 283
white dwarf (star), 104
White House, 9, 16, 39, 48, 75, 151, 184, 211, 299
White House science adviser, 91
White Nile, 41
White Sands Missile Range, N.M., 47, 95, 227, 229, 279, 288, 297, 301
White Sea, 283
Whitecloud (Navy surveillance system), 99
Whitten, Les, 78
Wichita, Kans., 81
wideband communications, 263
widebody jet. See jet aircraft.
Wieland, Richard J., 288
Wilcox Electric Co., 257
wildlife, 256

395

Wilhelm, John, 293
William Research Corp., 53
Williams, Darrel, 269
Williams, Dr. Gareth P., 189
Williams, Dr. Robert E., 82
Wilmington, Del., 36
Wilmington, N.C., 119
Wilson, E. Bright, Jr., 254
wind, 5, 63, 89, 120, 125, 140, 146, 209, 248, 277, 280, 300
wind-operated power generators, 14, 67, 102, 125, 167, 176, 181, 280
wind tunnels, 5, 24, 33, 95, 130, 168, 258
windmills, 67, 102, 167, 181, 280
windshield, 172
wing, aircraft, 87, 144, 231, 248, 280
Winkfield, England, 46
wireless, 143
Wisconsin, Univ. of, 254
Woiceshyn, Peter M., 76
Wolfe, Dr. John H., 58
women in space, 46, 59, 95, 142, 250, 255, 270, 288
Wonder of Engineering award, 241
Woodfin, R.Adm. Kenneth L., 83
Woomera rocket range, Australia, 93
Worcester, Mass., 190
Worden, Col. Alfred M., 66
workshop, 81, 179
World Administrative Radio Conference, 10
world government, 176
World Health Organization, 146
world records, 40, 71, 96, 98, 109, 140, 164, 190, 242, 253, 256, 282
World War II, 3, 17, 28, 67, 92, 98, 143, 211, 256
World Wildlife Fund, 256
Wouch, Jerry, 297
Wright brothers, 40, 137, 229
Wright Patterson AFB, 29, 129
Wu, Chien-Hsung, 254
Wyle Laboratories, 227
Wyoming, 121
Wyoming, Univ. of, 17

X-15 (supersonic research aircraft), 137, 241
X-24A (lifting body), 44
X-24B, 44, 241
x-ray test facility, MSFC, 62
x-rays, 23, 45, 62, 88, 93, 101, 106, 131, 132, 155, 156, 157, 163, 188, 216, 217, 223-224, 229
xenon, 181, 200, 226
XV-15 (tilt-rotor research aircraft), 258

Yak-42 (Soviet transport plane), 286
Yakutsk, USSR, 272
Yalta, USSR, 195
Yankee Clipper. See space sailer.
Yardley, John F., 17, 91, 95, 243, 290
Yeliseyev, Aleksey, 221
Yemen, 176, 293
Yen, J.L., 302
YF-12A (lightweight fighter plane), 164
YF-16 (USAF fuselage design), 135
YF-17, 95
Ying Manufacturing Corp., 116
York Univ., Ontario, 304
Yost, Ed, 242
Young, A, Thomas, 100, 154, 245, 271
Young, Sheila, 23
Yugoslavia, 143, 207, 214, 238, 245
Yuri Gagarin (training) Center, USSR, 219, 304
Yuri Gagarin medal, 282

Zaire, 73, 293
Zamengoe, Cameroon, 181
Zeiss MKF-6 (multispectral camera), 220-221
zero gravity, 95-96, 107, 140, 172, 249, 256, 270, 279
Zholobov, Vitaly, 140, 177, 190, 193, 204, 255, 342
zinc, 213
Zinno, Joseph A., 79
zipper, 60
Zirker, Jack, 179
zodiacal light, 8
Zolotov, Aleksey, 252-253
Zond 2 (USSR spacecraft), 15
Zook, Herbert A., 167
Zudov, Lt. Col. Vyacheslav, 250, 255, 342

ERRATA
In Earlier Volumes of Astronautics and Aeronautics

Astronautics and Aeronautics, 1966
- p. 331: In top line, first two words, substitute "October 26" for "October 27."
  At beginning of next item, insert "October 27:"
- p. 410: Appendix A, launch table, top item: Change date to Oct. 26 for *Intelsat II-A*.
- p. 421: Appendix B, NASA launch table, top item: Change launch date to Oct. 26 for *Intelsat II-A*.

Astronautics and Aeronautics, 1967
- p. 432: Appendix B, NASA launch table: In Feb. 4 entry, change launch vehicle to Atlas–Agena D, instead of B.

Astronautics and Aeronautics, 1969
- p. 110: In second sentence from the bottom of the page, delete "for Space Science and Applications."

Astronautics and Aeronautics, 1973
- p. 289: Substitute for first sentence at top: "An early explorer, probably Lucas Vasques de Ayllon, gave the name Cape Canaveral some time before 1536. It appears on a map dated 1536 and on most maps thereafter." In next to last line of item (sources), insert "4/17/73, S7555;" following "H8880;".

Astronautics and Aeronautics, 1974
- p. 146: In lines 6 and 8 from bottom, change "Castro II" to "Castor II." (Index entry is correct.)
- p. 203: In third line of last paragraph, change "Office" to "Organization."
- p. 314: Index, under U.S. Air Force, change "Space and Missiles Systems Office" to "Space and Missile Systems Organization."

Astronautics and Aeronautics, 1975
- p. 110: Last item on page, second and seventh lines, change "Kramer" to "Kraemer."
- p. 313: Index, change "Kramer, Robert S." to "Kraemer, Robert S."

www.ingramcontent.com/pod-product-compliance
Lightning Source LLC
Chambersburg PA
CBHW081716170526
45167CB00009B/3599